Strengthening, Corrosion and Protection of High Temperature Structural Materials

Strengthening, Corrosion and Protection of High Temperature Structural Materials

Guest Editors

Yingyi Zhang
Shahid Hussain
Awais Ahmad

Basel • Beijing • Wuhan • Barcelona • Belgrade • Novi Sad • Cluj • Manchester

Guest Editors

Yingyi Zhang
School of Metallurgical Engineering
Anhui University of Technology
Maanshan
China

Shahid Hussain
Department of Physics
Jiangsu University
Jiangsu
China

Awais Ahmad
Department of Chemistry
The University of Lahore
Lahore
Pakistan

Editorial Office
MDPI AG
Grosspeteranlage 5
4052 Basel, Switzerland

This is a reprint of the Special Issue, published open access by the journal *Coatings* (ISSN 2079-6412), freely accessible at: www.mdpi.com/journal/coatings/special_issues/high_temp_struct_mater.

For citation purposes, cite each article independently as indicated on the article page online and using the guide below:

Lastname, A.A.; Lastname, B.B. Article Title. *Journal Name* **Year**, *Volume Number*, Page Range.

ISBN 978-3-7258-2482-3 (Hbk)
ISBN 978-3-7258-2481-6 (PDF)
https://doi.org/10.3390/books978-3-7258-2481-6

© 2024 by the authors. Articles in this book are Open Access and distributed under the Creative Commons Attribution (CC BY) license. The book as a whole is distributed by MDPI under the terms and conditions of the Creative Commons Attribution-NonCommercial-NoDerivs (CC BY-NC-ND) license (https://creativecommons.org/licenses/by-nc-nd/4.0/).

Contents

About the Editors . . . . **vii**

Preface . . . . **ix**

Yingyi Zhang
Strengthening, Corrosion and Protection of High-Temperature Structural Materials
Reprinted from: *Coatings* 2022, *12*, 1136, https://doi.org/10.3390/coatings12081136 1

Yan Liu, Dingguo Zhao, Yue Li and Shuhuan Wang
Effects of Nano TiC on the Microhardness and Friction Properties of Laser Powder Bed Fusing Printed M2 High Speed Steel
Reprinted from: *Coatings* 2022, *12*, 825, https://doi.org/10.3390/coatings12060825 4

Tayyaba Ashfaq, Mariam Khan, Ifzan Arshad, Awais Ahmad, Shafaqat Ali and Kiran Aftab et al.
Electro-Oxidation of Metal Oxide-Fabricated Graphitic Carbon Nitride for Hydrogen Production via Water Splitting
Reprinted from: *Coatings* 2022, *12*, 548, https://doi.org/10.3390/coatings12050548 19

Tao Fu, Fuqiang Shen, Yingyi Zhang, Laihao Yu, Kunkun Cui and Jie Wang et al.
Oxidation Protection of High-Temperature Coatings on the Surface of Mo-Based Alloys—A Review
Reprinted from: *Coatings* 2022, *12*, 141, https://doi.org/10.3390/coatings12020141 32

Kai Liu, Shusen Cheng and Yaqiang Li
Study of Crack Sensitivity of Peritectic Steels
Reprinted from: *Coatings* 2021, *12*, 15, https://doi.org/10.3390/coatings12010015 56

Abduladheem Turki Jalil, Shameen Ashfaq, Dmitry Olegovich Bokov, Amer M. Alanazi, Kadda Hachem and Wanich Suksatan et al.
High-Sensitivity Biosensor Based on Glass Resonance PhC Cavities for Detection of Blood Component and Glucose Concentration in Human Urine
Reprinted from: *Coatings* 2021, *11*, 1555, https://doi.org/10.3390/coatings11121555 66

Sadaf Tariq, Sobia Tabassum, Sadia Aslam, Mika Sillanpaa, Wahidah H. Al-Qahtani and Shafaqat Ali
Detection of Virulence Genes and Biofilm Forming Capacity of Diarrheagenic *E. coli* Isolated from Different Water Sources
Reprinted from: *Coatings* 2021, *11*, 1544, https://doi.org/10.3390/coatings11121544 79

Kai Liu, Shusen Cheng, Jipeng Li and Yongping Feng
Effect of Solidifying Structure on Centerline Segregation of S50C Steel Produced by Compact Strip Production
Reprinted from: *Coatings* 2021, *11*, 1497, https://doi.org/10.3390/coatings11121497 89

Fuqiang Shen, Yingyi Zhang, Laihao Yu, Tao Fu, Jie Wang and Hong Wang et al.
Microstructure and Oxidation Behavior of Nb-Si-Based Alloys for Ultrahigh Temperature Applications: A Comprehensive Review
Reprinted from: *Coatings* 2021, *11*, 1373, https://doi.org/10.3390/coatings11111373 105

Qasim Abbas, Muhammad Sufyan Javed, Awais Ahmad, Sajid Hussain Siyal, Idrees Asim and Rafael Luque et al.
ZnO Nano-Flowers Assembled on Carbon Fiber Textile for High-Performance Supercapacitor's Electrode
Reprinted from: *Coatings* 2021, *11*, 1337, https://doi.org/10.3390/coatings11111337 . . . . **122**

Laihao Yu, Fuqiang Shen, Tao Fu, Yingyi Zhang, Kunkun Cui and Jie Wang et al.
Microstructure and Oxidation Behavior of Metal-Modified Mo-Si-B Alloys: A Review
Reprinted from: *Coatings* 2021, *11*, 1256, https://doi.org/10.3390/coatings11101256 . . . . **132**

Limin Zhang, Liguang Zhu, Caijun Zhang, Zhiqiang Wang, Pengcheng Xiao and Zenxun Liu
Physical Experiment and Numerical Simulation on Thermal Effect of Aerogel Material for Steel Ladle Insulation Layer
Reprinted from: *Coatings* 2021, *11*, 1205, https://doi.org/10.3390/coatings11101205 . . . . **155**

Laihao Yu, Yingyi Zhang, Tao Fu, Jie Wang, Kunkun Cui and Fuqiang Shen
Rare Earth Elements Enhanced the Oxidation Resistance of Mo-Si-Based Alloys for High Temperature Application: A Review
Reprinted from: *Coatings* 2021, *11*, 1144, https://doi.org/10.3390/coatings11091144 . . . . **173**

Lijie Zhang, Bing He, Shengnan Wang, Guangcun Wang and Xiaoming Yuan
Ecofriendly Ultrasonic Rust Removal: An Empirical Optimization Based on Response Surface Methodology
Reprinted from: *Coatings* 2021, *11*, 1127, https://doi.org/10.3390/coatings11091127 . . . . **193**

Umer Sharif, Beibei Sun, Md Shafiqul Islam, Kashif Majeed, Dauda Sh. Ibrahim and Orelaja Oluseyi Adewale et al.
Fracture Toughness Analysis of Aluminum (Al) Foil and Its Adhesion with Low-Density Polyethylene (LPDE) in the Packing Industry
Reprinted from: *Coatings* 2021, *11*, 1079, https://doi.org/10.3390/coatings11091079 . . . . **204**

Ali Dad Chandio, Iftikhar Ahmed Channa, Muhammad Rizwan, Shakeel Akram, Muhammad Sufyan Javed and Sajid Hussain Siyal et al.
Polyvinyl Alcohol and Nano-Clay Based Solution Processed Packaging Coatings
Reprinted from: *Coatings* 2021, *11*, 942, https://doi.org/10.3390/coatings11080942 . . . . **226**

Tao Fu, Kunkun Cui, Yingyi Zhang, Jie Wang, Xu Zhang and Fuqiang Shen et al.
Microstructure and Oxidation Behavior of Anti-Oxidation Coatings on Mo-Based Alloys through HAPC Process: A Review
Reprinted from: *Coatings* 2021, *11*, 883, https://doi.org/10.3390/coatings11080883 . . . . **237**

Xu Zhang, Tao Fu, Kunkun Cui, Yingyi Zhang, Fuqiang Shen and Jie Wang et al.
The Protection, Challenge, and Prospect of Anti-Oxidation Coating on the Surface of Niobium Alloy
Reprinted from: *Coatings* 2021, *11*, 742, https://doi.org/10.3390/coatings11070742 . . . . **248**

About the Editors

Yingyi Zhang

Yingyi Zhang is an Associate Professor at the Anhui University of Technology. He has presided over and participated in more than 20 research projects. He is mainly engaged in the research of the development of new ironmaking technology, the comprehensive utilization of metallurgical solid waste, the preparation of high-temperature ceramics, and the surface coating and corrosion protection of refractory metal alloys. With more than 100 publications, 36 patents, and 4 ESI Highly Cited Papers, Prof. Zhang is a leader in the hot dip silicon-plating technology field. In recent years, he has hosted two national key research and development programs, one national natural science foundation program, one Anhui Province science foundation program for excellent young scholars, and one China postdoctoral science foundation program. In addition, he has participated in two national science and technology support programs as the backbone, one international science and technology cooperation and exchange program, and six national natural science foundation programs.

Shahid Hussain

Prof. Dr. Shahid Hussain, born in Pakistan, is currently working as a Professor (Full) at the School of Materials Science and Engineering, Jiangsu University, China. He completed his PhD degree in Chongqing University, 2015, after starting his Post-Doctoral Research Fellowship 2015–2017. He joined Jiangsu University as an Associate Professor in July 2017, and based on his outstanding achievements and experiences, he was promoted to a Full Professor in July 2020 and was approved by the state Govt of China.

At present, Dr. Shahid Hussain and the project team have executed a lot of work in the field of metal oxide, sulfides, MXenes, metal–organic-based gas sensors, supercapacitors, and LiS Batteries. He has published high-quality research articles and has a wealth of experience, which laid a solid foundation for the project-related research. Dr Shahid Hussain has excellent working experience on gas sensors and has been working on sensor device fabrication since 2011. He has published more than 275+ SCI indexed journal research articles with an h-Index of 50 in Google Scholar with 7750 citations (as of Jan 2024) and 11 book chapters, including in *Nano Energy*, *Chemical Engineering Journal*, *Journal of Hazardous Materials*, *Applied Materials and Interfaces*, *Journal of Materials Chemistry A*, *Sensors and Actuators B*, *Chemosphere*, *Inorganic Chemistry*, *Journal of Cleaner Production*, *Applied Surface Science*, *Electrochemica Acta*, *Materials Science and Engineering*, etc., and has an impact factor of ~IF>1700. He is also working as an Editor for 17 SCI indexed journals (Elsevier, Springer, Frontiers, Hindawi, American Scientific Publishers, and MDPI). Moreover, he has appeared as a Keynote speaker in 22 international conferences and as a Conference Chair in 3 international conferences.

Awais Ahmad

Awais Ahmad is currently working as a Lecturer within the Department of Chemistry, the University of Lahore, Lahore, Pakistan, where I am carrying out research on nanomaterials, composite materials, coated materials, heterogenous catalysis, MOF, self-sacrifice MOF, single atom catalysis, photocatalysis, and electrochemical sensor formulations. I use these materials for waste water treatment, the removal of heavy metals, the degradation of organic pollutants, supercapacitors, dye sensing solar cells, gel electrolytes, sensing of hazardous metals in waste water, conversion of hazardous gases into useful products, nitrogen fixation, catalysis for organic transformation, Li-S batteries, GCE nafion fibers, and water splitting.

Preface

The increasing demand for advanced structural materials in high-temperature applications has necessitated the development of new alloys and protection techniques to combat degradation under extreme conditions. From aerospace engines to power plants and petrochemical industries, materials operating at elevated temperatures face several challenges, including mechanical stresses, thermal fatigue, oxidation, and corrosion. These factors significantly impact their performance, reliability, and lifespan.

This reprint, entitled "Strengthening, Corrosion, and Protection of High-Temperature Structural Materials", is aimed at researchers, engineers, and professionals working in fields where high-temperature performance is crucial. It provides a comprehensive overview of the current advancements in the strengthening mechanisms of structural materials and explores the various methods used to mitigate corrosion and degradation in such harsh environments. Through detailed discussions on material science and metallurgy, readers will gain insights into how modern materials can be engineered to enhance both strength and corrosion resistance.

In addition to an in-depth analysis of the metallurgical principles governing the behavior of structural materials under high temperatures, this book highlights protective coating technologies, surface treatments, and alloying strategies that extend the operational life of these materials. Recent advancements in nanostructured coatings, corrosion inhibitors, and thermal barrier systems are examined to offer solutions that meet the growing industrial demands for efficiency and durability.

The chapters within this reprint cover a wide range of materials, including superalloys, stainless steels, and ceramics, and integrate real-world case studies, demonstrating how theoretical concepts are applied to practical challenges. This reprint also addresses emerging trends in the field, such as additive manufacturing and sustainable material design, which promise to revolutionize how we approach high-temperature corrosion and protection in the future.

I would like to express my gratitude to the numerous colleagues and collaborators who have contributed to this work, and to the research community whose continuous advancements inspire new investigations into material resilience. This reprint is a product of both the collaborative efforts of researchers worldwide and my own commitment to exploring the intricacies of high-temperature structural materials.

It is my hope that this text will serve as a valuable resource for those seeking to deepen their understanding of material behavior in extreme environments, and that it will inspire further research in the critical areas of material science and engineering.

Yingyi Zhang, Shahid Hussain, and Awais Ahmad
Guest Editors

Editorial

Strengthening, Corrosion and Protection of High-Temperature Structural Materials

Yingyi Zhang

School of Metallurgical Engineering, Anhui University of Technology, Maanshan 243002, China; zhangyingyi@ahut.edu.cn; Tel.: +86-173-7507-6451

Abstract: This Special Issue presents a series of research papers and reviews about the second-phase enhancement, surface coating technology, high-temperature corrosion, wear, erosion, and protection of high-temperature structural materials. The effects of alloying and surface coating technology on the microstructure, mechanical properties, and oxidation resistance of materials were systematically introduced. In addition, this Special Issue also summarizes the strengthening mechanism of the second relatively refractory metal alloy and carbonized ceramic materials, compares the advantages and disadvantages of different surface coating technologies, and analyzes the oxidation behavior and failure mechanism of the coating in order to provide valuable research references for related fields.

Keywords: refractory metal alloys; ceramics; second-phase enhancement; surface technology; coating; alloying; corrosion protection; mechanical properties; frictional wear; oxidation behavior; mechanism

Citation: Zhang, Y. Strengthening, Corrosion and Protection of High-Temperature Structural Materials. *Coatings* **2022**, *12*, 1136. https://doi.org/10.3390/coatings12081136

Received: 21 July 2022
Accepted: 31 July 2022
Published: 7 August 2022

Publisher's Note: MDPI stays neutral with regard to jurisdictional claims in published maps and institutional affiliations.

Copyright: © 2022 by the author. Licensee MDPI, Basel, Switzerland. This article is an open access article distributed under the terms and conditions of the Creative Commons Attribution (CC BY) license (https://creativecommons.org/licenses/by/4.0/).

High-temperature structural materials are characterized by high melting points, high-strength and high-temperature creep resistance, low thermal expansion coefficients, and excellent corrosion resistance. Such materials are widely used in metallurgy, chemical applications, aerospace, nuclear reactors, and other situations where extreme environments are encountered.

At present, the main high-temperature structural materials include nickel base superalloys, refractory metal alloys, carbides, and nitride ultra-high temperature ceramics. However, during high-temperature service, high-temperature structural materials are exposed to extremely harsh high-temperature environments, bear mechanical and thermal loads, and are exposed to high-temperature oxidation, erosion, corrosion, etc. Therefore, high stress can be easily concentrated at a defect site, especially near the phase interface [1–3]. Thermal expansion stress will drive the nucleation and propagation of cracks. At the same time, friction, oxidation, and corrosion will also aggravate crack propagation and material failure, which will pose a catastrophic threat to the high-temperature components [4,5]. Therefore, the characterization, strengthening, and corrosion protection of high-temperature structural materials are very important. This Special Issue focuses on second phase enhancement, surface coating technology, high temperature corrosion, wear, erosion, and protection with respect to high temperature structural materials.

Among these refractory metal alloys, Mo-based alloys and Nb-based alloys are considered to have the most potential for new ultra-high temperature structural material and are favored by researchers [6–8]. However, low-temperature oxidation pulverization and high-temperature oxidation volatilization limit their further application [9–12]. A large number of studies show that alloying is an effective way to solve this problem [13–15]. This Special Issue provided a comprehensive review for the microstructure and oxidation resistance of Mo-based alloys and Nb-based alloys. Moreover, the influence of metallic elements and rare earth elements on the microstructure, phase compositions, oxidation kinetics and behavior of refractory metal alloys were also studied systematically. Finally, the modification mechanism of metallic elements was summarized in order to obtain refractory metal alloys with superior oxidation performance.

It is worth noting that the surface-coating technology can improve the oxidation resistance of the alloy at high temperature with as little impact on the mechanical properties as possible [16–21]. Therefore, it is favored by the majority of researchers. This Special Issue provides a summary of surface modification techniques for Mo-based and Nb-based alloys under high-temperature aerobic conditions of nearly half a century, including slurry sintering technology, plasma spraying technology, chemical vapor deposition technology, and liquid phase deposition technology. The growth mechanism and micromorphology of the coatings access by different preparation methods are evaluated. In addition, the advantages and disadvantages of various coating oxidation characteristics and coating preparation approaches are summarized. Finally, the coating's oxidation behavior and failure mechanism are summarized and analyzed, aiming to provide valuable research references in related fields.

In addition, TiC ceramics have become one of the most potential ultra-high temperature structural materials because of its high melting point, low density, and low price [22,23]. However, the poor mechanical properties seriously limit its development and application. In this Special Issue, the mechanism of the second-phase (particles, whiskers, and carbon nanotubes) reinforced TiC ceramics was reviewed. In addition, the effects of the second phase on the microstructure, phase composition, and mechanical properties of TiC ceramics were systematically studied: the addition of carbon black effectively eliminates the residual TiO_2 in the matrix, and the bending strength of the matrix is effectively improved by the strengthening bond formed between TiC; SiC particles effectively inhibit the grain growth through pinning, the obvious crack deflection phenomenon is found in the micrograph; the smaller grain size of WC plays a dispersion strengthening role in the matrix and makes the matrix uniformly refined; the addition of WC forms (Ti, W) C solid solution, and WC has a solid solution strengthening effect on the matrix; SiC whiskers effectively improve the fracture toughness of the matrix through bridging and pulling out, and the microscopic diagram and mechanism diagram of the SiC whisker action process are shown in this paper. The effect of new material carbon nanotubes (CNTs) on the matrix is also discussed: the bridging effect of CNTs can effectively improve the strength of the matrix; during sintering, some CNTs were partially expanded into nano-graphene ribbons (GNR); and in the process of crack bridging and propagation, more fracture energy is consumed by flake GNR. Finally, the existing problems of TiC-based composites are pointed out, and the future development direction is prospected.

To conclude, this Special Issue focuses on the reinforcement, surface coating technology, high-temperature corrosion, wear, erosion and corrosion protection of high-temperature structural materials. The significance of this Special Issue is to provide some references for further research on high-temperature structural materials.

Funding: This research was funded by the Anhui Province Science Foundation for Excellent Young Scholars (No. 2108085Y19).

Acknowledgments: As Editor of this Special Issue, I would like to thank first of all the authors of the articles who have shown an interest in these research topics, but also the reviewers, editors, and all those who have contributed to the publication of this Special Issue.

Conflicts of Interest: The authors declare no conflict of interest.

References

1. Cui, K.; Zhang, Y.; Fu, T.; Wang, J.; Zhang, X. Toughening Mechanism of Mullite Matrix Composites: A Review. *Coatings* **2020**, *10*, 672. [CrossRef]
2. Pan, Y.; Chen, S. Influence of alloying elements on the mechanical and thermodynamic properties of ZrB_2 boride. *Vacuum* **2022**, *198*, 110898. [CrossRef]
3. Cui, K.K.; Fu, T.; Zhang, Y.Y.; Wang, J.; Mao, H.B.; Tan, T.B. Microstructure and mechanical properties of $CaAl_{12}O_{19}$ reinforced Al_2O_3-Cr_2O_3 composites. *J. Eur. Ceram. Soc.* **2022**, *41*, 935–7945. [CrossRef]
4. Tu, L.Q.; Deng, Y.D.; Zheng, T.C.; Han, L.; An, Q.L.; Ming, W.W.; Chen, M. Wear and friction analysis of cubic boron nitride tools with different binders in high-speed turning of nickel-based superalloys. *Tribol. Int.* **2022**, *173*, 107659. [CrossRef]

5. Feng, K.L.; Shao, T.M. The evolution mechanism of tribo-oxide layer during high temperature dry sliding wear for nickel-based superalloy. *Wear* **2021**, *476*, 203747. [CrossRef]
6. Shen, F.Q.; Yu, L.H.; Fu, T.; Zhang, Y.Y.; Wang, H.; Cui, K.K.; Wang, J.; Hussain, S.; Akhtar, N. Effect of the Al, Cr and B elements on the mechanical properties and oxidation resistance of Nb-Si based alloys: A review. *Appl. Phys. A-Mater. Sci. Process.* **2021**, *127*, 852. [CrossRef]
7. Yu, L.H.; Shen, F.Q.; Fu, T.; Zhang, Y.Y.; Cui, K.K.; Wang, J.; Zhang, X. Microstructure and oxidation behavior of metal-modified Mo-Si-B alloys: A review. *Coatings* **2021**, *11*, 1256. [CrossRef]
8. Yu, L.H.; Zhang, Y.Y.; Fu, T.; Wang, J.; Cui, K.K.; Shen, F.Q. Rare earth elements enhanced the oxidation resistance of Mo-Si-based alloys for high temperature application: A review. *Coatings* **2021**, *11*, 1144. [CrossRef]
9. Zhang, Y.Y.; Li, Y.G.; Bai, C.G. Microstructure and oxidation behavior of Si-MoSi$_2$ functionally graded coating on Mo substrate. *Ceram. Int.* **2017**, *43*, 6250–6256. [CrossRef]
10. Fu, T.; Shen, F.; Zhang, Y.; Yu, L.; Cui, K.; Wang, J.; Zhang, X. Oxidation protection of high-temperature coatings on the surface of Mo-based alloys—A Review. *Coatings* **2022**, *12*, 141. [CrossRef]
11. Zhang, Y.Y.; Qie, J.M.; Cui, K.K.; Fu, T.; Fan, X.L.; Wang, J.; Zhang, X. Effect of hot dip silicon-plating temperature on microstructure characteristics of silicide coating on tungsten substrate. *Ceram. Int.* **2020**, *46*, 5223–5228. [CrossRef]
12. Zhang, Y.Y.; Cui, K.K.; Gao, Q.J.; Hussain, S.; Lv, Y. Investigation of morphology and texture properties of WSi$_2$ coatings on W substrate based on contact-mode AFM and EBSD. *Surf. Coat. Technol.* **2020**, *396*, 125966. [CrossRef]
13. Zhang, Y.; Fu, T.; Yu, L.; Cui, K.K.; Wang, J.; Shen, F.Q.; Zhang, X.; Zhou, K.C. Anti-Corrosion Coatings for Protecting Nb-Based Alloys Exposed to Oxidation Environments: A Review. *Met. Mater. Int.* **2022**, 1–17. [CrossRef]
14. Zhang, X.; Fu, T.; Cui, K.; Zhang, Y.; Shen, F.; Wang, J.; Yu, L.; Mao, H. The Protection, Challenge, and Prospect of Anti-Oxidation Coating on the Surface of Niobium Alloy. *Coatings* **2021**, *11*, 742. [CrossRef]
15. Pu, D.L.; Pan, Y. First-principles prediction of structure and mechanical properties of TM$_5$SiC$_2$ ternary silicides. *Vacuum* **2022**, *199*, 110981. [CrossRef]
16. Zhang, Y.Y.; Cui, K.K.; Fu, T.; Wang, J.; Shen, F.Q.; Zhang, X.; Yu, L.H. Formation of MoSi$_2$ and Si/MoSi$_2$ coatings on TZM (Mo-0.5Ti-0.1Zr-0.02C) alloy by hot dip silicon-plating method. *Ceram. Int.* **2021**, *47*, 23053–23065. [CrossRef]
17. Zhang, Y.; Yu, L.; Fu, T.; Wang, J.; Shen, F.; Cui, K. Microstructure evolution and growth mechanism of Si-MoSi$_2$ composite coatings on TZM (Mo-0.5Ti-0.1Zr-0.02 C) alloy. *J. Alloys Compd.* **2022**, *894*, 162403. [CrossRef]
18. Zhang, Y.; Yu, L.; Fu, T.; Wang, J.; Shen, F.; Cui, K.; Wang, H. Microstructure and oxidation resistance of Si-MoSi$_2$ ceramic coating on TZM (Mo-0.5Ti-0.1Zr-0.02C) alloy at 1500 °C. *Surf. Coat. Technol.* **2022**, *431*, 128037. [CrossRef]
19. Mao, H.; Shen, F.; Zhang, Y.; Wang, J.; Cui, K.; Wang, H.; Lv, T.; Fu, T.; Tan, T. Microstructure and mechanical properties of carbide reinforced TiC-based ultra-high temperature ceramics: A Review. *Coatings* **2021**, *11*, 1444. [CrossRef]
20. Zhang, Y.Y.; Fu, T.; Cui, K.K.; Shen, F.Q.; Wang, J.; Yu, L.H.; Mao, H.B. Evolution of surface morphology, roughness and texture of tungsten disilicide coatings on tungsten substrate. *Vacuum* **2021**, *191*, 110297. [CrossRef]
21. Zhang, Y.Y.; Cui, K.K.; Fu, T.; Wang, J.; Qie, J.M.; Zhang, X. Synthesis WSi$_2$ coating on W substrate by HDS method with various deposition times. *Appl. Surf. Sci.* **2020**, *511*, 145551. [CrossRef]
22. Cui, K.K.; Mao, H.B.; Zhang, Y.Y.; Wang, J.; Wang, H.; Tan, T.B.; Fu, T. Microstructure, mechanical properties, and reinforcement mechanism of carbide toughened ZrC-based ultra-high temperature ceramics: A review. *Compos. Interface* **2022**, *29*, 729–748. [CrossRef]
23. Mao, H.; Zhang, Y.; Wang, J.; Cui, K.; Liu, H.; Yang, J. Microstructure, Mechanical Properties, and Reinforcement Mechanism of Second-Phase Reinforced TiC-Based Composites: A Review. *Coatings* **2022**, *12*, 801. [CrossRef]

Article

Effects of Nano TiC on the Microhardness and Friction Properties of Laser Powder Bed Fusing Printed M2 High Speed Steel

Yan Liu, Dingguo Zhao *, Yue Li and Shuhuan Wang

College of Metallurgy and Energy, North China University of Science and Technology, Tangshan 063009, China; liuyan1342991123@163.com (Y.L.); liyue1593126@163.com (Y.L.); wshh88@ncst.edu.cn (S.W.)
* Correspondence: zhaodingguo@ncst.edu.cn

Abstract: In this work, TiC/M2 high speed steel metal matrix composites (MMCs) were prepared using the ball milling method and laser powder bed fusing process. By controlling the TiC content in TiC/M2HSS, the grain size, phase composition, and frictional wear properties of the samples were enhanced. The results showed that when TiC/M2HSS was supplemented with 1% TiC, the surface microhardness of the samples increased to a maximum value and the wear volume decreased by approximately 39%, compared to pure M2HSS. The hardness and friction wear properties of the TiC/M2HSS composites showed a decreasing trend as the TiC content increased, owing to an increase in internal defects in the samples, as a result of excess TiC addition. The physical phases of the TiC/M2HSS MMC samples prepared by LPBF were dominated by the BCC phase, with some residual FCC phases and carbide phases. This work explored the possibility of enhancing the frictional wear performance of TiC/M2HSS samples by controlling the TiC content.

Keywords: M2HSS; LPBF; Namo-TiC; friction properties

Citation: Liu, Y.; Zhao, D.; Li, Y.; Wang, S. Effects of Nano TiC on the Microhardness and Friction Properties of Laser Powder Bed Fusing Printed M2 High Speed Steel. *Coatings* **2022**, *12*, 825. https://doi.org/10.3390/coatings12060825

Academic Editor: Chang-Hwan Choi

Received: 14 May 2022
Accepted: 9 June 2022
Published: 12 June 2022

Publisher's Note: MDPI stays neutral with regard to jurisdictional claims in published maps and institutional affiliations.

Copyright: © 2022 by the authors. Licensee MDPI, Basel, Switzerland. This article is an open access article distributed under the terms and conditions of the Creative Commons Attribution (CC BY) license (https:// creativecommons.org/licenses/by/ 4.0/).

1. Introduction

W. Breelor developed M2 high speed steel high carbon content tool steel developed in 1937, which has been widely used in machining as a lathe sharpening material owing to its good wear resistance [1]. At present, the main production processes consist of melting and casting, electroslag remelting, and powder metallurgy. The traditional production method is mainly based on casting; however, its high alloy content will result in the solidification process of tissue segregation and carbide precipitation, resulting in the subsequent forging process of high-speed steel cracking, limiting the application of high-speed steel [2]. In the 1960s, researchers developed powder metallurgy techniques to improve the phenomenon of carbide segregation. However, M2 HSS prepared by powder metallurgy still requires complex machining to produce parts, making it difficult to machine complex structural parts such as ultra-thin-walled internal cavities. Thus, the further development of machining and forming techniques for M2 HSS is required.

Laser powder bed fusion (LPBF) technology consists of a metal additive manufacturing technique that can be used to prepare workpieces with complex geometries and fine grain structures [3–5]. In the LPBF process, the metal powder is laid flat on a powder bed and rapidly melted and solidified by a high-speed laser sweep, with melting and solidification occurring within 10^6–10^7 s [6–10]. Kempen et al. [11] successfully improved the warpage and cracking of SLM-prepared HSS by preheating the substrate temperature and investigated the microstructure of M2 HSS prepared by the SLM process. Liu et al. [12] demonstrated that the solidification process of the HSS melt pool was significantly higher in the center of the melt pool than at the edges, according to MATLAB simulations. The solidification of the melt pool in the center was found to be longer than the edge temperature, which was also conducive to the growth of crystals. Karolien. Kempen et al. [13]

prepared HSS lumps with a density of 98% using a substrate preheating temperature of 200 °C. Thus, the above researchers conducted a comprehensive study of the processing methods, microstructure, and physical phases of M2 HSS in the LPBF process.

Metal matrix composites (MMCs) have received widespread attention due to their excellent frictional wear properties [14–24], and TiC [25] is a very stable compound that has been commonly added to metal matrices to increase the wear resistance of composites due to its high melting point and strong chemical stability [26]. Li et al. [20] prepared TiC-containing 316 L stainless steel by LPBF and found that the addition of TiC refined the grain size of the sample, thus increasing sample strength. In addition, Luo et al. [16] used LPBF to prepare TiC-added magnesium alloys, which significantly strengthened the wear resistance of the magnesium alloy. However, no studies have been conducted on using LPBF to prepare MMCs with M2HSS as the matrix. High speed steel has mainly been used as a tool for machining equipment; therefore, it requires good wear resistance to enhance service life. In this study, we used LPBF to prepare TiC/M2HSS composites with different contents, to investigate the feasibility of using the LPBF process for preparing formed crack-free high density TiC/M2HSS composites, and we investigated the TiC/M2HSS material through microscopic characterization and friction wear performance testing.

2. Materials and Experiments
2.1. Preparation of Materials

The metal powders used for LPBF (range 15–53 μm) were obtained from M2 HSS ingots by Ar gas atomization and the powder chemical composition was determined by XRF (XRF-1800, Scimadzu, Kyoto, Japan), as listed in Table 1, and Learning ManagementSystem (LMS, Brno, Czech Republic), as shown in Figure 1a. The particle size distribution of the powders was determined using a laser diffraction particle size distribution meter and the results are shown in Figure 1b. The powder was mostly spherical with a Dv50 of 30.5 μm.

Irregularly shaped TiC nanopowders with an average size of 100–500 nm (purity of no less than 99%) and M2 HSS aerosolized powder were used as raw materials. The TiC nanopowders were added to the M2HSS powder in proportions of 1%, 3%, and 5%, and then ball milled and mixed in an Ar atmosphere using a planetary ball mill. The ball to powder weight ratio was 5:1, the ball mill speed was 240 r/min, and the ball milling time was set to 4 h, where the powder was cooled after each hour of ball milling for 15 min. After ball milling, the TiC nanopowders were uniformly distributed on the surface of the M2HSS powder without any obvious agglomeration. An energy dispersive X-ray spectroscopy (EDS, Oxford Swift 3000, Abingdon, UK) energy spectral surface scan of the prepared composite powder was obtained, as shown in Figure 1c, with different colors for Fe, Ti, and C, which showed that the TiC particles were uniformly distributed on the powder surface. The powders were preheated at an atmosphere temperature of 80 °C for 10 h to enhance flowability before LPBF.

Table 1. Chemical composition of the M2HSS powder (wt.%).

Composition	W	Mo	Cr	V	C	Mn	Fe
Nominal composition (wt.%)	5.7–6.0	4.6–4.9	3.6–3.9	1.6–1.9	0.7–0.9	0.1–0.3	Bal.
Actual composition (wt.%)	5.87	4.93	3.87	1.89	0.84	0.29	Bal.

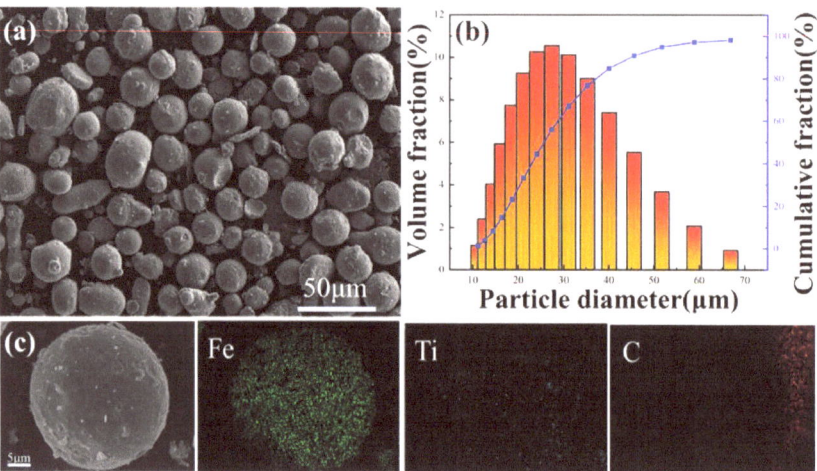

Figure 1. (a) Morphology of the TiC/M2HSS powder; (b) particle size distribution of the M2HSS powder; (c) elemental distribution of Fe, Ti, and C on a single M2 powder particle.

2.2. LPBF Process

All samples were LPBF-printed using laboratory-built equipment equipped with a ytterbium continuous single-mode fiber laser (maximum power 500 W, wavelength 1080 ± 5 nm) and focusing optical system using a 254 mm focal length F-θ lens, which produced a focused beam with a spot diameter of approximately 100 μm. The laboratory equipped the equipment with a substrate pre-heating device that allowed the stainless-steel substrate to be pre-heated up to 300 °C. During the LPBF process, Ar gas was used as the protective gas to prevent oxidation and the oxygen content in the forming cavity was below 100 ppm.

Figure 2a shows a schematic of the LPBF process, where a 67° oblique zoning scanning strategy was used between the adjacent layers in the forming process to reduce the thermal stresses [27], as shown in Figure 2b, where each zone was spaced 4 mm apart, using a laser to scan the adjacent zones. The printed samples are shown in Figure 2c, where two cubes, of 8 mm × 8 mm × 6 mm in size were used for microstructure, phase identification, and hardness testing, and a 15 mm × 15 mm × 5 mm sample was used for frictional wear testing, which was cut from the substrate after machining using electrical discharge machining (EDM, Jinli, Taizhou, China).

2.3. Characterization and Mechanical Testing

The densities of the samples were determined according to the Archimedes drainage method. The samples were polished with 240 to 5000 grit sandpaper and polished with 1 μm diamond polish for 20 min before they were observed under an optical microscope and by scanning electron microscope (SEM, Quanta 650 FEG, FEI, Madison, WI, USA) for defects such as holes and cracks.

For observations, the substrate was preheated at 300 °C, the scanning pitch (h) was fixed at 100 μm, the layer thickness was 30 μm, and the laser power values were 210, 240, 270, and 300 W, with scanning speeds of 500, 600, 700, and 800 mm/s, respectively. LPBF used η volumetric energy density as the evaluation criterion for the printing parameters [28]. The volumetric energy density η (J/mm^3) could be calculated by

$$\eta = \frac{p}{vhd} \qquad (1)$$

where p denotes the laser power, v is the scanning speed, h is the scanning interval, and d is the powder layer thickness.

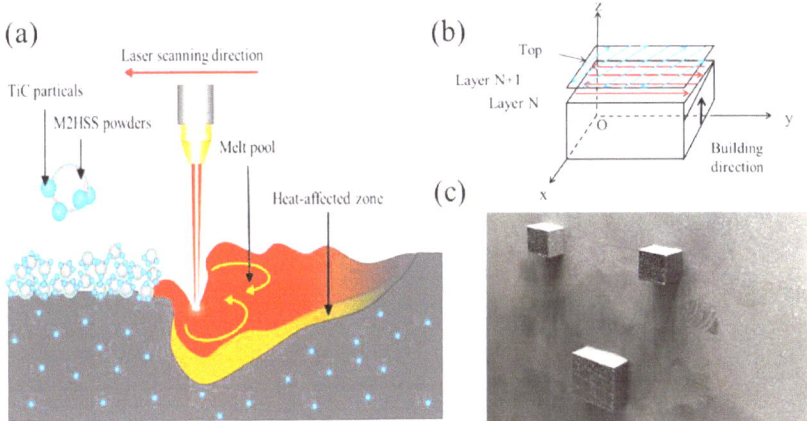

Figure 2. Schematic (**a**) diagram of LPBF TiC/M2HSS composite processing; (**b**) laser scanning strategy; (**c**) TiC/M2HSS samples printed by LPBF.

The samples were nano-indented using a Bruker TI 980 (Bruker Nano Surfaces, Madison, WI, USA) to obtain the micro-zone hardness and modulus of elasticity of the samples [29,30], using a Berkovich probe (Center for Tribology, Madison, WI, USA) and an experimental fixation depth of 1000 nm.

The relationship between the density and bulk energy density for 1% TiC/M2 is shown in Figure 3, which was divided into three regions based on the magnitude of the volumetric energy density, and typical light microscopy photographs, which were inserted in each region. The best combination of process parameters was obtained in region 2, due to the presence of pores and cracks on the surface of region 1, the presence of a large number of cracks on the surface of region 3, and the low number of defects on the surface of region 2. In this study, the optimal parameters were chosen as the laser power and scanning speed of 270 W and 700 mm/s, respectively. Figure 3 shows the relationship between density and volumetric energy density.

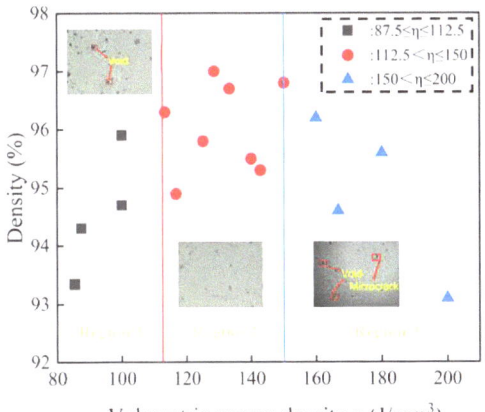

Figure 3. Relationship between density and the volumetric energy density.

The samples were subjected to metallographic etching using a composition of HNO_3 (5 mL) and anhydrous ethanol (45 mL). The surfaces of the samples were wiped with the etching solution for 180 s and rinsed clean using anhydrous ethanol, and the microstructure and wear morphology of the samples were observed under an SEM. The samples were scanned to determine their crystal structures using a Rigaku SmartLab SE (Japan) X-ray diffractometer (CuKα radiation, 40 KV, 40 mA, RIGAKU, Tokyo, Japan) with a scanning angle range of 10° to 80° and a scanning step angle of 0.013° for 1 s each time.

The friction and wear experiments were conducted using a Bruker (CETR, Center for Tribology, Wellesley, MA, USA) UMT-2 friction and wear tester with 6 mm Si_3N_4 ceramic balls, a normal load of 60 N, a friction stroke of 10 mm, and a sliding rate of 10 mm/s for 60 min to obtain the friction factor-time curves for the different samples. A laser confocal microscope (LSCM) (VK-X1000 of Keyence Corporation, Tokyo, Japan) with a wavelength of 661 nm was used to photograph the wear marks of the samples and to obtain the three-dimensional profiles and two-dimensional curves of the wear volume and wear marks [31,32]. The areas after the friction test were photographed using the SEM to analyze the frictional wear process. The hardness values of the polished samples were measured using a micro hardness tester (FM-800, FUTURE-TECH CORP, Toyko, Japan) at room temperature with a fixed loading force of 1 kg and a loading time of 15 s. Five randomly selected points on the surface of each sample were tested and the average value was taken as the hardness value of the sample.

3. Results and Analysis
3.1. TiC Action and Micro-Morphology Defects

Figure 4 shows the top surface of the LPBF-printed sample as observed under an optical microscope. Sample A0 had a smoother surface quality with no large holes or cracks, while sample A1 had micro cracks on the surface and grey precipitates were observed on the surfaces of some samples (magnified view). The precipitates were caused by the agglomeration of TiC in the matrix due to the addition of 1% TiC to sample A1 [33]. Sample A3 contained some circular holes on the surface of the sample, which were possibly caused by gas in the metal powder or the high purity argon atmosphere during the LPBF process [34]. Sample A5 had more holes distributed on the surface and the grey TiC precipitates were significantly denser (enlarged image), which was caused by the addition of more TiC and the fact that TiC in the melt pool reduced the viscosity of the melt pool [19], causing more defects on the surface and easier agglomeration.

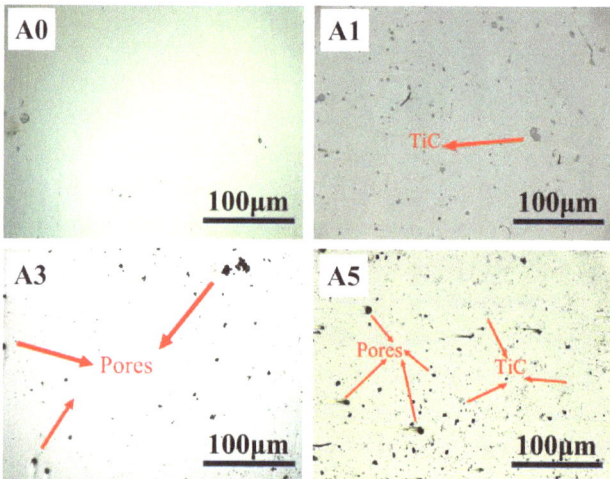

Figure 4. OM morphologies of the LPBF-printed TiC/M2HSS.

Figure 5a–d shows the different types of defects on the top surface of a selected A5 sample with 5% TiC addition, which was photographed using an SEM. Figure 5a shows a typical crack-like defect where the sample crack spread around along the unfused hole, eventually causing the sample surface to lose hardness. Figure 5b shows the crater-like defect and presence of TiC particles around the crater. This was due to the high viscosity of the melt, which allowed the melt to solidify without completely covering the crater. The crater shown in Figure 5c contained some spherical unmelted particles and a large amount of TiC around the crater, which was not enough to melt all of the powder due to the high energy absorption of TiC by the matrix. As the LPBF was built up layer by layer, defects were covered by a new layer when present, as shown in Figure 5d. The cause of these defects was generally considered to be due to the hole locking effect of the melt pool splash and metal vapor recoil [19,35].

During the LPBF-forming process when the laser irradiated the surface of the powder bed, micro-melt pools formed instantaneously, which overlapped together to form a dense structure, and the following relationship formed between the liquid phase viscosity μ of the pool, the pool temperature T, and surface tension γ [35]:

$$\mu = \frac{16}{15}\sqrt{\frac{m}{k_B T}}\gamma \qquad (2)$$

where m denotes the atomic mass; k_B is the Boltzmann constant, and γ is the surface tension of the liquid phase.

The liquid phase viscosity μ of the melt pool was proportional to γ and inversely proportional to the quadratic root of the melt pool temperature T when m and k_B were constant; thus, the melt pool viscosity decreased with an increasing temperature T when γ was constant. A moderate reduction in the melt pool viscosity was beneficial for improving the spatter and surface topography during LPBF; however, too low of a melt pool viscosity would cause the melt pool to become unstable and increase spatter. Owing to the high melting point of the titanium carbide nanoparticles, they were difficult to decompose during LPBF, and could only be retained in the melt pool as particles and the increase in the viscosity of the liquid phase μ, reduction in the liquid metal flow, and the melt pool solidification rate during LPBF were extremely fast (10^6–10^7 k/s) [36]. Therefore, a higher energy input was required to increase the melt temperature and reduce the melt viscosity to increase the wetting of the melt. As the energy input gradually increased, the density of the LPBF molded portion was increased to a certain extent. However, when the energy input was too high, it caused the powder bed to absorb too much energy and although the melt pool surface tension γ decreased and the melt pool viscosity μ also decreased, the thermal stress generation and microcrack expansion led to a decrease in the densities of the finished products [37,38]. These observations were consistent with the results shown in Figure 3.

The differences in laser absorption between M2HSS and the TiC powders resulted in a large temperature gradient inside the micro-melt pool, which in turn induced a surface tension gradient and Marangoni convection, promoting the flow of the liquid phase and generating capillary tension [39]. Subsequently, when applied to the irregularly shaped TiC particles, this promoted their movement and redistribution in the pool. At high levels of TiC particles, the particles collided with each other and agglomerated, forming larger particles at the micron scale. The excessive increase in TiC particles also led to an increase in the viscosity of the melt pool and a decrease in fluidity. Severe spherification, which reacted to the macroscopic level by the presence of a large number of spherical holes on the polished surface, was a phenomenon consistent with the observations made by Gu et al. in the preparation of TiC/Ti composites [40]

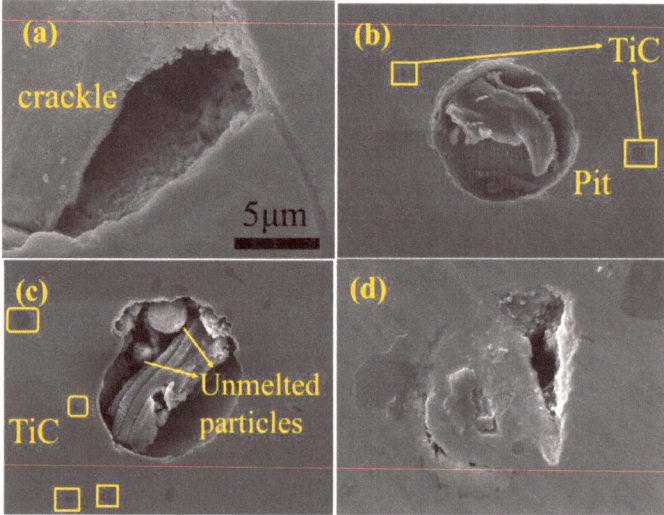

Figure 5. SEM images of the different defects on the sample surface: (**a**) Image of cracks containing TiC; (**b**) Image of a pore containing TiC; (**c**) Image of a hole containing unfused particles; (**d**) Typical SEM image of a defect.

3.2. Microstructure Characterization and Phase Analysis

Figure 6 shows the SEM images of the TiC/M2HSS composites with different TiC content. The organization of the LPBF-print sample with pure M2HSS was dominated by honeycomb isometric fine grains with a grain size of approximately 1–2 μm, as shown in Figure 6a, which was similar to the results reported by Liu et al. [8,9]. The organization of the TiC/M2HSS sample is shown in Figure 6b–d, which was dominated by ultra-fine isometric grains with a grain size that was almost always less than 1 μm, and the presence of grain boundaries, where the grain size was almost always less than 1 μm, with granular precipitates at the grain boundaries (enlarged image). The sample with 1% TiC showed a significant grain refinement compared to the original material; however, as the TiC content increased, the grains were not further refined.

The refinement could be attributed to the fast solidification process of the LPBF process, which resulted in a short grain growth time, the addition of TiC particles, which provided nucleation sites and provided a pegging effect, hindering grain growth, and the precipitates at the grain boundaries, which hindered grain growth [28], and acted as a refinement process.

Figure 7a shows the XRD pattern obtained in this experiment in the width range of 2θ (10° to 80°). Comparing the standard PDF card, the main phases of M2HSS prepared by LPBF were the FCC or BCC phases, and the phase of the MXCY-type carbides could be observed in the XRD pattern of the TiC/M2HSS composite with the addition of TiC. With an increasing amount of TiC, the intensity of the peaks of MXCY-type carbides also increased due to the precipitation of carbides induced by the addition of TiC [23].

According to Bragg's law, the 2θ angle between the X-rays and crystal plane was inversely related to the crystal plane spacing. Figure 7b focuses on intervals of 44°–45°, showing that as TiC content increased, the peak of BCC in the sample gradually moved to the right and the value of 2θ increased; thus, grain plane spacing d decreased and dislocations increased. This decrease in grain plane spacing was due to the aggregation and precipitation of TiC nanoparticles at the grain boundaries, which triggered lattice distortions and led to residual stress aggregation and increased dislocation density at the grain boundaries [17].

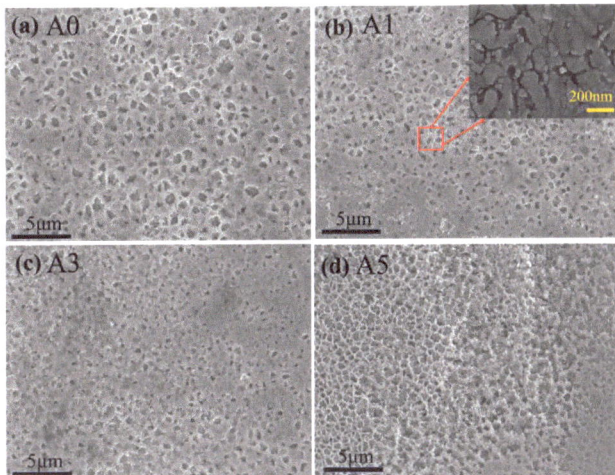

Figure 6. (a–d): SEM images of A0, A1, A3, A5 samples.

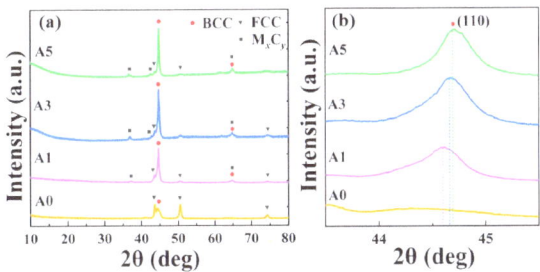

Figure 7. (a) XRD analysis of the A0, A1, A3, and A5 samples; (b) patterns at 41°–45° at a larger magnification.

To determine the effect of TiC addition on the stresses involved in the samples, all samples were characterized for residual stresses. Each sample was measured in three different areas of the surface and averaged. The test results for each sample are shown in Table 2.

Table 2. Samples residual stress.

Sample Number	A0	A1	A3	A5
Stress value	389.5 ± 29.6 MPa	−141.5 ± 10.2 MPa	−122.5 ± 10.9 MPa	−338.8 ± 32.6 MPa

From the test results, it can be seen that the residual stress is 389.5 MPa when the sample is not added with TiC, at which time the residual stress is positive, i.e., perpendicular to the sample surface. When TiC was added, all the formed stresses became shear stresses, i.e., parallel to the sample surface. The residual stress of the sample with 1% TiC addition is −141.5 MPa, and the re-residual stress of the sample with 3% TiC addition is −122.5 MPa, which was slightly lower than that of the sample with 1% TiC addition, while the residual stress of the sample with 5% TiC addition is −338.8 MPa, which is a significant improvement compared with the samples of A1, A3. This is due to the instability of the melt pool caused by the addition of too many TiC particles. The proper addition of TiC helps the transition from positive stress to shear stress, while when 5% TiC is added it causes a large increase in the residual stress of the sample.

3.3. Hardness and Friction Properties

The Vickers hardness of the A0–A5 samples is shown in Figure 8a, where the Vickers hardness of the A0 sample without TiC was approximately 683 HV. This was because the rapid melting and solidification process during the LPBF process not only resulted in grain refinement, but also heated up the matrix portion of the sample during the layer-by-layer scanning process; thus, a high hardness could be obtained without heat treatment [41].

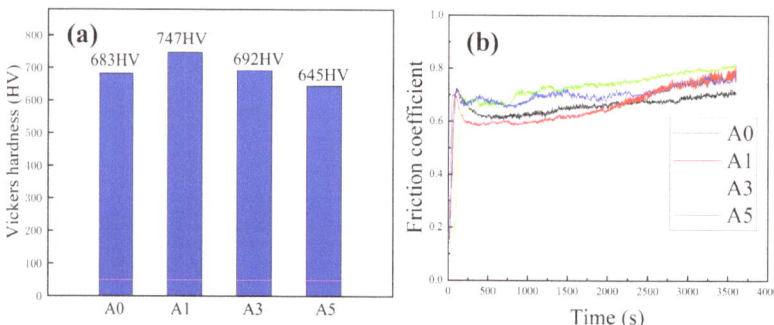

Figure 8. (**a**) Vickers hardness values of the A1, A2, A3, and A5 sample surfaces; (**b**) friction coefficient curve of M2HSS with different amounts of TiC after sliding for 60 min under 60 N.

The maximum Vickers microhardness of the TiC/M2HSS sample was 747 HV, and the A0 sample with 1% TiC addition reached peak hardness, with an increase of 9.37% over the pure M2HSS sample. The increase in hardness was due to the addition of TiC, which endowed the sample with a finer structure and higher energy absorption of the TiC particles, thus making the sample warmer during laser scanning. However, as TiC content increased, the Vickers hardness of the samples decreased, with Vickers microhardness measurements of 692 and 645 HV for samples A3 and A5, respectively. The decrease in hardness of the samples was due to the increase in surface defects such as cracks and holes due to the addition of excessive TiC (Figure 4).

Figure 8b shows the variations in the friction coefficient of the TiC/M2HSS composite with TiC content. The average friction coefficient of the samples decreased when 1% TiC was added and the time to reach the stabilization stage was slightly shorter. The friction factor was slightly higher when 3% and 5% TiC was added compared to the pure M2HSS samples. In general, the friction factor was related to the surface roughness of the samples [42], and the friction factor of the samples was slightly higher than the pure M2HSS samples due to the large number of defects such as pores and cracks on the surfaces of the A3 and A5 samples, in combination with the OM diagram shown in Figure 4.

In order to more accurately characterise the micro-zone hardness and modulus of elasticity of the material, the samples were tested using the nanoindentation method and the results are listed in Table 3. It can be seen from Table 3 that the micro-zone hardness of the samples follows the same trend as the macro hardness. This is due to the fact that the addition of TiC in the right amount can refine the structure and improve the hardness of the samples. However, when too much TiC is added, the defects in the sample increase and the hardness decreases. The modulus of elasticity of the samples also improves when 1% TiC is added due to the refinement of the structure, while it decreases when 3% and 5% TiC is added due to the increase in defects.

Table 3. Microzone hardness and microzone elastic modulus of different samples.

Sample Number	A0	A1	A3	A5
Microzone hardness (GPa)	5.66	7.17	6.30	6.02
Microzone modulus of elasticity (GPa)	158.33	163.89	140.54	102.47

The distribution of Figure 9a–d shows the shape of the 3D abrasion marks for samples A0, A1, A3, A5. The width and depth of the abrasion marks of sample A1 with 1% TiC addition were partially reduced, while the bottom of the abrasion marks of sample A0 without TiC addition was smoother and the bottom of the abrasion marks of sample A1 were rougher. This was due to the addition of TiC, which induced the precipitation and aggregation of carbide, and prevented contact between the grinding ball and sample matrix during the frictional wear process. Therefore, this improved the wear resistance and reduced the wear rate of the samples. The widths of the abrasion marks of the A3 and A5 samples with 3% and 5% addition were wider than the A0 samples. The wear volume of the A0–A5 samples is shown in Figure 9e. The wear volume of the samples without TiC addition was 3.1×10^6 μm^3. The A1 sample had the smallest volume loss, with approximately 39% less wear volume than the A0 sample, while the A3 sample had 16% more wear volume than the A0 sample, and the A5 sample had 35% more wear volume than the A0 sample. Therefore, adding too much TiC would lead to a decrease in the frictional wear performance of the samples, while adding the right amount of TiC would enhance the frictional wear performance of the samples and reduce the wear volume of the samples.

Figure 10a–d shows the color 2D plots of the abrasion marks on samples A0–A5, where the width of sample A1 with 1% TiC addition was clearly somewhat smaller than the widths of the other samples. The widest abrasion marks were found in sample A5 with 5% TiC addition, followed by sample A3. The middle sections of the sample abrasion marks were scanned using a laser scanning co-aggregation microscope, as shown by the arrows in Figure 9a–d, to obtain a two-dimensional profile of the abrasion marks. Figure 10e shows the two-dimensional view of the abrasion marks for samples A0–A5 in order from top to bottom, where A1 had the shallowest abrasion marks and sample A5 had the widest abrasion marks. By observing the contours of the abrasion marks, the bottom of the A0 sample without TiC addition was smooth and had no obvious bumps. The abrasion marks on samples A3 and A5 were deeper and sharper, with jagged bumps on both sides of the abrasion marks. This was due to an increase in internal defects in the samples due to the excess TiC, resulting in an increase in both the width and depth of the abrasion marks and a further increase in the wear volume.

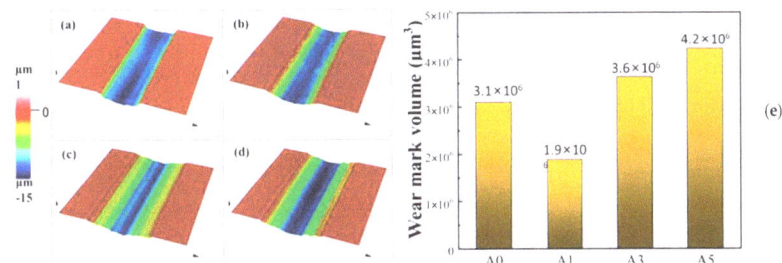

Figure 9. (a–d) The 3D surface morphologies of the wear marks on samples A0, A1, A3, and A5; (e) sample wear volume.

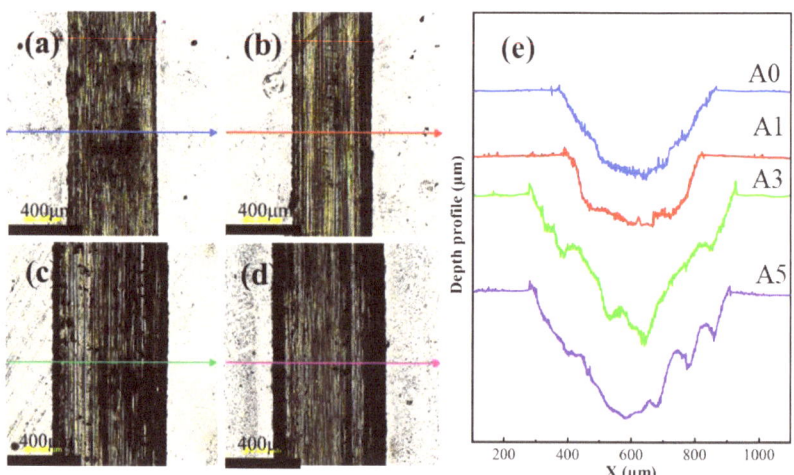

Figure 10. (**a**–**d**) Two-dimensional diagram of the sample wear marks; (**e**) cross-sections of the samples worn under 70 N for 60 min.

The microstructure diagrams of the A0–A5 sample abrasion marks are shown in Figure 11a–d, and Figure 10a shows the abrasion marks of the A0 sample of pure M2HSS. As shown in the figure, the abrasion marks on the surface of the sample were relatively wide with more scratches and pits. It was generally believed that the abrasion marks on the M2HSS sample were due to the plastic deformation of the grains, while the pits were caused by brittle cracking due to dislocation slip and stacking, which were more brittle and less plastic. The A1 sample with 1% TiC contained some dark grey TiC particles and cracks on the surface; however, the wear marks were narrower than those in the A0 sample, owing to the addition of TiC refining the microstructure of M2HSS and enhancing the plasticity of the sample. This resulted in fewer wear marks and aggregated TiC particles prevented the Si3N4 grinding balls from eroding the matrix. The A3 sample with 3% TiC had wider wear marks and a large number of pits and cracks, and the increase in defects within the sample allowed the grains to break away from the matrix during plastic deformation. The A5 sample with 5% TiC contained alternating black and grey abrasion marks, and a large number of pits were observed by magnifying some areas, which was due to the decrease in hardness of the sample. The process of grinding ball rubbing was caused by excessive defects in some areas of the sample surface, resulting in brittle cracking and consequent denting under the normal load of the grinding ball [43].

The addition of 3% and 5% TiC increased the internal defects of the sample and increased the surface cracks and porosity, which broke directly when the grinding ball was applied with a normal load and shear force, and the carbide particles were distributed within the matrix and acted as frictional substrates and participated in the frictional wear process. This increased the friction factor, making the volume of frictional loss larger and the wear marks wider and deeper. Therefore, the addition of TiC in the right amount refined the grains and precipitated carbides, enhancing the wear resistance of the sample. However, excess TiC addition would reduce the wear resistance of the sample.

The addition of 1% TiC helped to improve the frictional wear properties of the samples. The frictional wear mechanism of pure M2HSS with different TiC addition was described as follows. The wear resistance of the pure M2HSS material relied mainly on fine carbide grains such as WC, MoC, and VC, which were diffusely distributed in the matrix [44–53] with a grain size in the matrix ranging from 1 to 2 μm. When the grinding ball was applied to the sample surface with a normal load and shear force, the grains of M2HSS were pulled off, and the fragmentation is shown in Figure 12a. The addition of 1% TiC refined the grains of M2HSS to less than 1 µm and dispersed the internal stresses. When the grinding ball applied a normal load and shear force to the surface of the sample, grain refinement better dispersed the forces and TiC precipitated and collected in the matrix, preventing further contact between the friction substrate and the substrate. Therefore, the M2HSS friction coefficient increased and the wear volume decreased, as shown in Figure 12b.

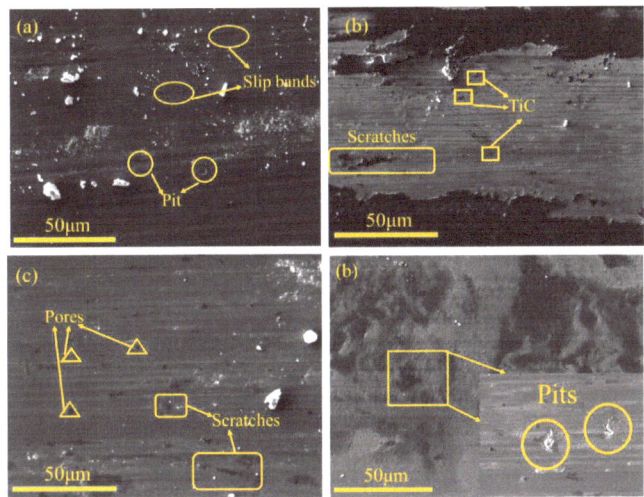

Figure 11. SEM images of the worn microstructures of the (**a**) A0, (**b**) A1, (**c**) A3, and (**d**) A5 samples.

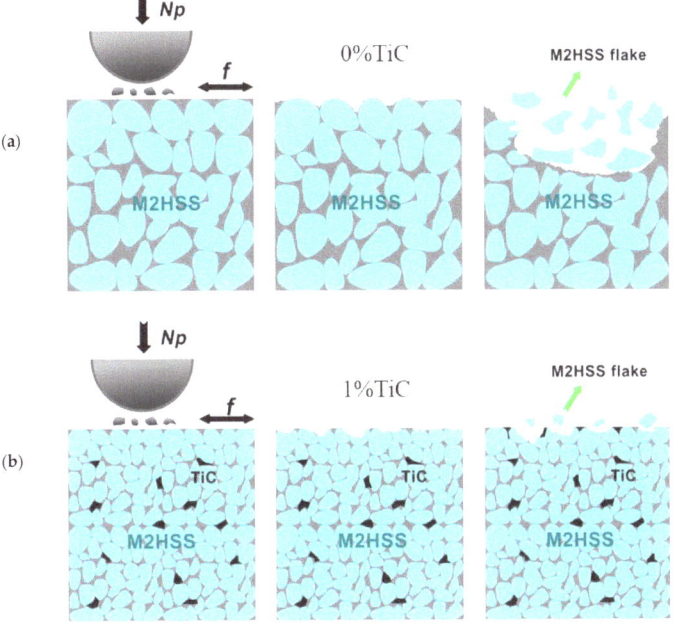

Figure 12. Schematic of the friction and wear mechanisms of the different alloys: (**a**) M2 HSS, (**b**) 1%TiC/M2HSS.

4. Conclusions

In this study, TiC/M2HSS composites were successfully prepared using physical mixing and LPBF processes. High density (97%) TiC/M2HSS samples were successfully prepared by investigating the relationship between the volumetric energy density, as well as the densities and surface morphologies of the samples. The effects of TiC content (0%, 1%, 3%, and 5%) on the microstructure, phase composition, defects, and frictional wear properties of the M2HSS samples were investigated. The following results were obtained.

1. The addition of TiC had a significant effect on the grain size of the TiC/M2HSS samples and promoted grain refinement of the samples.
2. The phase composition of the TiC/M2HSS samples consisted mainly of residual FCC and BCC phases and carbides, whereas the pure M2HSS samples consisted mainly of residual FCC and BCC phases.
3. As the TiC content increased, the hardness of the composite showed a tendency to increase and then decrease, and the maximum microhardness of the sample was 747 HV at 1% TiC content. Thus, the increase in hardness was mainly due to grain refinement.
4. The strengthening mechanism of the TiC/M2HSS samples was mainly the precipitation of carbide induced by TiC in the matrix, which prevented further erosion of the matrix by the grinding balls, and grain refinement also contributed to the frictional wear performance.

Author Contributions: Conceptualization, Y.L. and D.Z.; methodology, Y.L. (Yan Liu); software, Y.L. (Yan Liu); validation, Y.L. (Yan Liu), Y.L. (Yue Li) and D.Z.; formal analysis, Y.L. (Yan Liu); investigation, D.Z.; resources, Y.L. (Yan Liu); data curation, Y.L.(Yue Li); writing—original draft preparation, Y.L. (Yan Liu); writing—review and editing, D.Z.; visualization, Y.L. (Yan Liu); supervision, S.W.; project administration, D.Z.; funding acquisition, D.Z. All authors have read and agreed to the published version of the manuscript.

Funding: The work was supported by the National Natural Science Foundation of China (No. 52074128), Natural Science Foundation of Hebei (E2020209014, E2021209146).

Institutional Review Board Statement: Not applicable.

Informed Consent Statement: Not applicable.

Data Availability Statement: Not applicable.

Conflicts of Interest: The authors declare no conflict of interest.

References

1. Grinder, O. The HIP way to make cleaner, better steels. *Met. Powder Rep.* **2007**, *62*, 16–22. [CrossRef]
2. Peng, H.; Hu, L.; Li, L.; Zhang, L.; Zhang, X. Evolution of the microstructure and mechanical properties of powder metallurgical high-speed steel S390 after heat treatment. *J. Alloys Compd.* **2018**, *740*, 766–773. [CrossRef]
3. Li, J.; Cheng, T.; Liu, Y.; Yang, Y.; Li, W.; Wei, Q. Simultaneously enhanced strength and ductility of Cu-15Ni-8Sn alloy with periodic heterogeneous microstructures fabricated by laser powder bed fusion. *Addit. Manuf.* **2022**, *54*, 102726. [CrossRef]
4. Chen, H.; Cheng, T.; Li, Z.; Wei, Q.; Yan, W. Is high-speed powder spreading really unfavourable for the part quality of laser powder bed fusion additive manufacturing? *Acta Mater.* **2022**, *231*, 117901. [CrossRef]
5. Geenen, K.; Roettger, A.; Feld, F.; Theisen, W. Microstructure, mechanical, and tribological properties of M3: 2 high-speed steel processed by selective laser melting, hot-isostatic pressing, and casting. *Addit. Manuf.* **2019**, *28*, 585–599. [CrossRef]
6. Chen, H.; Yan, W. Spattering and denudation in laser powder bed fusion process: Multiphase flow modelling. *Acta Mater.* **2020**, *196*, 154–167. [CrossRef]
7. Badrossamay, M.; Childs, T.H.C. Further studies in selective laser melting of stainless and tool steel powders. *Int. J. Mach. Tools Manuf.* **2007**, *47*, 779–784. [CrossRef]
8. Liu, Z.H.; Chua, C.K.; Leong, K.F.; Thijs, L.; VanHumbeeck, J.; Kruth, J.P. Microstructural Investigation of M2 high speed steel produced by selective laser melting: Microstructural investigation of M2 high speed steel. In Proceedings of the 2012 Symposium on Photonics and Optoelectronics, Shanghai, China, 21–23 May 2012.
9. Liu, Z.H.; Chua, C.K.; Leong, K.F.; Kempen, K.; Thijs, L.; Yasa, E.; Kruth, J. A preliminary investigation on selective laser melting of M2 high speed steel. In Proceedings of the 5th International Conference on Advanced Research and Rapid Prototyping, Leiria, Portugal, 28 September–1 October 2011; Taylor & Francis Group: London, UK, 2011; pp. 339–346.

10. Liu, Z.H.; Zhang, D.Q.; Chua, C.K.; Leong, K.F. Crystal structure analysis of M2 high speed steel parts produced by selective laser melting. *Mater. Charact.* **2013**, *84*, 72–80. [CrossRef]
11. Kempen, K.; Vrancken, B.; Buls, S.; Thijs, L.; Van Humbeeck, J.; Kruth, J.P. Selective laser melting of crack-free high density M2 high speed steel parts by baseplate preheating. *J. Manuf. Sci. Eng.* **2014**, *136*, 4028513. [CrossRef]
12. Liu, Z.H.; Zhang, D.Q.; Chua, C.K.; Leong, K.F. Phase evolution of MHigh speed steel during selective laser melting: Experimental investigation and modelling. In Proceedings of the 1st International Conference on Progress in Additive Manufacturing, Singapore, 26–28 May 2014; pp. 151–157.
13. Chen, H.; Zhu., W.; Tang, H.; Yan, W. Oriented structure of short fiber reinforced polymer composites processed by selective laser sintering: The role of powder-spreading process. *Int. J. Mach. Tools Manuf.* **2021**, *163*, 103703. [CrossRef]
14. AlMangour, B.; Grzesiak, D.; Yang, J.M. In situ formation of TiC-particle-reinforced stainless steel matrix nanocomposites during ball milling: Feedstock powder preparation for selective laser melting at various energy densities. *Powder Technol.* **2018**, *326*, 467–478. [CrossRef]
15. Song, Q.; Zhang, Y.; Wei, Y.F.; Zhou, X.Y.; Shen, Y.F.; Zhou, Y.M.; Feng, X.M. Microstructure and mechanical performance of ODS superalloys manufactured by selective laser melting. *Opt. Laser Technol.* **2021**, *144*, 107423. [CrossRef]
16. Luo, X.; Zhao, K.; He, X.; Bai, Y.; De Andrade, V.; Zaiser, M.; An, L.; Liu, J. Evading strength and ductility trade-off in an inverse nacre structured magnesium matrix nanocomposite. *Acta Mater.* **2022**, *228*, 117730. [CrossRef]
17. Shi, Y.; Lu, Z.; Yu, L.; Xie, R.; Ren, Y.; Yang, G. Microstructure and tensile properties of Zr-containing ODS-FeCrAl alloy fabricated by laser additive manufacturing. *Mater. Sci. Eng. A* **2020**, *774*, 138937. [CrossRef]
18. Yang, M.; Wang, L.; Yan, W. Phase-field modeling of grain evolution in additive manufacturing with addition of reinforcing particles. *Addit. Manuf.* **2021**, *47*, 102286. [CrossRef]
19. Xue, F.; Shi, F.; Zhang, C.; Zheng, Q.; Yi, D.; Li, X.; Li, Y. The Microstructure and Mechanical and Corrosion Behaviors of Thermally Aged Z3CN20-09M Cast Stainless Steel for Primary Coolant Pipes of Nuclear Power Plants. *Coatings* **2021**, *11*, 870. [CrossRef]
20. Wang, X.; Jha, A.; Brydson, R. In situ fabrication of Al3Ti particle reinforced aluminium alloy metal–matrix composites. *Mater. Sci. Eng. A* **2004**, *364*, 339–345. [CrossRef]
21. Tjong, S.C. Novel nanoparticle-reinforced metal matrix composites with enhanced mechanical properties. *Adv. Eng. Mater.* **2007**, *9*, 639–652. [CrossRef]
22. Tjong, S.C.; Ma, Z.Y. Microstructural and mechanical characteristics of in situ metal matrix composites. *Mater. Sci. Eng. R Rep.* **2000**, *29*, 49–113. [CrossRef]
23. Liu, Z.Y.; Loh, N.H.; Khor, K.A.; Tor, S.B. Mechanical alloying of TiC/M2 high speed steel composite powders and sintering investigation. *Mater. Sci. Eng. A* **2001**, *311*, 13–21. [CrossRef]
24. Pagounis, E.; Lindroos, V.K. Processing and properties of particulate reinforced steel matrix composites. *Mater. Sci. Eng. A* **1998**, *246*, 221–234. [CrossRef]
25. Tan, T. Microstructure and Mechanical Properties of Carbide Reinforced TiC-Based Ultra-High Temperature Ceramics: A Review. *Coatings* **2021**, *11*, 1444.
26. Lu, Y.; Su, S.; Zhang, S.; Huang, Y.; Qin, Z.; Lu, X.; Chen, W. Controllable additive manufacturing of gradient bulk metallic glass composite with high strength and tensile ductility. *Acta Mater.* **2021**, *206*, 116632. [CrossRef]
27. Pham, M.S.; Dovgyy, B.; Hooper, P.A.; Gourlay, C.M.; Piglione, A. The role of side-branching in microstructure development in laser powder-bed fusion. *Nat. Commun.* **2020**, *11*, 749. [CrossRef] [PubMed]
28. Li, J.; Qu, H.; Bai, J. Grain boundary engineering during the laser powder bed fusion of TiC/316L stainless steel composites: New mechanism for forming TiC-induced special grain boundaries. *Acta Mater.* **2022**, *226*, 117605. [CrossRef]
29. Enneti, R.K.; Morgan, R.; Atre, S.V. Effect of process parameters on the Selective Laser Melting (SLM) of tungsten. *Int. J. Refract. Met. Hard Mater.* **2018**, *71*, 315–319. [CrossRef]
30. Ghidelli, M.; Sebastiani, M.; Collet, C.; Guillemet, R. Determination of the elastic moduli and residual stresses of freestanding Au-TiW bilayer thin films by nanoindentation. *Mater. Des.* **2016**, *106*, 436–445. [CrossRef]
31. Ast, J.; Ghidelli, M.; Durst, K.; Göken, M.; Sebastiani, M.; Korsunsky, A.M. A review of experimental approaches to fracture toughness evaluation at the micro-scale. *Mater. Des.* **2019**, *173*, 107762. [CrossRef]
32. Zhang, Y.; Yu, L.; Fu, T.; Wang, J.; Shen, F.; Cui, K.; Wang, H. Microstructure and oxidation resistance of Si-MoSi2 ceramic coating on TZM (Mo-0.5Ti-0.1Zr-0.02C) alloy at 1500 °C. *Surf. Coat. Technol.* **2022**, *431*, 128037. [CrossRef]
33. Zhang, Y.; Fu, T.; Yu, L.; Shen, F.; Wang, J.; Cui, K. Improving oxidation resistance of TZM alloy by deposited Si-MoSi2 composite coating with high silicon concentration. *Ceram. Int.* **2022**, *48*, 20895–20904. [CrossRef]
34. Cai, C.; Qiu, J.C.D.; Shian, T.W.; Han, C.; Liu, T.; Kong, L.B.; Srikanth, N.; Sun, C.-N.; Zhou, K. Laser powder bed fusion of Mo2C/Ti-6Al-4V composites with alternately laminated α'/β phases for enhanced mechanical properties. *Addit. Manuf.* **2021**, *46*, 102134. [CrossRef]
35. Dadbakhsh, S.; Speirs, M.; Kruth, J. Effect of SLM Parameters on Transformation Temperatures of Shape Memory Nickel Titanium Parts. *Adv. Eng. Mater.* **2014**, *16*, 1140–1146. [CrossRef]
36. Qiu, C.; Panwisawas, C.; Ward, M.; Basoalto, H.C.; Brooks, J.W.; Attallah, M.M. On the role of melt flow into the surface structure and porosity development during selective laser melting. *Acta Mater.* **2015**, *96*, 72–79. [CrossRef]
37. Carter, L.N.; Withers, P.J.; Martin, C. The influence of the laser scan strategy on grain structure and cracking behaviour in SLM powder-bed fabricated nickel superalloy. *J. Alloys Compd.* **2014**, *615*, 338–347. [CrossRef]

38. Cui, X.; Xue, Y.; Zhao, D.; Wang, S.; Guo, F. Physical modeling of bubble behaviors in molten steel under high pressure. *High Temp. Mater. Processes* **2021**, *40*, 471–484. [CrossRef]
39. Tang, D.; Pistorius, P.C. Isotope Exchange Measurements of the Interfacial Reaction Rate Constant of Nitrogen on Fe-Mn alloys and an Advanced High-Strength Steel. *Metall. Mater. Trans. B* **2020**, *52*, 51–58. [CrossRef]
40. Gu, D.; Wang, H.; Dai, D.; Yuan, P.; Meiners, W.; Poprawe, R. Rapid fabrication of Al-based bulk-form nanocomposites with novel reinforcement and enhanced performance by selective laser melting. *Scr. Mater.* **2015**, *96*, 25–28. [CrossRef]
41. Gu, D.; Meng, G.; Li, C.; Meiners, W.; Poprawe, R. Selective laser melting of TiC/Ti bulk nanocomposites, Influence of nanoscale reinforcement. *Scr. Mater.* **2012**, *67*, 185–188. [CrossRef]
42. Sander, J.; Hufenbach, J.; Giebeler, L.; Wendrock, H.; Kühn, U.; Eckert, J. Microstructure and properties of FeCrMoVC tool steel produced by selective laser melting. *Mater. Des.* **2016**, *89*, 335–341. [CrossRef]
43. Wen, S.; Hu, H.; Zhou, Y.; Chen, Z.; Wei, Q.; Shi, Y. Enhanced hardness and wear property of S136 mould steel with nano-TiB2 composites fabricated by selective laser melting method. *Appl. Surf. Sci.* **2018**, *457*, 11–20. [CrossRef]
44. Wu, M.; Huang, H.; Luo, Q.; Wu, Y. A novel approach to obtain near damage-free surface/subsurface in machining of single crystal 4H-SiC substrate using pure metal mediated friction. *Appl. Surf. Sci.* **2022**, *588*, 152963. [CrossRef]
45. Qiankun, Z.; Yao, J.; Weijun, S.; Huibin, Z.; Yuehui, H.; Nan, L.; Xiaolin, H. Direct fabrication of high-performance high speed steel products enhanced by LaB6. *Mater. Des.* **2016**, *112*, 469–478. [CrossRef]
46. Gimenez, S.; Zubizarreta, C.; Trabadelo, V.; Iturriza, I. Sintering behaviour and microstructure development of T42 powder metallurgy high speed steel under different processing conditions. *Mater. Sci. Eng. A* **2008**, *480*, 130–137. [CrossRef]
47. Ding, P.; Shi, G.; Zhou, S. As-cast carbides in high-speed steels. *Metall. Mater. Trans. A* **1993**, *24*, 1265–1272. [CrossRef]
48. Zhang, X. Microstructure and Oxidation Behavior of Metal-Modified Mo-Si-B Alloys: A Review. *Coatings* **2021**, *11*, 1256.
49. Aguirre, I.; Gimenez, S.; Talacchia, S.; Gomez-Acebo, T.; Iturriza, I. Effect of nitrogen on supersolidus sintering of modified M35M high speed steel. *Powder Metall.* **1999**, *42*, 353–357. [CrossRef]
50. Bolton, J.D.; Gant, A.J. Fracture in ceramic-reinforced metal matrix composites based on high-speed steel. *J. Mater. Sci.* **1998**, *33*, 939–953. [CrossRef]
51. Bolton, J.D.; Gant, A.J. Microstructural development and sintering kinetics in ceramic reinforced high speed steel metal matrix composites. *Powder Metall.* **1997**, *40*, 143–151. [CrossRef]
52. Gordo, E.; Velasco, F.; Antón, N.; Torralba, J.M. Wear mechanisms in high speed steel reinforced with (NbC) p and (TaC) p MMCs. *Wear* **2000**, *239*, 251–259. [CrossRef]
53. Sobczak, J.; Drenchev, L.B. *Metal Based Functionally Graded Materials*; Bentham Science Publishers: Sharjah, Emirate of Sharjah, 2010.

Article

Electro-Oxidation of Metal Oxide-Fabricated Graphitic Carbon Nitride for Hydrogen Production via Water Splitting

Tayyaba Ashfaq [1], Mariam Khan [2], Ifzan Arshad [3], Awais Ahmad [4], Shafaqat Ali [5,6,*], Kiran Aftab [1,*], Abdullah A. Al-Kahtani [7] and Ammar Mohamed Tighezza [7]

1. Department of Chemistry, Government College University, Faisalabad 38000, Pakistan; tayyabaa961@gmail.com
2. School of Applied Sciences & Humanities (NUSASH), National University of Technology, Islamabad 44000, Pakistan; mariamkhan@nutech.edu.pk
3. Department of Chemistry, University of Management and Technology, Sialkot Campus, Sialkot 51310, Pakistan; ifzan_ifzan@yahoo.com
4. Departamento de Quimica Organica, Universidad de Cordoba, Edificio Marie Curie (C-3), Ctra Nnal IV-A, Km 396, E14014 Cordoba, Spain; awaisahmed@gcuf.edu.pk
5. Department of Environmental Sciences and Engineering, Government College University, Allama Iqbal Road, Faisalabad 38000, Pakistan
6. Department of Biological Sciences and Technology, China Medical University, Taichung City 40402, Taiwan
7. Department of Chemistry, College of Science, King Saud University, Riyadh 11451, Saudi Arabia; akahtani@ksu.edu.sa (A.A.-K.); ammar@ksu.edu.sa (A.M.T.)
* Correspondence: shafaqataligill@yahoo.com (S.A.); drkiranaftab@gcuf.edu.pk (K.A.)

Citation: Ashfaq, T.; Khan, M.; Arshad, I.; Ahmad, A.; Ali, S.; Aftab, K.; Al-Kahtani, A.A.; Mohamed Tighezza, A. Electro-Oxidation of Metal Oxide-Fabricated Graphitic Carbon Nitride for Hydrogen Production via Water Splitting. *Coatings* 2022, *12*, 548. https://doi.org/10.3390/coatings12050548

Academic Editor: Peng Yu

Received: 26 November 2021
Accepted: 21 February 2022
Published: 19 April 2022

Publisher's Note: MDPI stays neutral with regard to jurisdictional claims in published maps and institutional affiliations.

Copyright: © 2022 by the authors. Licensee MDPI, Basel, Switzerland. This article is an open access article distributed under the terms and conditions of the Creative Commons Attribution (CC BY) license (https://creativecommons.org/licenses/by/4.0/).

Abstract: Hydrogen is a great sourcez of energy due to having zero emission of carbon-based contents. It is found primarily in water, which is abundant and renewable. For electrochemical splitting of water molecules, it is necessary to use catalytic materials that minimize energy consumption. As a famous carbon material, graphitic carbon nitride, with its excellent physicochemical properties and diversified functionalities, presents great potential in electrocatalytic sensing. In the present work, graphitic carbon nitride-fabricated metal tungstate nanocomposites are synthesized by the hydrothermal method to study their applications in catalysis, electrochemical sensing, and water splitting for hydrogen production. Nanocomposites using different metals, such as cobalt, manganese, strontium, tin, and nickel, were used as a precursor are synthesized via the hydrothermal process. The synthesized materials (g-C_3N_4/NiWO$_4$, g-C_3N_4/MnWO$_4$, g-C_3N_4/CoWO$_4$, g-C_3N_4/SnWO$_4$, g-C_3N_4/SrWO$_4$) were characterized using different techniques, such as FTIR and XRD. The presence of a functional groups between the metal and tungstate groups was confirmed by the FTIR spectra. All the nanocomposites show a tungstate peak at 600 cm^{-1}, while the vibrational absorption bands for metals appear in the range of 400–600 cm^{-1}. X-ray diffraction (XRD) shows that the characteristic peaks matched with the JCPDS in the literature, which confirmed the successful formation of all nanocomposites. The electrochemical active surface area is calculated by taking cyclic voltammograms of the potassium–ferrocyanide redox couple. Among the entire series of metal tungstate, the g-C_3N_4/NiWO$_4$ has a large surface area owing to the high conductive properties towards water oxidation. In order to study the electrocatalytic activity of the as-synthesized materials, electrochemical water splitting is performed by cyclic voltammetry in alkaline medium. All the synthesized materials proved to be efficient catalysts with enhanced conductive properties towards water oxidation. Among the entire series, g-C_3N_4/NiWO$_4$ is a very efficient electrocatalyst owing to its higher active surface area and conductive activity. The order of electrocatalytic sensing of the different composites is: g-C_3N_4-NiWO$_4$ > g-C_3N_4-SrWO$_4$ > g-C_3N_4-CoWO$_4$ > g-C_3N_4-SnWO$_4$ > g-C_3N_4-MnWO$_4$. Studies on electrochemically synthesized electrocatalysts revealed their catalytic activity, indicating their potential as electrode materials for direct hydrogen evolution for power generation.

Keywords: hydrothermal; cyclic voltammetry; surface area; electrochemical; sensor

1. Introduction

The rapid growth of energy use demands more sustainable and renewable energy production. The primary renewable energy resources such as solar and wind power are ecofriendly, but they possess seasonal intermittence and variabilities due to regional differences. To avoid such problems, such unstable energy sources should be replaced with the stable hydrogen energy [1]. The production of hydrogen is becoming increasingly popular due to its environment-friendly qualities and potential as a source of energy that is clean, non-polluting, and recyclable. In coping with the future energy crisis, hydrogen can play an important role [2]. In view of hydrogen generation, water electrolysis is a common and efficient method. Generally, water splitting is based on the two half processes: the cathodic oxygen evolution reaction (OER) and the anodic hydrogen evolution reaction (HER). Although water electrocatalysis produces very pure and ecofriendly hydrogen, due to the sluggish kinetics of OER and HER and large overpotential value, the practice of energy generation via water splitting is limited [3]. To solve this problem, the use of electrocatalysts should be implemented in order to minimize the overpotential value for cathodic hydrogen evolution (HER) and anodic oxygen evolution (OER) [4]. The electrocatalysts make the water splitting an energy-efficient process [5]. Recently, research on electrochemical water splitting using various electrocatalysts has been conducted to achieve this objective [4]. The noble metals such as platinum-, iridium-, and ruthenium-based nanomaterials proved to be efficient regarding HER and ORE processes [6,7]. The existence of such catalysts makes them indispensable for many technologically significant chemical processes, although their spontaneous aggregation and growth limit the lifetime and efficiency. Additionally, the ultrahigh price of noble metals has severely limited their future use. Due to the rapid development of modern industry, a critical need of today is highly active, stable, low-cost, and recyclable nanomaterial alternatives to these noble metals, designed for water splitting [8]. Recently, the conductive polymers (CPs) have been shown as a potential alternative to Pt-based materials due to their high electrical properties, large surface area, and greater physical and chemical stability [9–11]. Among the 2D CPs, graphitic carbon nitride (g-C_3N_4) is a graphene-like material having excellent structural features, such as high porosity, large surface area, and greater content of nitrogen. g-C_3N_4 has wide applications in electrochemical sensing, optoelectronics, electrochemical oxidations, and energy storage and conversion devices [12,13]. g-C_3N_4 can be used as a good supportive material for manufacturing 3D nanomaterials for different electrochemical applications. Doping of different nanomaterials with g-C_3N_4 generates highly active electrode materials by connection within hybrids [14]. In the past few years, scientists have enhanced the properties of CPs by doping with transition metals/metal oxides [15]. Recently, a series of low-cost metal tungstates, such as Bi_2WO_6, $CuWO_4$, $NiWO_4$, and $CoWO_4$, have been extensively studied as efficient dopants, with g-C_3N_4 showing various applications because of its outstanding optical, magnetic, and catalytic properties [16]. Hence, from these studies, it seems that by doping g-C_3N_4 with $CoWO_4$ [17] and $NiWO_4$ [18], its photocatalytic activity is enhanced. Similarly, strontium, tin, manganese, zirconium, and cobalt are the most promising transition metals and are reported as dopants with g-C_3N_4 for numerous applications [15,19,20]. These metals have a large bandgap, rendering them with weak ion transport kinetics as semiconductors or even insulators [19], and electrode film pulverization as a result of the pronounced expansion and contraction of volume during the charging/discharging processes [20].

Motivated by these results, we have designed this research work to explore the catalytic activity of transition metal tungstate-fabricated graphitic carbon nitride for electrochemical applications. In this work, the nanocomposites (g-C_3N_4-$NiWO_4$, g-C_3N_4-$MnWO_4$, g-C_3N_4-$SnWO_4$, g-C_3N_4-$CoWO_4$, g-C_3N_4-$SrWO_4$) were synthesized via the hydrothermal method and characterized by XRD, FTIR, and electrochemical cyclic voltammetry. The study of the electrocatalytic splitting of water for hydrogen production was carried out in alkaline medium via cyclic voltammetry. All the synthesized nanocomposites represent excellent catalytic activity towards water oxidation.

2. Materials and Methods

2.1. Materials

Melamine (purity: 99.95%), cyanuric acid (purity: >99%), sodium tungstate (Sigma Aldrich, St. Louis, MO, USA), sodium hydroxide (Sigma Aldrich), potassium hexacyanoferrate (III) ($K_4[Fe(CN)_6]$), alumina powder, nafion, and different metal precursors were used. All chemicals were purchased from Sigma Aldrich (St. Louis, MO, USA).

2.2. Preparation of g-C_3N_4

Following the protocol of Vilian et al., C_3N_4 was prepared [21]. The melamine and cyanuric acid were dissolved in water (75 mL) and ultrasonically maintained for 4 h. An oven (Memmert, Germany) was used at 200 °C for 12 h to heat the solution after sonication in a 100 mL stainless Teflon lining. Centrifugation, followed by cleaning with ethanol and water, was performed to eradicate any remaining contaminations from the precipitate, then centrifuging was carried out at 5000 RPM for 30 min (Figure 1). After drying at 100 °C for 24 h, the precipitate was removed. Consequently, g-C_3N_4 (Vilian et al., 2020) was synthesized [21].

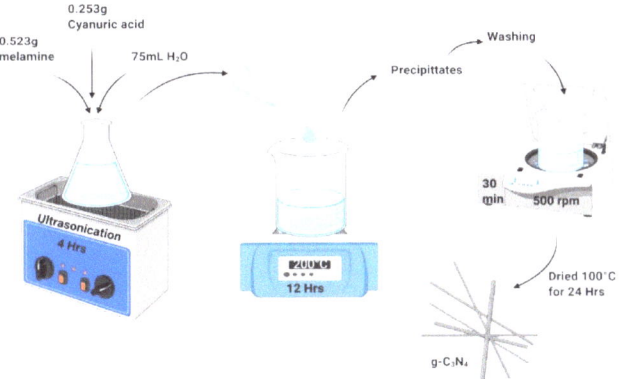

Figure 1. Schematic representation of the formation of graphitic carbon nitride (g-C_3N_4).

2.3. Preparation of g-C_3N_4/Metal Tungstate Composite

The graphitic carbon nitride g-C_3N_4/MetalWO_4 nanocomposite was synthesized by utilizing a simple hydrothermal procedure without any surfactants. Initially, a solution of sodium tungstate and graphitic carbon nitride was made. For this purpose, 0.01 g of g-C_3N_4 and 15 mM of $Na_2WO_4 \cdot 2H_2O$ were placed in 37 mL of deionized water. This solution was then kept under sonication for 20 min. The temperature was set at room temperature. After sonication, the mixture was placed aside. Then, 15 mM of salt solution in deionized water was prepared and added into the above mixture to form a homogeneous solution, and 1 mL of aqueous NaOH solution (0.1 M) was gradually added to the homogeneous solution. The mixture solution was then dispensed into an autoclave and heated at 180 °C for 12 h. After autoclaving, the resulting mixture was centrifuged to obtain the maximum analyte. To extract the excess unreacted starting materials, the obtained precipitate was collected and thoroughly rinsed three times with ethanol and water and desiccated at 60 °C. Finally, the resulting precipitation was strengthened in a vacuum oven at 300 °C for five hours to attain the g-C_3N_4/MetalWO_4 nanocomposites. The obtained g-C_3N_4/MetalWO_4 nanocomposite was circulated in ethanol to perform electrochemical measurements in Figure 2.

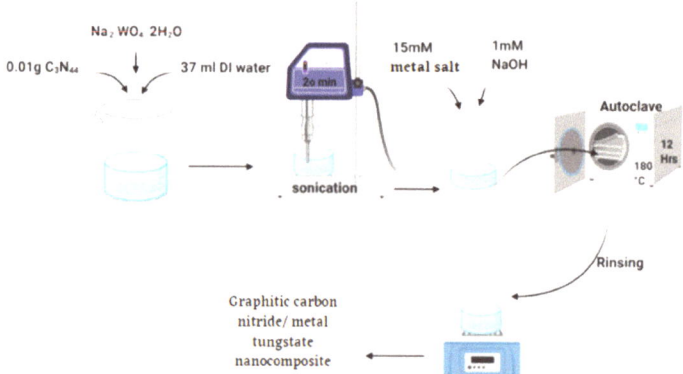

Figure 2. Schematic synthesis of graphitic carbon nitride-fabricated metal tungstate nanocomposite.

2.4. Electrochemical Studies

The electrochemical water oxidation of all synthesized nanocomposites was studied by using the Gamry Potentiostat interface 1000 (Gamry, Warminster, PA, USA) with a 3-electrode cell (100 mL) assembly in alkaline medium. The working electrode was glassy carbon (GC), the counter electrode was platinum wire (Pt), and the reference electrode was silver/silver chloride (Ag/AgCl). Before modification, the surface of glassy carbon was cleaned with alumina slurry by polishing the electrode in a figure eight motion for 5 min. Then, it was rinsed with deionized water and sonicated for 1 min in deionized water to remove remnants of alumina powder. It was dried at room temperature. The catalyst was then deposited on the GC surface by the drop casting method, where 5% nafion was used as a binder.

3. Characterization Studies

3.1. Fourier-Transform Infrared (FTIR) Spectroscopy

The chemical and structural characteristics of nanocomposites, as well as their atomic and molecular vibrations, were investigated using FTIR (Nicolet 5PC, Nicolet Analytical Instrument (Protea, Cambridgeshire, UK) in a 500 to 4000 cm^{-1} range. The FTIR spectrum of the cobalt tungstate-fabricated nanocomposite is depicted in Figure 3, where the two prominent peaks appeared at 1627.5 and 3350 cm^{-1}, representing N-H bond and hydroxyl bond stretching vibration, respectively [22]. The conjugated g-C_3N_4-based heterocycles are represented by several absorption peaks in the 1600–1200 cm^{-1} region [21], which appeared in nearly all spectra of g-C_3N_4/MetalWO$_4$. The band at 611.8 cm^{-1} corresponds to Co-O-W symmetric vibration [23].

Figure 4 presents the FTIR spectrum of g-C_3N_4/MnWO$_4$. There should be obvious peaks at 875, 827, 710, 605, and 512 cm^{-1} for pristine MnWO$_4$. The peaks at 870, 704, 621, and 514 cm^{-1} correspond to the W–O (symmetric), W–O (asymmetric), Mn–O, and Mn–O–Mn stretching vibrations, respectively, which confirms the formation of manganese tungstate [21]. In Figure 4, two strong peaks appeared at 794 and 670 cm^{-1}, demonstrating W–O (symmetric) and Mn–O–Mn stretching vibrations, respectively. Hence, Figure 4 confirms the successful preparation of g-C_3N_4/MnWO$_4$.

The FTIR spectrum of graphitic carbon nitride fabricated on nickel tungstate is shown in Figure 5. Strong peaks appeared at 1621 and 3322 cm^{-1}, showing conjugation between graphitic carbon nitride and metal oxide, as in previous spectra. Peaks in the series below 500 cm^{-1} are due to the vibrations of the NiO_6 polyhedron [24], which is not observed in this figure as the spectrum range is from 500 to 1500 cm^{-1}.

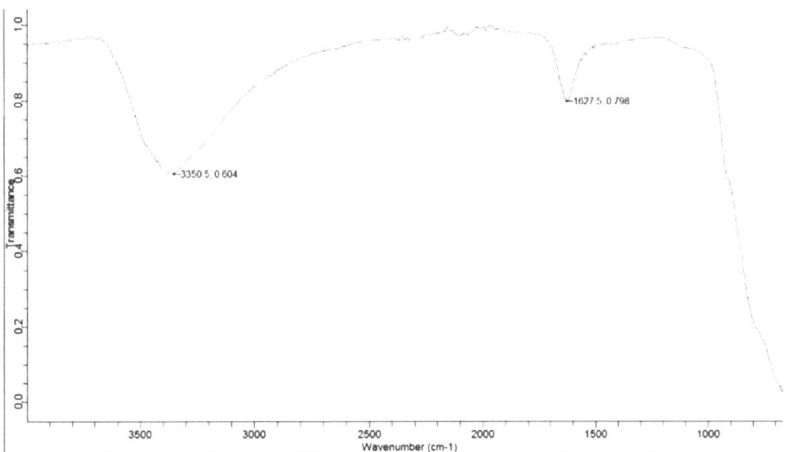

Figure 3. FTIR spectrum of g-C_3N_4-$CoWO_4$.

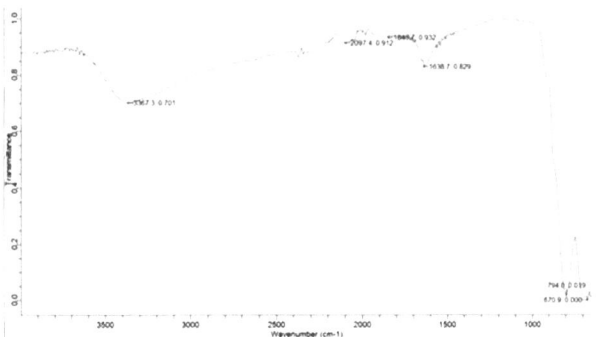

Figure 4. FTIR spectrum of g-C_3N_4-$MnWO_4$.

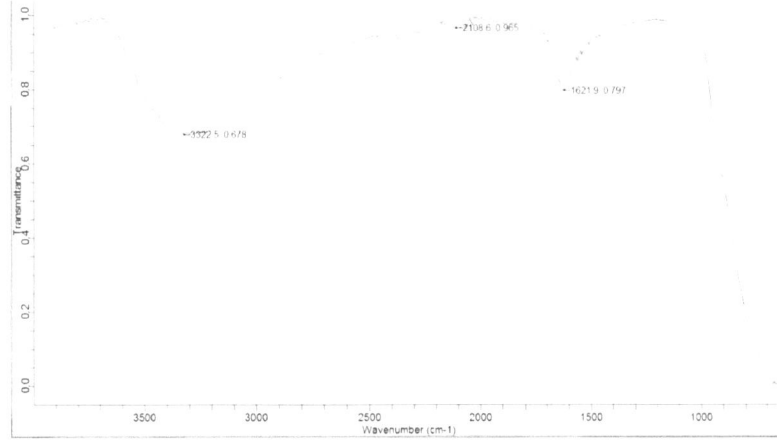

Figure 5. FTIR spectrum of g-C_3N_4-$NiWO_4$.

The FTIR of the tin tungstate–graphitic carbon nitride sample in Figure 6 depicts a band at 1000 cm^{-1} which is due to the W=O bond. The band at 650 cm^{-1} corresponds to Sn-O vibrational modes, which slightly appeared in Figure 6.

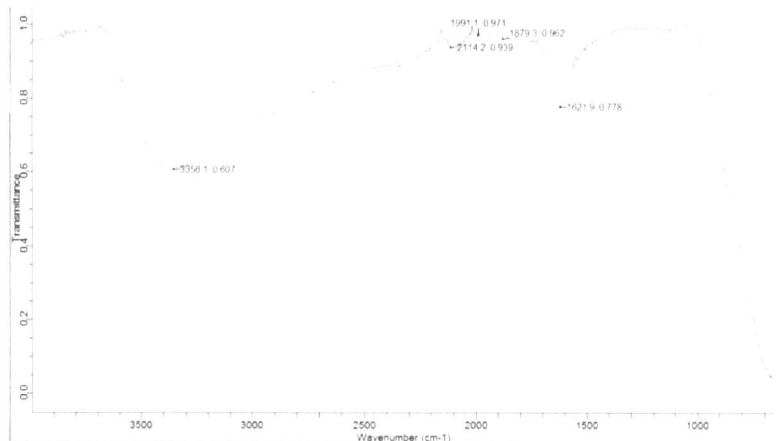

Figure 6. FTIR spectrum of g-C$_3$N$_4$-SnWO$_4$.

In Figure 7, the strong peak at 768 cm^{-1} is due to the W-O bond in the tetrahedral. The peak at 3350 cm^{-1} is due to OH stretching vibrations. Absorption peaks at 471 cm^{-1} are due to the bond between strontium and oxygen [25], which did not appear here due to the selected frequency range, which is 500–1500 cm^{-1}.

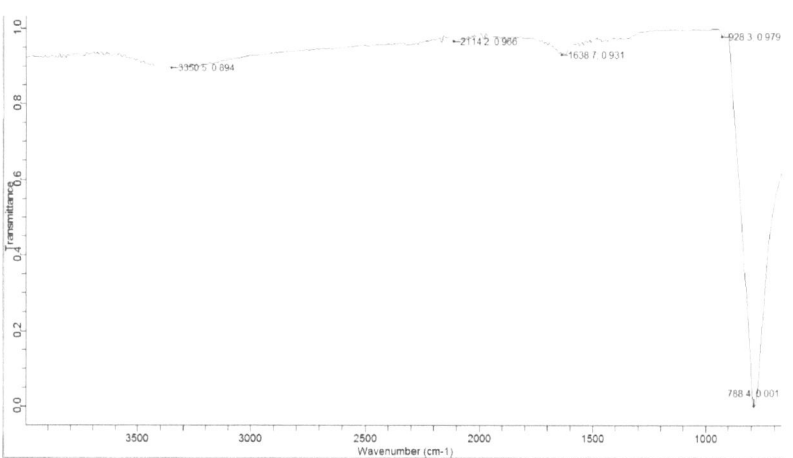

Figure 7. FTIR spectrum of C$_3$N$_4$-SrWO$_4$.

3.2. X-ray Diffraction (XRD)

X-ray diffraction (PANalytical X'PERT High Score's diffractometer, Malvern, UK) patterns of nanocomposites prepared by the hydrothermal method are shown in Figure 8. All the diffraction peaks in Figure 8 are sharp, indicating the good crystallinity of the material. Figure 8a represents the peaks of 2θ for g-C$_3$N$_4$/SnWO$_4$ at 28.3° and 53.1° with crystallographic plans of (121) and (161), with JCPDS card No. 29-1354 [26], which confirms the formation of orthorhombic α-SnWO$_4$. The spectrum in Figure 8b shows the XRD of g-C$_3$N$_4$/MnWO$_4$, which shows intensity peaks for MnWO$_4$ at 30.1°, 30.57°, 36.25°, and

67.1°, with crystal lattice (−111), (111), (120), and (140), respectively, with JCPDS card No. 72-0478 [27], along with the low-intensity peak of g-C_3N_4 at 26.7° (002). This spectrum in Figure 8b depicts the formation of g-C_3N_4/$MnWO_4$. The XRD spectrum in Figure 8c of g-C_3N_4/$CoWO_4$ shows the characteristic peak of g-C_3N_4 at 13.1°, indexed as (100) with JCPDS card No. 50-1512 [28], and this is due to the in-plane structural packing motif. The peaks for $CoWO_4$ appeared at 30.8°, 36.4°, and 54.3°, indexed as (111), (200), and (202), respectively, with JCPDS card No. 72-0479 [29]. The peaks of g-C_3N_4/$SrWO_4$ presented in Figure 8d are 33.6° and 55.8°, with corresponding miller indices (204) and (312) and JCPDS card No. 08-0490, which agree well with the reported data of $SrWO_4$ for Scheelite phase [30]. Figure 8e represents the XRD spectrum of g-C_3N_4/$NiWO_4$, having peaks at 31.1°, 36.2°, 55.4°, and 65.2° for $NiWO_4$ with JCPDS card No. 15-0755 [31]; thus, this confirmed the synthesis of g-C_3N_4/$NiWO_4$. The intensity of the peaks in many spectra decreased and the peak width increased, which indicated the interaction between metal tungstate and C_3N_4 nanoparticles, and thus confirmed the successful formation of nanocomposites. No characteristic peaks for other impurity phases were observed in all five XRD patterns, showing that the selected synthetic method is a feasible route to prepare pure phases of catalysts.

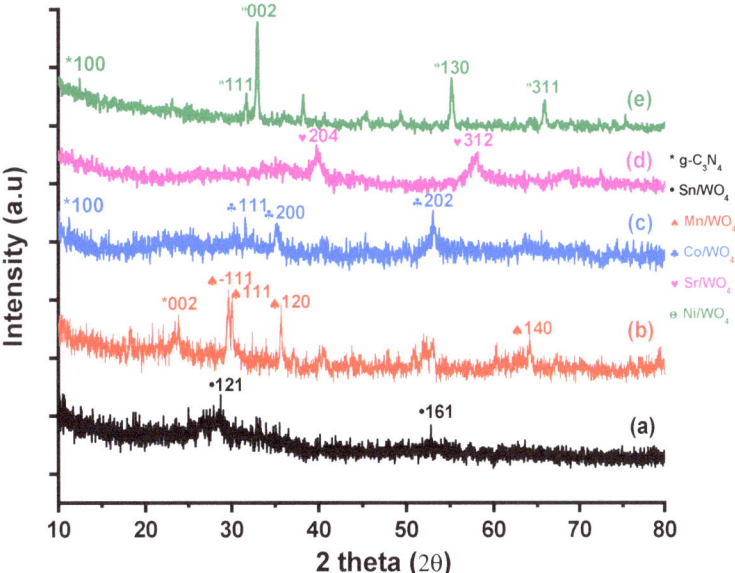

Figure 8. XRD patterns of (**a**) g-C_3N_4, (**b**) g-C_3N_4/$MnWO_4$, (**c**) g-C_3N_4/$CoWO_4$, (**d**) g-C_3N_4/$SrWO_4$ and (**e**) g-C_3N_4/$NiWO_4$ nanocomposites.

The size of the prepared materials was calculated by using the Debye–Scherrer equation [32]:

$$D_{avrg} = \frac{k\lambda}{\beta \cos\theta} \quad (1)$$

where D_{avrg} is equal to the average crystallite size, λ is the X-ray wavelength, θ is the Bragg angle of the desired peak, and β is the line broadening evaluated from the peak width at half height. The average crystalline size of all synthesized particles is presented in Table 1. In XRD patterns, the peak width is inversely proportional to the crystallite size. Normally, nanoparticles of small-sized peaks have a small particle size, while those of sharp peaks have large particle sizes. The small particle size is responsible for more catalytic activity [33].

Table 1. Calculated values of D_{avrg}.

Samples	D_{avrg} (XRD) (nm)
g-C_3N_4/$SnWO_4$	9.0
g-C_3N_4/$MnWO_4$	16.3
g-C_3N_4/$CoWO_4$	52.6
g-C_3N_4/$SrWO_4$	84.4
g-C_3N_4/$NiWO_4$	30.3

3.3. Estimation of Active Surface Area of Modified Electrode

To determine the active surface area of electrodes, which is altered by catalysts having diverse loadings of nanocomposites, cyclic voltammograms were recorded at a scan rate of 100 mV s^{-1} in a 0.1 M KCl solution, with 5 mM of potassium ferrocyanide as a model redox mediator (Figure 9). The peak currents show that the highest value corresponds to the $NiWO_4$-modified electrode. This means that the sample $NiWO_4$-presents better conductive properties towards the redox probe as related to the rest of the catalysts.

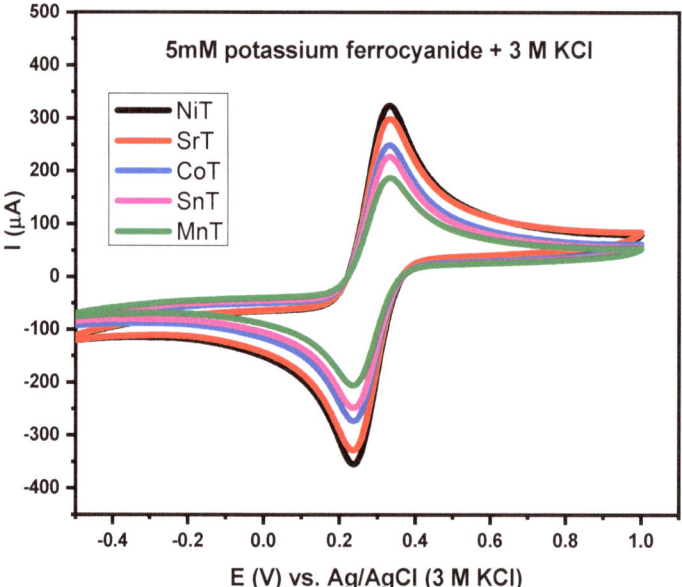

Figure 9. Cyclic voltammogram in 5 mM of potassium Ferrocyanide redox couple, with 0.1 M KCl as the supporting electrolyte at a 100 mV s^{-1} sweep rate at the surface of the modified GC.

The active surface area was estimated using the Randles–Sevcik equation for reversible reactions [34]:

$$Ip_a = 2.69 \times 10^5 n^{3/2} A D^{1/2} v^{1/2} C \tag{2}$$

where D is the diffusion coefficient (0.76×10^{-5} cm^{2}s^{-1}), n = 1, and C is the concentration of analytes (5 mM).

From Table 2, it can be seen that g-C_3N_4 with nickel tungstate presented a higher surface area as compared to the other metal tungstates. This means that the nickel tungstate-fabricated graphitic carbon nitride is an efficient electrocatalyst towards an electrochemical redox reaction.

Table 2. Active surface area calculated for modified electrodes.

Samples	Active Surface Area (cm^{-2})
g-C$_3$N$_4$/SnWO$_4$	0.061
g-C$_3$N$_4$/MnWO$_4$	0.050
g-C$_3$N$_4$/CoWO$_4$	0.068
g-C$_3$N$_4$/SrWO$_4$	0.081
g-C$_3$N$_4$/NiWO$_4$	0.088

4. Electrochemical Water Oxidation

CV (Gamry Interface 1000) was used to perform electrochemical studies, such as water oxidation. Three electrodes were utilized for the experiment: modified GC as a working electrode, *Pt-wire* as a counter electrode, and silver-silver chloride as a reference electrode. Water oxidation was carried out in a 1 M KOH solution in a potential window ranging from 0.5 to 1.8 V at different scan rates (10 to 100 mV s^{-1}). The cyclic voltammetry behaviors of nanocomposites were used to investigate the redox kinetics of the electrodes (Figure 10a–e). The anodic peak current with a growing scan rate from 10 to 100 mV s^{-1} was represented, which shows that the current increased with the increasing sweep rate, from 1.25 to 4.46 mA for g-C$_3$N$_4$-CoWO$_4$/GC, from 0.17 to 0.45 mA for g-C$_3$N$_4$/MnWO$_4$/GC, from 0.2 to 1.7 mA for g-C$_3$N$_4$-SnWO$_4$/GC, from 2.5 to 6.5 mA for g-C$_3$N$_4$-NiWO$_4$/GC, and from 2.0 to 5.5 mA for g-C$_3$N$_4$-SrWO$_4$/GC, and the peak potential shifted towards higher (anodic) potentials. The onset potential for all five nanocomposites is presented in Table 3. The reported onset potential for water oxidation on CoWO$_4$ was ~1.54 V (NHE) [35], while CoWO$_4$-fabricated g-C$_3$N$_4$ was 1.12 V (NHE). An ideal potential value for the water oxidation reaction is 1.23 V (NHE) [36–38]. From Table 3, it can be seen that all synthesized electrocatalysts presented lower onset potential values (NHE) for OER, which revealed that all the nanocomposites performed well in cyclic voltammetry, inherently possessing superior oxygen evolution activity. The peak current of different synthesized nanocomposites is shown in Figure 10f and Table 3, which was in the order: g-C$_3$N$_4$/NiWO$_4$ > g-C$_3$N$_4$/SrWO$_4$ > g-C$_3$N$_4$/CoWO$_4$ > g-C$_3$N$_4$/SnWO$_4$ > g-C$_3$N$_4$/MnWO$_4$. Hence, these synthesized nanocomposites are favorable for being excellent electrochemical catalysts towards water oxidation.

Table 3. Kinetics parameters for water oxidation at g-C$_3$N$_4$/MetalWO$_4$/GC.

Samples	E_{onset} (V)	Ip$_a$ (mA)	α	$D^o/10^{-6}$ cm^{2}s^{-1}
g-C$_3$N$_4$/SnWO$_4$	1.08 (NHE =1.23)	1.5	0.2	40.72
g-C$_3$N$_4$/MnWO$_4$	0.999 (NHE = 1.19)	0.45	0.3	14.09
g-C$_3$N$_4$/CoWO$_4$	0.92 (NHE = 1.12)	4.5	0.2	130.00
g-C$_3$N$_4$/SrWO$_4$	0.86 (NHE = 1.06)	5.3	0.2	132.06
g-C$_3$N$_4$/NiWO$_4$	0.80 (NHE = 1.01)	6.5	0.1	133.3

The linear relationship between peak current (Ip$_a$) and scan rates ($v^{1/2}$) is presented in Figure 11 and illustrates that the water oxidation on the surface of metal oxide catalysts is diffusion-controlled.

Determination of Diffusion Coefficient (D^o)

The diffusion coefficient was determined by using the Randles–Sevcik formula for irreversible reactions [32]:

$$Ip_a = 2.69 \times 10^5 \, n^{3/2} \, AC \times (1-a) \, n \, D^{1/2} \, v^{1/2} \tag{3}$$

where A is the surface area of the electrode (0.07 cm^2), C is the concentration of KOH (1 M), N is the number of electrons for OER (2), and α is the transfer coefficient (0.2–0.7). From the slope of the plot of I_{p_a} vs. $\nu^{1/2}$, we obtained the diffusion coefficients tabulated in Table 3. Table 3 depicts that g-C$_3$N$_4$/NiWO$_4$ has a higher value of the diffusion coefficient as compared to the other catalysts, showing that g-C$_3$N$_4$/NiWO$_4$ nanoparticles are an excellent candidate for rapid redox reactions towards water oxidation.

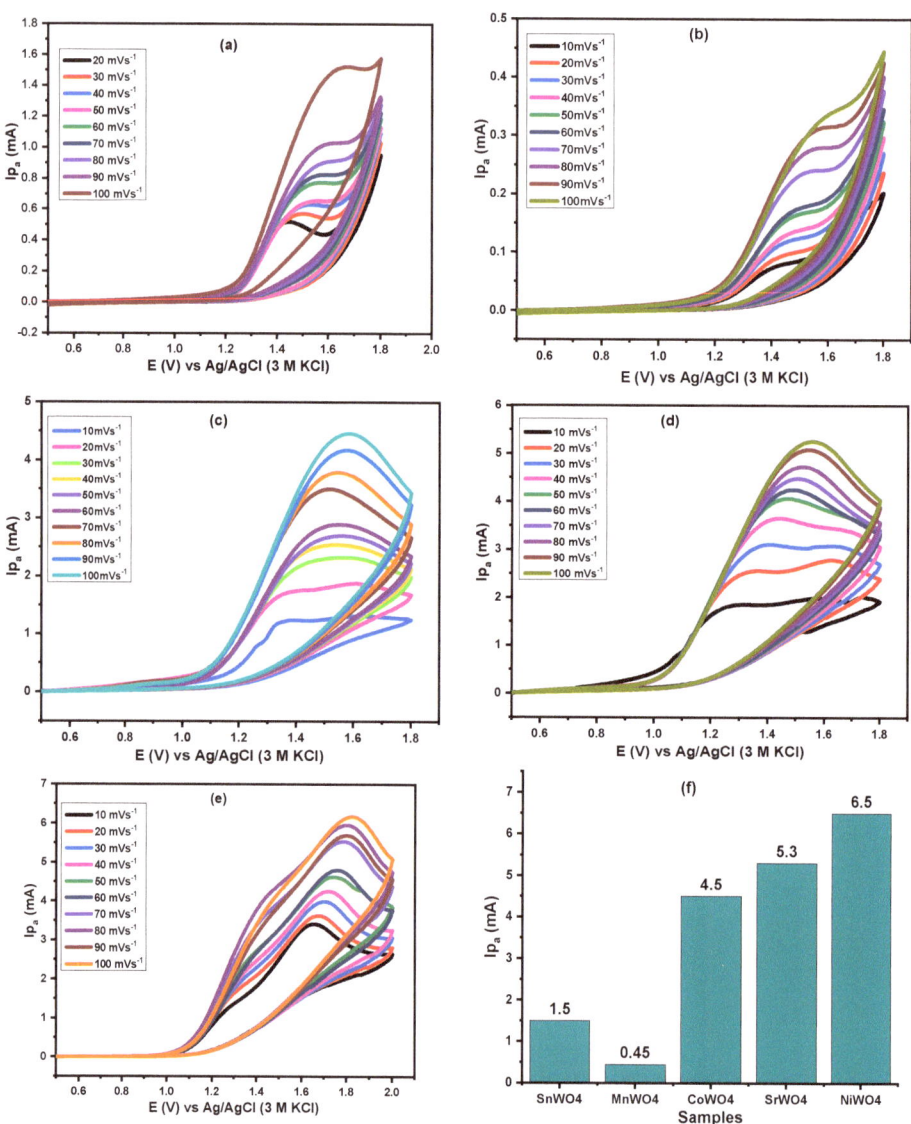

Figure 10. CV curves of water oxidation in 1 M KOH at different sweep rates (10–100 mV s^{-1}) of: (**a**) g-C$_3$N$_4$-SnWO$_4$, (**b**) g-C$_3$N$_4$-MnWO$_4$, (**c**) gC$_3$N$_4$-CoWO$_4$, (**d**) g-C$_3$N$_4$-SrWO$_4$, and (**e**) g-C$_3$N$_4$-NiWO$_4$. (**f**) Comparative peak current of all nanocomposites at 100 mV s^{-1} in 0.1 M KOH.

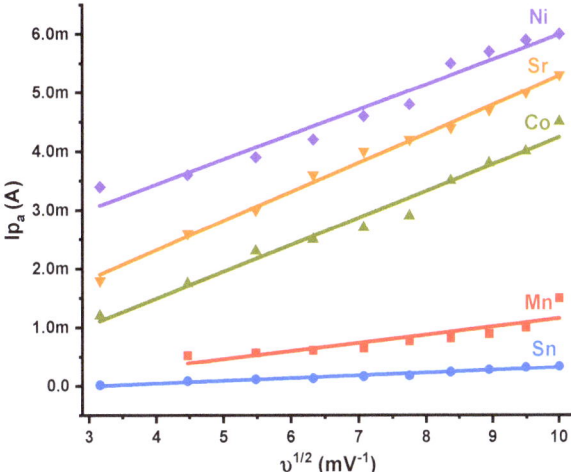

Figure 11. Dependence of I_{p_a} on $v^{1/2}$ showing linear fitting for scan rates for g-C$_3$N$_4$/MetalWO$_4$.

5. Conclusions

The present study presented a simple fabrication strategy for the fabrication of g-C$_3$N$_4$ on metal tungstates as a potential electrode material for the electrochemical splitting of water samples. Metal tungstates (nickel tungstate, copper tungstate, stannous tungstate, strontium tungstate, and manganese tungstate) were distributed on the graphitic carbon nitride surface. The composites were successfully formed, as revealed by XRD and FTIR analysis. In FTIR spectroscopy for functional group detection, all the composites showed a tungstate peak at 600 cm^{-1}. X-ray diffraction (XRD) was performed at 2 theta, showing characteristic peaks according to their JCPDS card numbers. The electrochemical properties of the electrodes were confirmed via cyclic voltammetry. The g-C$_3$N$_4$-NiWO$_4$ composite electrode displayed a unique electrochemical sensing behavior compared to the other composite electrodes, exhibiting a high peak current and low peak potential. The order of electrocatalytic sensing of different composites towards water oxidation was: g-C$_3$N$_4$-NiWO$_4$ > g-C$_3$N$_4$ SrWO$_4$ > g-C$_3$N$_4$-CoWO$_4$ > g-C$_3$N$_4$-SnWO$_4$ > g-C$_3$N$_4$-MnWO$_4$. Based on the results from the fabricated electrochemical sensor, the g-C$_3$N$_4$/NiWO$_4$ composite displayed the highest active surface area at 0.088 cm^{-2}. In this case, the peak current was higher and the sensitivity was also high. Aside from this, g-C$_3$N$_4$/NiWO$_4$ can also be employed in the manufacture of electrodes used in clinical, pharmaceutical, and medical applications. As a result, the proposed modified electrode can be employed in the long term for electrocatalytic sensing, with good selectivity and reproducibility. The findings indicated that the investigated approach could be useful in regular analytical applications.

Author Contributions: T.A.: Writing & Methodology. M.K.: Writing—review & editing. A.A.: Methodology. I.A.: Investigation. S.A.: Conceptualization. K.A.: Supervision. A.A.A.-K.: Data curation. A.M.T.: Funding acquisition. All authors have read and agreed to the published version of the manuscript.

Funding: This work was funded by the Researchers Supporting Project Number (RSP-2021/266) King Saud University, Riyadh, Saudi Arabia.

Institutional Review Board Statement: Not applicable.

Informed Consent Statement: Not applicable.

Conflicts of Interest: The authors declare no conflict of interest.

References

1. Ibrahim, H.; Ilinca, A.; Perron, J. Energy storage systems—Characteristics and comparisons. *Renew. Sustain. Energy Rev.* **2008**, *12*, 1221–1250. [CrossRef]
2. Zinatloo-Ajabshir, S.; Salehi, Z.; Salavati-Niasari, M. Synthesis of dysprosium cerate nanostructures using Phoenix dactylifera extract as novel green fuel and investigation of their electrochemical hydrogen storage and Coulombic efficiency. *J. Clean. Prod.* **2019**, *215*, 480–487. [CrossRef]
3. Zeng, K.; Zhang, D. Recent progress in alkaline water electrolysis for hydrogen production and applications. *Prog. Energy Combust. Sci.* **2010**, *36*, 307–326. [CrossRef]
4. Khan, S.B.; Kamal, T.; Asiri, A.M.; Bakhsh, E.M. Iron doped nanocomposites based efficient catalyst for hydrogen production and reduction of organic pollutant. *Colloids Surf. A Physicochem. Eng. Asp.* **2021**, *608*, 125502. [CrossRef]
5. Walter, M.G.; Warren, E.L.; McKone, J.R.; Boettcher, S.W.; Mi, Q.; Santori, E.A.; Lewis, N.S. Solar water splitting cells. *Chem. Rev.* **2010**, *110*, 6446–6473. [CrossRef]
6. Tian, J.; Liu, Q.; Asiri, A.M.; Sun, X. Self-supported nanoporous cobalt phosphide nanowire arrays: An efficient 3D hydrogen-evolving cathode over the wide range of pH 0–14. *J. Am. Chem. Soc.* **2014**, *136*, 7587–7590. [CrossRef]
7. Lee, Y.; Suntivich, J.; May, K.J.; Perry, E.E.; Shao-Horn, Y. Synthesis and activities of rutile IrO2 and RuO2 nanoparticles for oxygen evolution in acid and alkaline solutions. *J. Phys. Chem. Lett.* **2012**, *3*, 399–404. [CrossRef]
8. Ray, C.; Pal, T. Retracted article: Recent advances of metal–metal oxide nanocomposites and their tailored nanostructures in numerous catalytic applications. *J. Mater. Chem. A* **2017**, *5*, 9465–9487. [CrossRef]
9. Seo, C.U.; Yoon, Y.; Kim, D.H.; Choi, S.Y.; Park, W.K.; Yoo, J.S.; Baek, B.; Bin Kwon, S.; Yang, C.-M.; Song, Y.H.; et al. Fabrication of polyaniline–carbon nano composite for application in sensitive flexible acid sensor. *J. Ind. Eng. Chem.* **2018**, *64*, 97–101. [CrossRef]
10. Muthusankar, E.; Ragupathy, D. Supercapacitive retention of electrochemically active phosphotungstic acid supported poly (diphenylamine)/MnO$_2$ hybrid electrode. *Mater. Lett.* **2019**, *241*, 144–147. [CrossRef]
11. Das, T.K.; Prusty, S. Review on conducting polymers and their applications. *Polym. Technol. Eng.* **2012**, *51*, 1487–1500. [CrossRef]
12. Su, F.; Mathew, S.C.; Lipner, G.; Fu, X.; Antonietti, M.; Blechert, S.; Wang, X. mpg-C3N4-catalyzed selective oxidation of alcohols using O$_2$ and visible light. *J. Am. Chem. Soc.* **2010**, *132*, 16299–16301. [CrossRef]
13. Ramalingam, M.; Ponnusamy, V.K.; Sangilimuthu, S.N. A nanocomposite consisting of porous graphitic carbon nitride nanosheets and oxidized multiwalled carbon nanotubes for simultaneous stripping voltammetric determination of cadmium(II), mercury(II), lead(II) and zinc(II). *Mikrochim. Acta* **2019**, *186*, 69. [CrossRef]
14. Eswaran, M.; Dhanusuraman, R.; Tsai, P.-C.; Ponnusamy, V.K. One-step preparation of graphitic carbon nitride/Polyaniline/Palladium nanoparticles based nanohybrid composite modified electrode for efficient methanol electro-oxidation. *Fuel* **2019**, *251*, 91–97. [CrossRef]
15. Qamar, M.A.; Shahid, S.; Javed, M.; Sher, M.; Iqbal, S.; Bahadur, A.; Li, D. Fabricated novel g-C$_3$N$_4$/Mn doped ZnO nanocomposite as highly active photocatalyst for the disinfection of pathogens and degradation of the organic pollutants from wastewater under sunlight radiations. *Colloids Surf. A Physicochem. Eng. Asp.* **2021**, *611*, 125863. [CrossRef]
16. Dao, V.D.; Nguyen, T.D.; Van Noi, N.; Ngoc, N.M.; Pham, T.D.; Van Quan, P.; Trang, H.T. Superior visible light photocatalytic activity of g-C$_3$N$_4$/NiWO$_4$ direct Z system for degradation of gaseous toluene. *J. Solid State Chem.* **2019**, *272*, 62–68. [CrossRef]
17. Mousavi, M.; Habibi-Yangjeh, A. Decoration of Fe$_3$O$_4$ and CoWO$_4$ nanoparticles over graphitic carbon nitride: Novel visible-light-responsive photocatalysts with exceptional photocatalytic performances. *Mater. Res. Bull.* **2018**, *105*, 159–171. [CrossRef]
18. Mohamed, M.M.; Ahmed, S.A.; Khairou, K.S. Unprecedented high photocatalytic activity of nanocrystalline WO$_3$/NiWO$_4$ hetero-junction towards dye degradation: Effect of template and synthesis conditions. *Appl. Catal. B Environ.* **2014**, *150*, 63–73. [CrossRef]
19. Shanker, G.S.; Panchal, R.A.; Ogale, S.; Nag, A. g-C$_3$N$_4$:Sn-doped In$_2$O$_3$ (ITO) nanocomposite for photoelectrochemical reduction of water using solar light. *J. Solid State Chem.* **2020**, *285*, 121187. [CrossRef]
20. Kumar, A.; Chandel, M.; Thakur, M. Structural modifications of carbon nitride for photocatalytic applications. *Photocatal. Adv. Mater. React. Eng.* **2021**, *100*, 299–331.
21. Jiang, Y.; Virkar, A.V. Fuel composition and diluent effect on gas transport and performance of anode-supported SOFCs. *J. Electrochem. Soc.* **2003**, *150*, A942. [CrossRef]
22. George, J.M.; Antony, A.; Mathew, B. Metal oxide nanoparticles in electrochemical sensing and biosensing: A review. *Mikrochim. Acta* **2018**, *185*, 358. [CrossRef]
23. Zhou, D.; Qiu, C. Study on the effect of Co doping concentration on optical properties of g-C$_3$N$_4$. *Chem. Phys. Lett.* **2019**, *728*, 70–73. [CrossRef]
24. Vilian, A.; Oh, S.Y.; Rethinasabapathy, M.; Umapathi, R.; Hwang, S.-K.; Oh, C.W.; Park, B.; Huh, Y.S.; Han, Y.-K. Improved conductivity of flower-like MnWO$_4$ on defect engineered graphitic carbon nitride as an efficient electrocatalyst for ultrasensitive sensing of chloramphenicol. *J. Hazard. Mater.* **2020**, *399*, 122868. [CrossRef]
25. Adib, K.; Rahimi-Nasrabadi, M.; Rezvani, Z.; Pourmortazavi, S.M.; Ahmadi, F.; Naderi, H.R.; Ganjali, M.R. Facile chemical synthesis of cobalt tungstates nanoparticles as high performance supercapacitor. *J. Mater. Sci. Mater. Electron.* **2016**, *27*, 4541–4550. [CrossRef]
26. Pourmortazavi, S.M.; Rahimi-Nasrabadi, M.; Khalilian-Shalamzari, M.; Zahedi, M.M.; Hajimirsadeghi, S.S.; Omrani, I. Synthesis, structure characterization and catalytic activity of nickel tungstate nanoparticles. *Appl. Surf. Sci.* **2012**, *263*, 745–752. [CrossRef]

27. Hossainian, H.; Salavati-Niasari, M.; Bazarganipour, M. Photodegradation of organic dye using strontium tungstate spherical-like nanostructures; synthesis and characterization. *J. Mol. Liq.* **2016**, *220*, 747–754. [CrossRef]
28. Huang, R.; Ge, H.; Lin, X.; Guo, Y.; Yuan, R.; Fu, X.; Li, Z. Facile one-pot preparation of α-SnWO$_4$/reduced graphene oxide (RGO) nanocomposite with improved visible light photocatalytic activity and anode performance for Li-ion batteries. *RSC Adv.* **2013**, *3*, 1235–1242. [CrossRef]
29. Saranya, S.; Senthilkumar, S.T.; Sankar, K.V.; Selvan, R.K. Synthesis of MnWO$_4$ nanorods and its electrical and electrochemical properties. *J. Electroceram.* **2012**, *28*, 220–225. [CrossRef]
30. Zhou, M.; Hou, Z.; Zhang, L.; Liu, Y.; Gao, Q.; Chen, X. n/n junctioned g-C$_3$N$_4$ for enhanced photocatalytic H$_2$ generation. *Sustain. Energy Fuels* **2017**, *1*, 317–323. [CrossRef]
31. Viet, H.P.; Ngoc, A.D.T.; Minh, V.N.; Viet, H.T.T.; Do Van, D.; Thu, T.H.; Minh, P.N. Synthesis and characterization of Z-scheme heterostructure CoWO$_4$/g-C$_3$N$_4$ as a visible-light photocatalyst for removal of organic pollutant. *Vietnam. J. Catal. Adsorpt.* **2021**, *10*, 59–63. [CrossRef]
32. Singh, B.P.; Singh, J.; Singh, R.A. Luminescence properties of Eu^{3+}-activated SrWO$_4$ nanophosphors-concentration and annealing effect. *RSC Adv.* **2014**, *4*, 32605–32621. [CrossRef]
33. Ahmed, M.I.; Adam, A.; Khan, A.; Siddiqui, M.; Yamani, Z.; Qamar, M. Synthesis of mesoporous NiWO$_4$ nanocrystals for enhanced photoelectrochemical water oxidation. *Mater. Lett.* **2016**, *177*, 135–138. [CrossRef]
34. Khan, M.; Janjua, N.K.; Khan, S.; Qazi, I.; Ali, S.; Saad Algarni, T. Electro-oxidation of ammonia at novel Ag$_2$O−PrO$_2$/γ-Al$_2$O$_3$ catalysts. *Coatings* **2021**, *11*, 257. [CrossRef]
35. Smith, P.F.; Deibert, B.J.; Kaushik, S.; Gardner, G.; Hwang, S.; Wang, H.; Al-Sharab, J.F.; Garfunkel, E.; Fabris, L.; Li, J.; et al. Coordination geometry and oxidation state requirements of corner-sharing MnO6 octahedra for water oxidation catalysis: An investigation of manganite (γ-MnOOH). *ACS Catal.* **2016**, *6*, 2089–2099. [CrossRef]
36. Khan, S.; Shah, S.S.; Anjum, M.A.R.; Khan, M.R.; Janjua, N.K. Electro-oxidation of ammonia over copper oxide impregnated γ-Al2O3 nanocatalysts. *Coatings* **2021**, *11*, 313. [CrossRef]
37. AlShehri, S.M.; Ahmed, J.; Ahamad, T.; Arunachalam, P.; Ahmad, T.; Khan, A. Bifunctional electro-catalytic performances of CoWO$_4$ nanocubes for water redox reactions (OER/ORR). *RSC Adv.* **2017**, *7*, 45615–45623. [CrossRef]
38. Rani, B.J.; Ravi, G.; Ravichandran, S.; Ganesh, V.; Ameen, F.; Al-Sabri, A.; Yuvakkumar, R. Electrochemically active XWO$_4$ (X = Co, Cu, Mn, Zn) nanostructure for water splitting applications. *Appl. Nanosci.* **2018**, *8*, 1241–1258. [CrossRef]

Review

Oxidation Protection of High-Temperature Coatings on the Surface of Mo-Based Alloys—A Review

Tao Fu, Fuqiang Shen, Yingyi Zhang *[ ], Laihao Yu, Kunkun Cui, Jie Wang and Xu Zhang

School of Metallurgical Engineering, Anhui University of Technology, Maanshan 243002, China; ahgydxtaofu@163.com (T.F.); sfq19556630201@126.com (F.S.); aa1120407@126.com (L.Y.); 15613581810@163.com (K.C.); wangjiemaster0101@outlook.com (J.W.); zx13013111171@163.com (X.Z.)
* Correspondence: zhangyingyi@cqu.edu.cn

Abstract: Molybdenum and its alloys, with high melting points, excellent corrosion resistance and high temperature creep resistance, are a vital high-temperature structural material. However, the poor oxidation resistance at high temperatures is a major barrier to their application. This work provides a summary of surface modification techniques for Mo and its alloys under high-temperature aerobic conditions of nearly half a century, including slurry sintering technology, plasma spraying technology, chemical vapor deposition technology, and liquid phase deposition technology. The microstructure and oxidation behavior of various coatings were analyzed. The advantages and disadvantages of various processes were compared, and the key measures to improve oxidation resistance of coatings were also outlined. The future research direction in this field is set out.

Keywords: molybdenum alloys; coating; oxidation behavior; microstructure; high-temperature

Citation: Fu, T.; Shen, F.; Zhang, Y.; Yu, L.; Cui, K.; Wang, J.; Zhang, X. Oxidation Protection of High-Temperature Coatings on the Surface of Mo-Based Alloys—A Review. *Coatings* **2022**, *12*, 141. https://doi.org/10.3390/coatings12020141

Academic Editor: Fernando Pedraza

Received: 18 December 2021
Accepted: 21 January 2022
Published: 25 January 2022

Publisher's Note: MDPI stays neutral with regard to jurisdictional claims in published maps and institutional affiliations.

Copyright: © 2022 by the authors. Licensee MDPI, Basel, Switzerland. This article is an open access article distributed under the terms and conditions of the Creative Commons Attribution (CC BY) license (https://creativecommons.org/licenses/by/4.0/).

1. Introduction

With the rapid development of aerospace, national defense and the military industry, electronics, and so on, increasing attention has been paid to the research and application of refractory metals [1–4]. Molybdenum and molybdenum-based alloys have a high melting point (2620 °C), good high-temperature mechanical properties and high conductivity and thermal conductivity, and are widely used in high-temperature structures [5–10]. However, the alloys have a poor oxidation resistance, and the "Pesting oxidation" at 400–800 °C and oxidation decomposition above 1000 °C are the main factors that limit their application [11–14]. At present, the alloying and surface-coating technology are the main methods to increase the oxidation resistance of the basal materials [15,16]. The types of molybdenum alloys and the various surface coating technologies of Mo and its alloys are shown in Figure 1 [17–20]. It can been seen that the Ti, Zr, W, Re, Si, B, Hf, C and rare earth oxides are often added to pure Mo as beneficial elements to prepare molybdenum alloys. However, the result of alloying is not satisfactory when considering the mechanical properties and high-temperature oxidation resistance of the alloys [21,22]. For example, adding a certain amount Ti element to the alloy can enhance its strength, but it will further accelerate the oxidation of the alloy [23]. Mo–Si–B alloys have satisfactory high temperature oxidation resistance, but their fracture toughness is poor. Mo–Ti–Si–B alloys are considered as a promising ultra-high temperature material. However, their oxidation resistance and mechanical properties need to be further studied [24]. In contrast, the surface-coating technology can improve the oxidation resistance of the alloy at high temperature with as little impact on the mechanical properties as possible. Therefore, it is favored by the majority of researchers [25].

In past work, we discussed the composition, structure and oxidation characteristics of HAPC coating on the surface of molybdenum and its alloys in detail [26]. However, there are almost no reviews reporting on research about other methods in this field [27]. In this work, the latest research progress of high-temperature oxidation resistance coatings on the surface of molybdenum and its alloys is reviewed. The characteristics of different surface-coating preparation technologies are summarized and analyzed, including slurry sintering, plasma spraying, chemical vapor deposition, and liquid-phase deposition. As an important physical vapor deposition technology, magnetron sputtering technology is also widely used in metal surface coating [28]. In addition, the molten salt and laser cladding technologies are also mentioned [29–39]. The composition, structure and oxidation characteristics of all kinds of coatings have been given in relevant figures and tables [40]. Moreover, the process characteristics of various methods and key measures to improve the oxidation resistance of coatings are pointed out. The future research and development direction in this field will be outlined.

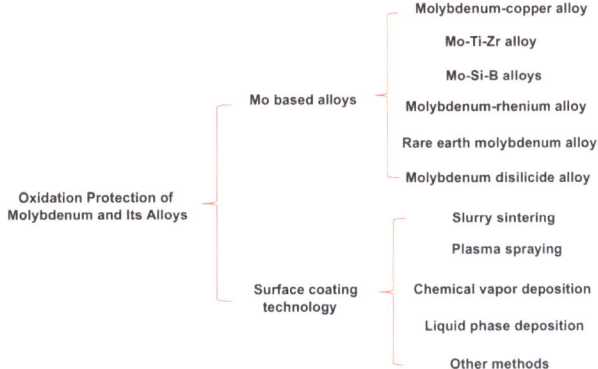

Figure 1. Overview of Mo alloy types and Mo and its alloy surface-coating technology.

2. Microstructure and Oxidation Behavior of Coatings

2.1. Coatings Prepared by Slurry Sintering (SS)

2.1.1. Microstructure and Growth Mechanism of SS Coatings

The slurry sintering (SS) method mixes alloy or silicide powder with binder in a certain proportion and then dissolves it in organic solvent to obtain the mixture. The mixture was evenly coated on the surface of the substrate, and then heated for a certain time in vacuum or Ar atmosphere, so that the substrate and mixture could be fully combined to form a coating on the surface [41,42]. As described in Table 1, the chemical composition and particle size of the mixture and process conditions have important effects on the composition, thickness, surface roughness and mechanical properties of silicide coatings [43–46]. Li et al. [43] investigated the influence of particle sizes of mixtures on surface grain sizes of the coatings. The reports show that the smaller the particle sizes of the mixture, the finer the grain size of the coatings, and they obtained coatings with surface grain size of only 1 to 5 μm by decreasing the particle size of the mixture. Similar results were reported by Wu et al. [44]. However, the surface roughness of the above two kinds of coating is still high. The surface roughness of most areas of the coating is above 15 μm, and the average roughness is 16.36 μm to 18.45 μm, as shown in Figure 2. The authors believe that the higher sintering temperature is the main reason for this result. It is worth noting that the interface layer of the two above coatings is relatively thin, only about 1 μm. In addition, the bonding strength and surface hardness of the coatings are not mentioned. Chakraborty et al. [45] successfully prepared a silicide coating on TZM surface with an interface thickness of 5 μm, a bonding strength of 25 MPa and a surface hardness of 2.00 GPa by slurry sintering technology, which has excellent mechanical properties. This is mainly due to the longer sintering time promoting the interdiffusion between the coating and the substrate, effectively. In addition, the surface quality of the coating will not decrease due to a sintering temperature that is too high.

Table 1. Summary of preparation process, coating composition and surface properties of TZM surface slurry coating.

Substrate	Slurry Composition and Particle Size		Process Conditions			Coating Composition and Thickness (μm)		Bond Strength (MPa)	Surface Hardness (GPa)	Grain Size (μm)	Refs.
	Composition (wt%)	Particle Size (μm)	Atmosphere	Treatment Time and Temperature		Outer Layer	Interface Layer				
TZM	75Si-10Mo-15Ti CN, EAC	1.00–3.00	Vacuum	1450 °C, 15.00 min		$MoSi_2$-(Mo,Ti) Si_2 (120.00)	$(Mo,Ti)_5Si_3$ (1.00)	-	-	1.00–5.00	[43]
	60Si-30Mo-10YSZ-SiO_2-PVB-NH_4F	1.00×10^{-1}	Ar	1450 °C, 1.00 h		$MoSi_2$-$ZrSi_2$-SiO_2 (120.00)	Mo_5Si_3 (1.00)	-	-	2.00–5.00	[44]
	MEK-PVB-10 to 20Si	45.00	Ar	1200 °C, 2.00 h		$MoSi_2$ (60.00)	Mo_5Si_3 (5.00)	25.00	2.00	10.00–20.00	[45]
	69.5Si-30Mo-0.5PVB-EA	-	Ar	1450 °C, 1.00 h		$MoSi_2$ (96.00)	Mo_5Si_3 (3.00)	-	-	2.00–4.00	[46]

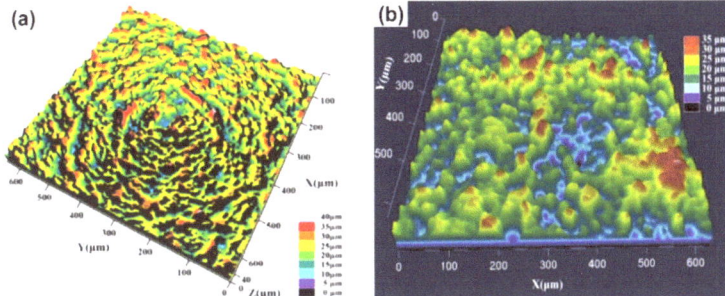

Figure 2. Roughness evolution of the Si-Mo coating (**a**) Reprinted with permission from [46]; reproduced from (Cai et al., 2017). surface roughness of the Si-Mo-5YSZ coating (**b**) Reprinted with permission from [44]; reproduced from (Cai et al., 2018).

The surface and corresponding cross-sectional images of the slurry coatings are shown in Figure 3. A great number of micro-cracks and holes are observed on the coating surface, and the volatilization of flux and binder during sintering is the main reasons for this result. Meanwhile, the high-temperature sintering shrinkage further aggravates the crack propagation and the increase of hole size [47], as shown in Figure 3a–d. However, the inner coatings are relatively dense and compact-bonded with the substrate. A thin interdiffusion zone (IDZ) can be clearly observed between the coating and the substrate. EDS analysis show that the atom ratio of Mo to Si is close to 5:3, which indicate that the inner coatings are a Mo_5Si_3 layer, as shown in Figure 3e,f. This is due to the decrease of diffusion rate with the decrease of silicon concentration during coating preparation, and finally a Mo_5Si_3 layer with lower Si concentration forms at the interface [48].

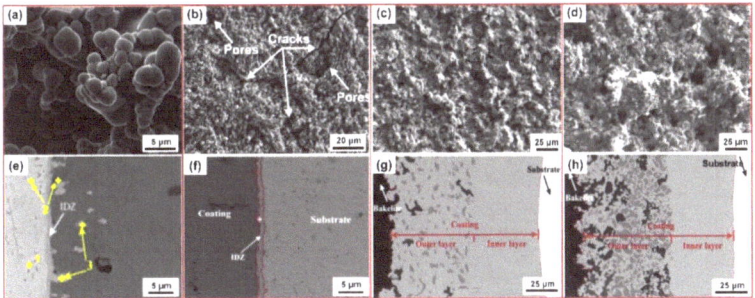

Figure 3. The images of surface and cross-sectional with different coating compositions. (**a,e**) $MoSi_2$/(Mo, Ti)Si_2 coating, Reprinted with permission from [43]; reproduced from (Li et al., 2018). (**b,f**) Si-Mo coating, Reprinted with permission from [46]; Reproduced from (Cai et al., 2017). (**c,g**) Si-Mo-5YSZ coating, (**d,h**) Si-Mo-10YSZ coating, Reprinted with permission from [44]; reproduced from (Cai et al., 2018).

The growth mechanism of the SS coatings and the main equations involved in the reaction process are shown in Figure 4. With the mutual diffusion between mixture and substrate, a dense $MoSi_2$ layer formed on the molybdenum alloys substrate. The content of Si element gradually decreased in the process of diffusion into substrate, and finally a thin interface layer (Mo_5Si_3 and Mo_3Si) formed with low silicon concentration between $MoSi_2$ and substrate. The results show that the growth mechanism of the interface layer in the $MoSi_2$ coating system is the same as that in $MoSi_2$/Mo diffusion couple, but the growth rate of Mo_5Si_3 is much higher than that of Mo_3Si [49,50]. In addition, an appropriate amount of beneficial elements (M), such as Zr, Ti and Y are usually added to the slurry mixture to optimize coating structure [51–53].

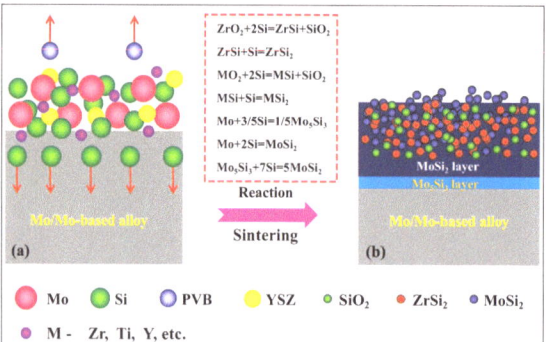

Figure 4. The diagram of growth mechanism of slurry sintering (SS) coating on molybdenum and Mo-based alloys. (**a**) Mixture composition and its diffusion law; (**b**) Structure of the coating after sintering reaction.

2.1.2. Oxidation Behavior and Mechanism of SS Coatings

The microstructure evolution and mass gain of the SS coatings before and after oxidation are shown in Table 2. It is observed that an oxide layer forms on the surface after oxidation, which is mainly composed of SiO_2, TiO_2, Mo_5Si_3, etc. Compared with the original coating, the thickness of the oxidized coating increases significantly, which is due to the volume of the coating expanding and the interface migration caused by the inter-diffusion reaction. However, the thickness of the $MoSi_2$ layer decreases significantly due to the growth of the oxide film and the migration of the interface layer. By contrast, the interdiffusion between the coating and the substrate becomes more sufficient with the increase of exposure time, resulting in a significant increase in the thickness of the interface layer dominated by Mo_5Si_3 [43–46]. The micro-structure and phase composition evolution during the high-temperature (above 1400 °C) oxidation is shown Figure 5. Relevant scholars believe that the integrity of the coating structure and the compactness of the oxide film are the key factors affecting the oxidation service life of the coating [48].

Table 2. Overview of the microstructure evolution and mass gain of SS coating on TZM alloy before and after oxidation.

Substrate	Composition and Thickness of Coatings (µm)		Exposure	Composition and Thickness of Oxidized Coatings (µm)			Mass Gain (mg·cm^{-2})	Refs.
	Outer Layer	Interface Layer		Oxide Layer	Intermediate Layer	Interface Layer		
TZM	$MoSi_2$-(Mo,Ti)Si_2 (120.00)	(Mo,Ti)$_5Si_3$ (1.00)	1600 °C, 5.00 h	SiO_2, Mo_5Si_3, TiO_2 (20.00–30.00)	$MoSi_2$-(Mo,Ti)Si_2 (70.00–75.00)	Mo_5Si_3 (53.00)	4.00	[43]
	$MoSi_2$-$ZrSi_2$-SiO_2 (120.00)	Mo_5Si_3 (1.00)	1725 °C, 6.00 h	SiO_2, ZrO_2, $ZrSiO_4$ (77.00)	$MoSi_2$ (41.00)	Mo_5Si_3 (37.00)	1.00	[44]
	$MoSi_2$ (60.00)	Mo_5Si_3 (5.00)	1000 °C, 5.00 h	-	-	-	Negligible	[45]
	$MoSi_2$ (96.00)	Mo_5Si_3 (3.00)	1650 °C, 4.00 h	SiO_2 (24.00)	$MoSi_2$ (41.00)	Mo_5Si_3 (44.00)	5.00×10^{-1}	[46]

The BSE images of the oxidized SS coatings are shown in Figure 6. The oxidation resistance of pure $MoSi_2$ coating is obviously poorer than that of composite coatings. We can clearly see from Figure 6a,d that the oxide layer of the pure $MoSi_2$ coating is very rough with a high porosity, and a large number of holes are observed. The addition of Ti can replace Mo atoms in the $MoSi_2$ coating, which changes the crystal structure of $MoSi_2$ and improves the coating density, as shown in Figure 6e. However, obvious cracks are still observed on the surface of the oxidized $MoSi_2$-(Mo,Ti)Si_2 composite coating [44]. This is due to the generation of SiO_2 partially crystallized as cristobalite during oxidation, and its phase transformation is accompanied by a volume change, which reduces the adhesion of SiO$_2$ [47], as shown in Figure 6b. Wu et al. [46] report a Si-Mo-10YSZ coating with excellent oxidation performance at high temperature, and the mass gain after oxidation at 1725 °C for 6 h was only 1.00 mg·cm^{-2}. As Figure 6c reveals, ZrO$_2$ and ZrSiO$_4$ oxide particles are dispersed in the oxide film on the surface of the coating, which optimizes the structure of the oxide film and enhances its compactness.

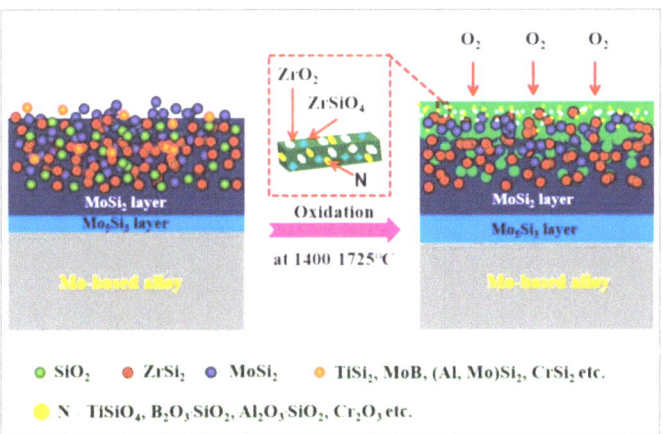

Figure 5. The diagram of oxidation mechanism of the SS coatings on molybdenum and its alloys.

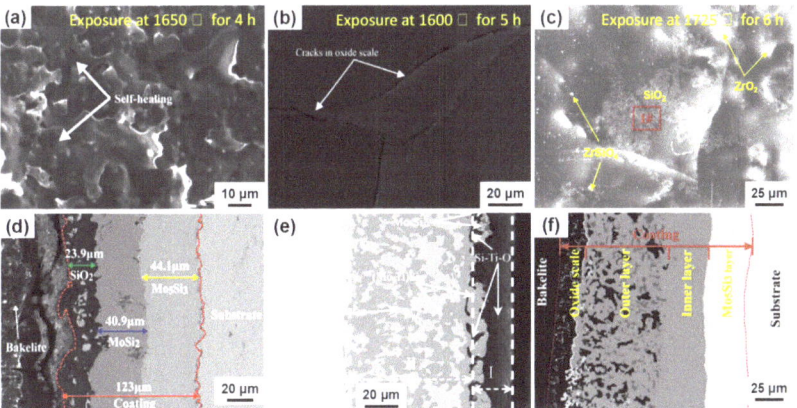

Figure 6. Surface and cross-sections images of SS coatings under different exposure conditions; (**a**,**d**) pure MoSi$_2$ coating, reprinted with permission from [45]; reproduced from (Chakraborty et al., 2016). (**b**,**e**) MoSi$_2$-(Mo, Ti)Si$_2$ coating, reprinted with permission from [46]; reproduced from (Cai et al., 2017). (**c**,**f**) Si-Mo-10YSZ coating, reprinted with permission from [44]; reproduced from (Cai et al., 2018).

2.2. Coatings Prepared by Plasma-Spraying Technique

2.2.1. Microstructure and Growth Mechanism of Plasma-Spraying Coatings

The plasma-spraying technique is one of the most widely used coating preparation in thermal-spraying technology. Its principle is heating and ionizing a certain gas (N$_2$, H$_2$, Ar, He or their mixture) by an electric arc. The generated high-energy plasma arc can heat powdery materials to molten or semi-molten state and spray them onto the substrate surface at high speed to form a coating [54–56]. Among them, the air plasma spraying technique (APS), plasma-transferred arc (PTA) and spark plasma sintering (SPS) are widely used in the surface oxidation protection of Mo and its alloys. The growth mechanism and main reaction equations involved in the preparation of silicide coatings by the plasma-spraying technique are shown in Figure 7 [57–61]. Generally, the spraying material consists of Si, Mo, MoSi$_2$, MoB, B$_4$C, and ZrO$_2$, etc. The MoSi$_2$ particles are formed by silicon powder and molybdenum powder at high temperature, which are attached to the surface of the substrate. At high temperature, the Si element in MoSi$_2$ further diffuses into the substrate, and finally an interface layer dominated by Mo$_5$Si$_3$ is formed between the coating and the substrate [62,63].

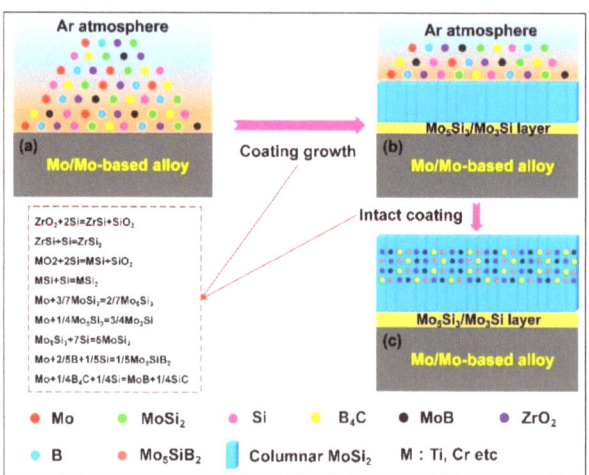

Figure 7. Diagram of the growth mechanism in plasma-spraying coating on molybdenum and its alloys. (**a**) Before spraying reaction (**b**) Coating structure at the initial stage of spraying (**c**) Structure of the coating after spraying.

Table 3 summarizes the process conditions and characteristics of antioxidant coatings prepared by the plasma-spraying technique. The microstructure and characteristics of coatings are mainly affected by powder feeding rate, spray gun power, vacuum degree, carrier gas flow rate, spraying distance, plasma gas composition [57–61]. Wang et al. [57] obtained a MoSi$_2$ coating on Mo substrate surface by the APS method, which had a high porosity and poor mechanical properties. The surface hardness, bonding strength, and porosity were 1.00 GPa, 10.00 MPa, and 28.73%, respectively. However, Deng et al. [58] prepared Mo–Si–B composite coating by the PTA process, and the surface hardness and porosity of the coating were 9.00 GPa and 18.00%, respectively. This was mainly due to the addition of element B, which improved the fluidity of Si and reduced the porosity of the coating. However, the coatings above have almost no interface layer, which results in very low bonding strength between the coating and the substrate. Chakraborty et al. [60] prepared a MoSi$_2$/Mo$_5$Si$_3$ gradient coating with a bonding strength of 40.00 MPa on TZM substrate by APS technology. A thicker interface layer formed between the coating and the substrate, which was due to the longer sintering time that makes the mutual diffusion between the coating and the substrate more sufficient. Baris et al. [61] obtained a Mo$_2$BC/MoB coating on TZM with surface hardness of 21.00 GPa by SPS technology. A large number of fine granular boron and carbide dispersed phases were generated during sintering, which was the main reason for the increase of coating hardness.

Table 3. Summary of process, composition and properties of plasma spraying coatings on molybdenum and its alloys.

Substrate	Spraying Material	Process Conditions						Composition and Thickness of Coatings (µm)		Bond Strength (MPa)	Surface Hardness (GPa)	Porosity (%)	Refs.
		Gas Flow (L·min^{-1})	Powder (kW)	Distance (mm)	Treatment Temperature and Time	Pressure (MPa)		Outer Layer	Interface Layer				
Mo	$MoSi_2$	Ar: 40.00 H_2: 5.00	32.00	80.00	-	-		$MoSi_2$, Mo_5Si_3 (600.00)	0.00	10.00	1.00	29.00	[57]
	Si, Mo, B	Ar: 6.00	47.00	20.00	-	-		Mo_3Si- Mo_5Si_3- Mo_5SiB_2 (6000.00)	0.00	-	9.00	18.00	[58]
	$MoSi_2$	-	-	-	1500 °C, 5.00 min	30.00		$MoSi_2$ (500.00)	Mo_5Si_3 (20.00)	-	10.00	-	[59]
	$MoSi_2$, ZrO_2, MoB	-	-	-	1500 °C, 5.00 min	30.00		$MoSi_2$, ZrO_2, MoB, Mo_5Si_3 (300.00)	Mo_5Si_3 (10.00)	-	11.00	-	
TZM	Si	-	15.00	100.00	1100–1300 °C, 3.00 h	-		$MoSi_2$ (150.00)	Mo_5Si_3 (10.00)	40.00	1.00	-	[60]
	B_4C	-	-	-	1420 °C, 10.00 min	60.00		Mo_2BC (214.00)	MoB (12.00)	-	21.00	-	[61]

The SEM images and the process conditions of plasma-spraying coatings are shown in Figure 8 and Table 4, respectively [52]. It is obvious that the coatings have a very rough surface, with a large number of pores and a wide diameter. In addition, the coating surface contains a great deal of spherical particles, which is the result of "splashing" liquid droplets generated when the mist particles collide with the coating surface during the spraying process, then deposited on the coating surface again and cooled, as shown in Figure 8a–d. The densification of the cross-sectional coatings is very poor with many pores and uneven distribution, as shown in Figure 8e–h. The uneven particle size and melting degree of sprayed powder during spraying are the main reasons for this result. Meanwhile, the carrier gas remaining inside the coating also exacerbates this result [64,65]. The research shows that the microhardness and bonding strength of the coating can be improved by appropriately increasing the spray gun power or reducing the Ar gas flow rate [57].

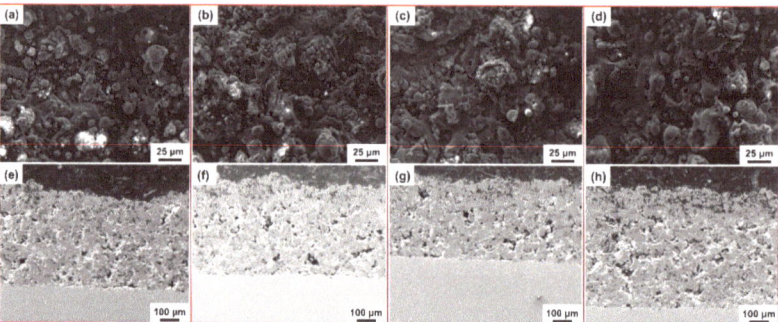

Figure 8. Surface (**a–d**) and cross-sectional (**e–h**) scanning electron microscopy (SEM) images of the MoSi$_2$ coatings on Mo with different process conditions. (**a–d**) represent MSi-1 to MSi-4, respectively. Reprinted with permission from [57]; reproduced from (Wang et al., 2013).

Table 4. Plasma-spraying conditions and the results of the characterization.

Sample No.	Power (kW)	Primary Gas (Ar) Flow (L/min)	Second Gas (H$_2$) Flow (L/min)	Powder Feed Rate (g·min^{-1})	Distance (mm)	Hardness (HV$_{50}$)	Porosity (%)
MSi-1	30.00	40.00	5.00	32.00	120.00	1302.00	30.00
MSi-2	30.00	50.00	5.00	32.00	120.00	1264.00	33.00
MSi-3	32.00	40.00	5.00	32.00	120.00	1303.00	30.00
MSi-4	32.00	50.00	5.00	32.00	120.00	1228.00	34.00

2.2.2. Oxidation Behavior and Mechanism of Plasma-Spraying Coatings

The micro-structure evolution and oxidation behavior of plasma-spraying coatings before and after oxidation are shown in Table 5. It should be noted that except for Mo$_2$BC coating, the mass of the other coatings increases compared with that before oxidation. This is mainly due to the strong affinity force between C and oxygen. During oxidation, the volatilization rate of CO is greater than the formation rate of B$_2$O$_3$, resulting in the reduction of the overall quality of the coating. The oxidation mechanism and structural evolution of the plasma-spraying coatings is shown in Figure 9. The coating surface is gradually covered by a layer of SiO$_2$ with the volatilization of MoO$_3$. However, the strong oxidizing volatilization and volume expansion will further aggravate the surface defects. Finally, the coating fails due to the rapid consumption of the main part. The images of plasma-sprayed coatings after oxidation are shown in Figure 10 [57–59]. It can be seen that the surface of the oxidized coating is very smooth without obvious cracks and holes, which is mainly composed of SiO$_2$ protective film, as shown in Figure 10a–c. Meanwhile, the results of

Figure 10d–f show that the holes and cracks in the coating were filled by SiO$_2$ [57,58]. The 3MoSi$_2$-MoB-3ZrO$_2$ composite coating shows good oxidation resistance compared to pure MoSi$_2$ coating. The analysis shows that the oxidized composite coating has dense structure, uniform composition and obvious layered structure, as shown in Figure 10f [59]. The mass gain (Δ m/S) of the coating is only 4.00×10^{-2} mg·cm^{-2} after oxidizing at 1400 °C for 80 h. By contrast, the MoSi$_2$ coating has failed to oxidize under the same conditions. This is due to the fact that a crack-free oxide film composed of ZrSiO$_4$ and SiO$_2$ phases forms on the surface of the composite coating, which effectively prevents further diffusion of oxygen, as shown in Figure 10c [59]. In addition, during coating preparation, the process parameters, surface roughness and substrate temperature will also have certain effects on the oxidation performance of the coating [66–68].

Table 5. Micro-structure evolution and mass gain of spark plasma sintering (SPS) coatings on molybdenum and its alloys before and after oxidation.

Substrate	Composition and Thickness of Coatings (μm)		Exposure	Composition and Thickness of Oxidized Coatings (μm)			Mass Gain (mg·cm^{-2})	Refs.
	Outer Layer	Interface Layer		Oxide Layer	Intermediate Layer	Interface Layer		
Mo	MoSi$_2$, Mo$_5$Si$_3$ (600.00)	-	1200 °C, 25.00 h	SiO$_2$	MoSi$_2$ (215.00)	Mo$_5$Si$_3$ (10.00)	2.00	[57]
	Mo$_3$Si, Mo$_5$Si$_3$, Mo$_5$SiB$_2$ (6000.00)	Mo$_3$Si, Mo$_5$SiB$_2$ (80.00)	1300 °C, 30.00 h	SiO$_2$, B$_2$O$_3$, MoO$_2$ (30.00)	Mo$_3$Si, SiO$_2$ (15.00)	Mo$_3$Si, Mo$_5$SiB$_2$	8.00	[58]
	MoSi$_2$ (500.00)	Mo$_5$Si$_3$ (20.00)	1400 °C, 80.00 h	SiO$_2$	MoSi$_2$	Mo$_5$Si$_3$	Failure	
	MoSi$_2$, ZrO$_2$, MoB, Mo$_5$Si$_3$ (300.00)	Mo$_5$Si$_3$ (10.00)	1400 °C, 80.00 h	SiO$_2$, ZrSiO$_4$ (2.00)	MoSi$_2$, ZrO$_2$ (396.00)	Mo$_5$Si$_3$, MoB (88.00)	4.00×10^{-2}	[59]
TZM	MoSi$_2$ (150.00)	Mo$_5$Si$_3$ (10.00)	1000 °C, 50.00 h	SiO$_2$ (10.00)	MoSi$_2$ (100.00)	Mo$_5$Si$_3$ (68.00)	1.00	[60]
	Mo$_2$BC (214.00)	MoB, Mo$_2$B (12.00)	1000 °C, 1.00 h	B$_2$O$_3$ (10.00)	Mo$_2$BC	MoB, Mo$_2$B	−12.00	[61]

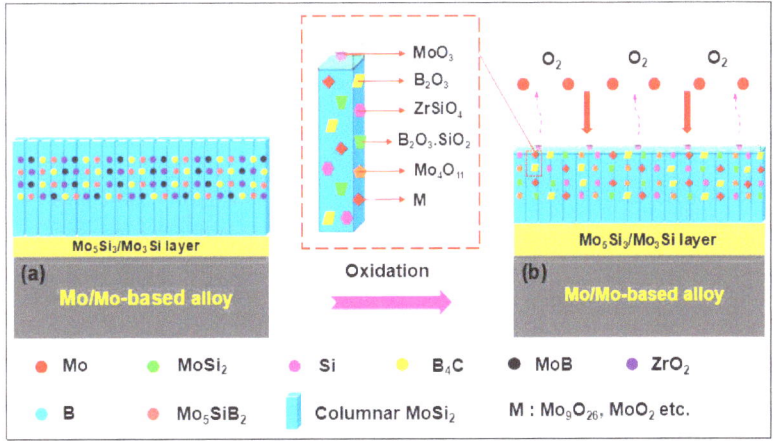

Figure 9. The diagram of oxidation mechanism of the plasma-spraying coatings on molybdenum and its alloys. (**a**) Coating structure before oxidation (**b**) Oxidized coating structure.

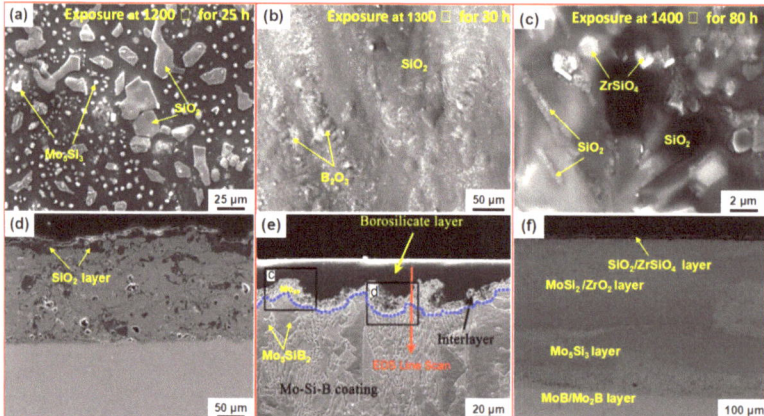

Figure 10. SEM image of surface and cross-section plasma spraying coatings after oxidation under different conditions. (**a**,**d**) Pure MoSi$_2$ coating, reprinted with permission from [57]; reproduced from (Deng et al., 2019). (**b**,**e**) Mo–Si–B coating, reprinted with permission from [58]; reproduced from (Zhu et al., 2019). (**c**,**f**) MoSi$_2$-MoB-ZrO$_2$ coating [59]. Reprinted with permission from [59]; reproduced from (Chakraborty et al., 2011).

2.3. Coatings Prepared by Chemical Vapor Deposition (CVD) Technology

2.3.1. Microstructure and Growth Mechanism of CVD Coatings

The principle behind chemical vapor deposition (CVD) technology is the process of using gaseous substances reacting with a solid substrate to generate solid deposits [69,70]. The process conditions and mechanical properties of the oxidation-resistant coatings prepared on molybdenum by the CVD technique as shown in Table 6 [71–75]. It is obvious that H$_2$ is often used as a carrier gas in the preparation of coatings. The images of the CVD coatings are shown in Figure 11 [71,72,75]. It can be seen that the coatings have a dense and homogeneous surface morphology with a granular structure, as shown in Figure 11a,b. However, the mismatch of thermal expansion coefficients (CTE) between MoSi$_2$ coating and Mo substrate makes obvious vertical cracks sprout inside the coating, as shown in Figure 11c,d. Huang et al. [73] prepared a TiB$_2$ coating on an Mo substrate with an average surface hardness of 28 GPa by CVD technology. However, the average surface hardness of MoSi$_2$ coating obtained under similar conditions is only 13 GPa. The author holds that the finer grains and lower surface roughness of the TiB$_2$ coating are the main reasons for this result.

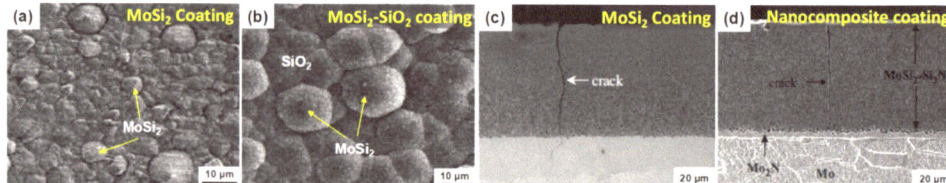

Figure 11. SEM images of different chemical vapor deposition (CVD) coatings on surface of Mo substrate. (**a**,**b**) Reprinted with permission from [71]; reproduced from (Nyutu et al., 2006). (**c**) Reprinted with permission from [72]; reproduced from (Yoon et al., 2005). (**d**) Reprinted with permission from [75]; reproduced from (Yoon et al., 2004).

Table 6. Summary of process, composition and properties of CVD coatings on Mo surface.

Substrate	Composition of Gas Mixture	Process Conditions			Composition and Thickness of Coatings (μm)		Bond Strength (MPa)	Hardness (GPa)	Surface Grain Size (μm)	Refs.
		Gas Flow Rate (ml·min^{-1})	Deposition Temperature (°C)	Deposition Time (h)	Outerlayer	Interface Layer				
Mo	$SiCl_4$, H_2	$SiCl_4$: 50.00 H_2: 100.00	620.00	3.00	SiO_2 (3.00)	$MoSi_2$ (5.00)	-	-	15.00	[71]
	NH_3, $SiCl_4$, H_2	NH_3: 100.00 H_2: 990.00 $SiCl_4$: 10.00	1100.00	NH_3: 2.00 $SiCl_4$: 5.00	$MoSi_2$, Si_3N_4 (72.00)	Mo_2N (5.00)	-	-	3.00×10^{-1}	[72]
	BCl_3, $TiCl_4$, H_2	BCl_3: 195.00 $TiCl_4$: 130.00 H_2: 635.00	1000.00	2.00	TiB_2 (13.00)	-	7.00	28.00	2.00	[73]
	WCl_2, H_2	-	1800.00	2.00	W (160.00)	-	-	-	20.00	[74]
	CH_4, $SiCl_4$, H_2	CH_4, H_2: 200.00 $SiCl_4$: 10.00 H_2: 990.00	1200.00, 1100.00	CH_4: 65.00 $SiCl_4$: 10.00	SiC, $MoSi_2$ (60.00)	Mo_2C (25.00)	-	-	3.00×10^{-1}	[75]

The growth mechanism of the oxidation-resistant protective coatings on molybdenum and its alloys prepared by CVD technology is summarized as shown in Figure 12. Si element decomposed from the mixed gas is deposited on substrate and reacts to generate a $MoSi_2$ coating at high temperature [71]. In order to enhance the low-temperature cyclic oxidation resistance of $MoSi_2$ coating, nitriding or carburizing treatment is usually carried out on the substrate before Si deposition. NH_3, CH_4, etc. are often used as nitrogen sources and carbon sources in this process to deposit on the substrate surface [76,77]. Then, a thinner Mo_2N or Mo_2C layer forms on the substrate surface, as shown in Figure 12b. The Mo_2N or Mo_2C layer will gradually consumed in the process of silicon deposition, most of them are replaced by a $MoSi_2$ layer with dispersed Si_3N_4 and SiC particles on the outer layer. The dispersed phase particles (such as Si_3N_4, SiC, etc.) can refine the grain size of $MoSi_2$, which significantly improves the mechanical properties and oxidation resistance of the coating, as shown in Figure 12c,d [72,77].

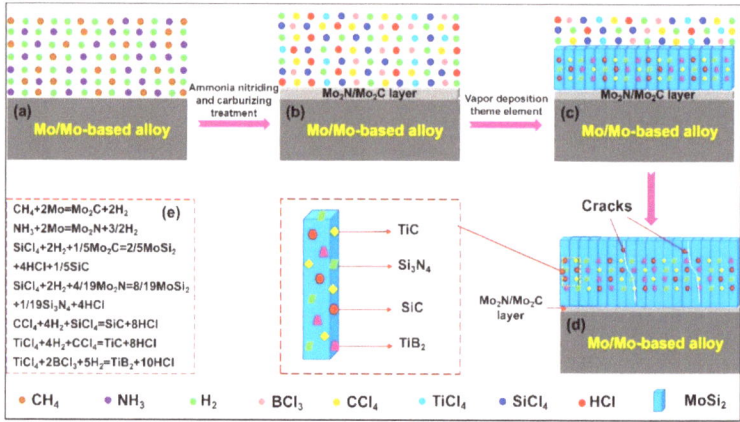

Figure 12. The diagram of growth mechanism of CVD coating on molybdenum and Mo-based alloys. (**a**–**d**) are the different stages of diffusion reaction, (**e**) is the equations involved in the reaction.

2.3.2. Oxidation Behavior and Mechanism of CVD Coatings

Table 7 shows the microstructure evolution and mass gain of CVD coatings before and after oxidation under different conditions. Obviously, researchers mainly reported the oxidation of the coating at low temperature (500 °C to 1000 °C), and the oxidized coatings mainly consist of an oxide layer, intermediate layer and interface layer [71–75,78]. The images of CVD coatings after low-temperature cyclic oxidation are shown in Figure 13. At the initial oxidation stage, oxygen reacts violently with $MoSi_2$ at the crack and holes on the coating surface, and the generated granular SiO_2 spreads along the crack, as shown in Figure 13a. The coating surfaces are gradually covered by SiO_2 with the increase of cycle number. The thickness of oxide layer reaches 90 μm with a high porosity, as shown in Figure 13b,c. However, the oxidized $MoSi_2/β$-SiC nanocomposite coating remains intact, with only a small amount of SiO_2 attached to the coating surface. The oxide layer thickness is only 2 to 3 μm on average, as shown in Figure 13e,f respectively. This is mainly due to the preferential oxidation of SiC particles, which inhibits the oxidation of $MoSi_2$ and reduces the generation of volatile MoO_3. Meanwhile, CO generated during the oxidation process reduces the oxidation pressure in the system, which further reduces the oxidation rate [72]. The author believes that volume expansion caused by low-temperature oxidation causes the failure of the coating. Figure 14 shows the oxidation behavior and mechanism of CVD coating. It can be seen that the longitudinal cracks inside the coating further increase and expand due to the mismatch of thermal expansion coefficient between coating and substrate during the oxidation process [75]. In addition, Anton et al. Prepared Mo–Si thin film coating on the surface of Mo–Si–B alloy by magnetron sputtering. The oxidation test shows that the coating can be used for 300 h at 1200 °C, while the inhibition time of medium temperature pulverization can reach 100 h at 800 °C. Moss- Mo_3Si-Mo_5SiB_2 phase is formed on the coating surface, which significantly improved its antioxidation activity [79–81].

Table 7. Microstructure evolution and mass gain of CVD coatings on molybdenum before and after oxidation.

Substrate	Composition and Thickness of Coatings (μm)		Exposure	Comments	Composition and Thickness of Oxidized Coatings (μm)		Mass Gain (mg·cm^{-2})	Refs.
	Outer Layer	Interface Layer			Oxide Layer	Intermediate Layer		
Mo	SiO_2 (3.00)	$MoSi_2$ (5.00)	1000 °C, 3.00 h	-	SiO_2, MoO_3	$MoSi_2$-Mo_5Si_3	12.00	[71]
	$MoSi_2$-Si_3N_4 (72.00)	Mo_2N (5.00)	500 °C, 1492.00 h	1.00 h cycles	Si_2ON_2, SiO_2, $MoO_3Mo_4O_{11}$, Mo_9O_{26}, (3.00)	$MoSi_2$-Si_3N_4 (100.00)	5.00×10^{-1}	[72]
	TiB_2 (13.00)	-	900 °C, 6.00 h	-	TiO_2, B_2O_3	-	8.00×10^{-2}	[73]
	TiB_2 (13.00)	-	450 °C, 5.00 h	-	TiO_2, B_2O_3	-	3.00×10^{-2}	[78]
	W (160.00)	W/Mo (2.00)	-	-	-	-	-	[74]
	$MoSi_2$-SiC (60.00)	MO_2C (25.00)	500 °C, 1492.00 h	1.00 h cycles	SiO_2, $MoO_3Mo_4O_{11}$, Mo_9O_{26} (8.00)	$MoSi_2$-SiC (80.00)	1.00×10^{-2}	[75]

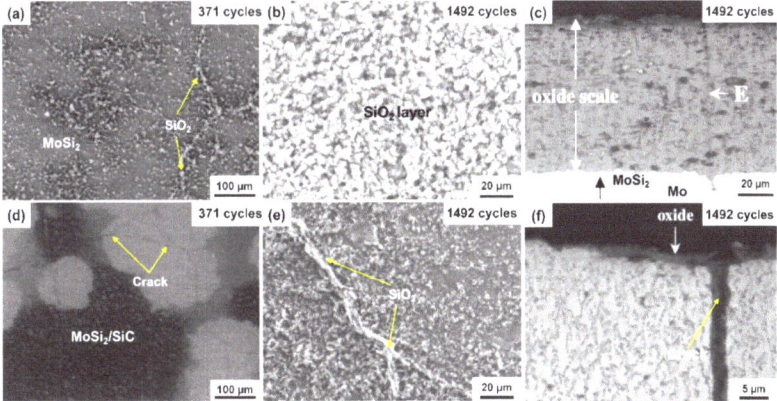

Figure 13. Surface and cross-sections images of oxidized coatings with different types and different cyclic oxidation times at 500 °C. (a–c) $MoSi_2$ coating; (d–f) $MoSi_2/\beta$-SiC nanocomposite coating. Reprinted with permission from [75]; reproduced from (Yoon et al., 2004).

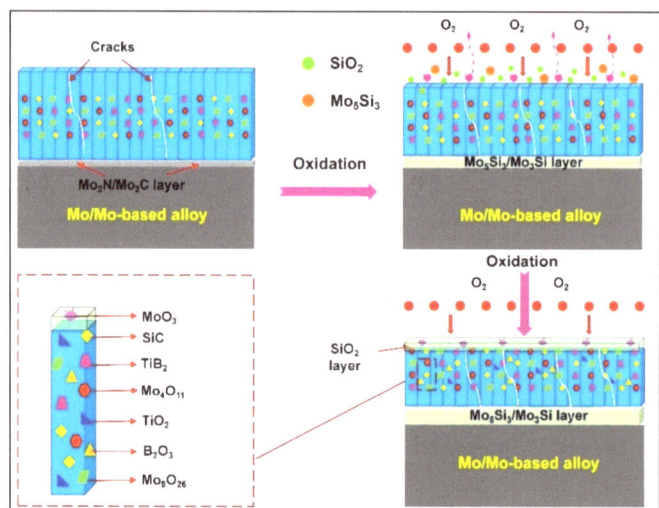

Figure 14. The mechanism diagram of oxidation of CVD coatings on Mo and Mo-based alloys.

2.4. Coatings Prepared by LiquidPhase Deposition Technology

2.4.1. Microstructure and Growth Mechanism of Liquid-Phase Deposition Coatings

Liquid-phase deposition technology inserts the refractory metal alloy into the alloy melt and prepares the intermetallic compound coating by a thermal diffusion reaction in a vacuum or inert gas atmosphere [82,83]. It is considered to be a promising surface coating technology for the oxidation protection of refractory metals [84]. The process details of liquid deposition technology for the preparation of Mo surface Si/Al coatings are shown in Table 8. It can be seen that hot dip time and temperature have important effects on coating composition, thickness and surface grain size [85–89]. Under high temperature and an Ar protection atmosphere, an intense diffusion reaction occurs between liquid silicon and substrate, and columnar $MoSi_2$ grains rapidly grow on the substrate's surface, as shown in Figure 15b. The thickness and surface grain size of the coating increase gradually with the increase of hot dip temperature and time. A thin interface layer (Mo_5Si_3/Mo_3Si layer) with low silicon concentration is observed at the bottom of $MoSi_2$ grain, as shown in Figure 15c. Zhang et al. [90]. reported a $Si-MoSi_2$ coating on Mo substrate by the liquid deposition Si technology. The coating mainly consists of a $MoSi_2$ outer layer and Mo_5Si_3/Mo_3Si interface layer, and the outer layer coating has a high surface silicon concentration and low roughness surface. It is worth noting that the surface of the coating has no cracks, holes and other defects. They find that the surface silicon concentration and grain size increase with the increase of deposition temperature and holding time, and the same conclusion has been reached by Wang et al. [89]. They report Al–Mo coatings with excellent oxidation resistance on the surface of Mo by the liquid-phase deposition Al technology. The Al–Mo coatings are mainly composed by out layer (Al-Al_{12}Mo or Al-Al_4Mo layer) and interface layer (Al_8Mo_3-Al_4Mo or Al_8Mo_3 layer), and the interface layer is thicker than the outer layer. The SEM images of the liquid-phase deposition coatings are shown in Figure 16 [87]. The coatings surface are very smooth and dense, almost no cracks and holes are observed. In addition, high silicon concentration has been observed at $MoSi_2$ grain boundaries, as shown in Figure 16a–c. The cross-sectional morphology shows that the silicide coatings are composed of $MoSi_2$ columnar crystals with a thin transition layer (Mo_5Si_3 and Mo_3Si), as shown in Figure 16d–f.

Table 8. Overview of process, composition and surface properties of hot-dip coating on a molybdenum surface.

Substrate	Osmotic Source and Purity		Process Conditions		Composition and Thickness of Coatings (µm)		Si/Al Content on Coating Surface (wt%)	Coating Surface Grain Size (µm)	Refs.
	Infiltratesource	Purity (wt%)	Atmosphere	Hot Dip Temperature and Time Min	Outer Layer	Interface Layer			
Mo	Si	99.00	Ar	1460 °C, 20 min	Si-MoSi$_2$ (20.00)	Mo$_5$Si$_3$-Mo$_3$Si (3.00)	47.00	-	[85]
				1500 °C, 20 min	Si-MoSi$_2$ (22.00)	Mo$_5$Si$_3$-Mo$_3$Si (2.00)	-	-	[86]
	Si	99.00		1460 °C, 15 min	Si-MoSi$_2$ (15.00)	Mo$_5$Si$_3$-Mo$_3$Si (2.00)	45.00	9.00	[87]
				1520 °C, 15 min	Si-MoSi$_2$ (20.00)	Mo$_5$Si$_3$-Mo$_3$Si (4.00)	56.00	7.00	
	Si	99.00		1490 °C, 5 min	Si-MoSi$_2$ (12.00)	Mo$_5$Si$_3$-Mo$_3$Si (1.00)	42.00	12.00	[88]
				1490 °C, 15 min	Si-MoSi$_2$ (19.00)	Mo$_5$Si$_3$-Mo$_3$Si (4.00)	56.00	7.00	
	Al	99.00	No oxygen	710 °C, 3 min	Al-Al$_{12}$Mo (30.00)	Al$_8$Mo$_3$-Al$_4$Mo (41.00)	90.00	30.00	[89]
				750 °C, 3 min	Al-Al$_4$Mo (35.00)	Al$_8$Mo$_3$ (51.00)	81.00	55.00	

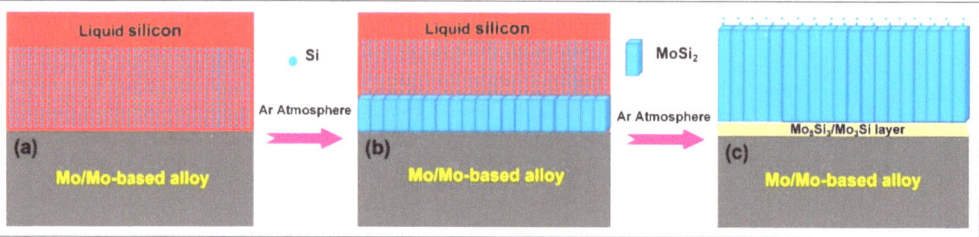

Figure 15. The growth mechanism diagram of liquid-phase deposition coating on molybdenum and its alloys. (**a**) is before reaction, (**b**) is the initial stage of the reaction, (**c**) is the late stage of reaction.

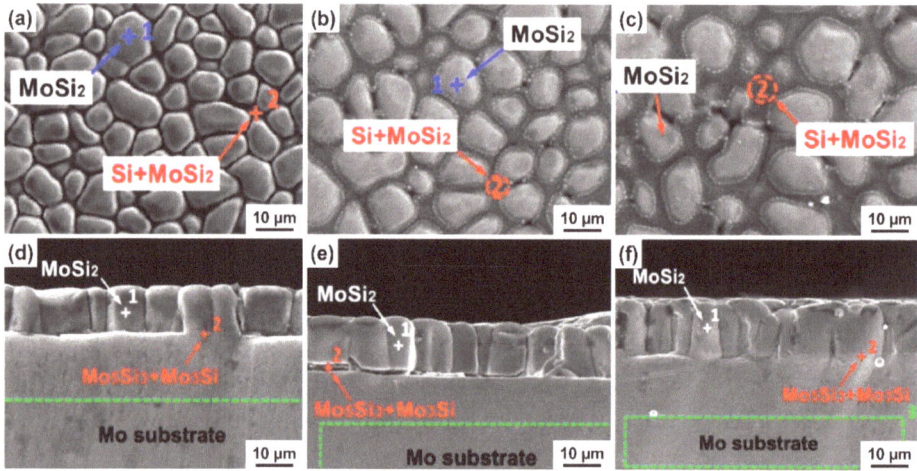

Figure 16. The surface and cross-section micrographs of Si-MoSi$_2$ functionally coatings obtained at different temperature for 15 min. 1460 °C (**a**,**d**), 1490 °C (**b**,**e**), 1520 °C (**c**,**f**). Reprinted with permission from [87]; reproduced from (Zhang et al., 2019).

In addition, Zhang et al. [91] also report a liquid deposition Si coating on TZM substrate. The results show that the surface roughness of the coating does not simply decrease with the increase of deposition time, as shown in Figure 17e–h. The surface of the sample deposited for 10 min is the roughest, and its Sa and Sq are 0.498 and 0.676 µm, respectively, as shown in Figure 17i. In addition, the surface roughness of the samples with deposition time of 15 min and 20 min is relatively low and close to each other, as shown in Figure 17j,k. However, their surface morphology are very different, as shown in Figure 17f,g. During the hot-dipping process, a great deal of molten silicon penetrated into the gap between the MoSi$_2$ particles, and it covered the grain surface during the cooling process. This gives the coatings obtained a smoother and more compact surface structure. It is worth noting that many banded textures were observed on the surface of the samples deposited for 25 min. This is due to the long deposition time reducing the Si viscosity, and the flow direction of the surface silicon changes during the extraction and cooling process of the sample, as shown in Figure 17d.

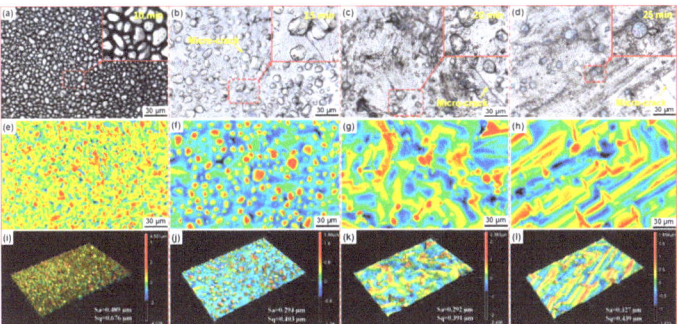

Figure 17. The CLSM (**a–d**), height distribution (**e–h**) and 3D images (**i–l**) of the coating obtained at 1480 °C for different times. Reprinted with permission from [91]; reproduced from (Zhang et al., 2021).

2.4.2. Oxidation Behavior and Mechanism of Liquid-Phase Deposition Coatings

The images of oxidized liquid-phase deposition coatings on Mo substrate are shown in Figure 18. After oxidation at 1200 °C for 2 h, the surface of Si-MoSi$_2$ coating is relatively rough with a small amount of pores, composed of SiO$_2$, Mo$_5$Si$_3$ and MoSi$_2$, as shown in Figure 18a [85]. However, a smooth and dense SiO$_2$ protective film forms on the coating surface when the oxidation temperature is 1600 °C, as shown in Figure 18d [89]. This is due to the good fluidity of SiO$_2$ at high temperature (above 1400 °C), which can fill the defects on the coating surface [90]. The oxidation mechanism and microstructure evolution of the coatings are shown in Figure 19. Compared with MoSi$_2$ phase, Si has a stronger affinity with oxygen in a high-temperature oxidation environment. Therefore, the silicon preferentially is oxidized to SiO$_2$, reducing the formation of volatile MoO$_3$. In order to improve the oxidation resistance of the Al–Mo coating, the coatings obtained at different hot dip temperatures are subjected to micro-arc oxidation (MAO) treatment (pre-oxidation). There are a lot of holes on the oxidized coating surface, which has typical MAO process characteristics, as shown in Figure 18b,c. The outer layer of the coatings is composed of Al$_2$O$_3$, the middle layer is an unoxidized aluminum dipping layer, and the inner layer is a diffusion layer of hot-dip aluminum, as shown in Figure 18e,f. The formation of the structure has great significance for delaying the diffusion of oxygen and prolonging the oxidation service life of the coating [88].

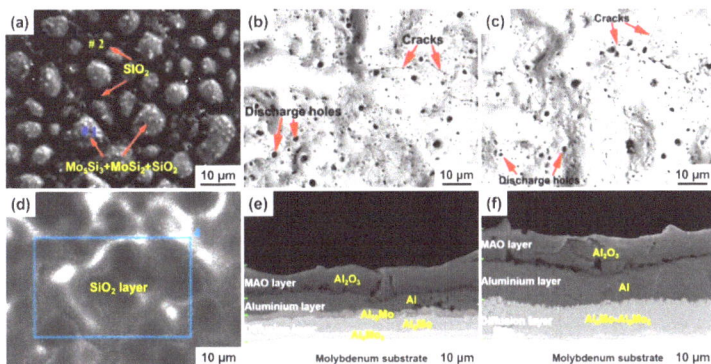

Figure 18. Surface topography of the oxidized Si-MoSi$_2$ coatings at different temperatures for 2 h, (**a**) 1200 °C, (**d**) 1600 °C. Reprinted with permission from [85]; reproduced from (Zhang et al., 2017). Images of Al–Mo coatings prepared at different temperature after MAO for 20 min; (**b,e**) 710 °C, (**c,f**) 750 °C. Reprinted with permission from [88]; reproduced from (Wang et al., 2020).

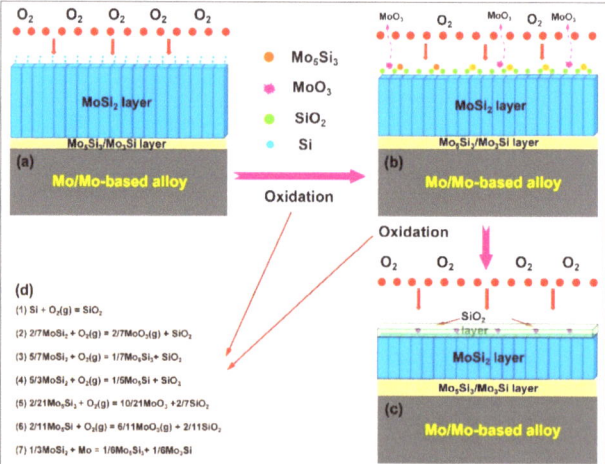

Figure 19. The diagram of oxidation mechanism of liquid-phase deposition coatings on Mo and its alloys. (**a**–**c**) are the coating structures before, during and after the oxidation reaction. (**d**) is the reaction equation involved in the oxidation process.

3. Conclusions and Prospects

As an important high-temperature structural material, the oxidation protection of Mo and its alloys has been of wide interest to relevant scholars. In this paper, the applications of various surface-coating preparation technologies in this field are reviewed, and the composition and oxidation characteristics of the coatings are shown in Figure 20. In addition, the characteristics of different coating preparation processes have also been analyzed and compared, and the details are shown in Table 9. During slurry sintering, due to volatilization of solvent and binder, the prepared coating has poor surface quality and high porosity. Reducing sintering temperature and prolonging sintering time can optimize coating structure and improve coating quality to a certain extent. The lower process temperature of CVD makes the preparation efficiency of the coating low and the preparation time long. However, the technology is suitable for workpieces with complex shapes and the coatings obtained have a good low temperature oxidation resistance. In contrast, plasma spraying and the hot-dip silicon method presented a high deposition efficiency due to high diffusion temperature. After 5 to 25 min of treatment, coatings several tens to several hundred microns thick can be obtained on the substrate surface. However, the plasma-spraying coatings have a high surface roughness and porosity because the spraying material is still mixed with a small quantity of residual gas and solid particles. It is worth noting that liquid-phase deposition coatings have a dense and smooth surface. This is conducive to the formation of protective oxide film on the coating surface in the oxidation process. However, the structure of the coatings are relatively simple, and the oxidation resistance of the coatings needs to be further studied. In addition, the molten salt method and laser-cladding technology have also been widely applied in the preparation of oxidation protective coatings on Mo and Mo-based alloys.

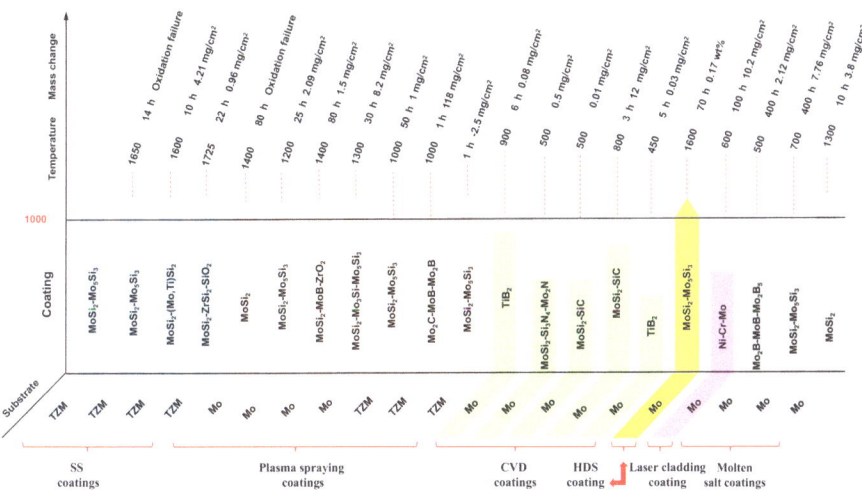

Figure 20. Overview of the composition and oxidation characteristics of silicide coating on molybdenum and its alloys.

Table 9. Summary of preparation methods and process characteristics of oxidation resistance coatings on molybdenum and its alloys.

Method Category	Process Temperature	Time	Advantages	Disadvantages	Refs.
SS method	1200–1450 °C	15.00–120.00 min	1. Simple preparation process, easy operation and low production cost. 2. That process adaptability is strong and the source materials are widely source. 3. That composition of the obtain coating is uniform.	1. The surface quality of the coating is poor, and there are many cracks and holes on the surface of the coating.	[43–46]
SPS method	>10,000 °C	5.00–10.00 min	1. High spraying temperature 2. That operation is simple and the application range is wide. 3. That deposition rate is high, and the coat preparation cost is low	1. That bond strength between the coating and the substrate is low. 2. High porosity of that coat	[57–61]
CVD method	500–1000 °C	2.00–10.00 h	1. The application range is wide and is not limited by the shape of the substrate. 2. The coating composition has uniform thickness and good bonding with the substrate.	1. The deposition temperature is low and the reaction time is long.	[71–75]
HD method	1430–1560 °C	10.00–25.00 min	1. High hot dip temperature, short permeation time and high deposition efficiency. 2. The surface of the coating is smooth, the density is high, and the adhesion between the coating and the substrate is good.	1. The structure of the coating is simple, and research on the oxidation resistance of the coating is relatively rare.	[85–91]

The addition of appropriate beneficial elements in $MoSi_2$ coating has a great significance for improving the antioxidant properties of the coating. Ti element can replace Mo in $MoSi_2$ to form $TiSi_2$ solid solution, which improves the strength and hardness of the coating. In addition, a continuous and dense Si–Ti–O protective film is easy to form on the coating surface during the oxidation process. B element can not only reduce the viscosity of SiO_2

at high temperature (above 1400 °C) and improve the self-healing ability of the coating, it can also combine with Si element to form Mo_5SiB_2 with a lower diffusion coefficient and maintain the coating structure. N element is dispersed in the coating with a granular Si_3N_4, which improves its mechanical properties. In addition, Si_3N_4 distributed on the coating surface is preferentially oxidized, which reduces the oxygen partial pressure in the system and alleviates the oxidation of $MoSi_2$. Similarly, element C exists in the coating in the form of SiC and plays a similar role. With the addition of YSZ/Y, a continuous and dense SiO_2 protective film with ZrO_2 and $ZrSiO_4$ particles forms in the outer layer of the coating, which is conducive to stabilizing the oxide film structure and prolongs the oxidation service life of the coating.

Therefore, oxidation resistance and mechanical properties of the coating can be advanced in the following two ways. On the one hand, the preparation process of the coating should be optimized to ensure the coatings obtained have uniform compositions, compact structures and smooth surfaces. On the other hand, by introducing appropriate amounts of modified elements and the second phases, the structure of the coating can be optimized, the consumption of the coating can be slowed down, and the formation of the continuous and uniform protective oxide film formed on its surface can be accelerated. Furthermore, we can also organically combine the preparation process of a single coating to overcome the problems existing in its single application. This will be the future research and development direction in this field.

Author Contributions: The manuscript was written through contributions of all authors. Y.Z.: Conceptualization, Investigation, and Supervision. Y.Z. and T.F.: Writing original draft and image processing. T.F., K.C. and J.W.: Validation, Resources, Investigation, Writing—review and editing. X.Z., L.Y. and F.S.: Visualization, Writing—review and editing. All authors have read and agreed to the published version of the manuscript.

Funding: This work was supported by the Anhui Province Science Foundation for Excellent Young Scholars (2108085Y19) and the National Natural Science Foundation of China (No.51604049).

Institutional Review Board Statement: Not applicable for studies not involving humans or animals.

Informed Consent Statement: Not applicable.

Data Availability Statement: Not applicable.

Conflicts of Interest: The authors declare no conflict of interest.

References

1. Chaudhuri, A.; Behera, A.N.; Sarkar, A.; Kapoor, R.; Ray, R.K.; Suwas, S. Hot deformation behaviour of Mo-TZM and understanding the restoration processes involved. *Acta Mater.* **2019**, *164*, 153–164. [CrossRef]
2. Tuzemen, C.; Yavas, B.; Akin, I.; Yucel, O.; Sahin, F.; Goller, G. Production and characterization of TZM based TiC or ZrC reinforced composites prepared by spark plasma sintering (SPS). *J. Alloys Compd.* **2019**, *781*, 433–439. [CrossRef]
3. Pint, B.A. Critical Assessment 4: Challenges in developing high temperature materials. *Mater. Sci. Technol.* **2013**, *30*, 1387–1391. [CrossRef]
4. Majumdar, S.; Kale, G.B.; Sharma, I.G. A study on preparation of Mo–30W alloy by aluminothermic co-reduction of mixed oxides. *J. Alloys Compd.* **2005**, *394*, 168–175. [CrossRef]
5. Xu, J.J.; Yang, T.T.; Yang, Y.; Qian, Y.H.; Li, M.S.; Yin, X.H. Ultra-high temperature oxidation behavior of micro-laminated $ZrC/MoSi_2$ coating on C/C composite. *Corros. Sci.* **2018**, *132*, 161–169. [CrossRef]
6. Khlyustova, A.; Sirotkin, N.; Titov, V.; Agafonov, A. Comparison of two types of plasma in contact with water during the formation of molybdenum oxide. *Curr. Appl. Phys.* **2020**, *20*, 1396–1403. [CrossRef]
7. Kong, H.; Kwon, H.S.; Kim, H.; Jeen, G.S.; Lee, J.; Lee, J.; Heo, Y.S.; Cho, J.; Jeen, H. Reductive-annealing-induced changes in Mo valence states on the surfaces of MoO_3 single crystals and their high temperature transport. *Curr. Appl. Phys.* **2019**, *19*, 1379–1382. [CrossRef]
8. Bahamonde, J.P.; Wu, C.Z.; Louie, S.M.; Bao, J.M.; Rodrigues, D.F. Oxidation state of Mo affects dissolution and visible-light photocatalytic activity of MoO_3 nanostructures. *J. Catal.* **2020**, *381*, 508–519. [CrossRef]
9. Smolik, G.R.; Petti, D.A.; Schuetz, S.T. Oxidation and volatilization of TZM alloy in air. *J. Nucl. Mater.* **2000**, *283-287*, 1458–1462. [CrossRef]

10. Majumdar, S.; Kapoor, R.; Raveendra, S.; Sinha, H.; Samajdar, I.; Bhargava, P.; Chakravartty, J.K.; Sharma, I.G.; Suri, A.K. A study of hot deformation behavior and microstructural characterization of Mo–TZM alloy. *J. Nucl. Mater.* **2009**, *385*, 545–551. [CrossRef]
11. Amakawa, K.; Wang, Y.Q.; Kröhnert, J.; Schlögl, R.; Trunschke, A. Acid sites on silica-supported molybdenum oxides probed by ammonia adsorption: Experiment and theory. *Mol. Catal.* **2019**, *478*, 110–580. [CrossRef]
12. Mannheim, R.L.; Garin, J.L. Structural identification of phases in Mo–Re alloys within the range from 5 to 95% Re. *J. Mater. Process. Technol.* **2003**, *143-144*, 533–538. [CrossRef]
13. Cui, K.K.; Zhang, Y.Y.; Fu, T.; Hussain, S.; Algarni, T.S.; Wang, J.; Zhang, X.; Ali, S. Effects of Cr_2O_3 Content on Microstructure and Mechanical Properties of Al_2O_3 Matrix Composites. *Coatings* **2021**, *11*, 234. [CrossRef]
14. Cui, Y.C.; Derby, B.; Li, N.; Misra, A. Fracture resistance of hierarchical Cu–Mo nanocomposite thin films. *Mater. Sci. Eng. A* **2021**, *799*, 139–891. [CrossRef]
15. Pan, Y.; Zhang, J. Influence of noble metals on the electronic and optical properties of the monoclinic ZrO_2: A first-principles study. *Vacuum* **2021**, *187*, 110112. [CrossRef]
16. Li, J.C.; Wei, L.L.; He, J.; Chen, H. The role of Re in improving the oxidation-resistance of a Re modified PtAl coating on Mo-rich single crystal superalloy. *J. Mater. Sci. Technol.* **2020**, *58*, 63–72. [CrossRef]
17. Paul, B.; Kishor, J.; Majumdar, S.; Kain, V. Studies on growth mechanism of intermediate layer of $(Mo,W)_5Si_3$ and interdiffusion in the (Mo,W)-$(Mo,W)Si_2$ system prepared by pack cementation coating. *Surf. Interfaces* **2020**, *18*, 100–458. [CrossRef]
18. Pan, Y. The structural, mechanical and thermodynamic properties of the orthorhombic TMAl (TM=Ti, Y, Zr and Hf) aluminides from first-principles calculations. *Vacuum* **2020**, *181*, 109742. [CrossRef]
19. Lu, Y.; Watanabe, M.; Miyata, R.; Nakamura, J.; Yamada, J.; Kato, H.; Yoshimi, K. Microstructures and mechanical properties of TiC-particulate-reinforced Ti–Mo–Al intermetallic matrix composites. *Mater. Sci. Eng. A* **2020**, *790*, 139–523. [CrossRef]
20. Cui, K.K.; Fu, T.; Zhang, Y.Y.; Wang, J.; Mao, H.B.; Tan, T.B. Microstructure and mechanical properties of $CaAl_{12}O_{19}$ reinforced Al_2O_3-Cr_2O_3 composites. *J. Eur. Ceram. Soc.* **2021**, *41*, 7935–7945. [CrossRef]
21. Niu, F.X.; Wang, Y.X.; Wang, Y.Y.; Ma, L.R.; Liu, J.J.; Wang, C.G. A crack-free SiC nanowire-toughened Si-Mo-W-C coating prepared on graphite materials for enhancing the oxidation resistance. *Surf. Coat. Technol.* **2018**, *344*, 52–57. [CrossRef]
22. Zhang, Y.Y.; Fu, T.; Cui, K.K.; Shen, F.Q.; Wang, J.; Yu, L.H.; Mao, H.B. Evolution of surface morphology, roughness and texture of tungsten disilicide coatings on tungsten substrate. *Vacuum* **2021**, *191*, 110297. [CrossRef]
23. Jiang, C.; Mariani, R.D.; Adkins, C.A. Ab initio investigation and thermodynamic modeling of the Mo–Ti–Zr system. *Materialia* **2020**, *10*, 100–701. [CrossRef]
24. Zhao, M.; Xu, B.Y.; Shao, Y.M.; Zhu, Y.; Wu, J.; Wu, S.S.; Yan, Y.W. Microstructure and oxidation mechanism of multiphase Mo–Ti–Si–B alloys at 800 °C. *Corros. Sci.* **2021**, *187*, 109518. [CrossRef]
25. Huang, L.; Pan, Y.F.; Zhang, J.X.; Du, Y.; Luo, F.H.; Zhang, S.Y. CALPHAD-type modeling of the C–Hf–Mo system over the whole composition and temperature ranges. *Thermochim. Acta* **2020**, *692*, 178–716. [CrossRef]
26. Fu, T.; Cui, K.K.; Zhang, Y.Y.; Wang, J.; Zhang, X.; Shen, F.Q.; Yu, L.H.; Mao, H.B. Microstructure and Oxidation Behavior of Anti-Oxidation Coatings on Mo-Based Alloys through HAPC Process: A Review. *Coatings* **2021**, *11*, 883. [CrossRef]
27. Hu, P.; Zhou, Y.H.; Chang, T.; Yua, Z.T.; Wang, K.S.; Yang, F.; Hua, B.L.; Cao, W.C.; Yu, H.L. Investigation on compression behavior of TZM and La_2O_3 doped TZM Alloys at high temperature. *Mater. Sci. Eng. A* **2017**, *687*, 276–280. [CrossRef]
28. Lange, A.; Braun, R.; Heilmaier, M. Oxidation behavior of magnetron sputtered double layer coatings containing molybdenum, silicon and boron. *Intermetallics* **2014**, *48*, 19–27. [CrossRef]
29. Suzuki, R.O.; Ishikawa, M.; Ono, K. $MoSi_2$ coating on molybdenum using molten salt. *J. Alloys Compd.* **2000**, *306*, 285–291. [CrossRef]
30. Kuznetsov, S.A.; Kuznetsova, S.V.; Rebrov, E.V.; Mies, M.J.M.; de Croon, M.H.J.M.; Schouten, J.C. Synthesis of molybdenum borides and molybdenum silicides in molten salts and their oxidation behavior in an air–water mixture. *Surf. Coat. Technol.* **2005**, *195*, 182–188. [CrossRef]
31. Anton, R.; Laska, N.; Schulz, U.; Obert, S.; Heilmaier, M. Magnetron Sputtered Silicon Coatings as Oxidation Protection for Mo-Based Alloys. *Adv. Eng. Mater.* **2020**, *22*, 7. [CrossRef]
32. Mao, H.; Shen, F.; Zhang, Y.; Wang, J.; Cui, K.; Wang, H.; Lv, T.; Fu, T.; Tan, T. Microstructure and Mechanical Properties of Carbide Reinforced TiC-Based Ultra-High Temperature Ceramics: A Review. *Coatings* **2021**, *11*, 1444. [CrossRef]
33. Huang, C.; Zhang, Y.Z.; Vilar, R. Microstructure and anti-oxidation behavior of laser clad Ni–20Cr coating on molybdenum surface. *Surf. Coat. Technol.* **2010**, *205*, 835–840. [CrossRef]
34. Zhang, H.A.; Lv, J.X.; Wu, Y.H.; Gu, S.Y.; Huang, Y.; Chen, Y. Oxidation behavior of $(Mo,W)Si_2$–Si_3N_4 composite coating on molybdenum substrate at 1600 °C. *Ceram. Int.* **2015**, *41*, 14890–14895. [CrossRef]
35. Yoon, J.K.; Lee, J.K.; Byun, J.Y.; Kim, G.H.; Paik, Y.H.; Kim, J.S. Effect of ammonia nitridation on the microstructure of $MoSi_2$ coatings formed by chemical vapor deposition of Si on Mo substrates. *Surf. Coat. Technol.* **2002**, *160*, 29–37. [CrossRef]
36. Pu, R.; Sun, Y.A.; Xu, J.W.; Zhou, X.J.; Li, S.; Zhang, B.; Cai, Z.Y.; Liu, S.N.; Zhao, X.J.; Xiao, L.R. Microstructure and properties of Mo-based double-layer $MoSi_2$ thick coating by a new two-step method. *Surf. Coat. Technol.* **2020**, *394*, 125840. [CrossRef]
37. Zhang, Y.Y.; Yu, L.H.; Fu, T.; Wang, J.; Shen, F.Q.; Cui, K.K. Microstructure evolution and growth mechanism of Si-$MoSi_2$ composite coatings on TZM (Mo-0.5Ti-0.1Zr-0.02 C) alloy. *J. Alloys Compd.* **2021**, *894*, 162403. [CrossRef]
38. Fu, T.; Cui, K.K.; Zhang, Y.Y.; Wang, J.; Shen, F.Q.; Yu, L.H.; Qie, J.M.; Zhang, X. Oxidation protection of tungsten alloys for nuclear fusion applications: A comprehensive review. *J. Alloys Compd.* **2021**, *884*, 161057. [CrossRef]

39. Pint, B.A. Invited Review Paper in Commemoration of Over 50 Years of Oxidation of Metals: Addressing the Role of Water Vapor on Long-Term Stainless Steel Oxidation Behavior. *Oxid. Met.* **2021**, *95*, 335–357. [CrossRef]
40. Cui, K.K.; Zhang, Y.Y.; Fu, T.; Wang, J.; Zhang, X. Toughening Mechanism of Mullite Matrix Composites: A Review. *Coatings* **2020**, *10*, 672. [CrossRef]
41. Zheng, X.Q.; Liu, Y. Slurry erosion–corrosion wear behavior in SiC-containing NaOH solutions of Mo_2NiB_2 cermets prepared by reactive sintering. *Int. J. Refract. Met. Hard Mater.* **2019**, *78*, 193–200. [CrossRef]
42. Gao, J.S.; Liu, Z.M.; Yan, Z.Q.; He, Y. A novel slurry blending method for a uniform dispersion of carbon nanotubes in natural rubber composites. *Results Phys.* **2019**, *15*, 102–720. [CrossRef]
43. Li, W.; Fan, J.L.; Fan, Y.; Xiao, L.R.; Cheng, H.C. $MoSi_2$/(Mo, Ti)Si_2 dual-phase composite coating for oxidation protection of molybdenum alloy. *J. Alloys Compd.* **2018**, *740*, 711–718. [CrossRef]
44. Cai, Z.Y.; Liu, S.N.; Xiao, L.R.; Fang, Z.; Li, W.; Zhang, B. Oxidation behavior and microstructural evolution of a slurry sintered Si-Mo coating on Mo alloy at 1650 °C. *Surf. Coat. Technol.* **2017**, *324*, 182–189. [CrossRef]
45. Chakraborty, S.P. Development of Protective Coating of $MoSi_2$ over TZM Alloy Substrate by Slurry Coating Technique. *Mater. Today Proc.* **2016**, *3*, 3071–3076. [CrossRef]
46. Cai, Z.Y.; Wu, Y.H.; Liu, H.Y.; Tian, G.Y.; Pu, R.; Piao, S.M.; Tang, X.Y.; Liu, S.N.; Zhao, X.J.; Xiao, L.R. Formation and oxidation resistance of a new YSZ modified silicide coating on Mo-based alloy. *Mater. Des.* **2018**, *155*, 463–474. [CrossRef]
47. Alam, M.Z.; Venkataraman, B.; Sarma, B.; Das, D.K. $MoSi_2$ coating on Mo substrate for short-term oxidation protection in air. *J. Alloys Compd.* **2009**, *487*, 335–340. [CrossRef]
48. Farje, J.A.V.; Matsunoshita, H.; Kishida, K.; Inui, H. Microstructure and mechanical properties of a $MoSi_2$-Mo_5Si_3 eutectic composite processed by laser surface melting. *Mater. Charact.* **2019**, *148*, 162–170. [CrossRef]
49. Yang, T.; Guo, X.P. Oxidation behavior of Zr-Y alloyed Mo-Si-B based alloys. *Int. J. Refract. Met. Hard Mater.* **2020**, *88*, 105–200. [CrossRef]
50. Alam, M.S.; Shafirovich, E. ShafirovichE. Mechanically activated combustion synthesis of molybdenum silicides and borosilicides for ultrahigh-temperature structural applications. *Proc. Combust. Inst.* **2015**, *35*, 2275–2281. [CrossRef]
51. Goodfellow, A.J.; Galindo-Nava, E.I.; Christofidou, K.A.; Jones, N.G.; Boyer, C.D.; Martin, T.L.; Bagot, P.A.J.; Hardy, M.C.; Stone, H.J. The effect of phase chemistry on the extent of strengthening mechanisms in model Ni-Cr-Al-Ti-Mo based superalloys. *Acta Mater.* **2018**, *153*, 290–302. [CrossRef]
52. Pint, B.A.; Unocic, K.A. Steam Oxidation Evaluation of Fe–Cr Alloys for Accident Tolerant Nuclear Fuel Cladding. *Oxid. Met.* **2017**, *87*, 515–526. [CrossRef]
53. Müller, F.; Gorr, B.; Christ, H.J.; Chen, H.; Kauffmann, A.; Laube, S.; Heilmaier, M. Formation of complex intermetallic phases in novel refractory high-entropy alloys NbMoCrTiAl and TaMoCrTiAl: Thermodynamic assessment and experimental validation. *J. Alloys Compd.* **2020**, *842*, 2515–5726. [CrossRef]
54. Vaunois, J.R.; Poulain, M.; Kanouté, P.; Chaboche, J.L. Development of bending tests for near shear mode interfacial toughness measurement of EB-PVD thermal barrier coatings. *Eng. Fract. Mech.* **2017**, *171*, 110–134. [CrossRef]
55. Gupta, M.; Li, X.H.; Markocsan, N.; Kjellman, B. Design of high lifetime suspension plasma sprayed thermal barrier coatings. *J. Eur. Ceram. Soc.* **2020**, *40*, 768–779. [CrossRef]
56. Gorr, S.M.B.; Christ, H.J.; Schliephake, D.; Heilmaier, M. Oxidation mechanisms of lanthanum-alloyed Mo–Si–B. *Corros. Sci.* **2014**, *88*, 360–371. [CrossRef]
57. Wang, Y.; Wang, D.Z.; Yan, J.H.; Sun, A.K. Preparation and characterization of molybdenum disilicide coating on molybdenum substrate by air plasma spraying. *Appl. Surf. Sci.* **2013**, *284*, 881–888. [CrossRef]
58. Deng, X.K.; Zhang, G.J.; Wang, T.; Ren, S.; Shi, Y.; Bai, Z.L.; Cao, Q. Microstructure and oxidation resistance of a multiphase Mo-Si-B ceramic coating on Mo substrates deposited by a plasma transferred arc process. *Ceram. Int.* **2019**, *45*, 415–423. [CrossRef]
59. Zhu, L.; Zhu, Y.S.; Ren, X.R.; Zhang, P.; Qiao, J.H.; Feng, P.Z. Microstructure, properties and oxidation behavior of $MoSi_2$-MoB-ZrO_2 coating for Mo substrate using spark plasma sintering. *Surf. Coat. Technol.* **2019**, *375*, 773–781. [CrossRef]
60. Chakraborty, S.P. Studies on the development of TZM alloy by aluminothermic coreduction process and formation of protective coating over the alloy by plasma spray technique. *Int. J. Refract. Met. Hard Mater.* **2011**, *29*, 5623–5630. [CrossRef]
61. Yavas, B.; Goller, G. A novel approach to boriding of TZM by spark plasma sintering method. *Int. J. Refract. Met. Hard Mater.* **2019**, *78*, 273–281. [CrossRef]
62. Shao, F.; Zhao, H.Y.; Liu, C.G.; Zhong, X.H.; Zhuang, Y.; Ni, J.X.; Tao, S.Y. Dense yttria-stabilized zirconia coatings fabricated by plasma spray-physical vapor deposition. *Ceram. Int.* **2017**, *43*, 2305–2313. [CrossRef]
63. Gao, L.H.; Guo, H.B.; Wei, L.L.; Li, C.Y.; Gong, S.K.; Xu, H.B. Microstructure and mechanical properties of yttria stabilized zirconia coatings prepared by plasma spray physical vapor deposition. *Ceram. Int.* **2015**, *41*, 8305–8311. [CrossRef]
64. Ganvir, A.; Calinas, R.F.; Markocsan, N.; Curry, N.; Joshi, S. Experimental visualization of microstructure evolution during suspension plasma spraying of thermal barrier coatings. *J. Eur. Ceram. Soc.* **2019**, *39*, 470–481. [CrossRef]
65. Ganvir, A.; Curry, N.; Markocsan, N.; Nylén, P.; Toma, F.L. Comparative study of suspension plasma sprayed and suspension high velocity oxy-fuel sprayed YSZ thermal barrier coatings. *Surf. Coat. Technol.* **2015**, *268*, 70–76. [CrossRef]
66. Yavas, B.; Goller, G. Investigation the effect of B 4 C addition on properties of TZM alloy prepared by spark plasma sintering. *Int. J. Refract. Met. Hard Mater.* **2016**, *58*, 182–188. [CrossRef]

67. Zhao, Y.L.; Wen, J.H.; Wen, J.H.; Peyraut, F.; Planche, M.P.; Misra, S.; Lenoir, B.; Ilavsky, J.; Liao, H.L.; Montavon, G. Porous architecture and thermal properties of thermal barrier coatings deposited by suspension plasma spray. *Surf. Coat. Technol.* **2020**, *386*, 125462. [CrossRef]
68. Gizynski, M.; Chen, X.; Dusautoy, N.; Araki, H.; Kuroda, S.; Watanabe, M.; Pakiela, Z. Comparative study of the failure mechanism of atmospheric and suspension plasma sprayed thermal barrier coatings. *Surf. Coat. Technol.* **2019**, *370*, 163–176. [CrossRef]
69. Zhang, Y.; Pint, B.A.; Cooley, K.M.; Haynes, J.A. Effect of nitrogen on the formation and oxidation behavior of iron aluminide coatings. *Surf. Coatings Technol.* **2005**, *200*, 1231–1235. [CrossRef]
70. Pochet, L.F.; Howard, P.; Safaie, S. Practical aspects of deposition of CVD SiC and boron silicon carbide onto high temperature composites. *Surf. Coat. Technol.* **1996**, *86-87*, 135–141. [CrossRef]
71. Nyutu, E.K.; Kmetz, M.A.; Suib, S.L. Formation of $MoSi_2$–SiO_2 coatings on molybdenum substrates by CVD/MOCVD. *Surf. Coat. Technol.* **2006**, *200*, 3980–3986. [CrossRef]
72. Yoon, J.K.; Kim, G.H.; Han, J.H.; Shon, I.J.; Doh, J.M.; Hong, K.T. Low-temperature cyclic oxidation behavior of $MoSi_2$/Si_3N_4 nanocomposite coating formed on Mo substrate at 773 K. *Surf. Coat. Technol.* **2005**, *200*, 2537–2546. [CrossRef]
73. Huang, X.X.; Sun, S.C.; Lu, S.D.; Li, K.H.; Tu, G.F.; Song, J.X. Synthesis and characterization of oxidation-resistant TiB_2 coating on molybdenum substrate by chemical vapor deposition. *Mater. Lett.* **2018**, *228*, 53–56. [CrossRef]
74. Du, J.H.; Li, Z.X.; Liu, G.J.; Zhou, H.; Huang, C.L. Surface characterization of CVD tungsten coating on molybdenum substrate. *Surf. Coat. Technol.* **2005**, *198*, 169–172. [CrossRef]
75. Yoon, J.K.; Lee, K.H.; Kim, G.H.; Han, J.H.; Doh, J.M.; Hong, K.T. Low-Temperature Cyclic Oxidation Behavior of $MoSi_2$/SiC Nanocomposite Coating Formed on Mo Substrate. *Mater. Trans.* **2004**, *45*, 2435–2442. [CrossRef]
76. Céspedes, E.; Wirz, M.; Sánchez-García, J.A.; Fraga, L.A.; Galindo, R.E.; Prietoa, C. Novel Mo–Si_3N_4 based selective coating for high temperature concentrating solar power applications. *Sol. Energy Mater. Sol. Cells* **2014**, *122*, 217–225. [CrossRef]
77. Jung, Y.I.; Kim, S.H.; Kim, H.G.; Park, J.Y.; Kim, W.J. Microstructures of diffusion bonded SiC ceramics using Ti and Mo interlayers. *J. Nucl. Mater.* **2013**, *441*, 510–513. [CrossRef]
78. Huang, X.X.; Sun, S.C.; Tu, G.F. Investigation of mechanical properties and oxidation resistance of CVD TiB_2 ceramic coating on molybdenum. *J. Mater. Res. Technol.* **2020**, *9*, 282–290. [CrossRef]
79. Anton, R.; Hüning, S.; Laska, N.; Weber, M.; Schellert, S.; Gorr, B.; Christ, H.J.; Schulz, U. Graded PVD Mo-Si interlayer between Si coating and Mo-Si-B alloys: Investigation of oxidation behaviour. *Corros. Sci.* **2021**, *192*, 109–843. [CrossRef]
80. Lange, A.; Braun, R.; Schulz, U. PVD thermal barrier coating systems for M-Si-B alloys. *Mater. High. Temp.* **2017**, *35*, 195–203. [CrossRef]
81. Krüger, M.; Franz, S.; Saage, H.; Heilmaier, M.; Schneibel, J.H.; J´ehanno, P.; B¨oning, M.; Kestler, H. Mechanically alloyed Mo–Si–B alloys with a continuous α-Mo matrix and improved mechanical properties. *Intermetallics* **2008**, *16*, 933–941. [CrossRef]
82. Zhang, Y.Y.; Hussain, S.; Cui, K.K.; Fu, T.; Wang, J.; Javed, M.S.; Lv, Y.; Aslam, B. Microstructure and Mechanical Properties of $MoSi_2$ Coating Deposited on Mo Substrate by Hot Dipping Processes. *J. Nanoelectron. Optoelectron.* **2019**, *14*, 1680–1685. [CrossRef]
83. Zhang, Y.Y.; Cui, K.K.; Fu, T.; Wang, J.; Qie, J.M.; Zhang, X. Synthesis WSi_2 coating on W substrate by HDS method with various deposition times. *Appl. Surf. Sci.* **2020**, *511*, 145551. [CrossRef]
84. Zhang, Y.Y.; Qie, J.M.; Cui, K.K.; Fu, T.; Fan, X.L.; Wang, J.; Zhang, X. Effect of hot dip silicon-plating temperature on microstructure characteristics of silicide coating on tungsten substrate. *Ceram. Int.* **2020**, *46*, 5223–5228. [CrossRef]
85. Zhang, Y.Y.; Li, Y.G.; Bai, C.G. Microstructure and oxidation behavior of Si–$MoSi_2$ functionally graded coating on Mo substrate. *Ceram. Int.* **2017**, *43*, 6250–6256. [CrossRef]
86. Zhang, Y.Y.; Li, Y.G.; Qi, Y.H.; Zou, Z.S. The Characters of Mo-$MoSi_2$ Functionally Graded Coating. *High Temp. Mater. Process.* **2014**, *33*, 239–244. [CrossRef]
87. Zhang, Y.Y.; Zhao, J.; Li, J.H.; Lei, J. Effect of hot-dip siliconizing time on phase composition and microstructure of Mo–$MoSi_2$ high temperature structural materials. *Ceram. Int.* **2019**, *45*, 5588–5593. [CrossRef]
88. Wang, S.P.; Zhou, L.; Li, C.J.; Li, Z.X.; Li, H.Z.; Yang, L.J. Morphology of composite coatings formed on Mo1 substrate using hot-dip aluminising and micro-arc oxidation techniques. *Appl. Surf. Sci.* **2020**, *508*, 144–761. [CrossRef]
89. Zhang, Y.Y.; Cui, K.K.; Gao, Q.J.; Hussain, S.; Lv, Y. Investigation of morphology and texture properties of WSi_2 coatings on W substrate based on contact-mode AFM and EBSD. *Surf. Coat. Technol.* **2020**, *396*, 125966. [CrossRef]
90. Zhang, Y.Y.; Yu, L.H.; Fu, T.; Wang, J.; Shen, F.Q.; Cui, K.K.; Wang, H. Microstructure and oxidation resistance of Si-$MoSi_2$ ceramic coating on TZM (Mo-0.5Ti-0.1Zr-0.02C) alloy at 1500 °C. *Surf. Coat. Technol.* **2021**, *431*, 128037. [CrossRef]
91. Zhang, Y.Y.; Cui, K.K.; Fu, T.; Wang, J.; Shen, F.Q.; Zhang, X.; Yu, L.H. Formation of $MoSi_2$ and Si/$MoSi_2$ coatings on TZM (Mo–0.5Ti–0.1Zr–0.02C) alloy by hot dip silicon-plating method. *Ceram. Int.* **2021**, *47*, 23053–23065. [CrossRef]

Article

Study of Crack Sensitivity of Peritectic Steels

Kai Liu [1], Shusen Cheng [1,*] and Yaqiang Li [2,*]

[1] School of Metallurgical and Ecological Engineering, University of Science and Technology Beijing, Beijing 100083, China; beikedaliukai@163.com
[2] College of Metallurgy and Energy, North China University of Science and Technology, Tangshan 063210, China
* Correspondence: chengsusen@metall.ustb.edu.cn (S.C.); liyq@ncst.edu.cn (Y.L.)

Abstract: By comprehensively considering both the high temperature mechanical properties and peritectic transformation during peritectic steel solidification, the strain ε_C^{th} is proposed to evaluate the crack sensitivity of peritectic steels produced in the brittle temperature range in the present work. The zero ductility temperature (ZDT) and the zero strength temperature (ZST) of Fe–C–0.32Si–1.6Mn–0.01P–0.015S steel under nonequilibrium conditions by taking the effect of the peritectic transformation on the solute segregation into account were calculated by the CK microsegregation model (Clyne–Kurz model) and were compared with the measured data. The comparison results show that this model can well simulate the nonequilibrium solidification process of peritectic steel. Then, based on the calculation of the CK microsegregation model, the strain during the peritectic phase transformation in the brittle temperature range (ZDT < T_B < LIT) was calculated under nonequilibrium conditions. The results show that the calculated strain is in good agreement with the actual statistical longitudinal crack data indicating that the strain can therefore be used to predict the crack sensitivity of peritectic steels effectively.

Keywords: peritectic steels; strain; crack sensitivity; peritectic phase transformation; microsegregation

Citation: Liu, K.; Cheng, S.; Li, Y. Study of Crack Sensitivity of Peritectic Steels. *Coatings* **2022**, *12*, 15. https://doi.org/10.3390/coatings12010015

Academic Editor: Annalisa Fortini

Received: 13 November 2021
Accepted: 14 December 2021
Published: 24 December 2021

Publisher's Note: MDPI stays neutral with regard to jurisdictional claims in published maps and institutional affiliations.

Copyright: © 2021 by the authors. Licensee MDPI, Basel, Switzerland. This article is an open access article distributed under the terms and conditions of the Creative Commons Attribution (CC BY) license (https://creativecommons.org/licenses/by/4.0/).

1. Introduction

Generally, the C percent of peretectic steels ranges in 0.09~0.17 wt %, for which the surface longitudinal cracks are prone to appear during continuous casting process decreasing the quality of the cast product and hampering production rates [1–6]. The peritectic phase transformation always occurs during the solidification of peritectic steels, which may lead to uneven strand shell growth and produce thermal stress as well as phase transformation stress. In order to reduce crack defects in the casting slabs, metallurgists hope to adjust the composition of liquid steel to reduce the crack sensitivity of steel during steelmaking. Much research has been conducted to estimate the solidification crack sensitivity during continuous casting. Kim et al. [7] investigated the effects of carbon and sulfur on longitudinal surface cracks by calculating the nonequilibrium pseudo binary Fe–C phase diagram and introducing the strain in the brittle temperature range for the continuous casting of steels. They found that the carbon content at which longitudinal surface cracks was maximized decreased with increasing sulfur content, and the possibility of surface cracks increased with increasing sulfur content at a given carbon content. Harste et al. [8] used a mechanical model and a thermal model to predict the crack susceptibility, and they found that there was a shrinkage peak at about 0.1 wt % C. Won et al. [9] proposed the concept of "Specific Crack Susceptibility" to analyze the crack tendency based on the critical fracture stress and critical stain, and they found that the crack susceptibility of 0.12 wt % C steel is the strongest.

Obviously, the chemical composition would have a noticeable effect on the process of peritectic transition. From the perspective of solidification, the solute elements segregate at the grain boundaries of the primary solidified shell in a mold, so that a low-melting liquid film would form between the dendrites leading to a reduced resistance of the shell from deformation, during which the peritectic phase transformation could occur near the solidus temperature resulting in the fact that the total thermal stress and phase transformation stress are close to the fracture strength of the primary solidified shell, which is easy to cause cracks in the weak parts of the shell. Therefore, the peritectic phase transformation during solidification has an important influence on the longitudinal cracks on the surface of the cast slab, and once the stress on the shell exceeds the critical stress at high temperature, cracks would occur in weak parts of the shell [10–15].

So, it is important to predict the sensitivity of the cracks during peritectic steel production. The methods proposed to describe the sensitivity of such steels in the literature mainly focuses on the basis of high temperature mechanical properties of the steels only, and the influence of peritectic phase transformation accompanied with element segregation on longitudinal cracks during solidification is always ignored.

In the present study, the characterized method for the sensitivity of the peritectic steels combined with both the high temperature mechanical properties and the peritectic transformation during peritectic phase transformation in the brittle temperature range is proposed comprehensively and is validated by statistical longitudinal crack data from literature, which can be used to evaluate the crack sensitivity of peritectic steels more reasonably. Furthermore, the results of this calculation are advantageous for the chemical composition control and utilization of mold fluxes to avoid the cracks occurring when producing such steels.

2. Calculation Procedures

2.1. Segregation Model

The CK microsegregation model is used to quantitatively describe the relationship between the concentration of the solutes and the solid fraction during the solidification of peritectic steels. In the solid–liquid interface, the solute concentration is assumed to be in the local equilibrium. The equilibrium distribution coefficient of the solute elements at the solid–liquid interface and liquidus slope is assumed to be constant. Solute elements are assumed to diffuse completely in the liquid. The effects of the solute segregation is assumed to be superimposable. The equation of the CK model is shown as follows [16]:

$$C_{L,i} = C_{0,i}[1 - (1 - 2\beta k_i)f_s]^{(k_i-1)/(1-2\beta k_i)} \tag{1}$$

where f_s is the solid fraction, k is the equilibrium distribution coefficient, $C_{0,i}$ is the initial concentration of the solute element i, $C_{L,i}$ is the concentration of the solute element i in the solidification front, and β is the correction factor given as follows:

$$\beta = \alpha_i(1 - e^{-\frac{1}{\alpha_i}}) - \frac{1}{2}e^{-\frac{1}{2\alpha_i}} \tag{2}$$

$$\alpha_i = \frac{4D_s \times t_f}{\lambda^2} \tag{3}$$

$$t_f = \frac{T_L - T_S}{C_R} \tag{4}$$

where α_i is the Fourier number of the solute element i, D_s is the diffusion coefficient of the solute in the solid phase, t_f is the local solidification time, T_L is the liquidus temperature, T_S is the solidus temperature, C_R is cooling rate, and λ is the secondary dendrite arm spacing given as follows [17]:

$$\lambda = (169.1 - 720.9 \times [\%C_C]) \times R_C^{-0.4935} \quad [\%C_C] \leq 0.15 \tag{5}$$

$$\lambda = 143.9 \times R_C^{-0.3616} \times [\%C_C]^{(0.5501-1.996[\%C_C])} \quad [\%C_C] > 0.15 \tag{6}$$

where $\%C_C$ is the carbon content.

2.2. Model Calculation

2.2.1. Determination of Liquidus Temperature and Solidus Temperature

The liquidus temperature (T_L) and the temperature of the solidification front (T_{int}) at different solid fractions are calculated using the following equations [18]:

$$T_L = 1538 - \sum_i m_i C_{0,i} \tag{7}$$

$$T_{int} = 1538 - \sum_i m_i C_{L,i} \tag{8}$$

where m_i is the liquidus slope of the solute element i. When the solid fraction is 1, the T_{int} calculated by Equation (8) is the solidus temperature.

δ-ferrite and γ-austenite appear during the solidification of peritectic steels. When the carbon content in the solidification front is lower than the peritectic end point (C_L), the peritectic steels will solidify in the L→δ-ferrite mode until the peritectic reaction occurs. With the solidification proceeding, the carbon content in the solidification front is continuously enriched. Once the carbon content in the solidification front exceeds the peritectic end point (C_L), the peritectic phase transformation occurs, and the residual liquid phase will solidify in the γ-austenite mode [19]. Due to the significant difference between the diffusion coefficient and equilibrium distribution coefficient of the solute elements in the δ-ferrite phase and the γ-austenite phase [20], the peritectic phase transformation will have an important effect on the segregation of the solute elements. The solidification parameters of the solute elements in the δ-ferrite and γ-austenite are shown in Table 1 [21–26]:

Table 1. Solidification parameters (R = 8.314 J/mol/K) [21–26].

Element	$k_i^{\delta/L}$	$k_i^{\gamma/L}$	$D_{s,i}^{\delta} \times 10^4$ (m^2·s^{-1})	$D_{s,i}^{\gamma} \times 10^4$ (m^2·s^{-1})	m_i
C	0.19	0.34	0.0127 × Exp (81,379/RT)	0.15 × Exp (−143,511/RT)	78
Si	0.77	0.52	8 × Exp (−248,948/RT)	0.3 × Exp (−251,458/RT)	7.6
Mn	0.76	0.78	0.76 × Exp (−224,430/RT)	0.055 × Exp (−249,366/RT)	4.9
P	0.23	0.13	2.9 × Exp (−230,120/RT)	0.01 × Exp (−182,841/RT)	34.4
S	0.05	0.035	4.56 × Exp (−214,639/RT)	2.4 × Exp (−223,426/RT)	48

where $k_i^{\delta/L}$ and $k_i^{\gamma/L}$ are the equilibrium distribution coefficients of the solute element i at the δ/L interface and γ/L interface, respectively; $D_{s,i}^{\delta}$ and $D_{s,i}^{\gamma}$ are the diffusion coefficients of the solute element i in the δ-ferrite phase and the γ-austentite phase, respectively; m_i is the slope of the liquidus temperature in the Fe-i phase diagram.

The segregation model is initially calculated in the δ-ferrite solidification mode. When the carbon content in the residual liquid phase reaches C_L, the critical solid fraction (f_{sw}) and the content of the solute elements (C_{sw}) are recorded. The virtual initial composition, which ensures that the solute content obtained by solidifying to f_{sw} in the γ-austenite mode is equal to the solute content obtained by solidifying the actual initial composition to f_{sw} in the δ-ferrite mode, can be obtained according to Equation (9). Then the segregation model is calculated in the γ-austenite solidification mode.

$$C'_{0,i} = C_{sw,i}[1 - (1 - 2\beta k_i)f_{sw}]^{\frac{(1-k_i)}{(1-2\beta k_i)}} \tag{9}$$

where $C'_{0,i}$ is the virtual initial concentration of the solute element i; $C_{sw,i}$ is the concentration of the solute element i at f_{sw}. The iteration method was used to calculate T_S. The process of the calculation is shown in Figure 1.

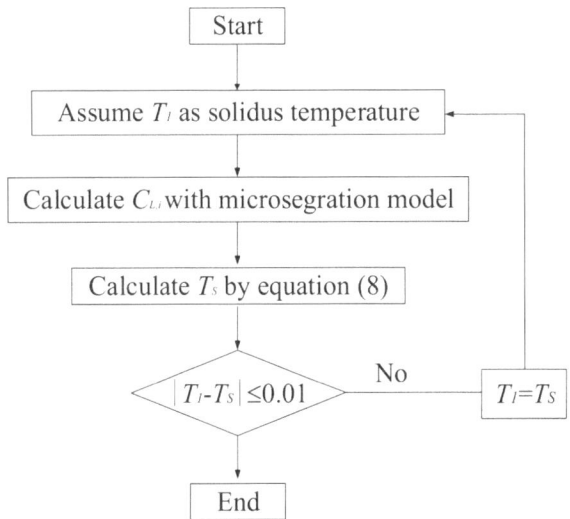

Figure 1. Flow chart for calculation of T_S.

2.2.2. Determination of Peritectic Phase Transformation Temperature during Solidification

When the carbon concentration at the solidification front reaches the carbon content of the peritectic end point (C_L), the T_{int} calculated by the segregation model is used as the starting temperature of the peritectic phase transformation (T_p), and the corresponding solid fraction is considered to be the starting solid fraction of the peritectic transformation (f_{sp}). The mass fraction of liquid phase (f_L) and solid phase (δ-ferrite or γ-austenite) at the starting or end temperature of the peritectic phase transformation (f_X) during solidification can be calculated as follows:

$$f_L = \frac{f_s \times \rho_L(T)}{(1-f_s) \times \rho_X(T) + f_s \times \rho_L(T)} \tag{10}$$

$$f_X = 1 - f_L \tag{11}$$

where $\rho_L(T)$ is the density of the liquid phase at the temperature of T, $\rho_X(T)$ is the density of the δ-ferrite or γ-austenite at the temperature of T, which is as follows [27]:

$$\rho_\delta(T) = 7875.96 - 0.297T - 5.62 \times 10^{-5}T^2. \tag{12}$$

$$\rho_\gamma(T) = 8099.79 - 0.506T \tag{13}$$

$$\rho_L(T) = 8319.49 - 0.835T \tag{14}$$

The ratio of the liquid phase to the δ-ferrite under the segregation condition is consistent with the equilibrium condition when the liquid phase and δ-ferrite transform completely into the γ-austenite in the same system [28]. This is determined by the conservation of mass and can be calculated using the thermodynamic Thermo-Calc software with the TCFE-7 database.

Based on the ratio and the mass fraction of the liquid phase and the δ-ferrite at T_p, the mass fraction of each phase at the end of the peritectic phase transformation during solidification can be obtained. When no liquid phase remains at the end of the peritectic phase transformation, the solid fraction at the end of the peritectic phase transformation (f_{sE}) is 1. When the liquid phase remains at the end of the peritectic phase transformation, f_{sE} is less than 1. Firstly, the mass fractions of the γ-austenite ($f_{\gamma'}$) and the liquid phase ($f_{L'}$)

at the end of the peritectic phase transformation were calculated, and then the iteration method was used to calculate f_{sE}. The process of calculation is shown in Figure 2.

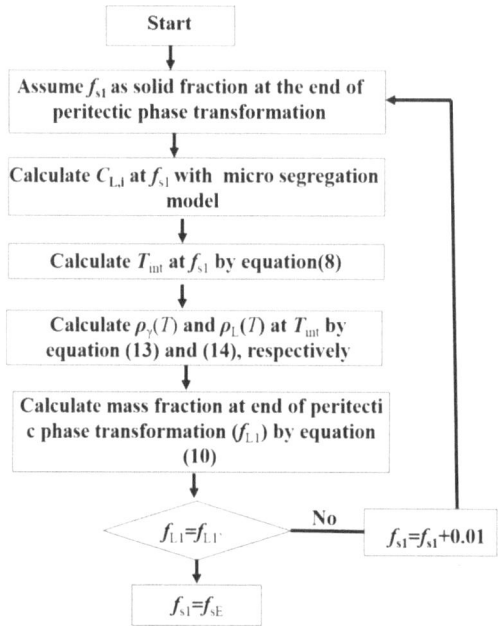

Figure 2. Flow chart for calculation of f_{sE}.

2.3. Model Validation

The nonequilibrium, pseudo-binary Fe–C phase diagram of the Fe–C–0.32Si–1.6Mn–0.01P–0.015S steel at a cooling rate of 0.17 K/s and the secondary dendrite arm spacing of 1000 μm were calculated using the microsegregation analysis. Schmidtman [29] measured the zero ductility temperature (ZDT) and zero strength temperature (ZST) of the steel as a function of the carbon content, which were used to be compared with the calculated nonequilibrium phase diagrams as shown in Figure 3.

In Figure 3, the complete solidification temperatures calculated are in good agreement with the ZDT measured by Schmidtman as well as the ZST calculated in the present work, and the solid fraction of the temperature, under which the measured ZDT and ZST agree well with the calculated results, is 0.7. Therefore, the results calculated by the microsegregation model proposed above can simulate the nonequilibrium solidification process of peritectic steel.

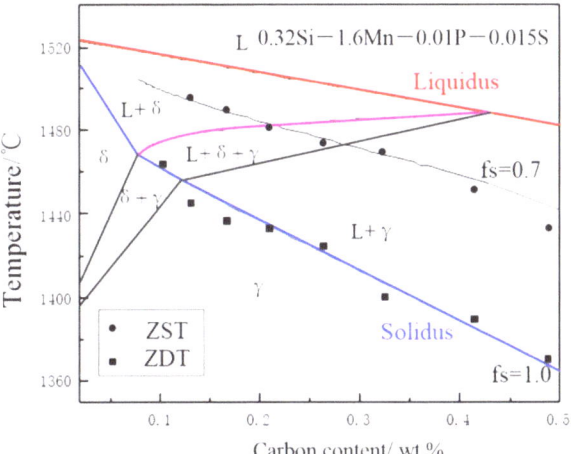

Figure 3. Nonequilibrium, pseudo binary Fe–C phase diagram of the Fe–C–0.32Si–1.6Mn–0.01P–0.015S steel.

3. Results and Discussion

3.1. Analysis of Crack Sensitivity

All cracks observed in the continuously cast steel originate and propagate along the interdendrites in the mushy zone except the transverse cracks. Clyne et al. [30] divided the mushy zone into the mass and liquid feeding zones and the crack zone. Cracks formed in the mass and liquid feeding zones are refilled with the surrounding liquid, whereas cracks formed in the crack zone can not be refilled with the liquid because the dendrite arms are compacted enough to resist the feeding of the liquid. They proposed the boundary between the two zones to be the liquid impenetrable temperature (LIT) and defined the brittle temperature range T_B as ZDT < T_B < LIT. Davies and Shin [31] used 0.9 as the solid fraction in the boundary. ZDT is defined as the temperature at which the solid fraction reaches unity ($f_s = 1$). The literature [32,33] reported that the peritectic transformation speed is so fast that the resulting shrinkage stress cannot be released in time. If the peritectic phase transformation occurs in the brittle temperature range (i.e., ZDT < T_B < LIT), the shrinkage stress would exceed the high temperature strength of the steel, easily leading to longitudinal cracks at the grain boundaries. Therefore, the strain ε_C^{th} during the peritectic phase transformation in the brittle temperature range (ZDT < T_B < LIT) can be used to evaluate the crack sensitivity of peritectic steels. ε^{th} is generally expressed as the sum of the strain caused by cooling and the strain caused by phase transformation as follows:

$$\varepsilon^{th} = \int_{T_{ref}}^{T} \alpha^* dT + \Delta\varepsilon^{\delta \to \gamma} \qquad (15)$$

where T is the temperature, T_{ref} is the reference temperature, α^* is the thermal expansion coefficient defined as $\alpha^* = d\varepsilon^{th}/dT$, and $\Delta\varepsilon^{\delta \to \gamma}$ is the strain due to the $\delta \to \gamma$ transformation. The strain ε_C^{th} during the peritectic phase transformation in the brittle temperature range (ZDT < T_B < LIT) could be expressed as:

$$\varepsilon_C^{th} = \int_{T_2}^{T_1} \alpha^* dT + \Delta\varepsilon_C^{\delta \to \gamma} = \varepsilon_C^* + \Delta\varepsilon_C^{\delta \to \gamma} \qquad (16)$$

where ε_C^* and $\Delta\varepsilon_C^{\delta \to \gamma}$ can be interpreted as the thermal strain and the strain induced by the $\delta \to \gamma$ transformation during the peritectic phase transformation in the brittle temperature range. T_1 is the initial temperature of the peritectic phase transformation in the brittle temperature range; T_2 is the end temperature of the peritectic phase transformation in the brittle temperature range. The strain ε_C^{th} can be simply expressed as a function of density as follows [7]:

$$\varepsilon_C^{th} = \sqrt[3]{\frac{\rho(T_2)}{\rho(T_1)}} - 1 \qquad (17)$$

where ρ_{T1} and ρ_{T2} are the density of the steel at T_1 and T_2, respectively. The density of the steel was obtained as follows [27]:

$$\frac{1}{\rho(T)} = \frac{f_\delta}{\rho_\delta} + \frac{f_\gamma}{\rho_\gamma} + \frac{f_L}{\rho_L} \qquad (18)$$

3.2. Validation of Crack Sensitivity

The comparison of the index of the longitudinal surface cracks of 0.14Si–0.36Mn–0.016P–0.013S steel under the different carbon content in the literature [34] with the calculated strain ε_C^{th} during the peritectic phase transformation in the brittle temperature range (ZDT < T_B < LIT) is shown in Figure 4. The index of the longitudinal surface cracks is a value expressing the longitudinal surface crack ratio. In view of a current slow cooling method used in continuous casting, the cooling rate is set as 10 °C/s. The carbon content at which the maximum number of longitudinal surface cracks appear is about 0.13~0.14 wt%, which agrees well with the experimental results in the literature [34].

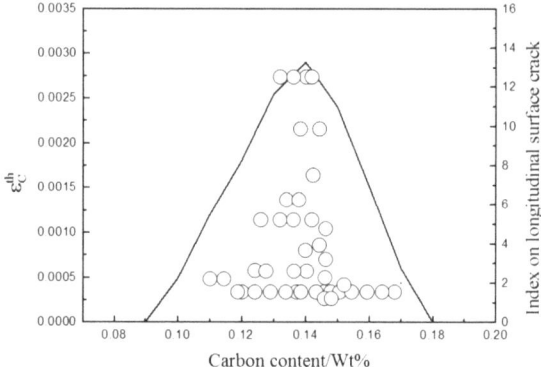

Figure 4. Measured index on longitudinal surface crack and calculated strain during the peritectic phase transformation in the brittle temperature range (ZDT < T_B < LIT).

In order to further verify the accuracy of the calculation method presented in the current work, the longitudinal crack ratio of the continuous casting slabs statistically obtained from an actual steel plant was used to compare with the calculation results. The comparison of the measured data and calculated data by present model is shown in Figure 5.

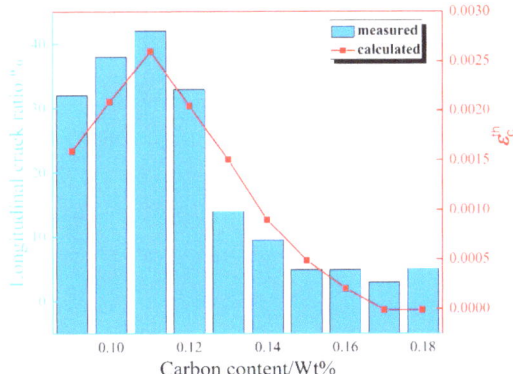

Figure 5. Comparison of measured data and calculated data by present model.

The steels obtained from the steel plant were divided into 10 parts according to the carbon content of 0.09~0.18 wt %, and the content of Si, Mn, P, and S was similar, and thus the average value of these steels were calculated in the 10 sections. The strain ε_C^{th} during the peritectic phase transformation in the brittle temperature range (ZDT < T_B < LIT) under different carbon content is also shown in Figure 5.

As shown in Figure 5, the strain ε_C^{th} during the peritectic phase transformation calculated in the brittle temperature range (ZDT < T_B < LIT) by the proposed model in this study is consistent with the measured data from the steel plant. The result calculated in the present work shows that the strain ε_C^{th} at 0.15~0.18 wt % of element C is apparently lower than that at 0.09~0.12 wt % meaning a lighter sensitivity in the range of element C 0.15~0.18 wt %. According to the results of comparison, the strain during the peritectic phase transformation in the brittle temperature range (ZDT < T_B < LIT) could characterize the tendency of the longitudinal surface crack during continuous casting perfectly, so the proposed model is reasonable to evaluate the sensitivity of cracks during peritectic steels.

4. Conclusions

The strain ε_C^{th} is proposed by comprehensively considering both the high temperature mechanical properties and the peritectic transformation during the peritectic steel solidification to evaluate the crack sensitivity of peritectic steels produced in the brittle temperature range (ZDT < T_B < LIT) in the present work, and the conclusions are as follows:

(1) Based on the analysis of the CK microsegregation model, which took the effect of the peritectic transformation on the solute segregation into account, the zero ductility temperature (ZDT) and zero strength temperature (ZST) of Fe–C–0.32Si–1.6Mn–0.01P–0.015S steel were determined. The calculated results are in good agreement with that measured in the literature.

(2) The model to calculate the strain ε_C^{th} by comprehensively considering both the high temperature mechanical properties and the peritectic transformation during peritectic steel solidification was proposed to evaluate the crack sensitivity of peritectic steels produced in the brittle temperature range (ZDT < T_B < LIT) in the present work, and the results were compared with the longitudinal crack ratio of peritectic steels in the literature showing a good agreement with the statistical longitudinal crack data from literature.

(3) Validated by the statistical data from steel plants and literature, the model proposed in the present work can be used to effectively predict the crack sensitivity of peritectic steels.

Author Contributions: Software, Y.L.; validation, Y.L. and K.L.; writing—original draft preparation, K.L.; writing—review and editing, K.L.; supervision, S.C. All authors have read and agreed to the published version of the manuscript.

Funding: This research received no external funding.

Institutional Review Board Statement: Not applicable.

Informed Consent Statement: Not applicable.

Data Availability Statement: Data is contained within the article.

Conflicts of Interest: The authors declare no conflict of interest.

References

1. Suzuki, M.; Yu, C.H.; Sato, H.; Tsui, Y.; Shibata, H.; Emi, T. Origin of heat transfer anomaly and solidifying shell deformation of peritectic steels in continuous casting. *Trans. Iron Steel Inst. Jpn.* **2007**, *36*, S171–S174. [CrossRef]
2. Saraswat, R.; Maijer, D.M.; Lee, P.D.; Mills, K.C. The Effect of Mould Flux Properties on thermomechanical behaviour during billet continuous casting. *ISIJ Int.* **2007**, *47*, 95–104. [CrossRef]
3. Konishi, J.; Militzer, M.; Samarasekera, I.V.; Brimacombe, J.K. Modeling the formation of longitudinal facial cracks during continu-ous casting of hypoperitectic steel. *Metall. Mater. Trans. B* **2002**, *33*, 413–423. [CrossRef]
4. Zhang, Y.; Yu, L.; Fu, T.; Wang, J.; Shen, F.; Cui, K. Microstructure evolution and growth mechanism of Si-MoSi$_2$ composite coatings on TZM (Mo–0.5Ti–0.1Zr–0.02 C) alloy. *J. Alloys Compd.* **2021**, *894*, 162403. [CrossRef]
5. Zhang, Y.; Fu, T.; Cui, K.; Shen, F.; Wang, J.; Yu, L.; Mao, H. Evolution of surface morphology, roughness and texture of tungsten disilicide coatings on tungsten substrate. *Vacuum* **2021**, *191*, 110297. [CrossRef]
6. Zhang, Y.; Cui, K.; Fu, T.; Wang, J.; Shen, F.; Zhang, X.; Yu, L. Formation of MoSi$_2$ and Si/MoSi$_2$ coatings on TZM (Mo–0.5 Ti–0.1 Zr–0.02 C) alloy by hot dip silicon-plating method. *Ceram. Int.* **2021**, *47*, 23053–23065. [CrossRef]
7. Kim, K.-H.; Yeo, T.-J.; Oh, K.H.; Lee, D.N. Effect of carbon and sulfur in continuously cast strand on longitudinal surface cracks. *ISIJ Int.* **1996**, *36*, 284–289. [CrossRef]
8. Harste, K.; Schwerdtfeger, K. Shrinkage of round iron-carbon ingots during solidification and subsequent cooling. *ISIJ Int.* **2003**, *43*, 1011–1020. [CrossRef]
9. Won, Y.M.; Han, H.N.; Yeo, T.-J.; Oh, K.H. Analysis of solidification cracking using the specific crack susceptibility. *ISIJ Int.* **2000**, *40*, 129–136. [CrossRef]
10. Cui, K.; Fu, T.; Zhang, Y.; Wang, J.; Mao, H.; Tan, T. Microstructure and mechanical properties of CaAl$_{12}$O$_{19}$ reinforced Al$_2$O$_3$-Cr$_2$O$_3$ composites. *J. Eur. Ceram. Soc.* **2021**, *41*, 7935–7945. [CrossRef]
11. Kim, K.-H.; Oh, K.H.; Lee, D.N. Mechanical behavior of carbon steels during continuous casting. *Scr. Mater.* **1996**, *34*, 301–307. [CrossRef]
12. Cui, K.; Zhang, Y.; Fu, T.; Hussain, S.; Saad AlGarni, T.; Wang, J.; Zhang, X.; Ali, S. Effects of Cr$_2$O$_3$ content on micro-structure and mechanical properties of Al$_2$O$_3$ matrix composites. *Coatings* **2021**, *11*, 234. [CrossRef]
13. Guo, L.; Li, W.; Bobadilla, M.; Yao, M.; Shen, H. High temperature mechanical properties of micro-alloyed carbon steel in the mushy zone. *Steel Res. Int.* **2010**, *81*, 387–393. [CrossRef]
14. Won, Y.M.; Yeo, T.-J.; Seol, D.J.; Oh, K.H. A new criterion for internal crack formation in continuously cast steels. *Met. Mater. Trans. A* **2000**, *31*, 779–794. [CrossRef]
15. Wang, J.; Zhang, Y.; Cui, K.; Fu, T.; Gao, J.; Hussain, S.; AlGarni, T.S. Pyrometallurgical recovery of zinc and valuable metals from electric arc furnace dust—A review. *J. Clean. Prod.* **2021**, *298*, 126788. [CrossRef]
16. Clyne, T.W.; Kurz, W. Solute redistribution during solidification with rapid solid state diffusion. *Metall. Trans. A* **1981**, *12*, 965–971. [CrossRef]
17. Won, Y.-M.; Thomas, B. Simple model of microsegregation during solidification of steels. *Met. Mater. Trans. A* **2001**, *32*, 1755–1767. [CrossRef]
18. Han, Z.Q.; Cai, K.K. Study on a mathematical model of microsegregation in continuously cast slab. *Acta Metall. Sin.* **2000**, *36*, 869–873.
19. Zhu, L.G.; Liu, Z.; Han, Y.H. A microsegregation model in the two-phase region of an ND steel continuous casting billet. *Chin. J. Eng.* **2019**, *41*, 461–469.
20. Cai, Z.Z.; Zhu, M.Y. Microsegregation of solute elements in solidifying mushy zone of steel and its effect on longitudinal sur-face cracks of continuous casting strand. *Acta Metall. Sin.* **2009**, *45*, 55–61.
21. Choudhary, S.K.; Ghosh, A. Mathematical model for prediction of composition of inclusions formed during solidification of liquid steel. *ISIJ Int.* **2009**, *49*, 1819–1827. [CrossRef]
22. Ueshima, Y.; Mizoguchi, S.; Matsumiya, T.; Kajioka, H. Analysis of solute distribution in dendrites of carbon steel with δ/γ transformation during solidification. *Met. Mater. Trans. A* **1986**, *17*, 845–859. [CrossRef]
23. Ma, Z.T.; Dieter, J. Characteristics of oxide precipitation and growth during solidification of deoxidized steel. *ISIJ Int.* **1998**, *38*, 46–52. [CrossRef]

24. Liu, Z.Z.; Wei, J.; Cai, K.K. A couple mathematical model of microsegregation and inclusion precipitation during solidification of silicon steel. *ISIJ Int.* **2002**, *42*, 958–963. [CrossRef]
25. Kobayashi, Y.; Liu, Z.; Nagai, K. Effect of phosphorus on sulfide precipitation in strip cast low carbon steel. *J. Jpn. Inst. Met.* **2006**, *70*, 440–446. [CrossRef]
26. Liu, Z.; Kobayashi, Y.; Yang, J.; Nagai, K.; Kuwabara, M. "In-situ" Observation of the δ/γ phase transformation on the sur-face of low carbon steel containing phosphorus at various cooling rates. *ISIJ Int.* **2006**, *46*, 847–853. [CrossRef]
27. Jablonka, A.; Harste, K.; Schwerdtfeger, K. Thermomechanical properties of iron and iron-carbon alloys: Density and thermal contraction. *Steel Res.* **1991**, *62*, 24–33. [CrossRef]
28. Xu, J.; He, S.; Jiang, X.; Wu, T.; Wang, Q. Analysis of crack susceptibility of regular carbon steel slabs using volume-based shrinkage index. *ISIJ Int.* **2013**, *53*, 1812–1817. [CrossRef]
29. Schmidtman, E.; Rakoski, F. Bnfluß des kohlenstoffgehaltes von 0.015 bis 1% und der gefügestruktur auf das hochtempera-turfestigkeits- und-zähigkeitsverhalten von baustählen nach der erstarrung aus der schmelze. *Steel Res. Int.* **1983**, *54*, 357–361.
30. Clyne, T.W.; Wolf, M.; Kurz, W. The effect of melt composition on solidification cracking of steel, with particular reference to continuous casting. *Met. Mater. Trans. A* **1982**, *13*, 259–266. [CrossRef]
31. Davies, G.J.; Shin, Y.K. *Solidification Technology in the Foundry and Cast House*; The Metal Society: London, UK, 1979; p. 517.
32. Fredriksson, H.; Stjerndahl, J. Microstructure evolution during the solidification of steel. *Met. Sci.* **1982**, *16*, 575–580. [CrossRef]
33. Shibata, H.; Arai, Y.; Suzuki, M.; Emi, T. Kinetics of peritectic reaction and transformation in Fe–C alloys. *Met. Mater. Trans. A* **2000**, *31*, 981–991. [CrossRef]
34. Blazek, K.E.; Saucedo, I.G.; Tsai, H.T. An investigation on mold heat transfer during continuous casting. *Steelmak. Conf. Proc.* **1988**, *71*, 411–421.

Article

High-Sensitivity Biosensor Based on Glass Resonance PhC Cavities for Detection of Blood Component and Glucose Concentration in Human Urine

Abduladheem Turki Jalil [1,2,3], Shameen Ashfaq [4,*], Dmitry Olegovich Bokov [5,6], Amer M. Alanazi [7], Kadda Hachem [8,*], Wanich Suksatan [9] and Mika Sillanpää [10,11]

1. Faculty of Biology and Ecology, Yanka Kupala State University of Grodno, 230023 Grodno, Belarus; abedalazeem799@gmail.com
2. College of Technical Engineering, The Islamic University, Najaf 54001, Wasit, Iraq
3. Department of Dentistry, Kut University College, Kut 52001, Wasit, Iraq
4. Women Medical Officer Punjab Health Department, Faisalabad 38000, Pakistan
5. Institute of Pharmacy, Sechenov First Moscow State Medical University, 8 Trubetskaya St., Bldg. 2, 119991 Moscow, Russia; fmmsu@mail.ru
6. Federal Research Center of Nutrition, Biotechnology and Food Safety, Laboratory of Food Chemistry, 2/14 Ustyinsky Pr., 109240 Moscow, Russia
7. Pharmaceutical Chemistry Department, College of Pharmacy, King Saud University, Riyadh 11451, Saudi Arabia; Amalanazi@ksu.edu.sa
8. Laboratory of Biotoxicology, Pharmacognosy and Biological Valorization of Plants (LBPVBP), Faculty of Sciences, University of Saida—Dr Moulay Tahar, Saida 20000, Algeria
9. Faculty of Nursing, HRH Princess Chulabhorn College of Medical Science, Chulabhorn Royal Academy, Bangkok 10210, Thailand; wanich.suk@pccms.ac.th
10. Department of Chemical Engineering, School of Mining, Metallurgy and Chemical Engineering, University of Johannesburg, P.O. Box 17011, Johannesburg 2028, South Africa; mikaesillanpaa@gmail.com
11. Department of Biological and Chemical Engineering, Aarhus University, Nørrebrogade 44, 8000 Aarhus, Denmark
* Correspondence: Saeed1wahla@gmail.com (S.A.); kadda46@hotmail.com (K.H.)

Citation: Jalil, A.T.; Ashfaq, S.; Bokov, D.O.; Alanazi, A.M.; Hachem, K.; Suksatan, W.; Sillanpää, M. High-Sensitivity Biosensor Based on Glass Resonance PhC Cavities for Detection of Blood Component and Glucose Concentration in Human Urine. *Coatings* **2021**, *11*, 1555. https://doi.org/10.3390/coatings11121555

Academic Editor: Awais Ahmad

Received: 17 November 2021
Accepted: 11 December 2021
Published: 17 December 2021

Publisher's Note: MDPI stays neutral with regard to jurisdictional claims in published maps and institutional affiliations.

Copyright: © 2021 by the authors. Licensee MDPI, Basel, Switzerland. This article is an open access article distributed under the terms and conditions of the Creative Commons Attribution (CC BY) license (https://creativecommons.org/licenses/by/4.0/).

Abstract: In this work, a novel structure of an all-optical biosensor based on glass resonance cavities with high detection accuracy and sensitivity in two-dimensional photon crystal is designed and simulated. The free spectral range in which the structure performs well is about FSR = 630 nm. This sensor measures the concentration of glucose in human urine. Analyses to determine the glucose concentration in urine for a normal range (0~15 mg/dL) and urine despite glucose concentrations of 0.625, 1.25, 2.5, 5 and 10 g/dL in the wavelength range 1.326404~1.326426 µm have been conducted. The detection range is RIU = 0.2×10^{-7}. The average bandwidth of the output resonance wavelengths is 0.34 nm in the lowest case. In the worst case, the percentage of optical signal power transmission is 77% with an amplitude of 1.303241 and, in the best case, 100% with an amplitude of 1.326404. The overall dimensions of the biosensor are 102.6 µm^2 and the sensitivity is equal to S = 1360.02 nm/RIU and the important parameter of the Figure of Merit (*FOM*) for the proposed biosensor structure is equal to *FOM* = 1320.23 RIU^{-1}.

Keywords: glucose concentration; optical biosensors; high-sensitivity; quality factor

1. Introduction

Photonic crystals consist of wave scatterers that are regularly arranged next to each other to form an interference pattern at more desirable angles [1–10]. Thus, there are many similarities between a photon crystal and an array of identical antennas, so that it can be said that photon crystals are an array of very small antennas in a regular arrangement of one, two, or three-dimensional next to each other [11–20]. If hypothetical isotropic antennas were used for this purpose, we would have a network of antennas that, due to the destructive and constructive interference of the internal members of the network at different angles of the environment, could input power like an array of antennas scattered at certain different angles. However, if antennas with anisotropic scattering patterns are used, the scattering and product scattering pattern will be the same for the pattern of each of the antennas and the network interference pattern. However, there are some differences. When an array of antennas is fed to send power at a certain angle in the source photon crystals, it is the input wave signal that is scattered around by the scatter-makers. In addition, due to the fact that the most important field of work on photon crystals is the range of optical and infrared waves. Millimeters, unlike antenna arrays, used metals due to their large losses in the frequency range not being very common. In this way, photonic crystals can be thought of as arrays of dielectrics with definite geometric shapes (such as cubes) arranged in a regular arrangement [21–30].

Research into refractive index-based (RI) photon sensors has been enhanced recently, with the goal of further engineering research into novel structures that are appropriate for detecting diverse physical features such as temperature and pressure. In particular, researchers have significantly considered waveguide structures and PhC cavities for bioassay uses [31–34]. Moreover, two-dimensional (2D) photonic crystal (PC) nanopores have shown exceptional optical intensifiers that have both small modal volumes (V) and quality factors (Q). Therefore, their very high Q/V values offer various advantages for optical devices such as high resolution, high sensitivity, low operating energy, and progress in linear and nonlinear optical phenomena. High-quality nanopores have been used for a variety of applications such as selective wavelength filters, biosensors, optical pulse manipulators, and solid-state quantum electrodynamics [35].

The structure of medical sensors is used to detect biochemical, biological or disease systems and food safety [36,37]. Authors adopted two general approaches to various important diagnoses: (1) The use of conventional labels (e.g., enzyme labels) to show analytes; and (2) the use of noninvasive unlabeled techniques so that they require no marker for detecting analytes. The latter can be desirable because labels are not attached to the molecules, and the actual nature and data of the desired biomaterials would not change. Protein changes in the cells and target cell lines may be analyzed by biosensors in cancer research [38]. When testing the blood glucose levels, diabetic cases experience bleeding that would be very uncomfortable and painful. According to the proposed biosensor structure in the present article, human urine could be used to test glucose due to its simplification and accessibility in comparison to the blood testing [39]. Moreover, the limit of such a diagnosis is very low because in human tears, glucose is 50 to 100 times less than in blood; by reducing the volume it can be measured in a shorter time [40]. Biosensors are a very promising platform for the development of on-chip laboratory devices. In addition, biosensors have been shown to be largely sensitive and may be readily constructed or integrated with other electrical components and can quickly and effectively detect DNA and biomolecules, as well as proteins for the early detection of disease. This article discusses the most common of these diseases and reduces mortality when diagnosed early.

The concentration of glucose in urine affects the physical properties of human urine, including surface tension, gravity, viscosity, and defect index. There is no glucose in the urine of the human body under normal conditions, or is in the range 0~15 mg/dL. On the one hand, glycosuria refers to the presence of glucose in the urine and the average level of blood glucose ranges from 165 to 180 mg/dL (mg/dL). Moreover, hypoglycemia represents a lower glucose concentration in the blood (<40 mg/dL) but hyperglycemia is a high level of glucose as well as a concentration of 270–360 mg/dL in blood sugar, which indicates diabetes. Photon measurement technology allows new measurements to be made by careful analyses to detect glucose concentration in urine. Therefore, the refractive index of urine samples and photonic crystal bandgap structures can be used to detect glucose [41].

To survive, human bodies need oxygen and food. However, just the presence of oxygen in the lungs and food in the stomach are not sufficient to survive. In fact, oxygen and food must be supplied to all parts of the body. The blood is responsible for delivering them to the whole body by flowing in a tube known as vein. Then, the heart pumps and circulates the blood in the vein. Almost 7%–8% of the weight of the body consists of blood so that, on average, five liters of the body's volume is made of blood in adults [42]. Blood resembles a red ink and includes different materials. WBCs, platelets, plasma, and red blood cells (RBCs) make up the components of human blood. However, the fluid part of the blood is not red. Blood redness is caused by the presence of RBC. RBC takes oxygen from the lung and carries it to other parts of the body. Furthermore, the fluid part of the blood is referred to as the plasma, most of which is formed of water. The basic function of plasma is to bring healthy food to each part of the body as well as eliminate the waste products. Blood disorders include leukemia, hemorrhage, lymphoma, anemia, and malaria. Blood counts, blood smears, bone marrow biopsies, and comb tests have been considered the most common blood tests [43,44].

2. Optical Biosensor

According to previous research, the rotation of the dielectric functions of a photonic crystal would be broken when there are several points and linear defects in its periodic lattice structures, which allow them to direct and limit light at several light wavelength scales. In particular, when defects enter a certain point in PhC-arranged networks, which possibly form a PhC nanopore enclosed by the reflection of the distributed boundaries. Actually, the new space provided in the center of the PhC network would support a light mode with a frequency in the photonic band (PBG). Hence, light may be "trapped" there for numerous field oscillation cycles, making it possible to confine stronger spatial and temporal light and extend the life of potentially long photons in a PhC cavity [45,46].

In this section, the proposed basic structure is designed and simulated. There is an optical gap created in the TM polarization mode. Therefore, we use the photon bandwidth diagram in terms of saturation r/a for the polarization mode TM using the flat wave clamp (PWE) method for obtaining the best radius of the dielectric rods, for which the photon band width is the maximum. The constant value of the network $n = 600$ nm is chosen and thus the radius of the dielectric rods employed in the structure is $n = 96$ nm. Moreover, the refractive index is $n = 3.4$ (Figure 1a). In addition, the widest band for the filling ratio is r/a = 0.2 in ranges a/λ = 0.285~0.446. In addition, the operating wavelength of the photon crystal at this bandwidth is λ = 1.12–1.75. Figure 1b depicts the input signal of the structure for several pulses that specifies the bandgap range for which the input light signal can propagate in the direction of the waveguides (Table 1).

Table 1. Major parameters of the biosensor computed for different elements of blood.

Analytic	RI	λ_0 (µm)	$\Delta\lambda$ (nm)	Q.f	TE (%)	S
plasma	1.35	1.326425	0.85	1560.5	99	Ref
WBC	1.36	1.312823	1	1312.82	80	1360.2
RBC	1.41	1.299492	0.34	3822.03	92	Ref
Biotin	1.45	1.286433	0.46	2796.59	93	326.47

We designed this new biosensor structure in two dimensions of photon crystal. Therefore, the grid structure was selected as a triangle. With regard to Figure 2, the sensor performance is determined according to the cavities filled with urine samples, all of which provide a specific resonant wavelength depending on the glucose concentration. The resonant cavity is made of glass with a refractive index of $n_G = 1.5$, which is marked in red in the center of the structure. The cavity radius equals $R_C = 200$ nm. Oval and blue dielectric rods with pink rods select the wavelength of intensification and a reduction of bandwidth, which naturally increases the biosensor sensitivity. Adjusting the radius of the red resonance cavity, depending on the application of the biosensor, causes the desired resonance wavelength to change and be transmitted to the output waveguides. Finally, the structure's width and length are the same, with a square structure, and its total area is 102.6 µm^2.

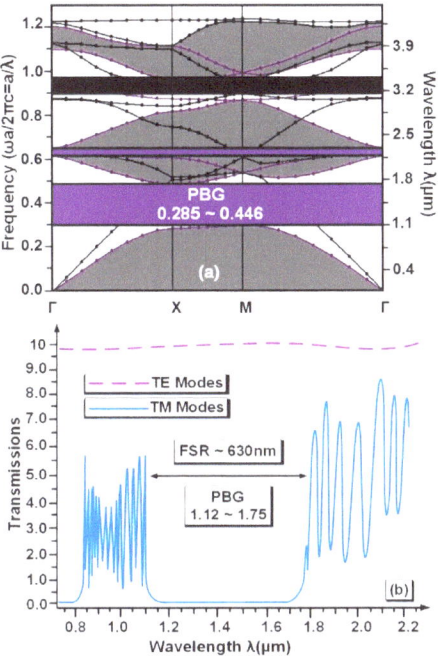

Figure 1. (a) Photon band diagram for polarization modes: TE and TM. (b) depicts the input pulse signal and determination of band gap range and resonant wavelengths.

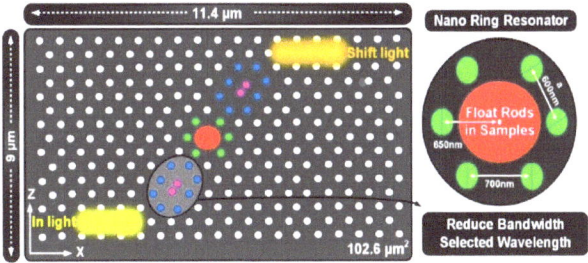

Figure 2. The new structure of an all-optical reflective index biosensor.

3. Materials and Methods

Simulations are related to the structure of the all-optical sensor using the finite difference method in the time domain for the two-dimensional model of R.soft software in a full-wave and band structure mode environment. The grid size (Δx, Δz) in the FDTD solution method is selected as an a/16 equal to 31.25 nm. The amount of time steps must be equal to the formula number for achieving a steady-state in simulating this structure (1). Therefore, the time step equals $\Delta t = 0.18125$ µs in this simulation, and the PML Width value would be selected to be 500 nm. The starting point in the FDTD solution method follows Maxwell's equations. Equation (1) represents the amount of the light emitted in a photon crystal that solves Maxwell's electro-magnetic equations.

$$\tilde{N} \times (\frac{1}{\varepsilon}\tilde{N} \times H) = (\frac{\omega}{C})^2 H, \tag{1}$$

where $\varepsilon = \eta^2$ and η is the refractive index and ω represents the frequency. As seen, the frequency "ω" has an inverse proportionality to the dielectric function. *FWHM* shows the average value of the width of the optical signal and *Q.f* implies the quality factor calculated in Equation (2). Equation (3) also shows the calculation of the sensitivity of the sensor that the rate of change of resonance wavelengths is obtained on the refractive index of the samples used inside the nanocavity.

$$Q.f = \frac{\lambda_0}{FWHM} \qquad (2)$$

$$S = \frac{\Delta \lambda_0}{\Delta RI} \frac{\text{nm}}{\text{RIU}}. \qquad (3)$$

λ_0 is the resonant frequency of the output signal of the structure. λ_0 indicates the difference observed between the resonant frequency of both output signals of the biosensor. ΔRI represents the differences between the samples' refractive index that were applied in the above structure.

One of the important parameters employed for describing the ability of the measurement of a sensor is found by considering the full width at half-maximum (*FWHM*) and the sensitivity parameter, which is an important parameter of the shape of the fit (*FOM*). It is calculated from (4).

$$FOM = \frac{S}{FWHM} \qquad (4)$$

The light signal entering the structure is entered into the structure by a tunable laser source with a continuous wave, and at the output of the structure, a time monitor receives and analyzes the light signal. To calculate the refractive index of the analytes that were applied in the structure, the refractometer with a measuring range of 1.3306 to 1.5284 is used [47]. A cytometer scan [48], scattering [49], and optical coherence tomography [50] have been introduced as some other methods for calculating refractive index. Finally, we placed water ($RI \frac{1}{4}$ 1.33) as a base element for setting 0 and, consequently, the corresponding analyzer on the refractometer prism to calculate the refractive index.

4. Results and Discussion

In the proposed biosensor, several elements of blood, such as RBC, blood plasma, WBC and Bovine Serum Albumin, as well as hemoglobin, would be detected and the detection is carried out by the resonant wavelength method. The process of biotin ling protein attachment to streptavidin would be used to study the interactions of the strongly sensitive proteins. Moreover, it is applied in biotechnology for detecting proteins [51]. It should be noted that the resulting streptavidin-biotin system shows the refractive index of 1.45. Furthermore, hemoglobin takes oxygen from the lung and carries carbon dioxide to the lung, which may be utilized to diagnose thalassemia with a fractured index of 1.38. Consequently, plasma, or blood fluid, is applied for transporting different materials via the blood that may be employed to diagnose cardiovascular disease (CVD) [52]. Its RI value equals 1.35. As mentioned earlier, RBC is largely utilized for carrying oxygen and detecting anemia with a refractive index of 1.40. Moreover, WBC, with a fractured index of 1.36, helps the protection of the body against several infections and is used for diagnosing infections, birth defects, medication, and bone marrow dysfunction. Considering Figure 3, the resonant wavelengths for the four different elements of human blood are obtained by our new biosensor.

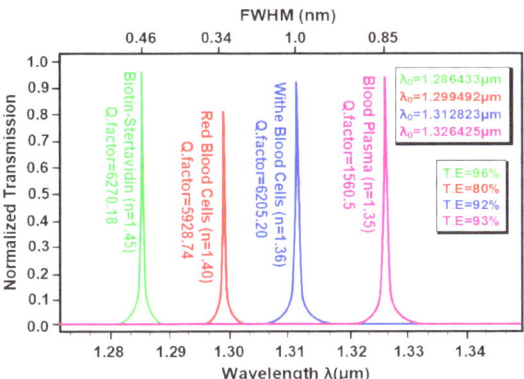

Figure 3. Show the resonance wave lengths of different elements of blood.

Here, the amount of glucose concentration would be determined in human urine. As seen in the table, the refractive index of the urine ranges from 1.332 to 1.340. The first-time urine prototype shows an average refractive index in the range 0.0019 ± 1.336 as well as the random samples 0.0017 ± 1.335. Moreover, the urine refractive index has shown high sensitivity to the changes in its concentration of glucose. A very high refractive index would be observed in a diabetic person [53]. The normal glucose level in normal persons ranges between 0 and 15 mg/dL. Table 2 reports the refractive index values for greater glucose concentration, respectively.

Table 2. The amount of glucose concentrations in human urine for sick and normal people.

Urine Sample with Glucose Concentration	Refractive Index
Normal (0~15 mg/dL)	1.335
0.625 gm/dL	1.336
1.25 gm/dL	1.337
2.5 gm/dL	1.338
5 gm/dL	1.341
10 gm/dL	1.347

In this section, a specific resonant wavelength is selected by the cavity in relation to the glucose density in the urine via putting the samples of human urine in the main glass resonance cavity of the proposed biosensor. With regard to Figure 4, both samples of human urine that had normal densities of 0 to 15 mg/dL and 0.625 mg/dL were detected in the structure so that all of them had a separate resonant wavelength that is very close to each other; the transmission power efficiency in the normal state is 91% higher than in the other state, 98%, which causes the separation of the normal person from the sick person. The sensitivity of this detection equals $S = 2$ nm/RIU. Finally, the figure of merit form equals $FOM = 1.913$ RIU^{-1}.

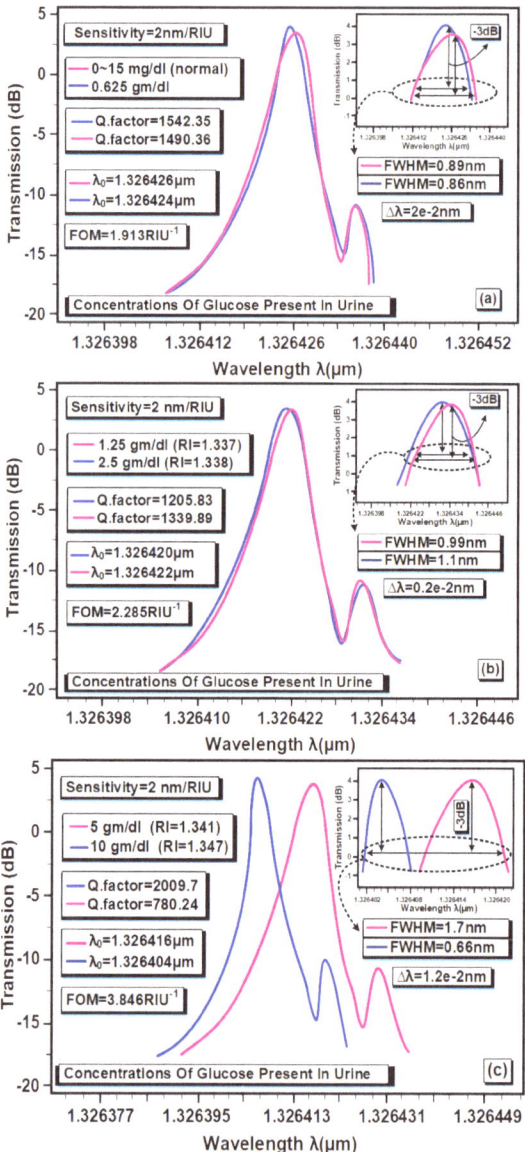

Figure 4. Detection of a normal person compared to a diseased person according to the glucose concentration in human urine in two cases (**a**) 0~15 mg/dL to 0.625 gm/dL, (**b**) 1.25 to 2.5 gm/dL, and (**c**) 5 to 10 gm/dL.

In Figure 4b, the detection of the two densities of glucose 2.5 and 1.25 mg/dL in human administration is shown. The sensitivity has been shown to be $S = 2$ nm/RIU and $FOM = 2.285$ RIU^{-1} for this case. Transmission power efficiency is 84% and 83% for the first and second cases, respectively. Figure 4c shows a difference between 5 mg/dL and 10 mg/dL glucose. The percentage of transfer power is equal to 81% for 5 mg/dL mode and 100% for 10 mg/dL mode. The parameters of the refractive index, quality coefficient, resonant wave-length, and sensitivity are calculated for several values of glucose concentration in human urine. Table 3 shown the parameters of refractive index, quality factor, sensitivity, as well as resonant wavelength for different values of glucose concentration in human urine.

Table 3. Parameters of refractive index, quality factor, sensitivity, as well as resonant wavelength for different values of glucose concentration in human urine.

Glucose	RI	λ_0 (μm)	$\Delta\lambda$ (nm)	Q.f	TE (%)	S
0~15 mg/dL	1.335	1.326426	0.89	1490.36	91	Ref.
0.625 gm/dL	1.336	1.326424	0.86	1542.35	98	2
1.25 gm/dL	1.337	1.326422	0.99	1339.89	84	Ref.
2.5 gm/dL	1.338	1.32642	1.1	1205.83	83	2
5 gm/dL	1.341	1.326416	1.7	780.24	81	Ref.
10 gm/dL	1.347	1.326404	0.66	2009.7	100	2

In this section, we will discuss the effects of the cavity resonance radius. These effects on transmission power efficiency and resonant wavelengths have been investigated. As shown in Figure 5a, when the R_C radius reaches 200 nm, the percentage of transmission power enhances, but if it reaches 300 nm R_C, the transmission power to the structure's output declines. Based on the above comparison, the most acceptable radius for the intensification of the cavity equals R_C = 200 nm and R_C = 300 nm. Enhancing the resonance radius of the cavity causes the resonance wavelength to reach a maximum of 200 nm and 300 nm. These changes are calculated for a refractive index of 1.36. In Figure 6b, the resonance wavelength has reached its maximum value at R_C = 300 nm. Moreover, the transmission power has been shown in the radius of R_C = 200 nm at best. The purpose of making such changes is to obtain the best radius for the cavity of the resonant. Put differently, it is possible to detect a sensor for several samples with the use of the transmission power efficiency and resonance wavelength.

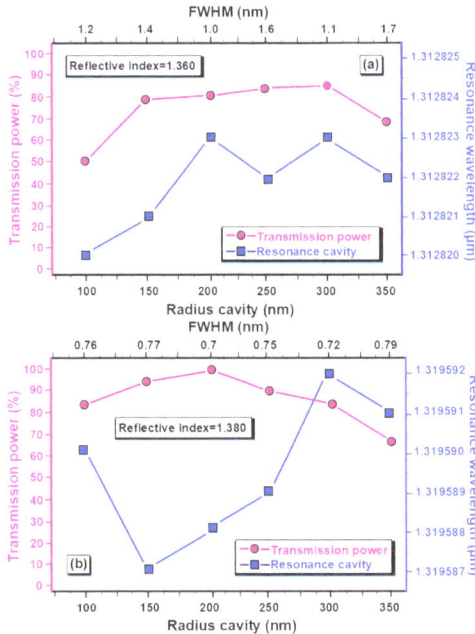

Figure 5. Calculation of the resonant wave-lengths as well as transmission power for several radius cavities, (**a**) for refractive index 1.36, (**b**) for refractive index 1.38.

Figure 6a the behavior of the biosensor structure at different refractive indices is calculated. With the enhanced refractive index, the resonant wavelength increases linearly and the transmitting power decreases linearly, too. The reason for using the cavitation refractive index changes and its impact on the sensor output is to detect the most acceptable performance of the structure in terms of transmission power. For example, when a sample with a refractive index of $RI = 1.41$ is detected by this sensor, we will reach the maximum resonance wavelength in the structure. If a normal person's blood sample with a refractive index of $RI = 1.35$ is used in this sensor, it will reach maximum transmitting power. Figure 6b the effects of changes in the reflective index of the resonance cavity on the parameters of FWHM as well as the quality factor are calculated. As can be seen in this figure, the changes of these two parameters are opposite to each other. Therefore, the quality factor parameter in the proposed sensor structure for the blood sample with the refractive index of 1.368 has the lowest value and reaches the maximum value at a refractive index of 1.38. Hence, the parameter changes of the FWHM with changes in the refractive index are the opposite of the changes of the quality factor. It reaches its maximum value (worst case). Therefore, for example, the blood sample with a reflective index of 1.368 is the best average value of the bandwidth of the intensified wavelength.

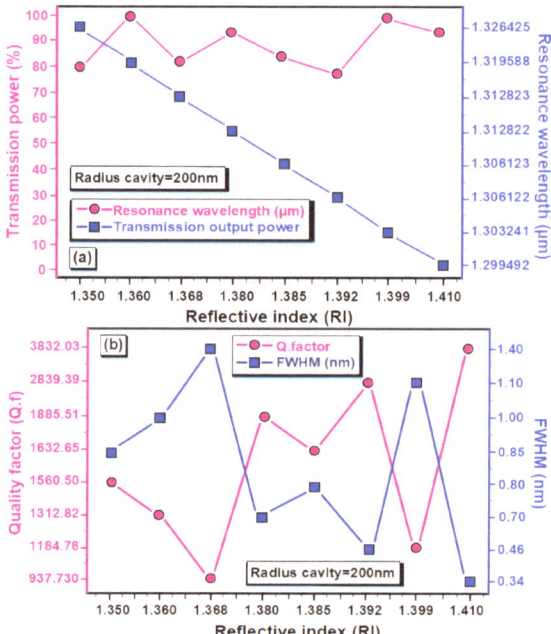

Figure 6. (**a**) Calculation of resonance wavelength parameters and transmission power efficiency for different refractive index. (**b**) Calculation of FWHM and Q factor parameters for different refractive index.

Different wavelengths are used in the input signal source to check the sensor performance in both active and inactive modes. Figure 7a the sensor operation in the inactive state is shown schematically and in the output diagram of the structure. For this case, for example, the refractive index of the resonant nano-cavity is set to $RI = 1.36$, and a wavelength other than the resonant wavelength is used for this case, which makes the sensor select no light signal transmission and does not amplify the wavelength and the sensor is inactive. Figure 7b shows the structure as fully active. To achieve this, place the wavelength of the input source exactly on the wave length of the central cavity, which enables 100% of the input light signal to be amplified by the cavity so that the rods around the cavity can be chosen and amplified, and finally transferred to the output wave-guide.

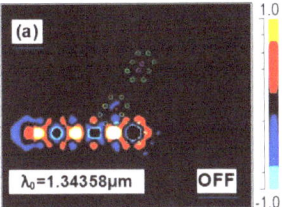 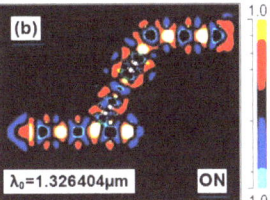

Figure 7. (a) Sensor performance in the passive mode, resonant wavelength λ_0 = 1.34358 nm. (b) Sensor function inactive mode, resonant wavelength λ_0 = 1.326404 nm.

Finally, to determine the superiority of the proposed structure over other similar structures, in Table 4, the major features of the all-optical biosensor structure include quality factor, transmission power, sensor sensitivity and figure of merit compared to the previous structure.

Table 4. Comparison of sample important characteristics of the proposed structure such as quality coefficient, transmission power, sensor sensitivity and figure of merit with the previously structures.

Ref	Detection Sample	Q.f	FOM	TE	S nm/RIU
Ref. [45]	Blood and Tears fluid	1082	-	-	6.5764
Ref. [52]	Glucose	-	-	86	422
Ref. [53]	Glucose	1.11×10^5	1117	92	462
Ref. [54]	Blood	262	-	100	-
Ref. [55]	-	-	88	98	263
Ref. [56]	-	1264	84	90	840
The present study	Urine, Tears fluid, Blood	3822.03	1320.23	100	1360.02

5. Conclusions

As mentioned earlier, our new sensor could detect a person with various ailments from a normal individual using blood and urine, as well as human tear fluid samples. Detectable diseases through the proposed sensor are blood cancer cells and the detection of different elements. In the blood, the distinction between the normal and diabetic cells is the concentration of glucose in human urine. Moreover, the key performance of nano cavitation detection is the resonance at the center of the structure. Due to the significance of sensitivity and accuracy features in designing the sensors, the average bandwidth of output intensification wavelengths in the lowest case equals 0.34 nm. In addition, the quality coefficient of this new sensor in the worst conditions equals 3822.03 and the power transmission percentage of the optical signal is between 77% to 100%. The parameter with the importance of sensitivity in the best case equals S60 = 1360.02 nm/RIU and the important parameter of the degree of suitability for the proposed sensor structure is equal to $FOM = 1320.23 \pm 118 \text{ RIU}^{-1}$. The sensor is designed as one of the most complete sensors in terms of scope of operation and high sensitivity, which can detect different patients from three types of blood, tears, and urine samples.

Author Contributions: Conceptualization, K.H.; methodology, S.A.; software, D.O.B.; validation, W.S.; formal analysis, A.T.J.; data curation, A.M.A.; writing—original draft preparation, S.A.; writing—review and editing, M.S.; supervision, M.S. All authors have read and agreed to the published version of the manuscript.

Funding: The Researchers Supporting Project Number (RSP-2021/261) King Saud University, Riyadh in Saudi Arabia funded the present study.

Institutional Review Board Statement: Not applicable.

Informed Consent Statement: Not applicable.

Data Availability Statement: Not applicable.

Conflicts of Interest: The authors declare no conflict of interest.

References

1. Lee, C.H.; Park, J.S. An SDN-Based Packet Scheduling Scheme for Transmitting Emergency Data in Mobile Edge Computing Environments. *Hum. Cent. Comput. Inf. Sci.* **2021**, *11*, 28. [CrossRef]
2. Baek, S.; Jeon, J.; Jeong, B.; Jeong, Y.S. Two-Stage Hybrid Malware Detection Using Deep Learning. Human-centric Computing and Information Sciences volume. *Hum. Cent. Comput. Inf. Sci.* **2021**, *11*, 27. [CrossRef]
3. Dai, H.; Li, J.; Kuang, Y.; Liao, J.; Zhang, Q.; Kang, Y. Multiscale Fuzzy Entropy and PSO-SVM Based Fault Diagnoses for Airborne Fuel Pumps. *Hum. Cent. Comput. Inf. Sci.* **2021**, *11*, 25. [CrossRef]
4. Maqsood, M.; Mehmood, I.; Kharel, R.; Muhammad, K.; Lee, J.; Alnumay, W.S. Exploring the Role of Deep Learning in Industrial Applications: A Case Study on Coastal Crane Casting Recognition. Human Centric Computing and Information Sciences. *Hum. Cent. Comput. Inf. Sci.* **2021**, *11*, 20.
5. Salim, M.M.; Shanmuganathan, V.; Loia, V.; Park, J.H. Deep Learning Enabled Secure IoT Handover Authentication for Blockchain Networks. *Hum. Cent. Comput. Inf. Sci.* **2021**, *11*, 21. [CrossRef]
6. Blundo, C.; De Maio, C.; Parente, M.; Siniscalchi, L. Targeted Advertising That Protects the Privacy of Social Networks Users. *Hum. Cent. Comput. Inf. Sci.* **2021**, *11*, 18. [CrossRef]
7. Liu, C.; Li, K.; Li, K. A Game Approach to Multi-Servers Load Balancing with Load-Dependent Server Availability Consideration. *IEEE Trans. Cloud Comput.* **2021**, *9*, 1–13. [CrossRef]
8. Liu, C.; Li, K.; Li, K.; Buyya, R. A New Service Mechanism for Profit Optimizations of a Cloud Provider and Its Users. *IEEE Trans. Cloud Comput.* **2021**, *9*, 14–26. [CrossRef]
9. Xiao, G.; Li, K.; Chen, Y.; He, W.; Albert, Y.; Zomaya, T.L. CASpMV: A Customized and Accelerative SpMV Framework for the Sunway TaihuLight. *IEEE Trans. Parallel Distrib. Syst.* **2021**, *32*, 131–146. [CrossRef]
10. Duan, M.; Li, K.; Li, K.; Tian, Q. A Novel Multi-Task Tensor Correlation Neural Network for Facial Attribute Prediction. *ACM Trans. Intell. Syst. Technol.* **2021**, *12*, 1–22. [CrossRef]
11. Chen, C.; Li, K.; Teo, S.G.; Zou, X.; Li, K.; Zeng, Z. Citywide Traffic Flow Prediction Based on Multiple Gated Spatio-temporal Convolutional Neural Networks. *ACM Trans. Knowl. Discov. Data* **2020**, *14*, 42:1–42:23. [CrossRef]
12. Zhou, X.; Li, K.; Yang, Z.; Gao, Y.; Li, K. Efficient Approaches to k Representative G-Skyline Queries. *ACM Trans. Knowl. Discov. Data* **2020**, *14*, 58:1–58:27. [CrossRef]
13. Zhang, C.; Liu, X.; Liu, C.; Luo, X. Characterization of the Complete Mitochondrial Genome of Acanthacorydalis fruhstorferi van der Weele (Megaloptera: Corydalidae). *J. Kans. Èntomol. Soc.* **2020**, *93*, 267–281. [CrossRef]
14. Sun, J.; Du, H.; Chen, Z.; Wang, L.; Shen, G. MXene Quantum Dot within Natural 3D Watermelon Peel Matrix for Biocompatible Flexible Sensing Platform. *Nano Res.* **2021**, *3*, 1–7. [CrossRef]
15. Zou, Y.; Wu, H.; Guo, X.; Peng, L.; Ding, Y.; Tang, J.; Guo, F. MK-FSVM-SVDD: A Multiple Kernel-based Fuzzy SVM Model for Predicting DNA-binding Proteins via Support Vector Data Description. *Curr. Bioinform.* **2021**, *16*, 274–283. [CrossRef]
16. Yang, Y.; Sun, F.; Chen, H.; Tan, H.; Yang, L.; Zhang, L.; Xie, J.; Sun, J.; Huang, X.; Huang, Y. Postnatal Exposure to DINP Was Associated with Greater Alterations of Lipidomic Markers for Hepatic Steatosis Than Dehp in Postweaning Mice. *Sci. Total Environ.* **2020**, *758*, 143631. [CrossRef] [PubMed]
17. Xu, Y.; Meng, L.; Chen, X.; Chen, X.; Su, L.; Yuan, L.; Shi, W.; Huang, G. A Strategy to Significantly Improve the Classification Accuracy of LIBS Data: Application for the Determination of Heavy Metals in Tegillarca Granosa. *Plasma Sci. Technol.* **2021**, *23*, 085503. [CrossRef]
18. Yan, Y.; Feng, L.; Shi, M.; Cui, C.; Liu, Y. Effect of Plasma-Activated Water on the Structure and in Vitro Digestibility of Waxy and Normal Maize Starches during Heat-Moisture Treatment. *Food Chem.* **2019**, *306*, 125589. [CrossRef]
19. Shi, M.; Wang, F.; Lan, P.; Zhang, Y.; Zhang, M.; Yan, Y.; Liu, Y. Effect of Ultrasonic Intensity on Structure and Properties of Wheat Starch-Monoglyceride Complex and Its Influence on Quality of Norther-Style Chinese Steamed Bread. *LWT* **2020**, *138*, 110677. [CrossRef]
20. Prather, D.W.; Shi, S.; Murakowski, J.; Schneider, G.J.; Sharkawy, A.; Chen, C.; Miao, B.; Martin, R. Self-Collimation in Photonic Crystal Structures: A New Paradigm for Applications and Device Development. *J. Phys. D Appl. Phys.* **2007**, *40*, 2635–2651. [CrossRef]
21. Duan, M.; Li, K.; Ouyang, A.; Win, K.N.; Li, K.; Tian, Q. EGroupNet: A Feature-Enhanced Network for Age Estimation with Novel Age Group Schemes. *ACM Trans. Multim. Comput. Commun. Appl.* **2020**, *16*, 1–23. [CrossRef]
22. Yang, W.; Li, K.; Li, K. A Pipeline Computing Method of SpTV for Three-Order Tensors on CPU and GPU. *ACM Trans. Knowl. Discov. Data* **2019**, *13*, 1–27. [CrossRef]

23. Zhou, X.; Li, K.; Yang, Z.; Xiao, G.; Li, K. Progressive Approaches for Pareto Optimal Groups Computation. *IEEE Trans. Knowl. Data Eng.* **2019**, *31*, 521–534. [CrossRef]
24. Mei, J.; Li, K.; Tong, Z.; Li, Q.; Li, K. Profit Maximization for Cloud Brokers in Cloud Computing. *IEEE Trans. Parallel Distrib. Syst.* **2019**, *30*, 190–203. [CrossRef]
25. Chen, Y.; Li, K.; Yang, W.; Xiao, G.; Xie, X.; Li, T. Performance-Aware Model for Sparse Ma-Trix-Matrix Multiplication on the Sunway TaihuLight Supercomputer. *IEEE Trans. Parallel Distrib. Syst.* **2019**, *30*, 923–938. [CrossRef]
26. Chen, J.; Li, K.; Bilal, K.; Zhou, X.; Li, K.; Philip, S.Y. A Bi-layered Parallel Training Architecture for Large-Scale Convolutional Neural Networks. *IEEE Trans. Parallel Distrib. Syst.* **2019**, *30*, 965–976. [CrossRef]
27. Liu, C.; Li, K.; Liang, J.; Li, K. Service Reliability in an HC: Considering from the Perspective of Scheduling with Load-Dependent Machine Reliability. *IEEE Trans. Reliab.* **2019**, *68*, 476–495. [CrossRef]
28. Chen, C.; Li, K.; Ouyang, A.; Li, K. FlinkCL: An OpenCL-Based in-Memory Computing Architecture on Heteroge-neous CPU-GPU Clusters for Big Data. *IEEE Trans. Comput.* **2018**, *67*, 1765–1779. [CrossRef]
29. Duan, M.; Li, K.; Li, K. An Ensemble CNN2ELM for Age Estimation. *IEEE Trans. Inf. Forensics Secur.* **2018**, *13*, 758–772. [CrossRef]
30. John, S.G.; Joannopoulos, D.; Winn, J.N.; Meade, R.D. *Photonic Crystals: Molding the Flow of Light*; InPrinceton University of Press: Princeton, NJ, USA, 2008.
31. Kassa-Baghdouche, L.; Cassan, E. Mid-Infrared Refractive Index Sensing Using Optimized Slotted Photonic Crystal Waveguides. *Photon. Nanostruct. Fundam. Appl.* **2018**, *28*, 32–36. [CrossRef]
32. Nakamura, T.; Takahashi, Y.; Tanaka, Y.; Asano, T.; Noda, S. Improvement in the Quality Factors for Photonic Crystal Nanocavities via Visualization of the Leaky Components. *Opt. Express* **2016**, *24*, 9541–9549. [CrossRef] [PubMed]
33. Kassa-Baghdouche, L.; Boumaza, T.; Cassan, E.; Bouchemat, M. Enhancement of Q-Factor in SiN-Based Planar Photonic Crystal L3 Nanocavity for Integrated Photonics in the Visible-Wavelength Range. *Optik* **2015**, *126*, 3467–3471. [CrossRef]
34. Sani, M.H.; Khosroabadi, S. A Novel Design and Analysis of High-Sensitivity Biosensor Based on Nano-Cavity for Detection of Blood Component, Diabetes, Cancer and Glucose Concentration. *IEEE Sens. J.* **2020**, *20*, 7161–7168. [CrossRef]
35. Jannesari, R.; Rancher, C.; Consani, C.; Grille, T.; Jakoby, B. Sensitivity Optimization of a Photonic Crystal Ring Resonator for Gas Sensing Applications. *Sens. Actuators A Phys.* **2017**, *264*, 347–351. [CrossRef]
36. Sharma, P.; Sharan, P.; Deshmukh, P. A Photonic Crystal Sensor for Analysis and Detection of Cancer Cells. In Proceedings of the 2015 IEEE International Conference on Pervasive Computing (ICPC), St. Louis, MO, USA, 23–27 March 2015; pp. 1–5.
37. Zhang, J.; Hodge, W.; Hutnick, C.; Wang, X. Noninvasive Diagnostic Devices for Diabetes through Measuring Tear Glucose. *J. Diabetes Sci. Technol.* **2011**, *5*, 166–172. [CrossRef]
38. Makaram, P.; Owens, D.; Aceros, J. Trends in Nanomaterial-Based Non-Invasive Diabetes Sensing Technologies. *Diagnostics* **2014**, *4*, 27–46. [CrossRef] [PubMed]
39. Yan, Q.; Peng, B.; Su, G.; Cohan, B.E.; Major, T.C.; Meyerhoff, M.E. Measurement of Tear Glucose Levels with Amperometric Glucose Biosensor/Capillary Tube Configuration. *Anal. Chem.* **2011**, *83*, 8341–8346. [CrossRef] [PubMed]
40. Beck, H.C.; Overgaard, M.; Rasmussen, L.M. Plasma Proteomics to Identify Biomarkers—Application to Cardiovascular Diseases. *Transl. Proteom.* **2015**, *7*, 40–48. [CrossRef]
41. Wolf, D.J.; Zitelli, J.A. Surgical Margins for Basal Cell Carcinoma. *Arch. Dermatol.* **1987**, *123*, 340–344. [CrossRef] [PubMed]
42. Siyal, S.H.; Javed, M.S.; Ahmad, A.; Sajjad, M.; Batool, S.; Khan, A.J.; Akramh, S.; Alothman, A.A.; Alshgarii, R.A.; Najam, T. Free-standing 3D Co3O4@ NF micro-flowers composed of porous ultra-long nanowires as an advanced cathode material for supercapacitor. *Curr. Appl. Phys.* **2021**, *31*, 221–227. [CrossRef]
43. Kassa-Baghdouche, L. High-Sensitivity Spectroscopic Gas Sensor Using Optimized H1 Photonic Crystal Microcavities. *J. Opt. Soc. Am. B* **2020**, *37*, A277. [CrossRef]
44. Kassa-Baghdouche, L.; Cassan, E. Mid-Infrared Gas Sensor Based on High-Q/V Point-Defect Photonic Crystal Nanocavities. *Opt. Quantum Electron.* **2020**, *52*, 260. [CrossRef]
45. Ahmad, A.; Jini, D.; Aravind, M.; Parvathiraja, C.; Ali, R.; Kiyani, M.Z.; Alothman, A. A novel study on synthesis of egg shell based activated carbon for degradation of methylene blue via photocatalysis. *Arab. J. Chem.* **2020**, *13*, 8717–8722. [CrossRef]
46. Shvalov, A.N.; Soini, J.T.; Chernyshev, A.V.; Tarasov, P.A.; Soini, E.; Maltsev, V.P. Light-Scattering Properties of Individual Erythro-Cytes. *Appl. Opt.* **1999**, *38*, 230–235. [CrossRef]
47. Mroczka, J.; Wysoczanski, D.; Onofri, F. Optical Parameters and Scattering Properties of Red Blood Cells. *Opt. Appl.* **2002**, *32*, 691–700.
48. Xu, X.; Wang, R.K.; Elder, J.B.; Tuchin, V. Effect of Dextran-Induced Changes in Refractive Index and Aggregation on Optical Properties of Whole Blood. *Phys. Med. Biol.* **2003**, *48*, 1205–1221. [CrossRef] [PubMed]
49. Cailleau, R.; Young, R.; Olivé, M.; Reeves, W.J., Jr. Breast Tumor Cell Lines from Pleural Effusions 2. *J. Natl. Cancer Inst.* **1974**, *53*, 661–674. [CrossRef]
50. Yaroslavsky, A.N.; Patel, R.; Salomatina, E.; Li, C.; Lin, C.; Al-Arashi, M.; Neel, V. High-Contrast Mapping of Basal Cell Carcinomas. *Opt. Lett.* **2012**, *37*, 644–646. [CrossRef]
51. Gharsallah, Z.; Najjar, M.; Suthar, B.; Janyani, V. High Sensitivity and Ultra-Compact Optical Biosensor for Detection of UREA Concentration. *Opt. Quantum Electron.* **2018**, *50*, 249. [CrossRef]

52. Kashif, M.; Jafaar, E.; Sahari, S.K.; Low, F.W.; Hoa, N.D.; Ahmad, A.; Abbas, A.; Ngaini, Z.; Shafa, Z.; Qurashi, A. Organic sensitization of graphene oxide and reduced graphene oxide thin films for photovoltaic applications. *Int. J. Energy Res.* **2021**, *45*, 9657–9666. [CrossRef]
53. Arafa, S.; Bouchemat, M.; Bouchemat, T.; Benmerkhi, A.; Hocini, A. Infiltrated Photonic Crystal Cavity as a Highly Sensitive Platform for Glucose Concentration Detection. *Opt. Commun.* **2017**, *384*, 93–100. [CrossRef]
54. Arunkumar, R.; Suaganya, T.; Robinson, S. Design and Analysis of 2D Photonic Crystal Based Biosensor to Detect Different Blood Components. *Photon. Sens.* **2018**, *9*, 69–77. [CrossRef]
55. Almpanis, E.; Papanikolaou, N. Dielectric Nanopatterned Surfaces for Subwavelength Light Localization and Sensing Applications. *Microelectron. Eng.* **2016**, *159*, 60–63. [CrossRef]
56. Lu, X.; Zhang, T.; Wan, R.; Xu, Y.; Zhao, C.; Guo, S. Numerical Investigation of Narrowband Infrared Absorber and Sensor Based on Dielectric-Metal Metasurface. *Opt. Express* **2018**, *26*, 10179–10187. [CrossRef] [PubMed]

Article

Detection of Virulence Genes and Biofilm Forming Capacity of Diarrheagenic *E. coli* Isolated from Different Water Sources

Sadaf Tariq [1], Sobia Tabassum [2,*], Sadia Aslam [3], Mika Sillanpaa [4,5], Wahidah H. Al-Qahtani [6] and Shafaqat Ali [7,8,*]

1. Department of Biochemistry, Government College University Faisalabad, Faisalabad 38000, Pakistan; sadaftariq787@gmail.com
2. Department of Biological Sciences, International Islamic University, H-10 Campus, Islamabad 44000, Pakistan
3. Department of Botany, Government College University Faisalabad, Faisalabad 38000, Pakistan; sadiaaslam646@gmail.com
4. Department of Chemical Engineering, School of Mining, Metallurgy and Chemical Engineering, University of Johannesburg, P. O. Box 17011, Doornfontein 2028, South Africa; mikaesillanpaa@gmail.com
5. Department of Biological and Chemical Engineering, Aarhus University, Nørrebrogade 44, 8000 Aarhus, Denmark
6. Department of Food Sciences & Nutrition, College of Food & Agriculture Sciences, King Saud University, Riyadh 11351, Saudi Arabia; wahida@ksu.edu.sa
7. Department of Environmental Sciences and Engineering, Government College University, Allama Iqbal Road, Faisalabad 38000, Pakistan
8. Department of Biological Sciences and Technology, China Medical University, Taichung 40402, Taiwan
* Correspondence: sobia.tabassum@iiu.edu.pk (S.T.); shafaqataligill@yahoo.com (S.A.)

Citation: Tariq, S.; Tabassum, S.; Aslam, S.; Sillanpaa, M.; Al-Qahtani, W.H.; Ali, S. Detection of Virulence Genes and Biofilm Forming Capacity of Diarrheagenic *E. coli* Isolated from Different Water Sources. *Coatings* **2021**, *11*, 1544. https://doi.org/10.3390/coatings11121544

Academic Editor: Teen-Hang Meen

Received: 17 November 2021
Accepted: 8 December 2021
Published: 16 December 2021

Publisher's Note: MDPI stays neutral with regard to jurisdictional claims in published maps and institutional affiliations.

Copyright: © 2021 by the authors. Licensee MDPI, Basel, Switzerland. This article is an open access article distributed under the terms and conditions of the Creative Commons Attribution (CC BY) license (https://creativecommons.org/licenses/by/4.0/).

Abstract: Diarrheagenic *Escherichia coli* (DEC) are associated with frequent incidences of waterborne infections and pose health risk to individuals who contact water for domestic or recreational uses. Detection of DEC pathotypes in drinking water can be used as an indicator of fecal contamination. This study aimed to investigate the occurrence of DEC pathotypes and their capacity to form biofilms in drinking water samples collected from different water sources. In this study, PCR analysis was used to determine the occurrence of four clinically significant virulence genes of diarrheagenic *E. coli*, *eaeA* (Enteropathogenic *E. coli*), *stx1*, *stx2* (Enterohemorrhagic *E. coli*) and *sth* (Enterotoxigenic *E. coli*), in drinking water samples (n = 35) by using specific primers and conditions. PCR amplicons were visualized by using agarose gel electrophoresis. A total of 12/35 (34%) samples were detected as positive for at least one of the four DEC virulence genes and 11/12 (91%) *E. coli* isolates harbored virulence gene while 1/12 (8%) *E. coli* isolates harbored none. The *eaeA* and *sth* genes were the most detected genes (75%), while *stx1* and *stx2* genes were least detected genes (66%). Biofilm assay confirmed that ETEC pathotypes can cause damage in enteric walls by attaching and effacing to persist diarrheal conditions. This study indicated that drinking water of different sources is contaminated with potential DEC pathotypes and it can be a source of diarrheal diseases. The amplification of four virulence genes associated with DEC pathotypes (EPEC, EHEC and ETEC) in drinking water demonstrates that potentially virulent DEC pathotypes are distributed in water sources and may be a cause of health concern. There is, therefore, an urgent need to monitor DEC pathotypes in drinking water.

Keywords: diarrheagenic *E. coli*; virulence genes; contamination; pathotypes; isolates; water

1. Introduction

In order to ensure good health, the availability of safe and good quality water is very essential [1]. Waterborne infections are major cause of high morbidity and mortality rate and have become a public health problem [2]. In developing countries, ground water is the sole elementary source of drinking water supply, which may contain several pathogens, viruses, bacteria and protozoa, causing 2.5 million deaths from enteric diarrheal diseases annually [3,4]. Diarrheal and other gastrointestinal illnesses are common infections among infants and children in developing countries [5]. Improper sanitation and hygiene cause comparatively greater health risk [6,7]. *Escherichia coli* are one of inhabitants of gut flora and most commonly used indicator for fecal contamination in different drinking water distribution systems [8]. *E. coli* can survive in drinking water for 4 to 12 weeks, depending upon various environmental factors. According to its biological significance, *E. coli* live safely in the intestine as harmless commensal [9] and pathogenic strains of bacteria cause intestinal and extra-intestinal infections both in healthy and immunocompromised individuals [2,10].

The accessory genome expresses pathogenic traits that are developed by horizontal gene transfer. The virulence genes are located on the accessory genome: a kind of genome plasticity formed by gain and loss of genetic information [1,11]. On the basis of virulence determinants (acquired), specific combinations and horizontal gene transfer, diarrheagenic *E. coli* pathotypes are classified into different pathotypes. These pathotypes are enteropathogenic *E. coli* (EPEC), enteroinvasive *E. coli* (EIEC), enterohemorrhagic *E. coli* (EHEC), enteroaggregative *E. coli* (EAEC), diffusely adherent *E. coli* (DAEC) and enterotoxigenic *E. coli* (ETEC) [11,12]. These pathotypes define different clinical features, phenotypic traits, epidemiological evidences and specific virulence factors of DEC pathotypes. These pathotypes vary for their clinical, epidemiological and pathogenic significance [2] (Table 1).

Table 1. Clinical manifestation, intestinal pathology and epidemiological features of epidemics caused by five basic pathotypes of diarrheagenic *E. coli*.

Pathotype	Clinical Manifestation	Pathology	Susceptible Population
ETEC	Watery diarrhea	Not any definite change	Children in developing areas, travelers to risky areas
EIEC	Bacillary diarrhea	Disrupted mucosa with inflammation of large intestine	Common in developing countries, affect all ages
EPEC	Gastroentritis	Attaching and effacing lesions	Infants and travelers to risky areas
EHEC	Bloody diarrhea	Attaching and effacing lesions in intestine followed by necrosis	Inhabitants of industrialized areas
EAEC	Persistent diarrhea	Inflammatory responses in intestine, cytotoxicity of enterocytes	Children in developing areas and travelers to those areas

DEC pathotypes form colonies on the epithelia of intestines by either inhabiting and invading the intestinal cells. They are specialized to form adhesions on the epithelia and known for their attaching and effacing ability to form lesions [9]. They possess resistance against immune system as the colonies target and respond in a similar manner. The virulence factors inject themselves in the cells by mimicking the cell ligands and disrupt plasticity of cytoskeleton [13], thereby reducing endocytic trafficking; they make themselves resistant to phagocytosis. The host cells respond and defend by inducing inflammatory reactions. It changes the intestinal fluid balance to excrete unwanted and harmful bacteria that cause diarrhea [14,15]. The extent of infection depends on the interaction between the cell and bacteria and also on the defense system [16–18].

DEC pathotypes have developed important structures for attaching and effacing the cells such as bundle forming pili (*bfp*), *EspA* filaments and *EAF* shiga-like toxins and intimin [19,20]. The *bfp* encoded by a ~80 kb adherence factor forms bacterial micro-colonies. Shiga-like toxin and *intimin* are encoded by *stx* and *eaeA* genes, respectively. The *eaeA* gene is an *E. coli* outer membrane protein. Intimin is known for its intimate adhesions. EHEC are shiga toxin producers posessing two main classes: *stx1* and *stx2*; they are further divided into many subtypes [20,21]. STEC detection methods mostly depend on serotype 0157H7. The EPEC secreted proteins are translocated through a type III secretion system [16,22]. It forms a pore inside the bacterial membrane. *EspA* filaments are parts of type III secretion system [23] and directly related to the epithelial cells due to its interaction in a manner functioning as adhesions [17,24]. *EAF* is important for its role in pathogenicity, biofilm production and colonization. *EspA* forms a filament-like translocation tube outside the bacteria [25], while *EspB* and *EspD* are incorporated inside the membrane of host cells [26]. It allows the movement of other proteins through the cell membrane, e.g., Tir an effector protein [23].

The presence of DEC pathotypes forming biofilms can cause a serious threat because it can cause difficulties in medical treatments [27,28]. DEC pathotypes, especially ETEC, either make biofilms or attach to already existing biofilms. Biofilms are bacterial surface groups surrounded by extracellular matrix [29,30]. It provides bacteria with shelter, prevents its desiccation, make it resistant to antibiotics and protects it from bacteriophages. Biofilms are rapidly developed on both abiotic and biotic surfaces that are in direct contact with water [18,27]. Bacterial biofilm production is highly complex and regulated. They have evolved a complex system of DNA, exopolysaccharides and proteins to protect themselves from environmental stress [31,32]. In developing countries, very few studies have been performed with respect to reporting the virulence genes of DEC isolated from drinking water. Therefore, this study aimed to detect virulence-associated genes of DEC pathotypes associated with human pathology and to assess their ability to form biofilms to determine the risk of drinking contaminated water.

2. Material and Method

2.1. Water Samples Collection

Drinking water samples were collected from different water sources: ten samples were collected from tank water, five samples were collected from industries, five samples were infused water, three samples were collected from standing water areas, five samples were collected from wells and seven samples were collected from tap labeled as 1–35, respectively. A total of 35 drinking water samples were collected for this study. An amount of 200 mL of drinking water sample was collected from each site in sterilized glass containers having lids according to the American Public Health Association 2001. These samples were transported to laboratory in a cooler and processed in 4 h.

2.2. Isolation and Biochemical Identification of E. coli

The collected samples were cultured on MacConkey agar (Oxoid) plates by swabbing sterilized cotton swab and incubated at 37 °C for 24 h. These plates were observed to detect grown bacterial colonies. Only suspected colonies were considered for Gram's reaction. Gram-negative bacillus colonies were detected. Each suspected sample was subjected to further biochemical testing. Triple Sugar Iron agar (TSI) test was used for biochemical characterization. The autoclaved TSI agar was allowed to set in a slanted position in sterilized test tubes. The slants were inoculated by picking a colony from the MacConkey agar plates and kept in an incubator at 37 °C overnight.

2.3. Extraction of DNA

The boiling centrifugation method was used to extract DNA from all isolated *E. coli* colonies as described by [33], with slight modifications. Isolated *E. coli* colonies were subcultured on nutrient agar at 37 °C. Isolated *E. coli* colonies were picked from a nutrient agar plate in a sterile Eppendorf tube and washed with 1 mL sterile normal saline (NS). It was centrifuged at 10,000 rev/min for 2 min. The pellet was recovered and re-suspended in 100 µL distilled water. Then, the tubes were boiled at 85 °C for 10 min. The lysate was centrifuged at 13,000 rev/min for 5 min. The supernatant containing template DNA stock was stored at 20 °C. The presence and concentration of extracted template DNA were measured by using agarose gel electrophoresis in the sample, and it was compared with a specific DNA marker of known concentration.

2.4. Polymerase Chain Reaction (PCR)

PCR (Thermal cycle, London, UK) was performed for the detection of target genes *stx1* and *stx2* for EHEC, *sth* for ETEC and *eaeA* for EPEC by using specific primers (Table 2). The reagents used for each reaction in volumes of 25 µL were as follows: 7 µL syringe water, 2.5 µL 100 mM KCl, 3 µL 3 mM MgCl2, 1 µL 400 µM dNTPs, 2.5 µL 20 mM tris-HCl, 0.5 µL forward primer, 0.5 µL reverse primer, 2.5 µL 0.1% gelatin, 5.0 µL DNA (Template DNA) and 0.5 µL 2.5 units Taq Polymerase. The PCR mixture was subjected to an initial denaturation step at 95 °C for 5 min, followed by 35 cycles of denaturation at 95 °C for 45 s, primer annealing at 57 °C for 45 s, extension at 72 °C for 45 s and the final extension at 72 °C for 10 min.

Table 2. Primers used in study.

Pathotype	Target Gene	Primer Sequence (5'–3')	Size of Product (kb)	References
EHEC	*stx1*	F:ACTTCTCGACTGCAAAGACGTATG R:ACAAATTATCCCCTGAGCCACTATC	132	[34]
EHEC	*stx2*	F:TTCCGGAATGCAAATCAG TC R:CGATACTCCGGAAGCACATTG	264	[35]
EPEC	*eaeA*	F:CCGATTCCTCTGGTGACGA R:CCACGGTTTATCAAACTGATAACG	105	[36]
ETEC	*sth*	F:TTCACCTTTCCCTCAGGATG R:CTATTCATGCTTTCAGGACCA	120	[36]

2.5. Visualization of PCR Products

Agarose (2%) was used to make agarose gel. The casting tray was flooded by 10× TBE buffer near the gel surface, and 5 µL PCR product of each sample along with DNA dye was loaded into each well. An amount of 3 µL of DNA ladder was injected in the first well of casting tray. Gel electrophoresis apparatus was then connected to a power supply of 100 V/cm for 30 min. The gel was removed from the holder and visualized by a UV illuminator.

2.6. Biofilm Assay

The PCR positive samples were further assayed to check biofilm production qualitatively. In order to assay biofilm formation, 3 mL of autoclaved Luria Bertani broth was taken in a test tube and incubated overnight at 37 °C at a 130 rpm shaker. An amount of 3 mL of LB was poured in each test tube, and three sets were made and they were autoclaved. An amount of 100 µL of culture was inoculated into each test tube of triplicate sets. These three sets of test tubes were incubated at 37 °C for different time intervals, i.e., one set for 24 h, the second set for 48 h and the third set for 72 h. After 24 h, cultured media were discarded from test tubes and stained with 0.5% crystal violet dye for 5 min. Each test tube was washed with distilled water. Ninety-five percent ethanol (4 mL) was used to dissolve the ring formed in test tubes after staining them with 0.5% crystal violet dye and optical density was measured at 570 nm of all the three sets of test tubes incubated at 37 °C for 24, 48 and 72 h [30,37].

2.7. Statistical Analysis

Statistical analysis for the prevalence of virulence genes of EPEC, ETEC and EHEC in drinking water samples was performed using Microsoft Excel spreadsheet and the result was presented as a percentage or graph.

3. Results and Discussion
3.1. Distribution of Isolated Organisms

All collected water samples (n = 35) were screened for the presence of DEC bacteria. Twelve out of thirty-five (34%) water samples were screened as positive for the presence of DAE; thus, 12 *E. coli* bacteria samples were isolated from collected drinking water. The distribution of water samples was as follows: 30% (3/10) were from tank water, 20% (1/5) were from industrial water, 20% (1/5) were from infused water, 66% (2/3) were from standing water, 20% (1/5) were from well water and 57% (4/7) were from tap water (Table 3).

Table 3. Virulence genes of *E. coli* isolates from different water sources.

Sample	*eaeA* (EPEC)	*stx1* (EHEC)	*stx2* (EHEC)	*sth* (ETEC)
1	-	-	-	-
2	-	-	-	-
3	-	-	-	-
4	+	+	+	-
5	-	+	+	+
6	-	-	-	-
7	-	-	-	-
8	+	+	+	+
9	-	-	-	-
10	-	-	-	-
11	-	-	-	-
12	+	+	+	+
13	-	-	-	-
14	-	-	-	-
15	-	-	-	-
16	-	-	-	-
17	-	-	-	-
18	-	-	-	-
19	-	-	-	-
20	-	-	-	-
21	+	+	+	+
22	+	+	+	+
23	-	-	-	-
24	-	-	-	-
25	-	-	-	-
26	+	-	-	+
27	-	-	-	-
28	-	-	-	-
29	+	+	+	+
30	-	+	+	-
31	+	-	-	+
32	-	-	-	-
33	-	-	-	-
34	-	-	-	-
35	+	-	-	+

3.2. Prevalence of Virulence Genes from E. coli Isolates

Amongst 35 collected water samples, twelve confirmed *E. coli* isolates were evaluated for various virulence genes *eaeA*, *stx1*, *stx2* and *sth*. Ninety-one percent (11/12) of the *E. coli* isolates harbored virulence gene while 8% (1/12) *E. coli* isolates harbored none. The most frequent virulence factor genes were *eaeA* and *sth*, each of which was detected in 75% of *E. coli* isolates. In contrast, four virulence genes, including *eaeA*, *stx1*, *stx2* and *sth*, were each observed in 41% of *E. coli* isolates (Table 3). In the current study, the frequencies of virulence genes were as follows: *eaeA* (75%), *stx1* (66%), *stx2* (66%) and *sth* (75%). These readings showed that the prevalence rate of virulence genes *eaeA* and *sth* was very high among DEC pathotypes isolated from drinking water. Therefore, the *E. coli* strains isolated from drinking water carried the virulence-associated *eaeA* and *sth* genes more frequently than compared to *stx1* and *stx2* (Figure 1). In addition, the results of the PCR for the identification of *eaeA*, *stx1*, *stx2* and *sth* genes are shown in Figure 2.

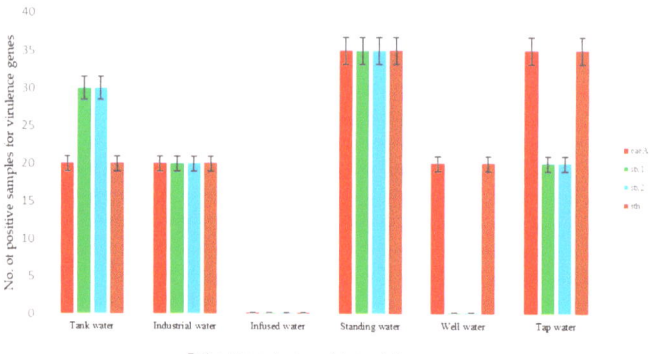

Figure 1. Comparative distribution of virulence genes by individual *E. coli* isolates.

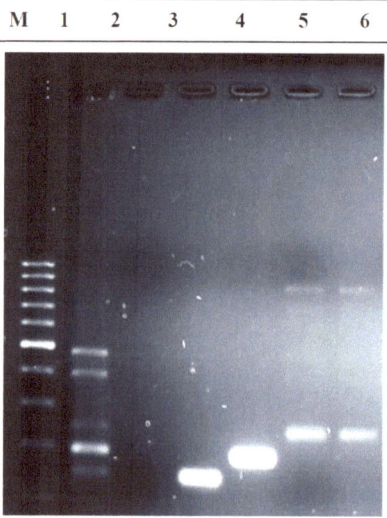

Figure 2. Result of PCR Assay for detection of virulence gene. M: DNA size ladder 1000 bp; number 1: positive control; 2: negative control; 3: *eaeA*; 4: *sth*; 5: *stx1*; 6: *stx2*.

3.3. Biofilm Assay

A biofilm assay was used to check the ability of biofilm production of EPEC, EHEC and ETEC pathotypes. The PCR positive samples were further assayed to check biofilm production. All PCR positive samples were incubated in LB media for different time intervals at 37 °C. Set A of test tubes containing only LB media was taken as control. Three sets of test tubes were prepared; B was incubated for 24 h, C was incubated for 48 h and D was incubated for 72 h. Zero-point five percent crystal violet dye was used to stain all the test tubes. After staining, the test tubes were washed with distilled water in order to remove excessive dye. The most prominent biofilm ring was observed in the test tube incubated for 24 h possessing the ETEC pathotype. These strains developed adhesion to the walls of test tubes and formed a ring. The ring formation around the walls of test tubes confirmed biofilm production by ETEC pathotype (Figure 3).

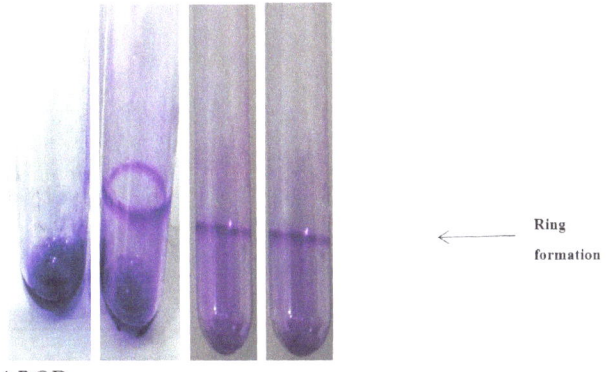

Figure 3. Biofilm assay. (**A**): Control; (**B**): *E. coli* culture inoculated for 24 h; (**C**): *E. coli* culture inoculated for 48 h; (**D**): *E. coli* inoculated culture for 72 h.

All the samples that showed positive results in the qualitative biofilm assay were further processed in a quantitative biofilm assay. The ring was dissolved by 95% ethanol, and optical density (OD) was measured in spectrophotometer at 570 nm (Table 4).

Table 4. Incidence of biofilm formation in each group classified by absorbance.

Samples	OD_{570}	Strength of Adhesion
1	0.2	Strongly adherent
2	0.15	Moderately adherent
3	0.1	Weakly adherent

The OD 0.2 of sample 1 reported that these strains can strongly adhere to biotic and abiotic surfaces. Sample 2 producing an OD of 0.15 showed that these strains are also involved in biofilm production with moderate adhesion to the surfaces. Sample 3 with OD 0.1 showed that the strains were involved in weak adhesion to the surfaces with biofilm production as shown by [37]. In current research, virulence genes were detected in DEC isolates to detect the presence of virulent *E. coli* strains in drinking water. Other studies have also reported the presence *E. coli* strains as fecal indicator bacteria originating from defective hygiene and poor sanitation and animal fecal contact in freshwater places in developing countries [2,17,38]. The presence of ETEC in drinking water is indicative of a high risk of contamination. In general, these results found that the risk of contamination may increase over time, and it is high time to follow appropriate preventive measures.

4. Conclusions

Water sources harbor different pathotypes of DEC, which can possibly be suspended by natural or synthetic events. In the present study, four virulence genes (*eaeA*, *stx1*, *stx2* and *sth*) were investigated pertaining to three DEC pathotypes (EPEC, EHEC and ETEC), which signify a potential health concern for individual drinking water from these sources. Approximately 34% of analyzed water samples were found positive for at least one of the four virulence genes. The most frequent pathotypes were EPEC and ETEC as *eaeA* and *sth* genes were detected in 75% of pathotypes. The frequencies of pathotypes were as follows: EPEC (75%), EHEC (66%) and ETEC (75%). These findings demonstrated that the rate of prevalence of EPEC and ETEC is very high among DEC pathotypes isolated from drinking water sources. The biofilm assay demonstrated that ETEC pathotypes form adhesions on the walls of test tubes. All ETEC positive samples showed different strength of adhesion from weak to strong. These findings suggest that these four virulence genes are responsible for waterborne infections. This study detected the presence of *E. coli* pathotypes carrying virulence genes isolated from different drinking water samples. Moreover, the biofilm forming capacity of ETEC pathotypes revealed that attaching and effacing ETEC to enteric walls can result in persistent infection. It can cause serious difficulties in medical treatments. In this manner, this study highlights the significance of sanitation and good hygiene in developing countries. Therefore, the detection of virulence genes of *E. coli* from drinking water directs us to the need to study its origin, reservoir and transmission pathway to create a better preventive and controlling plans. As for public health concern, this data will prove to be a better source for estimating the risk factors related to gastrointestinal infections, and the data will provide a better understanding about public health complications caused by *E. coli* pathotypes carrying virulence genes. In this study, a set of four virulence genes has been tested and linked to assign the definite pathotypes as a source of human diarrhea. Additional screening of other virulence genes along with serotyping and other assays may provide data on the pathogenicity of DEC isolates.

Author Contributions: Conceptualization, S.T. (Sobia Tabassum) and S.A. (Sadia Aslam); methodology, S.T. (Sadaf Tariq).; software, S.A. (Shafaqat Ali); writing—original draft preparation, S.T. (Sadaf Tariq); writing—review and editing, M.S.; visualization, W.H.A.-Q.; supervision, M.S. All authors have read and agreed to the published version of the manuscript.

Funding: This work was funded by the Researchers Supporting Project Number (RSP-2021/293) King Saud University, Riyadh, Saudi Arabia.

Institutional Review Board Statement: Not applicable.

Informed Consent Statement: Not applicable.

Data Availability Statement: Not applicable.

Conflicts of Interest: The authors declare no conflict of interest.

References

1. Ahmad, A.; Jini, D.; Aravind, M.; Parvathiraja, C.; Ali, R.; Kiyani, M.Z.; Alothman, A. A novel study on synthesis of egg shell based activated carbon for degradation of methylene blue via photocatalysis. *Arab. J. Chem.* **2020**, *13*, 8717–8722. [CrossRef]
2. Aravind, M.; Ahmad, A.; Ahmad, I.; Amalanathan, M.; Naseem, K.; Mary, S.M.M.; Zuber, M. Critical green routing synthesis of silver NPs using jasmine flower extract for biological activities and photocatalytical degradation of methylene blue. *J. Environ. Chem. Eng.* **2021**, *9*, 104877. [CrossRef]
3. Pervaiz, M.; Ahmad, I.; Yousaf, M.; Kirn, S.; Munawar, A.; Saeed, Z.; Rashid, A. Synthesis, spectral and antimicrobial studies of amino acid derivative Schiff base metal (Co, Mn, Cu, and Cd) complexes. *Spectrochim. Acta Part A Mol. Biomol. Spectrosc.* **2019**, *206*, 642–649. [CrossRef] [PubMed]
4. Hussain, S.; Khan, A.J.; Arshad, M.; Javed, M.S.; Ahmad, A.; Shah, S.S.A.; Qiao, G. Charge storage in binder-free 2D-hexagonal $CoMoO_4$ nanosheets as a redox active material for pseudocapacitors. *Ceram. Int.* **2021**, *47*, 8659–8667. [CrossRef]
5. Zhan, M.; Hussain, S.; AlGarni, T.S.; Shah, S.; Liu, J.; Zhang, X.; Liu, G. Facet controlled polyhedral ZIF-8 MOF nanostructures for excellent NO_2 gas-sensing applications. *Mater. Res. Bull.* **2021**, *136*, 111133. [CrossRef]

6. Kashif, M.; Ngaini, Z.; Harry, A.V.; Vekariya, R.L.; Ahmad, A.; Zuo, Z.; Alarifi, A. An experimental and DFT study on novel dyes incorporated with natural dyes on titanium dioxide (TiO_2) towards solar cell application. *Appl. Phys. A* **2020**, *126*, 1–13. [CrossRef]
7. Saleem, M.; Irfan, M.; Tabassum, S.; Albaqami, M.D.; Javed, M.S.; Hussain, S.; Zuber, M. Experimental and theoretical study of highly porous lignocellulose assisted metal oxide photoelectrodes for dye-sensitized solar cells. *Arab. J. Chem.* **2021**, *14*, 102937. [CrossRef]
8. Kashif, M.; Jaafar, E.; Bhadja, P.; Low, F.W.; Sahari, S.K.; Hussain, S.; Al-Tamrah, S.A. Effect of potassium permanganate on morphological, structural and electro-optical properties of graphene oxide thin films. *Arab. J. Chem.* **2021**, *14*, 102953. [CrossRef]
9. Zhang, X.Z.; Xu, P.H.; Liu, G.W.; Ahmad, A.; Chen, X.H.; Zhu, Y.L.; Qiao, G.J. Synthesis, characterization and wettability of Cu-Sn alloy on the Si-implanted 6H-SiC. *Coatings* **2020**, *10*, 906. [CrossRef]
10. Fallah, Z.; Zare, E.N.; Ghomi, M.; Ahmadijokani, F.; Amini, M.; Tajbakhsh, M.; Varma, R. Toxicity and remediation of pharmaceuticals and pesticides using metal oxides and carbon nanomaterials. *Chemosphere* **2021**, *275*, 130055. [CrossRef]
11. Bibi, S.; Ahmad, A.; Anjum, M.A.R.; Haleem, A.; Siddiq, M.; Shah, S.S.; Al Kahtani, A. Photocatalytic degradation of malachite green and methylene blue over reduced graphene oxide (rGO) based metal oxides (rGO-Fe_3O_4/TiO_2) nanocomposite under UV-visible light irradiation. *J. Environ. Chem. Eng.* **2021**, *9*, 105580. [CrossRef]
12. Ahmad, I.; Jamal, M.A.; Iftikhar, M.; Ahmad, A.; Hussain, S.; Asghar, H.; Khan, S. Lanthanum-zinc binary oxide nanocomposite with promising heterogeneous catalysis performance for the active conversion of 4-nitrophenol into 4-aminophenol. *Coatings* **2021**, *11*, 537. [CrossRef]
13. Javed, M.S.; Khan, A.J.; Ahmad, A.; Siyal, S.H.; Akram, S.; Zhao, G.; Alfakeer, M. Design and fabrication of bimetallic oxide nanonest-like structure/carbon cloth composite electrode for supercapacitors. *Ceram. Int.* **2021**, *47*, 30747–30755. [CrossRef]
14. Javed, M.S.; Najim, T.; Hussain, I.; Batool, S.; Idrees, M.; Mehmood, A.; Shah, S.S.A. 2D V_2O_5 ultrathin nanoflakes as a binder-free electrode material for high-performance pseudocapacitor. *Ceram. Int.* **2021**, *47*, 25152–25157. [CrossRef]
15. Beena, V.; Rayar, S.L.; Ajitha, S.; Ahmad, A.; Albaqami, M.D.; Alsabar, F.A.A.; Sillanpää, M. Synthesis and characterization of Sr-doped ZnSe nanoparticles for catalytic and biological activities. *Water* **2021**, *13*, 2189. [CrossRef]
16. Siyal, S.H.; Javed, M.S.; Ahmad, A.; Sajjad, M.; Batool, S.; Khan, A.J.; Najam, T. Free-standing 3D Co_3O_4@ NF micro-flowers composed of porous ultra-long nanowires as an advanced cathode material for supercapacitor. *Curr. Appl. Phys.* **2021**, *31*, 221–227. [CrossRef]
17. Syah, R.; Ahmad, A.; Davarpanah, A.; Elveny, M.; Ramdan, D.; Albaqami, M.D.; Ouladsmane, M. Incorporation of Bi_2O_3 residuals with metallic bi as high performance electrocatalyst toward hydrogen evolution reaction. *Catalysts* **2021**, *11*, 1099. [CrossRef]
18. Abbas, Q.; Javed, M.S.; Ahmad, A.; Siyal, S.H.; Asim, I.; Luque, R.; Tighezza, A.M. ZnO nano-flowers assembled on carbon fiber textile for high-performance supercapacitor's electrode. *Coatings* **2021**, *11*, 1337. [CrossRef]
19. Bibi, S.; Khan, A.; Khan, S.; Ahmad, A.; Sakhawat Shah, S.; Siddiq, M.; Al-Kahtani, A.A. Synthesis of Cr doped $LiMnPO_4$ cathode materials and investigation of their dielectric properties. *Int. J. Energy Res.* **2021**. [CrossRef]
20. Raees, A.; Jamal, M.A.; Ahmad, A.; Ahmad, I.; Saeed, M.; Habila, M.A.; Alomar, T.S. Synthesis and characterization of Ceria incorporated Nickel oxide nanocomposite for promising degradation of methylene blue via photocatalysis. *Int. J. Environ. Sci. Technol.* **2021**, 1–8. [CrossRef]
21. Beena, V.; Rayar, S.L.; Ajitha, S.; Ahmad, A.; Iftikhar, F.J.; Abualnaja, K.M.; Ali, S. Photocatalytic dye degradation and biological activities of Cu-doped ZnSe nanoparticles and their insights. *Water* **2021**, *13*, 2561. [CrossRef]
22. Mahmud, Z.H.; Kabir, M.H.; Ali, S.; Moniruzzaman, M.; Imran, K.M.; Nafiz, T.N.; Islam, M.S.; Hussain, A.; Hakim, S.A.I.; Worth, M.; et al. Extended-spectrum beta-lactamase-producing Escherichia coli in drinking water samples from a forcibly displaced, densely populated community setting in Bangladesh. *Front. Public Health* **2020**, *8*, 228. [CrossRef] [PubMed]
23. Ochien, G.; Atieno, L. Prevalence of Enterotoxigenic Escherichia coli among Children under Five Years in Siaya County, Western Kenya. Master's Thesis, Maseno University, Kisumu, Kenya, 2021.
24. Hassan, A.; Ojo, B.; Abdulrahman, A. Escherichia coli as a global pathogen. *Achiev. J. Sci. Res.* **2021**, *3*, 239–260.
25. Marquezini, M.G.; da Costa, L.H.; Bromberg, R. Occurrence of the seven most common serotypes of Shiga toxin-producing Escherichia coli in beef cuts produced in meat-processing plants in the state of São Paulo, Brazil. *J. Food Prot.* **2021**. [CrossRef] [PubMed]
26. Fonseca, T.G.; Motta, E.A.; Mass, A.P.; Fongaro, G.; Ramos, F.M.; Machado, M.S.; Bocchese, D.C.; Viancelli, A.; Michelon, W. Toxicity and enterobacteriaceae profile in water in different hydrological events: A case from south Brazil. *Water Air Soil Pollut.* **2021**, *232*, 1–12. [CrossRef]
27. Bel, J.S.; Khaper, N.; Kurissery, S.; Leung, K.T. A novel comparison of virulence genes, biofilm-forming capacity, antibiotic resistance, and level of reactive oxygen species of sediment, sewage, and O157 E. coli. *Water Air Soil Pollut.* **2021**, *232*, 1–19. [CrossRef]
28. Angulo-Zamudio, U.A.; Gutiérrez-Jiménez, J.; Monroy-Higuera, L.; Flores-Villaseñor, H.; Leon-Sicairos, N.; Velazquez-Roman, J.; Vidal, J.E.; Tapia-Pastrana, G.; Canizalez-Roman, A. Non-diarrheagenic and diarrheagenic E. coli carrying supplementary virulence genes (SVG) are associated with diarrhea in children from Mexico. *Microb. Pathog.* **2021**, *157*, 104994. [CrossRef]
29. Ravi, M.; Ngeleka, M.; Kim, S.-H.; Gyles, C.; Berthiaume, F.; Mourez, M.; Middleton, D.; Simko, E. Contribution of AIDA-I to the pathogenicity of a porcine diarrheagenic Escherichia coli and to intestinal colonization through biofilm formation in pigs. *Vet. Microbiol.* **2007**, *120*, 308–319. [CrossRef]

30. Corzo-Ariyama, H.A.; García-Heredia, A.; Heredia, N.; García, S.; León, J.; Jaykus, L.; Solís-Soto, L. Phylogroups, pathotypes, biofilm formation and antimicrobial resistance of Escherichia coli isolates in farms and packing facilities of tomato, jalapeño pepper and cantaloupe from Northern Mexico. *Int. J. Food Microbiol.* **2019**, *290*, 96–104. [CrossRef]
31. Safadi, R.A.; Abu-Ali, G.S.; Sloup, R.E.; Rudrik, J.T.; Waters, C.M.; Eaton, K.A.; Manning, S.D. Correlation between in vivo biofilm formation and virulence gene expression in *Escherichia coli* O104: H4. *PLoS ONE* **2012**, *7*, e41628. [CrossRef] [PubMed]
32. Sukkua, K.; Rattanachuay, P.; Sukhumungoon, P. Ex vivo adherence to murine ileal, biofilm formation ability and presence of adherence-associated of human and animal diarrheagenic *Escherichia coli*. *Southeast Asian J. Trop. Med. Public Health* **2016**, *47*, 40.
33. Al-Gallas, N.; Annabi, T.A.; Bahri, O.; Boudabous, A. Isolation and characterization of shiga toxin-producing Escherichia coli from meat and dairy products. *Food Microbiol.* **2002**, *19*, 389–398. [CrossRef]
34. Chandran, A.; Mazumder, A. Occurrence of diarrheagenic virulence genes and genetic diversity in Escherichia coli isolates from fecal material of various avian hosts in British Columbia, Canada. *Appl. Environ. Microbiol.* **2014**, *80*, 1933–1940. [CrossRef] [PubMed]
35. Huang, S.-W.; Hsu, B.-M.; Su, Y.-J.; Ji, D.-D.; Lin, W.-C.; Chen, J.-L.; Shih, F.-C.; Kao, P.-M.; Chiu, Y.-C. Occurrence of diarrheagenic *Escherichia coli* genes in raw water of water treatment plants. *Environ. Sci. Pollut. Res.* **2012**, *19*, 2776–2783. [CrossRef]
36. Chandran, A.; Mazumder, A. Prevalence of diarrhea-associated virulence genes and genetic diversity in Escherichia coli isolates from fecal material of various animal hosts. *Appl. Environ. Microbiol.* **2013**, *79*, 7371–7380. [CrossRef]
37. Wakimoto, N.; Nishi, J.; Sheikh, J.; Nataro, J.P.; Sarantuya, J.; Iwashita, M.; Manago, K.; Tokuda, K.; Yoshinaga, M.; Kawano, Y. Quantitative biofilm assay using a microtiter plate to screen for enteroaggregative *Escherichia coli*. *Am. J. Trop. Med. Hyg.* **2004**, *71*, 687–690. [CrossRef] [PubMed]
38. Moglad, E.H.; Jalil Adam, O.A.E.; Alnosh, M.M.; Altayb, H.N. Retracted: Detection of virulence genes of diarrheagenic Escherichia coli strains from drinking water in Khartoum State. *J. Water Health* **2020**. [CrossRef]

Article

Effect of Solidifying Structure on Centerline Segregation of S50C Steel Produced by Compact Strip Production

Kai Liu [1], Shusen Cheng [1,*], Jipeng Li [1,2] and Yongping Feng [2]

1. School of Metallurgical and Ecological Engineering, University of Science & Technology Beijing, 30 Xueyuan Road, Haidian Dsitrict, Beijing 100083, China; beikedaliukai@163.com (K.L.); lijipeng@jiugang.com (J.L.)
2. Jiuquan Iron & Steel (Group) Co., Ltd., Jiayuguan 735100, China; yp_0311@163.com
* Correspondence: chengsusen@metall.ustb.edu.cn

Citation: Liu, K.; Cheng, S.; Li, J.; Feng, Y. Effect of Solidifying Structure on Centerline Segregation of S50C Steel Produced by Compact Strip Production. *Coatings* **2021**, *11*, 1497. https://doi.org/10.3390/coatings11121497

Academic Editor: Awais Ahmad

Received: 6 November 2021
Accepted: 29 November 2021
Published: 5 December 2021

Publisher's Note: MDPI stays neutral with regard to jurisdictional claims in published maps and institutional affiliations.

Copyright: © 2021 by the authors. Licensee MDPI, Basel, Switzerland. This article is an open access article distributed under the terms and conditions of the Creative Commons Attribution (CC BY) license (https://creativecommons.org/licenses/by/4.0/).

Abstract: Medium-high carbon steels having a high quality are widely used in China. It is advantageous to produce high value-added hot-rolled plates with the crystal refined and chemical composition homogenized in the casting slabs. However, element segregation occurs easily during high-medium carbon steels' production. Generally, the centerline segregation is improved by enlarging the equiaxed zone with low-superheat casting and electromagnetic stirring (EMS). Studies were conducted on centerline segregation of S50C steel slabs with a thickness of 52 mm produced by the compact strip production (CSP) process in China without EMS equipped. By sampling along the width at different position, the secondary dendrite arm spacing (SDAS) was measured after etching and picture processing, based on which the cooling rate was calculated. It was found that the cooling rate increased from the center to the surfaces of the slabs ranging in 1~20 K/s, 10 times faster than that of a conventional process. The faster cooling rate led to a refined solidifying structure and columnar dendrite through the center of the slabs. The SDAS tended to increase from surfaces to the center, ranging only 32~120 μm smaller than that of a conventional process in 100~300 μm, indicating a finer solidifying structure by the CSP process. Results by EPMA indicated that elements C, Si, and Mn distribute in dispersed spots, increasing towards the center, and the centerline segregation changed in a narrow range: for C mainly in 1.0~1.1, Si in 0.98~1.08, Mn in 0.96~1.02, respectively, meaning a more chemical homogenization than that of thick slabs. Elements' segregation originated from solute redistribution between solid and liquid. According to thermodynamic calculation, δ region of S50C is so narrow that the solute redistribution mainly occurred between γ-Fe and liquid during solidification. As the equilibrium partition coefficient of element C was the smallest, it was easy for C to be rejected to the residual liquid in the inter-dendritic space, leading to obvious segregation, relatively. Besides, as a result of high-cooling intensity, the solidifying structure became so fine that the Fourier number increased and the volume of the residual liquid decreased, making centerline segregation alleviated effectively both in volume and degree. Although bulging was observed during the industrial experiment, the centerline segregation was still inhibited obviously as the refining solidifying structure with permeability ranged only in 0.1~2.3 μm^2 from the surfaces to centerline, which showed a good resistance on the residual flow towards the centerline.

Keywords: CSP process; S50C steel; solidifying structure; cooling rate; micro-segregation; centerline segregation

1. Introduction

In medium-high carbon steels' production, elements' segregation can occur easily as the solidifying range is relatively wide; thus, it has called more attention to be improved [1–3]. Usually, the micro-segregation can be reduced in the subsequent processing; however, macro-segregation [4–9], which may cause inconsistent transformation products, leading to failure of the finished products in service [10–14], is more harmful. Therefore, it is vital to control the macro-segregation at a desired level.

Actually, the segregation originates from redistribution of solutes during solidification in the solidifying front, leading to solute-enriched liquid in the inter-dendritic space, and then the liquid is driven, by some factors such as thermal shrinkage and bulging, to the centerline of slabs, forming centerline segregation finally. Therefore, the redistribution of solutes [15], solidifying structure, and the motion of the solute-enriched liquid in the inter-dendritic space would have an important effect on the centerline segregation; commonly, the element segregation degree is used to describe the extent of the element segregation, shown as follows [16]:

$$r_i = C_i/C_{i0} \tag{1}$$

where C_i is the concentration of solute i at a certain position of the slabs, and C_{i0} is the nominal concentration of solute i.

Currently, many actions have been taken to improve the internal quality of casting billets, such as electromagnetic stirring (EMS) [17–23], soft reduction, etc. Most of the actions taken are to enlarge the equiaxed zone and dismiss the shrinkage so that the segregation could be inhibited effectively [24–27]. Ma and Zhang et al. [28] simulated the center macroscopic segregation process of Q345 slabs, producing in a conventional process at a casting speed of 0.85 m/min with mechanical reduction, and the results showed that the segregation could be improved by an amount of 10 mm of reduction at the position with 0.4 of liquid fraction, which was verified by experiments, indicating that the segregation degree of C ranged from 0.7~1.2. However, in addition to the cost of such expensive equipment, it is always difficult to determine the reasonable position and operative parameters for the equipment, as the complexity of the industrial plants and the solidifying characteristics of a certain steel grade [29]. Additionally, according to Tsuchida and O. Haida [30,31], few effects have EMS and soft reduction on the volume of solute-enriched liquid transported to the centerline of the slabs to form the centerline segregation: EMS may improve the degree of segregation but enlarges the zone of segregation, and, though the thickness of segregation might be minimized by soft reduction, the degree of segregation becomes heavier.

With a higher cooling intensity, faster heat transfer, and thinner dimension, it is advantageous for the CSP process to improve the internal quality of slabs. The CSP process, as shown in Figure 1, is characterized by high cooling rate, compact layout, and lower energy consumption, making the casting speed rise to 4.0~6.0 m/min, and has been thriving for decades [32–34].

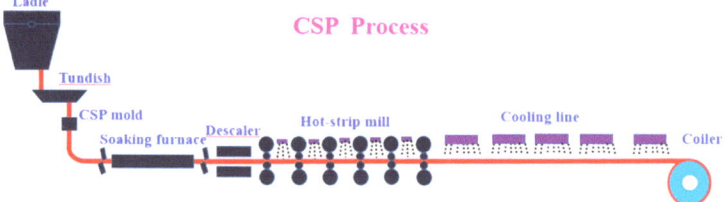

Figure 1. Schematic representation of CSP plant.

Connected with a rolled-hearth tunnel furnace from the caster to the hot rolling mill, in a CSP process the slabs, unlike the conventional process, have experienced only $L \rightarrow L + \delta\text{-Fe} \rightarrow L + \gamma\text{-Fe} \rightarrow \gamma\text{-Fe}$ before finishing rolling. As shown in Figure 2, $R(1)$ represents the solidification sequence of the CSP process and $R(2)$ and $R(3)$ represent the solidification sequence of conventional casting ($R(2)$ is reheated from 600 °C while $R(3)$ is from room temperature), respectively. From Figure 2, $R(1)$ indicates that the liquid steel solidifies with a relatively higher cooling rate and the surface temperature of the slab is above A_3 at about 1000 °C before entering the soaking furnace. Then, the slab is directly transported into the soaking furnace, preparing for finishing rolling. While both $R(2)$ and $R(3)$ solidify with lower cooling rates and were cooling down below A_1, and reheated over A_3 before rolling.

Consequently, solidifying structure of slabs produced by CSP is finer, resulting in inhibiting elements' segregation. Additionally, slabs produced by CSP remain above A$_3$, avoiding precipitation before rolling, which is of great advantage for chemical homogenization, such that the strips exhibit a greater uniformity as regards chemical composition, microstructure, mechanical properties, and dimensional accuracy [35,36].

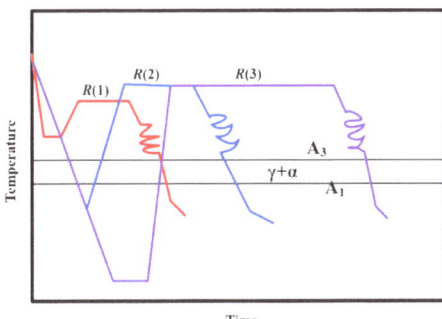

Figure 2. Comparison between conventional casting and compact strip processing.

However, for now, there is little literature, especially for producing medium-high carbon steels, reporting the solidifying structure, element segregation, and the mechanism for the formation of centerline segregation during a CSP process sufficiently and systematically. In view of this, the objective of present work was to investigate the solidifying structure, element segregation, and the improvement of element segregation with the solidifying structure refined for S50C steels produced by CSP process.

In this contribution, samples were collected from a CSP plant in China without electromagnetic stirring, and the studies on the solidifying structure and segregation of S50C steel were conducted systematically from micro scale to macro scale. The results of the present work will give a deep insight to a solidifying structure and the segregation for the CSP process.

2. Experimental Method

The present study mainly included several activities, as follows:

- Collection of samples from an industrial trial in a CSP plant in JISCO in Jiuquan, China.
- Measurement of SDAS by sampling from the transverse section of thin slab (with a dimension 1270 mm × 52 mm) samples.
- Micro-segregation tests for elements C, Si, and Mn by EPMA (Electron Probe Microanalysis, EPMA-1720H, Shimadzu, Kyoto, Japan) through mapping scanning mode.
- Macro-segregation tests by OES (Optical Emission Spectrometer-ARL8860, Thermo Fisher Scientific, Waltham, MA, USA).
- Calculation for macro-segregation degree for elements C, Si, and Mn.

The slab samples were collected from a CSP plant in China, together with the corresponding chemical composition, as shown in Table 1.

Table 1. Nominal Composition (mass percent, %).

C	Si	Mn	P	S
0.48	0.18	0.64	0.0082	0.0021

The CSP process is characterized by a high casting speed and high cooling rate, and the primitive process parameters are listed in Table 2 for the present work.

Table 2. The primitive process parameters in the present work.

Parameters	Value
Casting speed/m·min^{-1}	4.5
Superheat degree (K)	33
Specific water flow of second cooling zone (L·kg^{-1})	2.8
Liquidus temperature (K)	1763

After getting samples from the plant, one of the slabs was cut into a smaller size with a limited dimension, as shown in Figure 3a, where the label number started from the narrow side with 1 and ended with 9 at the center of the slab width. To reveal the details of SDAS, samples were fine-ground and polished so that a smooth surface was obtained. Then, the samples were etched with picric acid for 10 min at 70 °C. After being etched, the samples were cleaned by alcohol. Then, the morphology of the dendrite arms was observed through the whole sample surface (with an area about 4 mm × 4 mm) with LEXT OLS4100 Laser Scanning Confocal Microscope (Olympus Corporation, Tokyo, Japan). Finally, the image analyzer, Image Pro Plus 6.0, was employed to measure the SDAS, as shown in Figure 3b, and at least 30~50 measurements were taken for each zone. The data of the SDAS can be used for the calculation of the cooling rate ε during solidification, by Equation (2):

$$\lambda_s = \alpha\varepsilon^{-n} \qquad (2)$$

where λ_s is SDAS in μm, ε is the cooling rate in K·s^{-1}, and α and n are constants. According to Jacobi and Schwerdtfeger [16], the value of α is 109.2 and n is 0.44; so, Equation (2) can be expressed as:

$$\lambda_s = 109.2\varepsilon^{-0.44} \qquad (3)$$

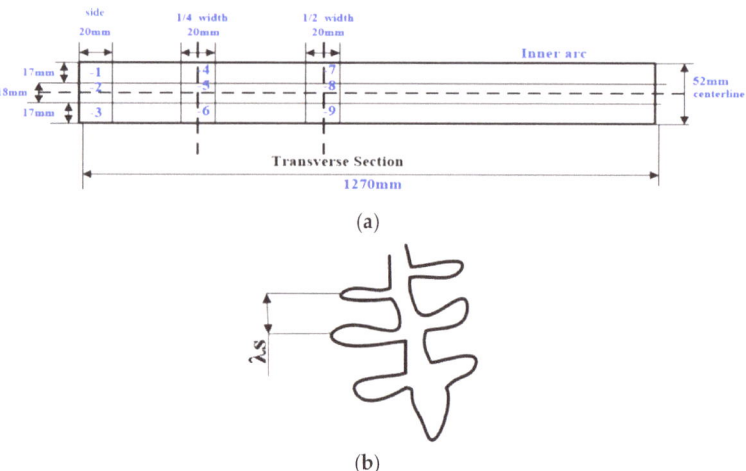

Figure 3. Sampling and measurement: (**a**) sampling for SDAS and EPMA; (**b**) measurement of SDAS.

Sampling for EPMA tests is also shown in Figure 3a. The samples were ground and polished to get a smooth surface, and the mapping scanning mode, set with a working distance of about 11 mm, an accelerating voltage of 15 kV, and a beam current of 15 nA, was used to obtain the distribution of elements C, Mn, and Si.

As shown in Figure 4, the macro-segregation was measured every 5-mm interval by OES, represented by the circles on the transverse section, and the calculation for the segregation degree of each measured element is given, as Equation (1).

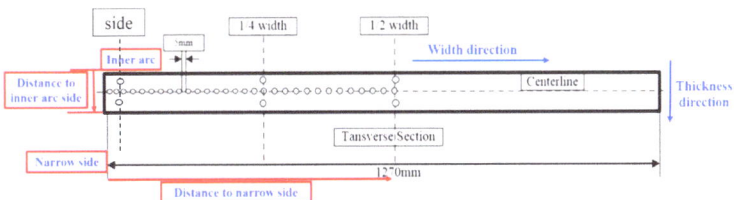

Figure 4. Testing position for OES.

3. Results and Discussion

3.1. Morphology of Solidifying Structure

Figure 5 shows the morphology of the solidifying structure produced by the CSP process. As shown in Figure 5, the solidifying structure consisted of a chill zone a few millimeters beneath the surface (Figure 5a,c) and a coarsened columnar dendrites' zone after the chill zone towards the center of the studied slabs (Figure 5b). It is obvious that, unlike attempting to enlarge the equiaxed zone during a conventional continuous cast process [17–23], the centerline of slabs produced by the CSP process was occupied by columnar dendrites (Figure 5b). The fine chill zone reflected the solidification with a rapid cooling rate in the meniscus, beyond which the columnar dendrites became coarsened towards the center of the slab. The orientation of the columnar dendrites illustrated that the direction of heat extraction was almost vertical to the surface of the slab; in other words, the solidification that occurred in the CSP process can be treated as one-dimension heat extraction.

The SDAS (λs) was measured from both surfaces towards the center of the slab, as shown in Figure 6. As can be seen from Figure 6, the SDAS tended to increase towards the center of the slabs, and the values of λs ranged from 32~120 μm, while λs of the thick slab (235 mm thick) varied over a wider range, from 100~300 μm, with bigger values [10,37]. Therefore, the solidifying structure produced by CSP was finer than that in a conventional thick slab process.

3.2. Micro-Segregation

The distribution of elements in micro scale was measured by EPMA using the mapping scanning mode, and the results are shown in Figure 7.

Figure 7 depicts the distribution of elements C, Si, and Mn in the inter-dendritic space in a microscopic scale. As can be seen, the micro-segregation distributed as dispersed spots, indicating a relatively homogeneous distribution of elements. This is because the size of the inter-dendritic space was small, resulting in the decreasing of the volume of residual liquid in it. But the micro-segregation tended to become serious from the surface to the center of the slabs, as the same trend of SDAS, among which the distribution of C was far more non-uniform compared with Si and Mn. This is because the value of partition coefficient of C (k_c) was smaller than that of Mn (k_{Mn}) and Si (k_{Si}) [15], indicating that there would be more element C in the residual liquid, which will be discussed later.

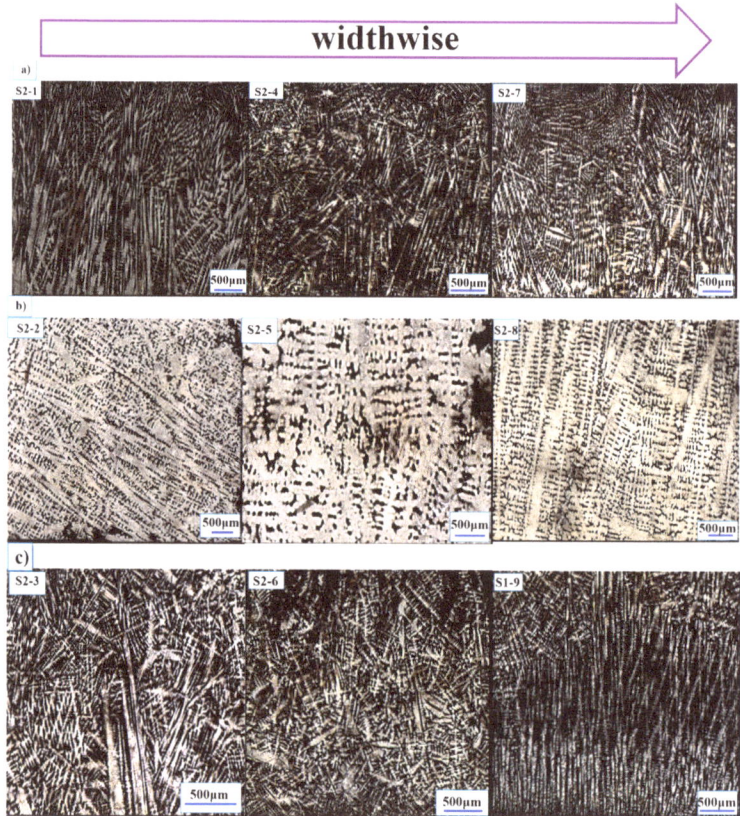

Figure 5. Morphology of Solidifying Structure by CSP: (**a**) near inner arc; (**b**) around centerline; (**c**) near outer arc.

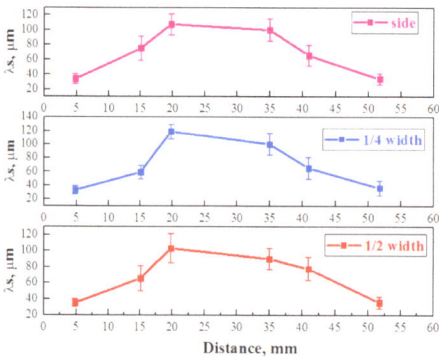

Figure 6. SDAS from surface to the center of slabs.

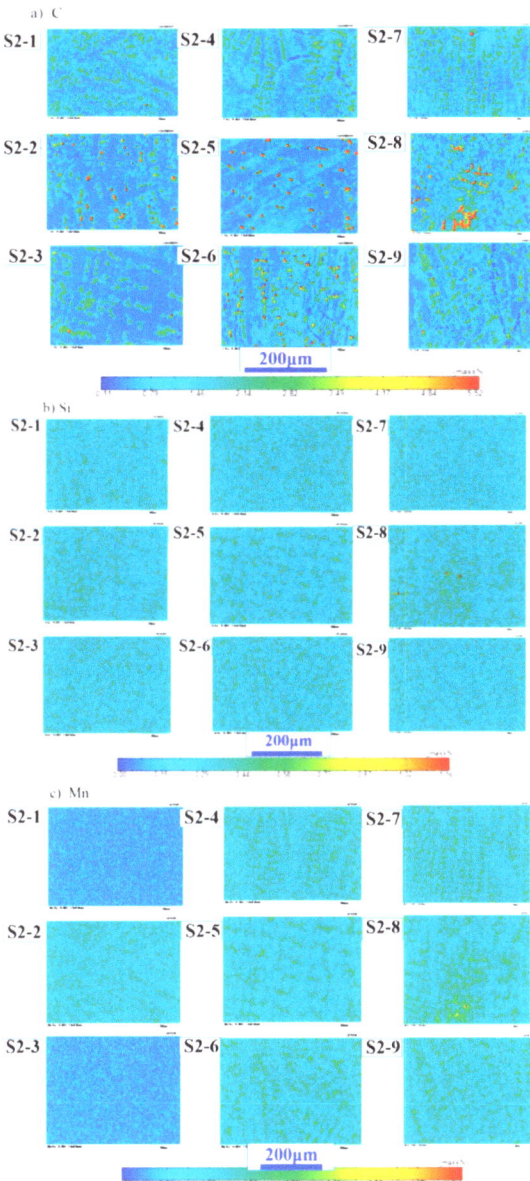

Figure 7. Results of EPMA: (**a**) C (**b**) Si (**c**) Mn.

3.3. Centerline Segregation

The centerline segregation is shown in Figure 8, in which the x-axis represents the distance to the narrow side along the centerline of the slabs' transverse section, schematically shown in Figure 4, and the y-axis indicates the extent of the element segregation calculated by Equation (1). Besides a few scattered points around 1.1, r_C was mainly in the range of 1.0 to 1.1, which shows the trend to increase towards the half width, while the segregation degree for Si and Mn was so small that the value of the segregation degree only varied mainly in the range of 0.98 to 1.08 and 0.96 to 1.02, respectively [28].

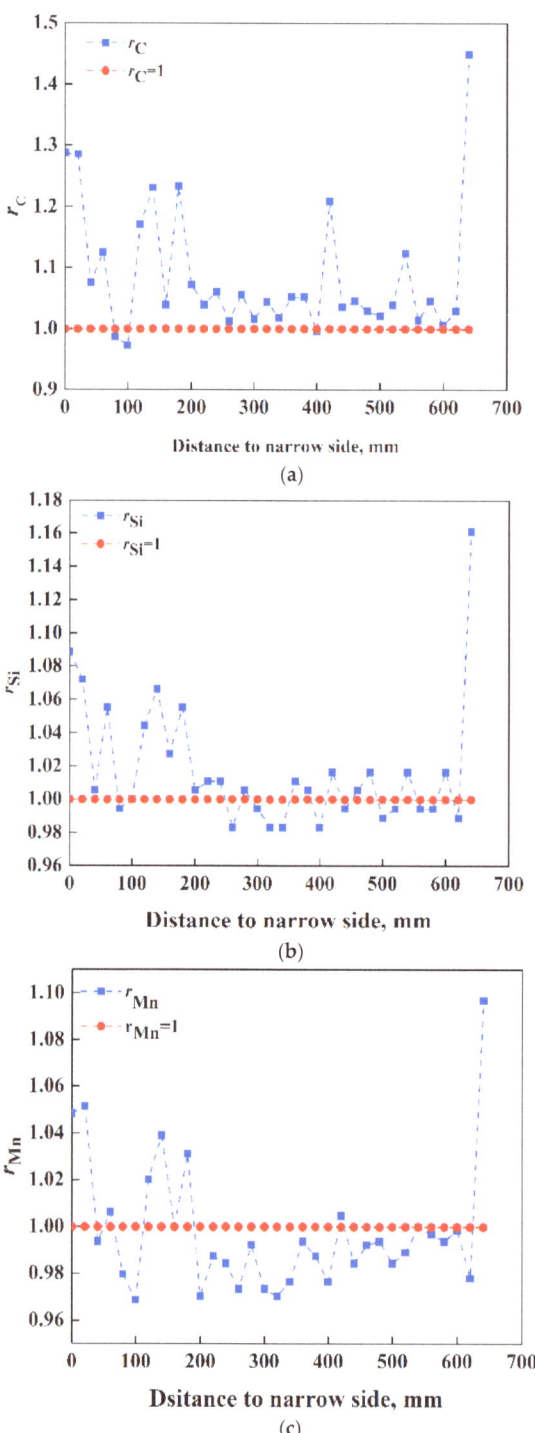

Figure 8. Centerline segregation degree (**a**) r_C, (**b**) r_{Si}, and (**c**) r_{Mn}.

Figure 9 illustrates the segregation along the thickness of the slabs. As shown in Figure 9, the segregation tended to be more serious towards the center of the slabs, but the value of each element was very small, especially for elements Si and Mn, both of which ranged very little around 1.0. The centerline segregation was relatively obvious for element C, and the values of element C segregation ranged from 1.0~1.1 at both one-quarter and one-half width of the slabs, meaning that the macro-segregation of S50C steel produced by the CSP process was homogeneous along the thickness.

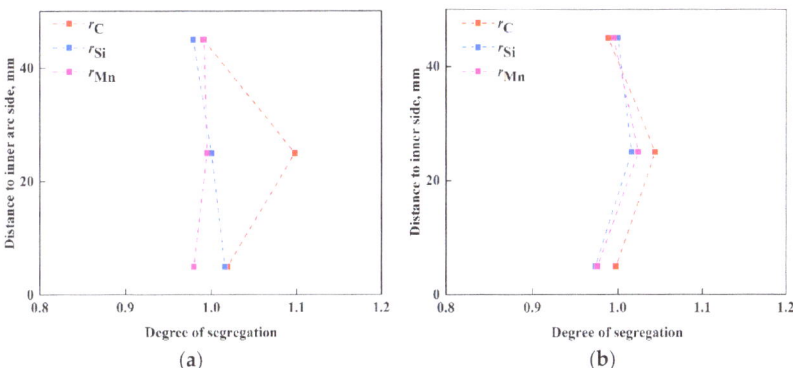

Figure 9. Macro-segregation along thickness direction: (**a**) one-quarter width, (**b**) one-half width.

According to literature [38], the macro-segregation depends on (1) partition coefficient (k) of solute elements and the parameter, R/k_m (R growth rate and k_m mass transfer coefficient), (2) morphology of the solidifying structure, (3) movement of solid and liquid phases during solidification, and (4) the extent of chemical reactions during freezing. In the current study, the (1), (2) and (4) were treated as the same for a certain element in the same position. Considering the bulging of the slab may have an influence on the movement of solid and liquid phases, i.e., factor (3), the thickness along the width direction was measured and is illustrated in Figure 10, where the x-axis represents the distance to the narrow side of the slabs along (as shown in Figure 4) the centerline, and the y-axis reflects the thickness of the slabs measured at intervals of a certain distance along the width direction of the slab.

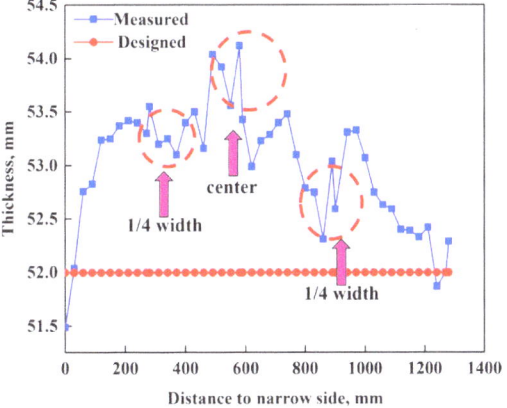

Figure 10. Thickness changes along the width direction.

As illustrated in Figure 10, the thickness of the slab increased (bulging) from both narrow sides towards the half width with the same trends of the centerline segregation degree, as shown in Figure 8. It is clear from Figure 10 that the slab, with the designed thickness of 52 mm, was about 54 mm around the center part. It is the bulging that caused the motion of the inter-dendritic liquid phase to the center of the slabs, forming the centerline segregation finally, and, because the bulging around the half width was heavier than that around the quarter width, resulting in more serious centerline segregation around the half width. Therefore, the formation of macro-segregation in the CSP process was determined mainly by bulging, which is also the key factor to control the macro-segregation during a CSP process.

3.4. Analysis on the Formation of Centerline Segregation during CSP Process

It is well known that the element segregation originates from the redistribution between the solid and liquid phases, around the solid–liquid interface. The partition coefficient determines the degree of the solutes rejected to the liquid. Usually, the partition coefficient of elements in the steel is smaller than 1.0, meaning that the solutes would be rejected from solid phase to liquid phase, forming the solute-enriched residual liquid. Then, the residual liquid remains in the inter-dendritic space. If the residual liquid solidified locally, the segregation might occur in a micro scale. In this case the degree of segregation was determined by: (1) the deviation of partition coefficient from 1.0 (the smaller the partition coefficient is, the heavier the segregation degree would be) and (2) the size of the inter-dendritic space (the finer the solidifying structure is, the smaller the amount of residual liquid remaining in the inter-dendritic space would be). Actually, the residual liquid in the inter-dendritic space may be driven to move towards a certain position (centerline) by some factors (such as bulging and thermal shrinkage) [27,39]. Then, the residual liquid converges in the very place, forming the macro-segregation (centerline segregation, as an example) finally. During this process, the motion of the residual liquid in the mushy zone plays an important role on the formation of macro-segregation [40].

According to literature [41], the cooling rate of thin slab casting and direct rolling technology is 10 times faster than that of a traditional thick slab process, resulting in a finer solidifying structure. To reveal the effects of the CSP process on the formation mechanism of S50C slabs, the main aspects, including (1) chemical composition, (2) cooling rate, and (3) the effects of solidifying structure on the flow of residual liquid, were analyzed subsequently.

(1) Chemical composition

The chemical composition (as listed in Table 1) determines the phase transformation process during solidification and the partition coefficient. By using Factsage7.0, the phase transformation process and equilibrium partition coefficient were calculated, as shown in Figures 11 and 12, respectively.

Connected with a rolled-hearth tunnel furnace from the caster to the hot rolling mill (Figure 1), the CSP process cancelled the reheating process, avoiding cooling down below $A_1 \rightarrow$ reheating to the A_3 process, so that the slabs would remain in the γ region after solidification before finishing rolling. Based on that, the results of phase transformation were calculated and are shown in Figure 11.

As shown in Figure 11, S50C steel experienced $L \rightarrow \delta \rightarrow \gamma$ transformation process before finishing rolling, and the region of δ was so narrow that it can be neglected. Therefore, within the mushy zone, the redistribution of solutes occurs mainly between γ and L, and the partition coefficient of solute i ($k_i{}^\gamma$) between γ and L was taken to characterize the redistribution during solidification.

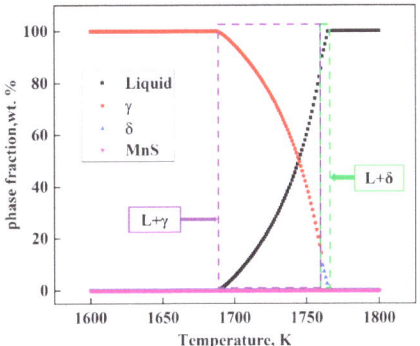

Figure 11. Phase transformation of S50C steel.

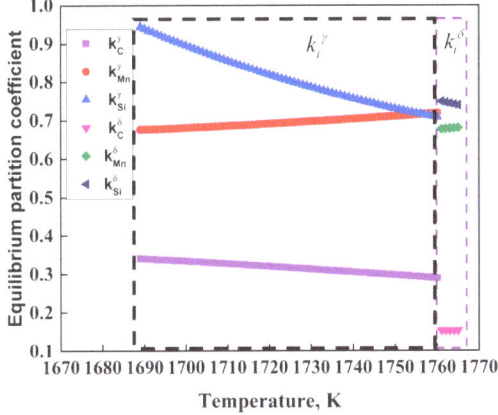

Figure 12. The equilibrium partition coefficient of S50C steel.

Figure 12 depicts the equilibrium partition coefficient of S50C during phase transformation. Both k_C^δ and k_{Mn}^δ increased to k_C^γ and k_{Mn}^γ around 1760 K, after which k_C^γ and k_{Mn}^γ changed a little with temperature, while k_{Si}^δ decreased to k_{Si}^γ around 1760 K, then increased with temperature decreasing. Therefore, with the temperature decreasing, δ-Fe emerged firstly in the liquid, and the solutes redistributed between δ-Fe and liquid. The δ region was so narrow that γ-Fe emerged in the liquid quickly, and then the redistribution occurred between γ-Fe and liquid over a long range. Because k_C^γ (0.34) was the smallest, there was more element C rejected to the residual liquid, resulting in the degree of micro-segregation of C being more serious than Mn and Si, as shown in Figure 7.

(2) Cooling rate

The relationship between the cooling rate and λs is expressed in Equation (2). The results of the cooling rate (ε) for present work are shown in Figure 13, where the x-axis represents the distance to the inner arc side along the thickness direction, as shown in Figure 4, and the y-axis is the cooling rates calculated by Equation (3). As can be seen from Figure 13, the cooling rate tended to decrease to be a constant from the surface to the center zone. The values of the cooling rate varied in the range of 1 to 20 K/s.

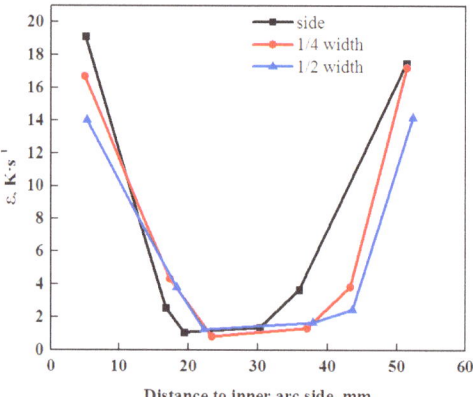

Figure 13. Results of cooling rate.

According to [42], within the mushy zone, the elements converged in the inter-dendritic space during solidification and can be described by:

$$C_L^* = C_0\left[1 - \frac{f_s}{\alpha k_0 + 1}\right]^{(k_0-1)} \quad (4)$$

where C_0 and C_L^* are the nominal concentration and concentration of solute in the liquid around the solid–liquid interface, respectively; f_s is the volume fraction of the solid phase in the mushy zone; k_0 is the partition coefficient of solute; and α is the diffusion parameter (Fourier number). Usually, $k_0 < 1$ and $f_s > 0$ during solidification; so Equation (4) indicates that the solute would decreases if α increase, and α could be written as:

$$\alpha = \frac{D_s}{\frac{\lambda_s^2}{4t_f}} \quad (5)$$

where D_s is the diffusivity of solute element and t_f is the local solidification time. It is obvious that Equation (5) indicates the accumulation of solutes in the inter-dendritic space within a length of $\lambda s/2$ during solidification. Therefore, with λs decreasing, the solute elements enriched in the inter-dendritic space would decrease. As shown in Figure 6, λs increased from the surface towards the center of the slab, so the solute enriched in the inter-dendritic also increased, leading to a heavier segregation degree, as shown in Figure 7.

However, it must be clarified that during the CSP process, the slabs remained in γ-Fe before finishing rolling, so that the segregation degree of element C could be improved by diffusion as the diffusivity of C was bigger [43,44], rather than the facts that the samples were taken by cooling down to the room temperature rapidly without enough time and temperature remained for elements' diffusion, which indicated that the CSP process was advantageous at inhibiting the elements' segregation.

(3) Effect of solidifying structure on the motion of liquid in mushy zone

Actually, the macro-segregation is always due to a relative velocity of the liquid of enriched solutes with respect to the solid phase, which can be induced by thermo-solute convection, forced convection, solidification shrinkage, transport and sedimentation of grains, or deformation of the solid skeleton [45,46], leading to centerline segregation in continuous casting of steel. Therefore, the factors that have a resistance on the flow can improve the centerline segregation during continuously casting slabs. Practically, the segregation originating within mushy zone consisted of dendritic arms, which can be treated as a porosity medium, and the flow in the porosity medium space can be described by D'Arcy's law, by which the velocity (v) of inter-dendritic liquid can be written as [47–49]:

$$v = -\left(\frac{K_p}{\mu g_L}\right)(\nabla P - \rho g) \quad (6)$$

where μ is the viscosity of the inter-dendritic liquid, N·s·m^{-2}; g_L is the volume fraction of the inter-dendritic liquid; P is the pressure, N·m^{-2}; ρ is the density of the inter-dendritic liquid, kg·m^{-3}; g is the gravitational acceleration, m·s^{-2}; and K_p is the specific permeability of the solid-liquid zone, m^2.

In different processes, K_p varies, which has an important influence on the flow in the inter-dendritic region. The relationship between K_P and λs is expressed as [50]:

$$K_P = \frac{\lambda_s^2 \times (1-g_s)^3}{180 \times g_s^2} \quad (7)$$

where g_s is the fraction of solid phase, usually g_s = 0.75 [51,52]; so K_P can be determined by λs.

Based on the measured data, K_p was calculated by Equation (7), and the results are shown in Figure 14, in which the x-axis is the distance to the inner arc side of the slab along the thickness direction, as shown in Figure 4, and the y-axis is the value of K_P calculated by Equation (7). K_p increased from surface to the center of the slabs, just the same as λs, and it changed from 0.1~2.3 μm^2, indicating an obvious resistance of inter-dendritic flows, because of which the centerline segregation was alleviated significantly.

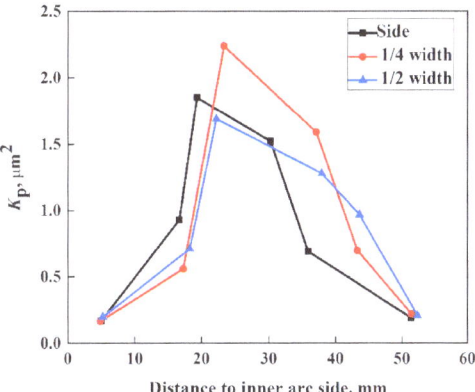

Figure 14. Results of permeability along slab thickness.

Based on what have been analyzed above, the advantages of a CSP process at alleviating segregation even without EMS are that as the cooling rate is high, as shown in Figure 13, the solidification proceeds immediately, leading to a refining of the solidifying structure, and the higher the cooling rate is, the finer the solidifying structure would be, as shown in Figure 6. Usually, the solidification of alloy is always accompanied with redistribution of solute as a result of the difference of solubility between solid and liquid, which could be characterized by equilibrium partition coefficient (k_i), as shown in Figure 12. But it is hard to solidify under equilibrium condition, which brings the assumption that the equilibrium was obtained only around the solid–liquid interface [28,42], from which the solutes were rejected to liquid phase (as k_i < 1), forming the solutes-enriched liquid among the inter-dendritic space, and at the same time the solutes could diffuse into the solid phase. The results of EPMA, shown in Figure 7, illustrated the elements' distribution throughout the dendrites qualitatively. Generally, the inter-dendritic liquid would be driven by thermal shrinkage and bulging (shown in Figure 10) to flow towards the centerline where centerline segregation came up. However, the flow would be resisted by a low permeability (K_P), as shown in Figure 14, so the centerline segregation was improved.

As is mentioned above, the bulging of the slabs drove the flow of the inter-dendritic liquid, but SDAS of slabs produced by CSP process was finer than that of conventional process; so, the centerline segregation was inhibited by the refining SDAS obviously, even though without EMS [17–23]. On a summary, unlike suppressing segregation by enlarging the equiaxed zone with EMS during conventional process, the CSP process is of great advantage for improving the segregation with a finer solidifying structure and higher cooling rate, even without EMS [30,31], and as the bulging had an obvious influence on the formation of centerline segregation, it needs more attention to be paid to control the bulging.

4. Conclusions

This study was mainly focused on the centerline segregation of S50C steel produced by CSP without EMS in China, and the effects of solidifying structure are analyzed on inhibition of centerline segregation. The conclusions are as follows:

(1) The cooling rate of CSP process increases from center to the surfaces of the slabs ranging from 1~20 K/s, 10 times faster than that of a conventional process. The faster cooling rate leads to refined solidifying structure and columnar dendrite through the center of the slabs. The SDAS tends to increase from surfaces to the center, ranging only from 32~120 μm, smaller than that of a conventional process, in 100~300 μm, indicating a finer solidifying structure by CSP process.

(2) According to thermodynamic calculation, δ region of S50C is so narrow that the solute redistribution mainly occurs between γ-Fe and liquid during solidification. As the equilibrium partition coefficient of element C is the smallest, it is easy for C to be rejected to the residual liquid in the inter-dendritic space leading to obvious segregation relatively.

(3) Results by EPMA indicate that elements C, Si, and Mn distribute in dispersed spots, increasing towards the center, and the centerline segregation changes in a narrow range: for C mainly in 1.0~1.1, Si in 0.98~1.08, and Mn in 0.96~1.02, respectively, meaning a more chemical homogenization than that of thick slabs.

(4) The segregation is suppressed mainly by refining the solidifying structure during CSP process. As a result of high-cooling intensity, the solidifying structure of slabs becomes so much finer that the Fourier number increases and the volume of the residual liquid decreases, making centerline segregation alleviated effectively both in volume and degree, even without EMS.

(5) Although bulging was observed during the industrial experiment, the centerline segregation was still inhibited obviously with permeability ranging only from 0.1~2.3 μm^2 from surfaces to centerline as a result of the refining solidifying structure, which showed a good resistance on the residual flow towards the centerline.

Author Contributions: Supervision, S.C.; Validation, J.L. and Y.F.; Writing—original draft, K.L. All authors have read and agreed to the published version of the manuscript.

Funding: This research was funded by National Natural Science Foundation of China No.: 62071034.

Institutional Review Board Statement: Not applicable.

Informed Consent Statement: Not applicable.

Data Availability Statement: Data is contained within the article.

Acknowledgments: The authors would like to thank National Natural Science Foundation of China No. 62071034, and the management of Jiuquan Iron & Steel (Group) Co., Ltd., China, for giving permission to publish this work.

Conflicts of Interest: The authors declare no conflict of interest.

References

1. Flemings, M.C. Our understanding of macro-segregation past and present. *ISIJ Int.* **2000**, *40*, 838–841. [CrossRef]
2. Brimacombe, J.K. The challenge of quality in continuous casting processes. *Met. Mater. Trans. A* **1999**, *30*, 553–566. [CrossRef]
3. Flemings, M.C. *Modeling of Casting, Welding and Advanced Solidification Processes-VIII*; Thomas, B.G., Beckermann, C., Eds.; TMS: Warrendale, PA, USA, 1998.
4. Cui, K.K.; Fu, T.; Zhang, Y.Y.; Wang, J.; Mao, H.B.; Tan, T.B. Microstructure and mechanical properties of $CaAl_{12}O_{19}$ rein-forced Al_2O_3-Cr_2O_3 composites. *J. Eur. Ceram. Soc.* **2021**, *41*, 7935–7945. [CrossRef]
5. Wang, J.; Zhang, Y.; Cui, K.; Fu, T.; Gao, J.; Hussain, S.; AlGarni, T.S. Pyrometallurgical recovery of zinc and valuable metals from electric arc furnace dust—A review. *J. Clean. Prod.* **2021**, *298*, 126788. [CrossRef]
6. Sung, P.K.; Poirier, D.R.; Yalamanchili, B. Segregation of carbon and manganese in continuously cast high carbon steel for wire rod. *Ironmak. Steelmak.* **1990**, *17*, 424–430.
7. Cui, K.K.; Zhang, Y.Y.; Fu, T.; Hussain, S.; Algarni, T.S.; Wang, J.; Zhang, X.; Ali, S. Effects of Cr_2O_3 content on microstructure and mechanical properties of Al_2O_3 matrix composites. *Coatings* **2021**, *11*, 234. [CrossRef]
8. Iwata, H.; Yamada, K.; Fujita, T.; Hayashi, K. Electromagnetic stirring of molten core in continuous casting of high carbon steel. *Trans. Iron Steel Inst. Jpn.* **1976**, *16*, 374–381. [CrossRef]
9. Mori, H.; Tanaka, N.; Sato, N.; Hirai, M. Macrostructure of and segregation in continuously cast carbon steel billets. *Trans. Iron Steel Inst. Jpn.* **1972**, *12*, 102–111. [CrossRef]
10. Zhou, T.; Zhang, P.; Kuuskman, K.; Cerilli, E.; Cho, S.-H.; Burella, D.; Zurob, H.S. Development of medium-to-high carbon hot-rolled steel strip on a thin slab casting direct strip production complex. *Ironmak. Steelmak.* **2017**, *45*, 603–610. [CrossRef]
11. Xiao, Y.; Li, W.; Zhao, H.; Lu, X.; Jin, X. Investigation of carbon segregation during low temperature tempering in a medium carbon steel. *Mater. Charact.* **2016**, *117*, 84–90. [CrossRef]
12. Lesoult, G. Macrosegregation in steel strands and ingots: Characterisation, formation and consequences. *Mater. Sci. Eng. A* **2005**, *413–414*, 19–29. [CrossRef]
13. Miyamura, K.; Taguchi, I.; Soga, H. New evaluation techniques of segregation in continuously cast steel. *Trans. Iron Steel Inst. Jpn.* **1984**, *24*, 883–890. [CrossRef]
14. Sirgo, J.A.; Campo, R.; Lopez, A.; Diaz, A.; Sancho, L. Measurement of centerline segregation in steel slabs. In Proceedings of the Conference Record of the 2006 IEEE Industry Applications Conference Forty-First IAS Annual Meeting, Tampa, FL, USA, 8–12 October 2006; Volume 1, pp. 516–520.
15. Ghosh, A. *Principles of Secondary Processing and Casting of Liquid Steel*; Oxford & IBH Publishing: New Delhi, India, 1990; p. 174.
16. Jacobi, H.; Schwerdtfeger, K. Dendrite morphology of steady state unidirectionally solidified steel. *Met. Mater. Trans. A* **1976**, *7*, 811–820. [CrossRef]
17. Ayata, K.; Mori, T.; Fujimoto, T. Improvement of macro-segregation in continuously cast bloom and billet by electromagnetic stirring. *Tran. Iron Steel Inst. Jpn.* **2006**, *24*, 931–939. [CrossRef]
18. Song, X.P.; Cheng, S.S.; Cheng, Z.J. Mathematical modelling of billet casting with secondary cooling zone electromagnetic stirrer. *Ironmak. Steelmak.* **2013**, *40*, 189–198. [CrossRef]
19. Trindade, L.; Vilela, A.; Filho, A.F.F.; Vilhena, M.; Soares, R. Numerical model of electromagnetic stirring for continuous casting billets. *IEEE Trans. Magn.* **2002**, *38*, 3658–3660. [CrossRef]
20. Javurek, M.; MBarna, M.; Gittler, P. Flow Modelling in round bloom strands with electromagnetical stirring. In Proceedings of the Steelsim, Graz, Austria, 1 January 2007.
21. Liu, H.; Xu, M.; Qiu, S.; Zhang, H. Numerical simulation of fluid flow in a round bloom mold with in-mold rotary electromagnetic stirring. *Met. Mater. Trans. A* **2012**, *43*, 1657–1675. [CrossRef]
22. Oh, K.S.; Chang, Y.W. Macro-segregation behavior in continuously cast high carbon steel blooms and billets at the final stage of solidification in combination stirring. *ISIJ Int.* **1995**, *35*, 866–875. [CrossRef]
23. Xiao, C.; Zhang, J.-M.; Luo, Y.-Z.; Wei, X.-D.; Wu, L.; Wang, S.-X. Control of macrosegregation behavior by applying final electromagnetic stirring for continuously cast high carbon steel billet. *J. Iron Steel Res. Int.* **2013**, *20*, 13–20. [CrossRef]
24. Zhang, Y.Y.; Yu, L.H.; Fu, T.; Wang, J.; Shen, Q.F.; Cui, K.K. Microstructure evolution and growth mechanism of Si-$MoSi_2$ composite coatings on TZM (Mo-0.5Ti-0.1Zr-0.02 C) alloy. *J. Alloys Compd.* **2022**, *894*, 162403. [CrossRef]
25. Zhang, Y.; Fu, T.; Cui, K.; Shen, F.; Wang, J.; Yu, L.; Mao, H. Evolution of surface morphology, roughness and texture of tungsten disilicide coatings on tungsten substrate. *Vacuum* **2021**, *191*, 110297. [CrossRef]
26. Zhang, Y.Y.; Cui, K.K.; Fu, T.; Wang, J.; Shen, F.Q.; Zhang, X.; Yu, L.H. Formation of $MoSi_2$ and Si/$MoSi_2$ coatings on TZM (Mo–0.5Ti–0.1Zr–0.02C) alloy by hot dip silicon-plating method. *Ceram. Int.* **2021**, *47*, 23053–23065. [CrossRef]
27. Rauter, W.; Reiter, J.; Srienc, K.; Brandl, W.; Erker, M.; Huemer, K.; Mair, A. Soft reduction at a round bloom caster: Implementation and results. *BHM Berg-Und Hüttenmänn. Mon.* **2014**, *159*, 454–460. [CrossRef]
28. Ma, H.; Zhang, J.; Cheng, R.; Wang, S. Numerical simulation study on quality control of the center macroscopic segregation in the continuous casting slab. *Trans. Indian Inst. Met.* **2019**, *72*, 825–835. [CrossRef]
29. Han, T.; Cheng, C.G.; Mei, J.X.; Zhu, J.F.; Jin, Y. Optimization of soft reduction process for continuous thick slab casting. *Adv. Mater. Res.* **2014**, *881–883*, 1558–1561. [CrossRef]
30. Tsuchida, Y.; Nakada, M.; Sugawara, I.; Miyahara, S.; Murakami, K.; Tokushige, S. Behavior of semi-macroscopic segregation in continuously cast slabs and technique for reducing the segregation. *Trans. Iron Steel Inst. Jpn.* **1984**, *24*, 899–906. [CrossRef]

31. Haida, O.; Kitaoka, H.; Habu, Y.; Kakihara, S.; Bada, H.; Shiraishi, S. Macro- and semi-macroscopic features of the centerline segregation in CC slabs and their effect on product quality. *Trans. Iron Steel Inst. Jpn.* **1984**, *24*, 891–898. [CrossRef]
32. Rosenthal, D.; Krämer, S.; Klein, C.; Geerkens, C.; Mueller, J. 20 years of CSP: Success story of an extraordinary technology. *Müller Stahl Und Eisen* **2009**, *129*, 73–86.
33. Hoen, K.; Klein, C.; Krämer, S.; Chung, J.S. Recent Development of Thin Slab Casting and Rolling Technology in a Challenging Market. In Proceedings of the 10th Int. Rolling Conf. and the 7th European Rolling Conference, Congress Graz, Graz, Austria, 6–9 June 2016; p. 1217.
34. Janssen, H.; Sowka, E. Seven years of progress in the casting-rolling plant at Thyssen Krupp Steel. *Stahl Und Eisen* **2007**, *127*, 27–33.
35. Klinkenberg, C.; Kintscher, B.; Hoen, K.; Reifferscheid, M. More than 25 Years of experience in thin slab casting and rolling current state of the art and future developments. *Steel Res. Int.* **2017**, *88*, 1700272. [CrossRef]
36. Fu, T.; Cui, K.; Zhang, Y.; Wang, J.; Shen, F.; Yu, L.; Qie, J.; Zhang, X. Oxidation protection of tungsten alloys for nuclear fusion applications: A comprehensive review. *J. Alloys Compd.* **2021**, *884*, 161057. [CrossRef]
37. Kaspar, R. Microstructural aspects and optimization of thin slab direct rolling of steels. *Steel Res. Int.* **2003**, *74*, 318–326. [CrossRef]
38. Choudhary, S.K.; Ganguly, S. Morphology and segregation in continuously cast high carbon steel billets. *ISIJ Int.* **2007**, *47*, 1759–1766. [CrossRef]
39. Saeki, T.; Imura, H.; Oonishi, Y.; Niimi, H.; Miwa, E.; Yoshida, T.; Igari, S.; Kitamine, S. Effect of bulging and solidification structure on segregation in continuously cast slab. *Trans. Iron Steel Inst. Jpn.* **1984**, *24*, 907–916. [CrossRef]
40. Pikkarainen, T.; Vuorenmaa, V.; Rentola, I.; Leinonen, M.; Porter, D. Effect of superheat on macrostructure and mac-ro-segregation in continuous cast low-alloy steel slabs. *Iop Conf.* **2016**, *117*, 012064.
41. Gadellaa, R.F.; Kreijger, P.J.; Cornelissen, M.C.M.; Donnay, B.; Herman, J.C.; Leroy, V. Metallurgical aspects of thin slab casting and rolling of low carbon steels. In Proceedings of the 2nd European Conference Continuous Casting (METEC 94), Dusseldorf, Germany, 20–22 June 1994; Volume 1, pp. 382–389.
42. Clyne, T.W.; Kurz, W. Solute redistribution during solidification with rapid solid state diffusion. *Met. Mater. Trans. A* **1981**, *12*, 965–971. [CrossRef]
43. Meng, Y.; Thomas, B. Heat-transfer and solidification model of continuous slab casting: CON1D. *Met. Mater. Trans. A* **2003**, *34*, 685–705. [CrossRef]
44. Ueshima, Y.; Mizoguchi, S.; Matsumiya, T.; Kajioka, H. Analysis of solute distribution in dendrites of carbon steel with δ/γ transformation during solidification. *Met. Mater. Trans. A* **1986**, *17*, 845–859. [CrossRef]
45. Kajatani, T.; Drezet, J.-M.; Rappaz, M. Numerical simulation of deformation-induced segregation in continuous casting of steel. *Met. Mater. Trans. A* **2001**, *32*, 1479–1491. [CrossRef]
46. Miyazawa, K.; Schwerdtfeger, K. Macrosegregation in continuously cast steel slabs: Preliminary theoretical investigation on the effect of steady state bulging. *Archiv Eisenhüttenwes.* **1981**, *52*, 415–422. [CrossRef]
47. Chakraborty, S.; Dutta, P. Effects of dendritic arm coarsening on macroscopic modelling of solidification of binary alloys. *Mater. Sci. Technol.* **2001**, *17*, 1531–1538. [CrossRef]
48. Poirier, D.R. Permeability for flow of interdendritic liquid in columnar-dendritic alloys. *Met. Mater. Trans. A* **1987**, *18*, 245–255. [CrossRef]
49. Yoo, H.; Viskanta, R. Effect of anisotropic permeability on the transport process during solidification of a binary mixture. *Int. J. Heat Mass Transf.* **1992**, *35*, 2335–2346. [CrossRef]
50. Ahmad, N.; Rappaz, J.; Desbiolles, J.-L.; Jalanti, T.; Rappaz, M.; Combeau, H.; Lesoult, G.; Stomp, C. Numerical simulation of macrosegregation: A comparison between finite volume method and finite element method predictions and a confrontation with experiments. *Met. Mater. Trans. A* **1998**, *29*, 617–630. [CrossRef]
51. Yamanaka, A.; Nakajima, K.; Yasumoto, K.; Kawashima, H.; Nakai, K. New evaluation of critical strain for internal crack formation in continuous casting. *Rev. Métall.* **1992**, *89*, 627–634. [CrossRef]
52. Cornelissen, M.C.M. Mathematical model for solidification of multi component alloys. *Ironmak. Steelmak.* **1986**, *13*, 204–212.

Review

Microstructure and Oxidation Behavior of Nb-Si-Based Alloys for Ultrahigh Temperature Applications: A Comprehensive Review

Fuqiang Shen [1], Yingyi Zhang [1,*], Laihao Yu [1], Tao Fu [1], Jie Wang [1], Hong Wang [1,2] and Kunkun Cui [1]

1. School of Metallurgical Engineering, Anhui University of Technology, Maanshan 243002, China; sfq19556630201@126.com (F.S.); aa1120407@126.com (L.Y.); ahgydxtaofu@163.com (T.F.); wangjiemaster0101@outlook.com (J.W.); 0531@126.com (H.W.); 15613581810@163.com (K.C.)
2. Beijing Metallurgical Equipment Research Design Institute Co. Ltd., Beijing 100029, China
* Correspondence: zhangyingyi@cqu.edu.cn

Abstract: Nb-Si-based superalloys are considered as the most promising high-temperature structural material to replace the Ni-based superalloys. Unfortunately, the poor oxidation resistance is still a major obstacle to the application of Nb-Si-based alloys. Alloying is a promising method to overcome this problem. In this work, the effects of Hf, Cr, Zr, B, and V on the oxidation resistance of Nb-Si-based superalloys were discussed. Furthermore, the microstructure, phase composition, and oxidation characteristics of Nb-Si series alloys were analyzed. The oxidation reaction and failure mechanism of Nb-Si-based alloys were summarized. The significance of this work is to provide some references for further research on high-temperature niobium alloys.

Keywords: multivariate alloy; microstructure; alloying; oxidation behavior; mechanism

Citation: Shen, F.; Zhang, Y.; Yu, L.; Fu, T.; Wang, J.; Wang, H.; Cui, K. Microstructure and Oxidation Behavior of Nb-Si-Based Alloys for Ultrahigh Temperature Applications: A Comprehensive Review. *Coatings* **2021**, *11*, 1373. https://doi.org/10.3390/coatings11111373

Received: 25 October 2021
Accepted: 7 November 2021
Published: 9 November 2021

Publisher's Note: MDPI stays neutral with regard to jurisdictional claims in published maps and institutional affiliations.

Copyright: © 2021 by the authors. Licensee MDPI, Basel, Switzerland. This article is an open access article distributed under the terms and conditions of the Creative Commons Attribution (CC BY) license (https:// creativecommons.org/licenses/by/ 4.0/).

1. Introduction

With the advancement of aerospace technology, the high thrust-to-weight ratio aeroengine puts forward higher requirements for the high-temperature resistance of structural materials. Due to the limitation of the melting point and high-temperature strength, the traditional Ni-based superalloys have gradually failed to meet the development needs of aerospace industry. Thus, there is a great need for a high-temperature structural material with better performance [1–3]. Nb-Si-based alloys have a higher melting point and lower density: an ideal candidate material [4–8]. The comparison of properties of Ni-based alloy and Nb-based alloy is shown in Table 1. The insufficient high-temperature oxidation resistance and high-temperature creep resistance of Nb-based alloys has greatly hindered its development and application [9,10]. The oxidation of Nb to Nb_2O_5 at high temperature leads to a sharp volume expansion, and the stress in the oxide layer increases gradually, resulting in cracks and spalling, which eventually leads to rapid oxidation failure of the alloy [11–13]. Various methods have been used to solve this problem, and one of the most common is alloying. Alloying elements can ameliorate the oxidation characteristics of the oxide layer, improve the density and integrity of oxide scales, and inhibit the permeation of O_2. Furthermore, alloying elements may also produce the intermediate phase with thermal expansion coefficient (hereinafter abbreviated as TEC) between the substrate and oxide layer, which can reduce the internal stress between the substrate and oxide layer, improve the bonding force of the oxide layer, reduce cracks, and optimize its antioxidant properties [14–20].

The weaker oxidation resistance of alloy is mainly due to the lack of protective oxide layer, and it also depends on the oxidation characteristics of the material itself. Although the metal surface coating technology can extend the oxidation life of the alloy effectively, once the metal coating fails, the antioxidant properties of the alloy become extremely important [21–23]. Therefore, studying uncoated alloys is essential to determine how the material will respond when the coating fails. In the past few decades, people have studied a variety of alloying elements such as Hf, Cr, Al, W, Mo, Ti, Zr, V, B, Ge, Sn, and Ta [22–26]. Studies show that Al and Cr can form the protective Al_2O_3 and $CrNbO_4$ oxide layers [27], and B and Ge can facilitate the formation of SiO_2 protective film. Moreover, W and Mo can generate volatile WO_3 and MoO_3 [28–30] to reduce the internal stress and increase the adhesion of the oxide scales. In addition, Ti, Hf, and Sn can increase the fracture toughness of the alloy. Meanwhile, Hf and Sn can also inhibit the diffusion of O_2 and enhance the antioxidant properties of the alloy.

Table 1. Comparison of properties of Ni base superalloy and Nb base superalloy.

Alloy Composition	Melting Point (°C)	Density (g/cm^3)	Operating Temperature (°C)	Reference
Ni-based superalloy	1350–1400	8.1–8.5	1150	[1,5,17,19]
Nb-based superalloy	>1700	6.6–7.2	1200–1400	[11,16,22,25]

At present, there are many related research studies on niobium alloy, but there is little summary or sorting of these studies. It is of great significance to summarize the effects of different alloy elements, temperature, element content, and other factors on the further study of high-temperature resistant alloy. In this work, the effects of Hf, Cr, Zr, B, and V on the oxidation resistance of multi-component Nb-Si-based superalloys were summarized. By analyzing the oxidation kinetics, oxide composition, and microstructure of the alloy, the oxidation and failure mechanisms of Nb-Si alloy were summarized, which have a certain reference significance for the further study of high-temperature niobium alloys.

2. Nb-Si-Based Alloys

The Si element has strong stability, and Nb-Si intermetallic compound is one of the most common high-temperature structural materials. After the oxidation of Si at high temperature, the flowing SiO_2 glass film is generated, which has an excellent repair and protection effect on the alloy substrate [31–34]. The phase composition of Nb-Si-based alloys mainly includes niobium solid solution (Nbss), Nb_3Si, and Nb_5Si_3. Nbss has excellent ductility, but its high-temperature oxidation resistance is poor. The pesting phenomenon occurs at medium-low temperature (700 °C); thus, the alloy is difficult to use at high temperatures [35–37]. Alloying can alleviate the problem of insufficient high-temperature oxidation resistance of Nb-Si alloys to a certain extent. Unfortunately, the oxidation resistance and mechanical properties of ternary Nb-Si-based alloys, such as Nb-Si-Al and Nb-Si-Cr, are still insufficient. Therefore, it is significant to optimize the comprehensive properties of Nb-Si series alloy by further alloying. Based on this, the focus of the research has been shifted to multi-component high-temperature-resistant niobium alloys.

3. Modified Nb-Si-Based Multi-Element Alloys

3.1. Nb-Si-Based Alloys Modified with Hf Elements

In order to promote the high-temperature oxidation resistance of Nb-Si-based alloys, researchers have tried to add different kinds of elements, and Hf is a common addition element. Geng et al. [38] produced the Nb-Si-Al-Cr alloy by a non-consumption arc-melting process and the effects of Ti, Mo, and Hf on the oxidation behavior of the alloy at different temperatures were studied. Figure 1 shows cross-sections of the oxidized alloys. The scales of the alloy were a complex mixture of Nb and Ti oxides; oxygen was dissolved in the niobium solid solution (Nbss) below the oxide scale of the alloy. The oxidation kinetics curve is shown in Figure 2a. It can be seen that the oxidation of these alloys includes two stages: the initial linear stage and the later parabolic stage. The weight gain rate of the alloy with the Hf element decreases gradually after 55 h of oxidation, while its oxidation resistance becomes worse after heat treatment. Furthermore, Geng et al. [39] have also studied the effects of Ti, Hf, and Sn on the oxidation resistance of the Nb-Si-Al-Cr-Mo alloy, and the oxidation weight gain curve is shown in Figure 2b. It is clear that the addition of Hf decreases the oxidation rate of as-cast alloy but increases the oxidation rate of heat-treated alloy. The oxidation behavior of these alloys is mainly controlled by the oxidation of Nbss, and the volume fraction of Nbss has a great impact on the oxidation resistance of the alloy.

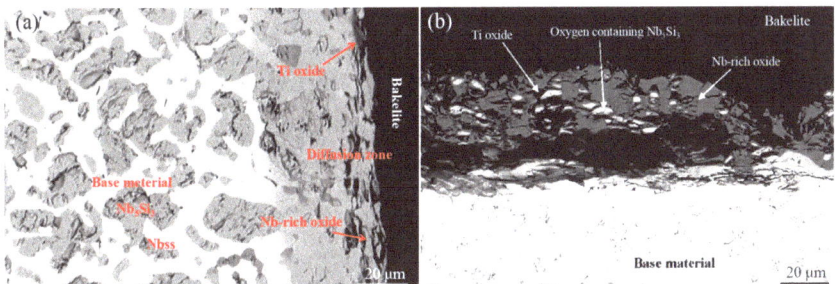

Figure 1. BSE images of the microstructures of cross-sections of the (**a**) Nb-24Ti-18Si-5Al-5Cr-2Mo and (**b**) Nb-24Ti-18Si-5Al-5Cr-2Mo-5Hf after oxidation at 800 °C. Copyright 2006 Elsevier.

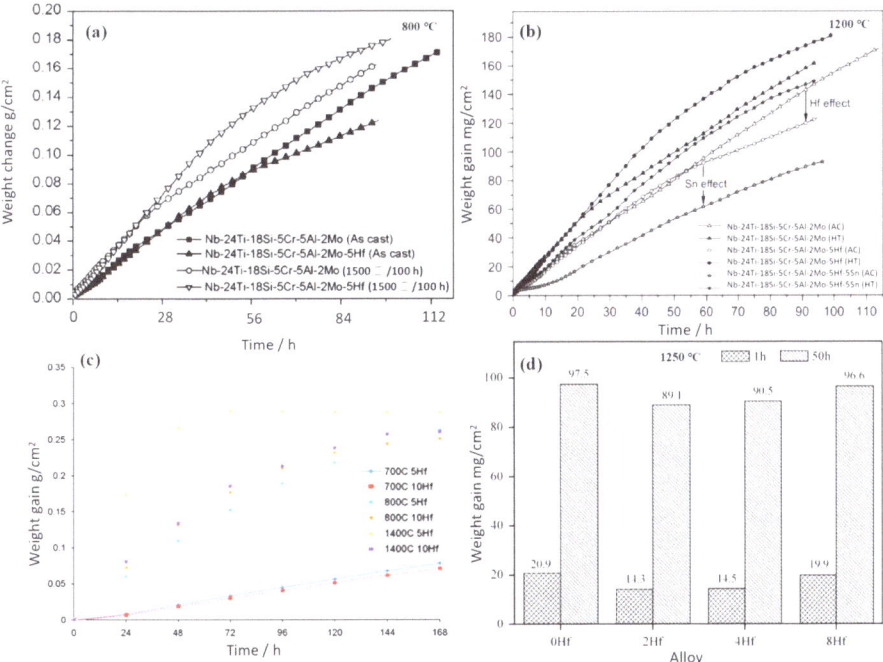

Figure 2. Weight change curves of Nb-Si-based alloys with Hf addition oxidized at (**a**) 800 °C and (**b**) 1200 °C. (**c**) Cyclic oxidation curves for Nb-20Si-20Cr-(5, 10 at.%) Hf alloys. (**d**) Oxidation weight gain histogram of Nb-22Ti-16Si-3Cr-Al-2B-xHf (x = 0 at.%, 2 at.%, 4 at.%, 8 at.%) alloys at 1250 °C. Copyright 2006 Elsevier, 2011 Elsevier and 2015 Elsevier.

In addition, Vazquez et al. [40] also studied the effects of Hf element on the antioxidant properties of Nb-Si-based alloys. The microscopic morphology of the alloy after oxidation is shown in Figure 3. Analysis shows that the products of oxidation are Nb_2O_5, HfO_2, and un-reacted $NbCr_2$. Nbss in the alloy is prone to selective oxidation. After long-term oxidation at 700 °C, there are a large number of powder oxides and cracks in the oxide layer, which has no protective effect. Figure 2c shows the oxidation curves for both alloys doped with 5Hf and 10Hf (at.%) at 700, 800, and 1400 °C, respectively. The oxidation experiments reveal that both alloys exhibit a good oxidation resistance at 700 °C, the 10Hf alloy has lower oxidation weight gain than 5Hf alloy at all temperatures. The researchers attribute the alloy's better oxidation resistance to the formation of a thin protective layer.

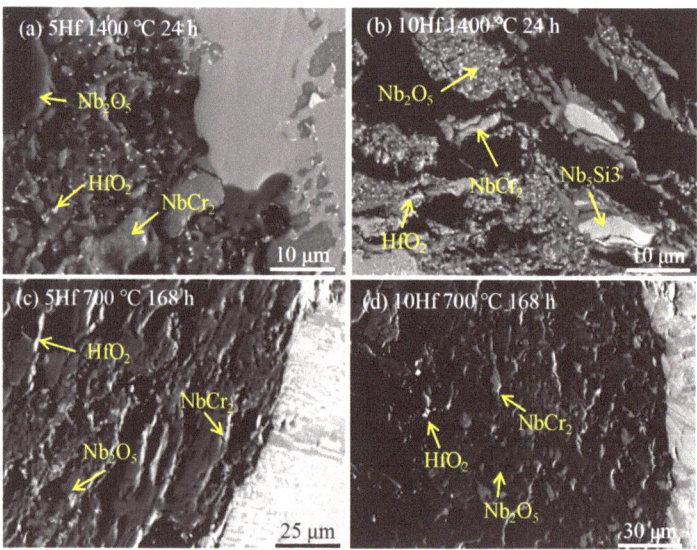

Figure 3. Microstructure of Nb-20Cr-20Si-5Hf alloy (**a**,**c**) and Nb-20Cr-20Si-10Hf alloy (**b**,**d**) after short-term oxidation at 1400 °C and long-term oxidation at 700 °C. Copyright 2011 Elsevier.

Zhang et al. [41] studied the effects of Hf, B, and Cr on the antioxidant properties of Nb-Si-based superalloys. The research shows that the weight gain per unit area of the alloy decreases from 157 to 139 mg/cm^2 after oxidation at 1250 °C for 50 h. Meanwhile, the synergistic addition of Hf, B, and Cr makes the oxidation resistance of the alloy more excellent. Zhang et al. [42]. also studied the antioxidant properties of Nb-Si-based alloys with different Hf contents and found that the oxide layers of these alloys fell off and had poor adhesion. The cross-sectional BSE image of the alloy after oxidation is shown in Figure 4. The oxides of the alloys are TiO_2 and HfO_2, as determined by EDS. HfO_2 is mainly found at the interface between Nbss and $(Nb,X)_5Si_3$ (rod-like morphology) or at the edges of silicide blocks (needle-like morphology). More severe internal oxidation occurs with Hf addition. With the addition of Hf, the formation of $\alpha(Nb,X)_5Si_3$ was inhibited, while the formation of $\gamma(Nb,X)_5Si_3$ was promoted, and the former has better oxidation resistance, since the former phase is not attacked, while the latter phase is partly oxidized. The histogram of oxidation weight gain is shown in Figure 2d. It is obvious that the weight change of the alloy decreases first and then increases with the increase in Hf, and the alloy with 2Hf (at.%) has the least weight gain. It may be due to the large size of Hf atoms, which inhibited the diffusion of other metal atoms. Although the addition of 8Hf (at.%) did not improve the oxidation resistance of the alloy, it increased the room temperature fracture toughness of the alloy, and the eutectic structure in the alloy has been refined significantly. It can be seen that the contents of Hf have a great influence on the comprehensive performance of the alloy.

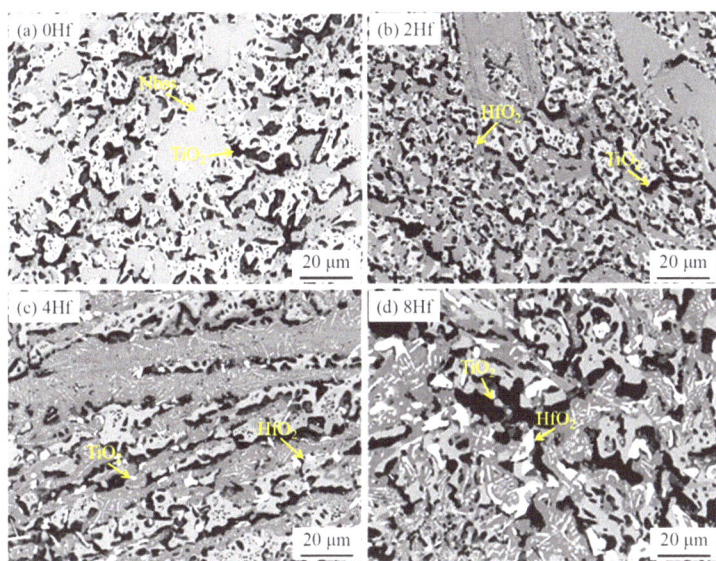

Figure 4. Cross-sectional BSE images of the internal oxidation zones of the Nb-22Ti-16Si-3Cr-3Al-2B-xHf. (x = 0 at.%, 2 at.%, 4 at.%, 8 at.%) alloys upon oxidation at 1250 °C for 50 h.(**a**) 0Hf alloy. (**b**) 2Hf alloy. (**c**) 4Hf alloy. (**d**) 8Hf alloy. Copyright 2015 Elsevier.

3.2. Nb-Si-Based Alloys Modified with Cr Elements

Chromium has the advantages of high melting point, excellent chemical stability, and easy alloying with other elements. It is widely used in Fe-based and Ni-based alloys. The Cr_2Nb phase formed after the addition of Cr element in niobium alloy has the advantages of high melting point and oxidation resistance [43–45]. The oxide $CrNbO_4$ has better oxidation resistance than Nb_2O_5, and it can also improve the adhesion between the oxide layer and substrate [46,47]. Zhang et al. [41] studied the effects of Hf, B, and Cr elements on the microstructure and properties of Nb-Si-based superalloys by a vacuum non-consumption arc-melting process. (The alloy with x at.% Hf, y at.% B, and z at.% Cr content is denoted by xHf-yB-zCr below.) The histogram of the oxidation weight gain of alloys with different compositions is shown in Figure 5a. Obviously, the 0Hf-0B-0Cr alloy (base alloy) has the most weight gain after oxidation (157 mg/cm^2 for 50 h), displaying the worst oxidation resistance. The oxidation weight gain decreases from 157 to 139 mg/cm^2 after Hf addition (4Hf-0B-0Cr alloy), and the oxidation resistance is slightly ameliorated. Furthermore, the synergistic effects of Hf, Cr, and B enhance the oxidation resistance of the alloy significantly, and the weight gain of the 4Hf-2B-5Cr alloy is 91 mg/cm^2 at 50 h. The room temperature fracture toughness of the alloy decreases with the addition of Cr, and the microhardness increases with the addition of Cr. In addition, Wang et al. [46] prepared the Nb-22Ti-14Si-2HF-2Al-xCr (x = 2 at.%, 6 at.%, 10 at.%, 14 at.%, 17 at %) alloys, and the effect of Cr on oxidation resistance and mechanical properties of the alloy were studied. Figure 5b shows the oxidation kinetics curve of these alloys at 1250 °C. It is obvious that with the increase in the contents of Cr, the oxidation weight gain decreases gradually, and the oxidation resistance of the alloy improves gradually. The 17Cr (at.%) alloy has the best antioxidant properties, which may be due to the formation of the protective Cr_2O_3/SiO_2 film. In addition, with the increase in Cr content, the fracture toughness of the alloy decreases, and the hardness of Nbss increases. Moreover, the studies of Zelenitsas [48] and Esparza [49] show that the synergy of Cr and Al can form the protective $Al_2O_3/CrNbO_4$ film and reduce the oxidation rate inside the alloy. The Cr element has stronger O affinity than Nb, and the formation of stable Cr_2O_3 can improve the compactness and stability of the oxide layer [50–52]; thus, the alloy has better oxidation resistance.

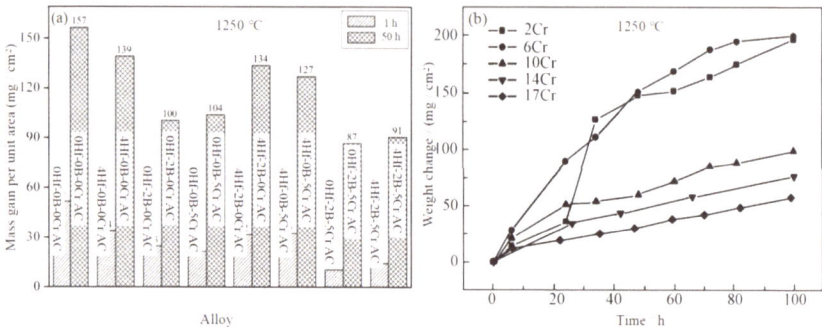

Figure 5. (a) Oxidative weight gain histogram of the Nb-Si-based alloys oxidized at 1250 °C. (b) The oxidation weight gain curve of Nb-Si alloy with different Cr content at 1250 °C. Copyright 2012 Elsevier and 2015 Elsevier.

3.3. Nb-Si-Based Alloys Modified with Zr Elements

Among the many types of alloy systems, Zr addition can refine the microstructure and promote the oxidation resistance of the alloy [53–56]. After the oxidation of Zr, ZrO_2 is formed, which has the characteristics of low thermal conductivity, excellent thermal vibration resistance, and oxidation resistance. In the work of Zhang et al. [57], the effect of V, Zr, and other elements on the antioxidant properties of Nb-15Si-24Ti-4Cr-2Al-2Hf (at %) alloy was investigated. The surface SEM images of the alloy oxide upon 1250 °C for 100 h are shown in Figure 6a–c; Figure 6(a1–c1) show the macroscopic morphology after oxidation. It is apparent that the alloy has a rough and porous surface, and the oxide layer is seriously peeled off. The condition has not been improved after adding V, but the oxide layer is relatively dense and complete after adding Zr. Figure 6d–f shows the SEM morphology of the oxidation cross-section, and the oxide layer consists of inner and outer layers. The surface of the base alloy and 1 V alloy after oxidation is loose and porous after peeling off. Although the oxide layer of 1Zr alloy falls off also, an oxide layer rich in silicon and oxygen is formed on the surface. Combined with the oxidation kinetics curve in Figure 7a, it can be inferred that a protective oxide layer is produced on the surface of the 1Zr alloy. The results of WDS and EPMA analysis show that an SiO_2 film is formed in the oxide outer layer, which inhibits the diffusion of O_2 and protects the alloy substrate.

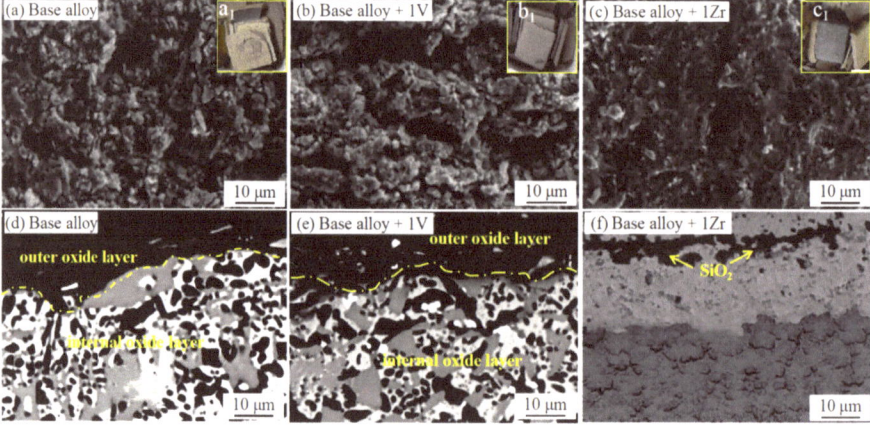

Figure 6. Surface SEM images of Nb-Si-based alloys after oxidation at 1250 °C for 100 h: (a) base alloy, (b) 1V alloy, (c) 1Zr alloy. a1–c1 are the corresponding macroscopic morphology. Cross-sectional SEM images of Nb-Si-based alloys after oxidation: (d) base alloy, (e) 1 V alloy, (f) 1Zr alloy. Copyright 2017 Springer Nature.

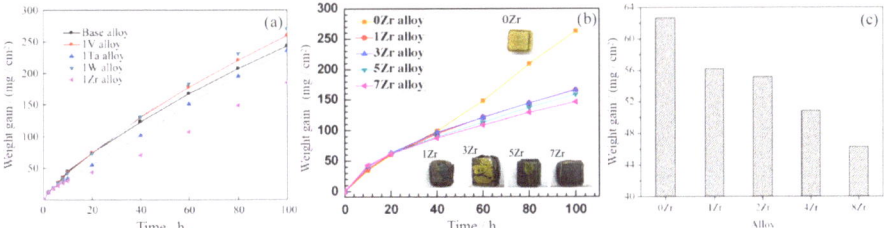

Figure 7. Weight gain versus time for Nb-Si-based alloys oxidized at 1250 °C for 100 h (**a**). Oxidation weight gain curve (**b**) and oxidation weight gain histogram (**c**) of Nb-Si-based alloys with different Zr contents. Copyright 2017 Springer Nature, 2019 Elsevier and 2017 Elsevier.

In order to ascertain the influence of Zr content on the oxidation resistance of Nb-Si-based alloy, Zhang et al. [58] designed the Nb-Si-based alloys with different Zr content, and the oxidation behavior of the alloy at 1300 °C was studied. The microstructure of the oxide layer on the alloy surface is shown in Figure 8a–d. It is apparent that the oxide on the surface of 0Zr and 3Zr alloys is rod-like, rough, and porous. When the Zr content reaches more than 5 at.%, the oxides have a polygonal structure and compact arrangement. Figure 8e–h show the oxidation fracture surface microstructures of 0Zr, 3Zr, 5Zr, and 7Zr, respectively, and the oxide layer is mainly composed of TiO_2, Nb_2O_5, $TiNb_2O_7$, and $Ti_2Nb_{10}O_{29}$. A small amount of ZrO_2 is detected in the 7Zr alloy. Meanwhile, no SiO_2 is detected in the oxide, which may be attributed to the amorphous state of SiO_2. Figure 7b shows the oxidation kinetics curve of the alloy. It is obvious that the weight gain of the alloy decreases with the increase in Zr. The weight gain curves of 1Zr and 3Zr alloys almost coincide, while the weight gain of 7Zr alloy is about 44% lower than that of 0Zr alloy. Combined with the macroscopic morphology of these alloys after oxidation, it can be seen that the 7Zr alloy has the best oxidation resistance.

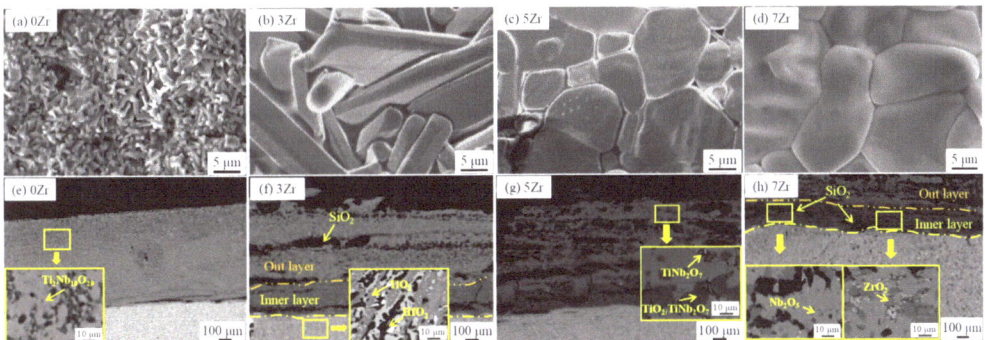

Figure 8. Surface morphologies and cross-sectional microstructures of Nb-15Si-24Ti-4Cr-2Al-2Hf-xZr (x = 0 at.%, 3 at.%, 5 at.%, 7 at.%) alloys after oxidation at 1300 °C for 100 h. (**a,e**) 0Zr; (**b,f**) 3Zr; (**c,g**) 5Zr; (**d,h**) 7Zr. Copyright 2019 Elsevier.

In addition, Qiao et al. [59] also carried out a similar study on the effect of Zr content on the microstructure and properties of Nb-Si-based alloys. Figure 9 shows the cross-section BSE image of the alloy after oxidation at 1250 °C for 20 h. It is evident that there are many cracks and holes in the oxide layer, and the thickness of the oxide layers ranges from 260 to 380 μm. The EPMA-WDS analysis shows that the phase composition of the outer layer is $TiNb_2O_7$ and TiO_2, and the inner layer is $Ti_2Nb_{10}O_{29}$ and Nb_2O_5. In addition, ZrO_2 was also detected in the scales of 4Zr and 8Zr alloys. Obviously, the weight gain of the sample decreases with the increase in Zr, as shown in Figure 7c. It can be seen that Zr plays an important role in advancing the antioxidant properties of Nb-Si-based alloys; meanwhile, the addition of Zr improves not only the room-temperature fracture toughness but also the high-temperature strength and compressive strength of the alloy. In addition, Ma et al. [60] showed that adding Zr or V alone can improve the oxidation resistance of Nb-Si series alloys, but the simultaneous addition of Zr and V will form the $CrVNbO_6/VNb_9O_{25}$ harmful oxides, which will reduce the service time of Nb-Si alloys. Moreover, the addition of Zr improves the room-temperature fracture toughness and compressive yield strength of the alloy, and the phase microhardness is also improved due to the solution strengthening effect of Zr.

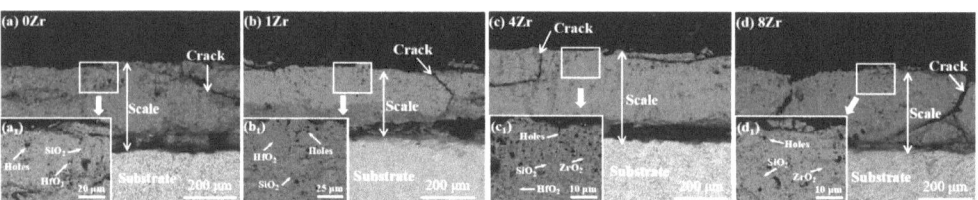

Figure 9. Cross-sectional BSE images of the Nb-22Ti-15Si-5Cr-3HF-3Al-xZr (x = 0 at.%, 1 at.%, 4 at.%, 8 at %) alloys after oxidation at 1250 °C for 20 h. (**a**) 0Zr; (**b**) 1Zr; (**c**) 4Zr; (**d**) 8Zr. The a_1, b_1, c_1, and d_1 are microarea images in figures (**a–d**), respectively. Copyright 2017 Elsevier.

3.4. Nb-Si-Based Alloys Modified with B Elements

B elements can refine the grain size and improve the toughness of materials. Studies have shown that the borosilicate glassy layer formed by B and Si in Mo-Si-B alloys has a good repairing effect on cracks and holes in oxide scales [61–66], which is of great significance to ameliorate the comprehensive properties. Zhang et al. [65] prepared the Nb-22Ti-16Si-5Cr-4Hf-3Al alloy with different contents of B and studied the effect of B on the microstructure and properties of the alloy. The layer is composed of $TiNb_2O_7$, (Ti, Nb, Cr)O_2, and Si, as shown in Figure 10. Unexpectedly, no silicon oxide was detected in the oxide layer, which may be due to the formation of amorphous silicate. The oxidation weight gain of 0B, 2B, and 5B alloys are 127.1, 90.5, and 67.6 mg/cm^2, respectively, and the thickness of the oxide layer decreases from 882 to 441 µm. Obviously, the antioxidant performance of the sample increases gradually with the increase in B content. The fracture toughness of the alloy increases first and then decreases, while the microhardness of Nbss increases slightly with the increase in B content. Furthermore, Zhang et al. [41] also studied the synergistic effects of B, Hf, and Cr, and the results showed that the synergy of B and Cr had a better oxidation resistance, while the addition of Hf was not conducive to enhancing the oxidation resistance of the alloy. The fracture toughness of the alloy decreases with the addition of B, while the macroscopic hardness does not change significantly. Moreover, in the work of Thomas et al. [67], the effect of different Si and B content (at %) on the oxidation resistance of Nb-25Cr-15Mo-xSi-yB alloy was studied (hereinafter, it is abbreviated as xSi-yB alloy). The oxidation products of 20Si-10B and 20Si-15B (at.%) alloys are mainly $CrNbO_4$ and SiO_2. The weight gain of 20Si-15B alloy is the minimum at 1200 and 1400 °C, while the weight gain of 15Si-15B alloy is the minimum at 900 °C, which may be attributed to the occurrence of pest oxidation and the spalling of the oxide layer, as shown in Figure 11. Therefore, the high content of Si and B is more beneficial to enhance the oxidation resistance of the Nb alloys. Moreover, the research of Su et al. [68] shows that the synergy of B and Ge can facilitate the production of continuous and dense oxide film in the alloy and reduce the penetration rate of O_2. It is worth noting that the dissolution of B_2O_3 and GeO_2 in silicide can also improve the TEC of the oxide scales and make it more compatible with the substrate so as to suppress the occurrence of cracks and improve the oxidation life of the alloy.

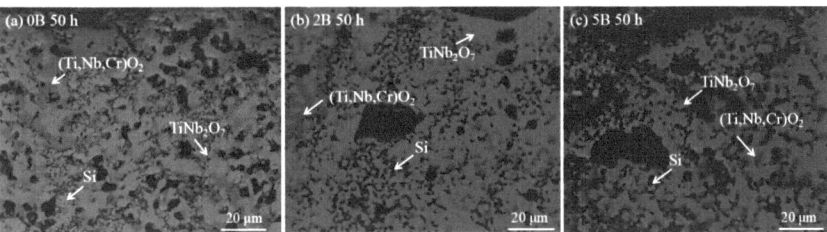

Figure 10. Cross-sectional BSE images of the Nb-22Ti-16Si-5Cr-4Hf-3Al-xB (x = 0 at.%, 2 at.%, 5 at.%) alloys after oxidation at 1250 °C: (**a**) 0B, (**b**) 2B, and (**c**) 5B. Copyright 2014 Elsevier.

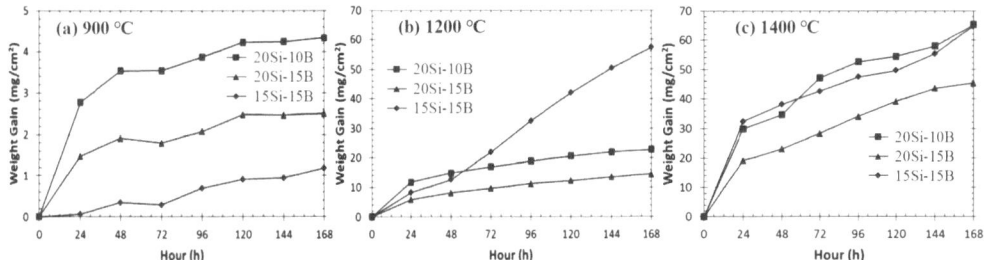

Figure 11. Long-term oxidation curves for alloys at (**a**) 900 °C, (**b**) 1200 °C, and (**c**) 1400 °C. Copyright 2015 Elsevier.

3.5. Nb-Si-Based Alloys Modified with V Elements

It is found that the V element has a positive effect on the oxidation resistance and mechanical properties of Nb-Si-based alloys [69,70]. Guo et al. [69] studied the effect of V (1 at.%) on the fracture toughness and oxidation resistance of the Nb-15Si-24Ti-4Cr-2Al-2Hf alloy. Compared with V free alloy, 1 V alloy has better fracture toughness. Figure 12. shows the cross-section morphology of the alloy oxidized at 1300 °C for 100 h. V is mainly distributed in Nbss, and V will be oxidized preferentially within Nbss due to its stronger O affinity than Nb. The oxidation kinetic curve of the alloy is shown in Figure 13a. After oxidation for 20 h, the V-free alloy increases linearly, while the 1 V alloy still increases its parabola. The author believes that although the microstructure of the alloy has no obvious change, the oxidation characteristics of Nbss have been ameliorated; thus, the antioxidant properties of the alloy have been enhanced. Ma et al. [60] studied the synergistic effect of V and Zr on the microstructures and properties of the Nb-22Ti-15Si-5Cr-3Al-2Hf alloy. Figure 13b shows the histogram of oxidation weight gain and macroscopic morphology of the alloy. The study shows that the addition of V or Zr alone can enhance the oxidation resistance of the alloy, while the effect of adding both at the same time is the opposite. The addition of V can reduce the difference in the oxidation rate of Nbss and silicide, which is beneficial to the formation of a continuous and dense oxide layer, thereby improving the oxidation resistance of the alloy. However, the addition of Zr has the opposite effect [71,72]. The addition of V can also improve the fracture toughness and compressive yield strength of the alloy. However, there are some different views about the effect of vanadium on the antioxidant performance of Nb-Si-based high temperature alloys.

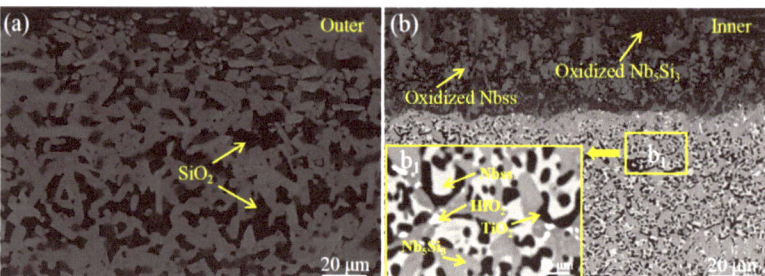

Figure 12. Cross-sectional microstructure of Nb-15Si-24Ti-4Cr-2Al-2HF-1V (at.%) alloy in outer layer (**a**) and inner layer (**b**) after oxidation, and the enlarged images of the yellow rectangular zones (b$_1$). Copyright 2017 Elsevier.

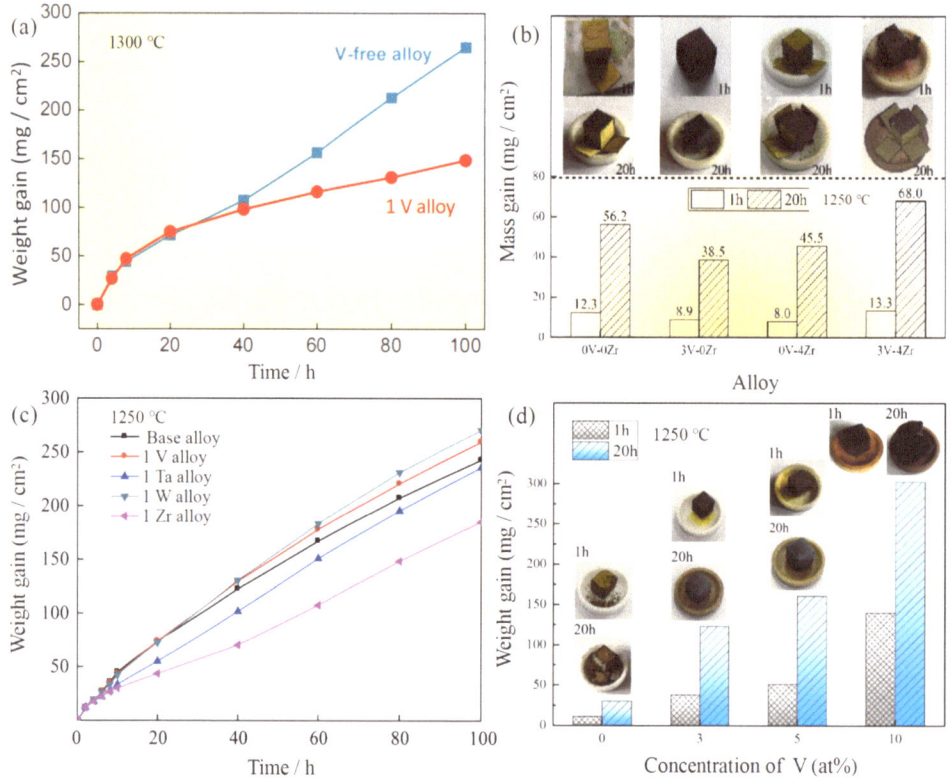

Figure 13. Weight gain curve of Nb-Si-based alloys after adding V. (**a**) Nb-15Si-24Ti-4Cr-2Al-2Hf-xV (x = 0, 1 at.%). (**b**) Nb-22Ti-15Si-5Cr-3Al-2Hf-xV-yZr ((x, y) = (0, 0), (3, 0), (0, 4), (3, 4) at.%). (**c**) Nb-15Si-24Ti-4Cr-2Al-2Hf alloys with separate V, Ta, W and Zr additions. (**d**) Nb-22Ti-15Si-5Cr-5Mo-4Zr-3Al-2Hf-xV (x = 0, 3, 5, 10 at.%). Copyright 2017 Elsevier, 2021 Elsevier, 2017 Springer Nature and 2020 Elsevier.

The study by Zhang et al. [57] showed that V reduces the oxidation resistance of Nb-Si-based alloys, and the oxidation weight gain curve of the Nb-15Si-24Ti-4Cr-2Al-2Hf alloy with different modification elements (V, Ta, W, and Zr) is shown in Figure 13c. The oxide layer composed of Nb_2O_5, $TiNb_2O_7$, $Ti_2Nb_{10}O_{29}$, and TiO_2 has no protective effect on the substrate, which may be the reason for the poor oxidation resistance of the alloy [73,74]; besides, volatile V_2O_5 produced after V oxidation will cause the oxidation holes and accelerate the oxidation failure process. In addition, Ma et al. [75] studied the oxidation behavior of Nb-22Ti-15Si-5Cr-5Mo-4Zr-3Al-2Hf-xV (x = 0 at.%, 3 at.%, 5 at.%, 10 at.%) alloys at 1250 °C. Figure 14 shows the cross-section morphology and the magnified BSE image of the alloy after oxidation at 1250 °C for 1 h. It is clearly that the oxide layer is divided into internal and external layers. The outer layer is composed of $TiNb_2O_7$, Nb_2O_5, TiO_2, SiO_2, $CrVNbO_6$, and Nb_9VO_{25}, while the inner layer is composed of Nbss and silicide. It can be seen from Figure 13d that with the increase in V content, the weight gain per unit area of oxidation increases gradually, and the thickness of the oxide layer increases gradually also. Meanwhile, the pores caused by V_2O_5 volatilization are also increasing. Although the pores are beneficial to relieve the internal stress of the oxide layer, they also accelerate the internal diffusion of O_2, thus worsening the antioxidant properties of the alloy. The fracture toughness of the alloy increases first and then decreases with the increase in V content, and the microhardness of the alloy increases with the increase in V content, while the compressive strength decreases. However, there is still a controversy about the effect of vanadium on the antioxidant performance of Nb-Si-based alloys, which may be related to the substrate composition, alloy preparation process, oxidation conditions, V content, and other factors.

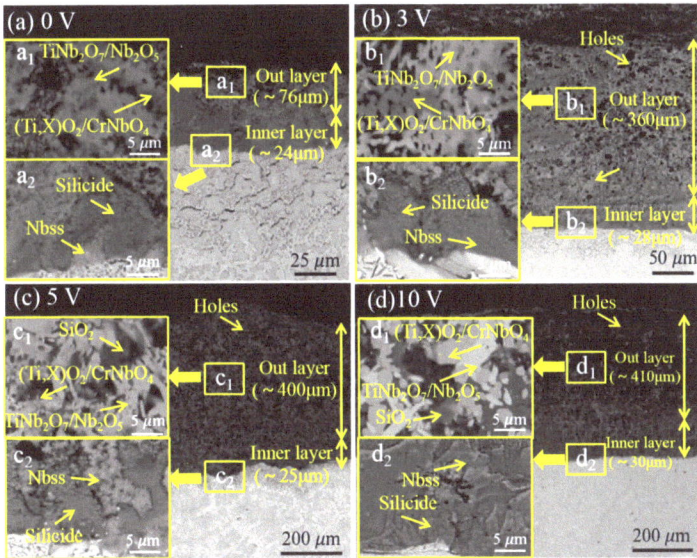

Figure 14. Cross-sectional and enlarged (a_1–d_1, a_2–d_2) BSE images of scales of the Nb-Si-based alloys with different V content oxidized at 1250 °C for 1 h: (**a**) (a1, a2) 0 at.%V. (**b**) (b1, b2) 3 at.% V. (**c**) (c1, c2) 5 at.%V. (**d**) (d1,d2) 10 at.%V. Copyright 2020 Elsevier.

3.6. Nb-Si-Based Alloys Modified with Other Elements

The oxidation and mechanical properties of the alloys are shown in Table 2. Apart from the alloying elements mentioned above, the researchers also studied the effects of Ti, Al, Mo, Ta, and other elements on the oxidation behavior of Nb-Si-based alloys. Geng et al. [38,39] found that Ti can effectively increase the oxidation resistance of the Nb-Si-Cr-Al alloy at 800 °C. It may be attributed to the preferential oxidation of Ti in Nbss, forming a large number of stable Ti oxide particles, which enhance the oxidation resistance of Nbss. This is consistent with Yonosuke et al. [76]. The Al_2O_3 protective film will be produced after the oxidation of Al, which has a good protective ability on the substrate. Esparza [49] and Su [52] et al. point out that a certain amount of Al is beneficial to prolong the service life of

Nb-Si-based superalloys. However, because Al reduces the melting point of the alloy, this alloy is not suitable for a high-temperature oxidation environment. Mo can improve the melting point and high-temperature strength of the alloy [77]. The research of Geng [38] and Ma et al. [78] shows that the low content of Mo (2 at.%) can form the volatile MoO_3 after oxidation, reduce the internal stress of oxide scales, and enhance the antioxidant capacity of the oxide layer. However, a large amount of MoO_3 volatilizes and causes holes in the oxide layer, which will deteriorate the oxidation resistance of the alloy. The study by Zelenitsas [48] and Zhang [57] et al. showed that Ta can slightly advance the oxidation properties of Nb-Si-based alloys. Moreover, other studies show that an appropriate amount of elements such as W, Y, Er, Ce, Sn, and Ge can also enhance the oxidation resistance of Nb-Si-based super alloys to some extent [39,57,68,79–81].

Table 2. Comparison between different alloys.

Alloy Composition (at.%)	Oxidation Conditions (°C and h)	Weight Gain (mg/cm^2)	Room Temperature Fracture Toughness ($MPa \cdot m^{1/2}$)	Reference
Nb-22Ti-16Si-3Cr-3Al-2B-8Hf	1250/50	96.6	14.1	[42]
Nb-22Ti-14Si-2Al-2Hf-17Cr	1250/100	60	8.7	[46]
Nb-22Ti-15Si-5Cr-3Hf-3Al-8Zr	1250/20	46.3	15.01	[59]
Nb-22Ti-16Si-5Cr-4Hf-3Al-5B	1250/50	67.6	11.0	[65]
Nb-15Si-24Ti-4Cr-2Al-2Hf-1V	1300/100	148.5	12.98	[69]
Nb-22Ti-15Si-5Cr-5Mo-4Zr-3Al-2Hf-10V	1250/20	302.8	6.42	[75]

The oxidation behavior of niobium alloy is affected not only by the type and amount of alloying elements but also by the alloy preparation process and oxidation conditions. For instance, the research of Guo et al. [82,83] shows that the alloys prepared by selective laser melting (SLM) have a finer microstructure, denser oxide layer, and better oxidation resistance than vacuum induction melting (VIM). The research of Mathieu [84] and Alvarez et al. [85] shows that Nb-based alloys are more prone to pesting oxidation at medium and low temperature (700 °C), resulting in the pulverization of the oxide layer. At high temperature, the oxidation weight gain is more severe, the outer oxide layer spalls gradually, and the alloy fails faster.

4. Mechanisms of Oxidation and Failure

The oxidation reaction of niobium alloy usually proceeds gradually from the outside to the inside. At the beginning of the oxidation reaction, O atoms dissolve and diffuse to Nbss to form an oxide layer firstly; then, O atoms diffuse to the substrate along the cracks and holes of the oxide layer, and oxidation occurs inside the alloy. The thickness of the oxide layer increases gradually with the increase in oxidation time and temperature. Due to the distinction of TEC between the oxide layer and substrate, the layer expands and falls off, loses its protective effect eventually. Thereby, it is vital to strengthen the oxidation performance of the oxide layer.

Figure 15 shows the oxidation behavior and failure mechanism of Nb-Si-based alloys. After Nbss is oxidized, Nb_2O_5 is formed as the volume sharply expands, and there is no protection. Si, Al, Zr, Cr, B, etc. can form the protective oxide films, SiO_2, Al_2O_3, borosilicate, and other fluid oxide films can fill the cracks and holes timely in the oxide scales, inhibit the permeation of O_2, and improve the oxidation resistance of the oxide layer. In addition, elements such as Al, Cr, Hf, Si, and Ti have a stronger O affinity than Nb, and the selective oxidation of these elements in Nbss can increase the oxidation resistance of Nbss. Furthermore, Si, Cr etc. can also reduce the diffusion rate and solubility of O atoms in Nbss and inhibit the oxidation reaction of Nbss, thereby improving the antioxidant performance of the alloy. In addition, an appropriate amount of V can enhance the adhesion of oxide scales, ameliorate the oxidation performance of the alloy. However, the excessive addition of V will cause a large number of oxide pores, which will deteriorate the antioxidant properties of the alloy. In addition, an appropriate amount of Hf can participate in the formation of amorphous silicate and improve the oxidation resistance, while an excessive amount of Hf will increase the brittleness of scales and deteriorate the oxidation performance of the alloy. Therefore, the content of modified elements also has a significant effect on the oxidation resistance of the alloy. The excellent oxidation characteristics of the oxide layer is essential for the oxidation resistance of the niobium alloy. A protective oxide layer can suppress the diffusion of O_2 effectively and strengthen the anti-oxidation performance of the alloy. The non-protective oxide layer is often loose and porous due to the discrepancy in the TEC of the oxides, and the oxide layer is prone to peeling and crushing. In addition, the volatile oxides in the oxide layer can also cause the cracks and holes, resulting in rapid oxidation failure of the alloy.

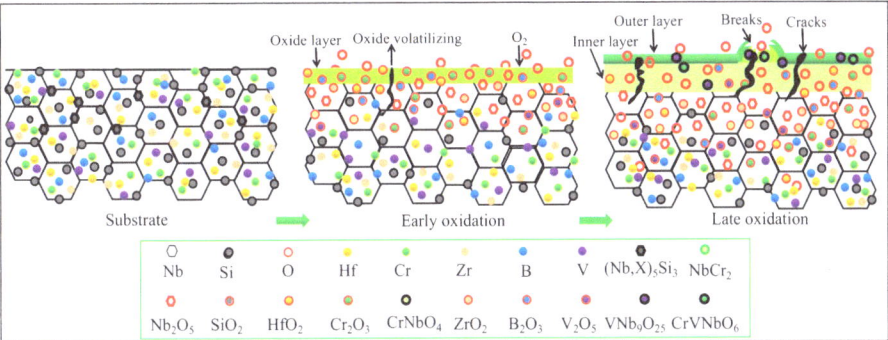

Figure 15. Schematic diagram of oxidation behavior and failure mechanism of Nb-Si-based superalloys.

5. Conclusions and Prospect

Niobium silicon series alloys are widely used in propulsion systems of aerospace and spacecraft, military weapons, and ultrahigh-temperature hot end parts, and they are the most promising new high-temperature-resistant structural materials. Studies show that some modifying elements such as Hf, Cr, Zr, B, and V can promote the oxidation resistance of Nb-Si-based alloys to some extent. Furthermore, the type and content of modified elements, alloy preparation process, and oxidation conditions will also have a great impact on the oxidation performance of materials.

Studies show that adding an appropriate amount of Hf can improve the antioxidant properties of Nb-Si-based alloys, but an excessive amount of Hf will aggravate the brittleness and porosity of the oxide layer, which is not conducive to the generation of a protective oxide layer, which will lead to the deterioration of the antioxidant properties of the alloy. Cr elements can reduce the diffusion rate of O atoms in Nbss and improve the compactness and stability of the oxide layer. In the Nb-Si-based alloys, an oxidation-resistant Nb_2Cr phase will be formed if enough Cr is added. With the increase in Cr content, the volume fraction of Nb_2Cr and Nb_5Si_3 increases, while the volume fraction of Nbss decreases. $CrNbO_4$ oxides can strengthen the adhesion of scales and have better oxidation resistance than Nb_2O_5. The addition of Zr can refine the microstructure and promote the generation of a glassy SiO_2 oxide film, which is beneficial to improve the compactness and integrity of the oxide scales and enhance the anti-oxidation performance of the alloy. The B elements can refine the grain, and the dissolution of B and Ge in the oxide layer will facilitate the formation of continuous SiO_2 film and inhibit the diffusion of O_2. In addition, B can also alleviate the problem of mismatch between the TEC of the oxide layer and substrate, enhance the adhesion of oxide scales, and reduce the cracks and spalling. The effect of V addition on the oxidation properties of Nb-Si-based alloy is still controversial. Some studies suggest that V will form volatile V_2O_5 in the alloy, which is not conducive to the formation of the protective oxide scales and deteriorates the oxidation resistance of the alloy. It is also considered that V has a stronger O affinity than Nb, which improves the oxidation characteristics of Nbss, reduces the oxidation rate difference between Nbss and silicide, and makes the oxide layer more compact and complete, thus promoting the oxide resistance of the alloy.

To improve the high-temperature oxidation resistance of metal by alloying, it is necessary to accurately control the content of modified elements. The oxidation resistance of the alloy cannot be improved effectively if the content of modified elements is insufficient, while excessive modified elements will reduce the mechanical properties such as the fracture toughness and high-temperature creep property of the substrates. Alloying makes it difficult to keep the balance between high-temperature oxidation resistance and mechanical properties. In recent years, the rapid development of metal surface coating technology can better solve this problem; high-temperature resistant coating is formed on the substrate surface of a layer of heat insulation or the oxygen isolation protective layer, with a certain fluidity and thickness. Among them, silicide ceramic high-temperature-resistant coating is the most widely used, its preparation process is relatively simple and low cost, and it has excellent high temperature resistance, giving it broad potential as a high-temperature coating. In addition, oxide coating, precious metal coating, composite coating, etc. also have their own advantages. High-temperature-resistant coating technology greatly makes up for the deficiency of the high-temperature oxidation resistance of metal substrates, and it is an excellent improved process, which is worthy of further study in the field of high-temperature-resistant materials.

Author Contributions: The manuscript was written through the contributions of all authors. Y.Z.: conceptualization, investigation, and supervision. Y.Z. and F.S.: writing—original draft and image processing. F.S., L.Y. and T.F.: validation, resources, investigation, writing—review and editing. K.C., H.W. and J.W.: visualization, writing—review and editing. All authors have read and agreed to the published version of the manuscript.

Funding: This work was supported by the Anhui Province Science Foundation for Excellent Young Scholars (Grant No. 2108085Y19) and National Natural Science Foundation of China (Grant No. 51604049).

Institutional Review Board Statement: Not applicable.

Informed Consent Statement: Not applicable.

Data Availability Statement: Data sharing is not applicable to this article.

Conflicts of Interest: The authors declare no conflict of interest.

References

1. Tsakiropoulos, P. Alloys for application at ultra-high temperatures: Nb-silicide in situ composites challenges, breakthroughs and opportunities. *Prog. Mater. Sci.* **2022**, *123*, 100714. [CrossRef]
2. Cheng, J.C.; Yi, S.H.; Park, J.S. Oxidation behaviors of Nb-Si-B ternary alloys at 1100 °C under ambient atmosphere. *Intermetallics* **2012**, *23*, 12–19. [CrossRef]
3. Murakamia, T.; Sasakia, S.; Ichikawaa, K.; Kitahara, A. Oxidation resistance of powder compacts of the Nb-Si-Cr system and $Nb_3Si_5Al_2$ substrate compacts prepared by spark plasma sintering. *Intermetallics* **2001**, *9*, 629–635. [CrossRef]
4. Al Zoubi, W.; Kamil, M.P.; Ko, Y.G. Synergistic influence of inorganic oxides (ZrO_2 and SiO_2) with N_2H_4 to protect composite coatings obtained via plasma electrolyte oxidation on Mg alloy. *Phys. Chem. Chem. Phys.* **2017**, *19*, 2372–2382. [CrossRef] [PubMed]
5. Li, Z.F.; Tsakiropoulos, P. Study of the effects of Ge addition on the microstructure of Nb-18Si in situ composites. *Intermetallics* **2010**, *18*, 1072–1078.
6. Sala, K.; MitraK, R. Microstructural evolution and mechanical properties of as-cast and annealed Nb-Si-Mo based hypoeutectic alloys with quaternary additions of Ti or Fe. *Mater. Sci. Eng. A* **2021**, *802*, 140663. [CrossRef]
7. Maji, P.; Mitra, R.; Ray, K. Effect of Cr on the evolution of microstructures in as-cast ternary niobium-silicide-based composites. *Intermetallics* **2017**, *85*, 34–47. [CrossRef]
8. Zoubi, W.A.; Ko, Y.G. Chemical stability of synergistic inorganic materials for enhancing electrochemical performance. *Compos. Sci. Technol.* **2020**, *199*, 108383. [CrossRef]

9. Wang, J.; Zhang, Y.; Cui, K.; Fu, T.; Gao, J.; Hussain, S.; AlGarni, T.S. Pyrometallurgical recovery of zinc and valuable metals from electric arc furnace dust—A review. *J. Clean. Prod.* **2021**, *298*, 126788. [CrossRef]
10. Liu, W.; Ren, X.Y.; Li, N.; Gao, C.; Xiong, H.P. Rapid directionally solidified microstructure characteristic and fracture be-haviour of laser melting deposited Nb-Si-Ti alloy. *Prog. Nat. Sci. Mater. Int.* **2021**, *31*, 113–120. [CrossRef]
11. Zhang, S.; Liu, W.; Sha, J. Microstructural evolution and mechanical properties of Nb-Si-Cr ternary alloys with a tri-phase Nb/Nb_5Si_3/Cr_2Nb microstructure fabricated by spark plasma sintering. *Prog. Nat. Sci.* **2018**, *28*, 626–634. [CrossRef]
12. Guo, Y.; Li, Z.; He, J.; Su, H.; Jia, L.; Zhang, J.; Liu, L.; Zhang, H. Surface microstructure modification of hypereutectic Nb-Si based alloys to improve oxidation resistance without damaging fracture toughness. *Mater. Charact.* **2020**, *159*, 110051. [CrossRef]
13. Al Zoubi, W.; Kamil, M.P.; Fatimah, S.; Nashrah, N.; Ko, Y.G. Recent advances in hybrid organic-inorganic materials with spatial architecture for state-of-the-art applications. *Prog. Mater. Sci.* **2020**, *112*, 100663. [CrossRef]
14. Shen, F.Q.; Yu, L.H.; Fu, T.; Zhang, Y.Y.; Wang, H.; Cui, K.K.; Wang, J.; Hussain, S.; Akhtar, N. Effect of the Al, Cr and B elements on the mechanical properties and oxidation resistance of Nb-Si based alloys: A review. *Appl. Phys. A Mater. Sci. Process.* **2021**, *127*, 852. [CrossRef]
15. Jiang, W.; Shao, W.; Sha, J.; Zhou, C. Experimental studies and modeling for the transition from internal to external oxidation of three-phase Nb-Si-Cr alloys. *Prog. Nat. Sci.* **2018**, *28*, 740–748. [CrossRef]
16. Ma, R.; Guo, X. Effects of Mo and Zr composite additions on the microstructure, mechanical properties and oxidation resistance of multi-elemental Nb-Si based ultrahigh temperature alloys. *J. Alloys Compd.* **2021**, *870*, 159437. [CrossRef]
17. He, J.-H.; Guo, X.-P.; Qiao, Y.-Q. Oxidation and hot corrosion behaviors of Nb−Si based ultrahigh temperature alloys at 900 °C. *Trans. Nonferrous Met. Soc. China* **2021**, *31*, 207–221. [CrossRef]
18. Yu, L.H.; Shen, F.Q.; Fu, T.; Zhang, Y.Y.; Cui, K.K.; Wang, J.; Zhang, X. Microstructure and oxidation behavior of metal-modified mo-si-b alloys: A review. *Coatings* **2021**, *11*, 1256. [CrossRef]
19. Zhang, G.P.; Sun, J.; F, Q.G. Effect of mullite on the microstructure and oxidation behavior of thermalsprayed $MoSi_2$ coating at 1500 °C. *Ceram. Int.* **2020**, *46*, 10058–10066. [CrossRef]
20. Liu, L.; Zhang, H.Q.; Lei, H.; Li, H.Q.; Gong, J.; Sun, C. Influence of different coating structures on the oxidation resistance of $MoSi_2$ coatings. *Ceram. Int.* **2020**, *46*, 5993–5997. [CrossRef]
21. Fu, T.; Cui, K.; Zhang, Y.; Wang, J.; Zhang, X.; Shen, F.; Yu, L.; Mao, H. Microstructure and oxidation behavior of anti-oxidation coatings on Mo-based alloys through hapc process: A review. *Coatings.* **2021**, *11*, 883. [CrossRef]
22. Zhang, Y.Y.; Yu, L.H.; Fu, T.; Wang, J.; Shen, F.Q.; Cui, K.K. Microstructure evolution and growth mechanism of Si-$MoSi_2$ composite coatings on TZM (Mo-0.5Ti-0.1Zr-0.02 C) alloy. *J. Alloys Compd.* **2022**, *894*, 162403. [CrossRef]
23. Zhang, H.; Lin, X.; Yu, Y.; Mei, F.; Zou, L.; Gao, J. Positive modification on the mechanical, tribological and oxidation properties of AlCrNbSiN coatings by regulating the Nb/Si-doping ratio. *Ceram. Int.* **2021**, *47*, 31603–31616. [CrossRef]
24. Zhang, Y.Y.; Cui, K.K.; Gao, Q.J.; Hussain, S.; Lv, Y. Investigation of morphology and texture pr1operties of WSi_2 coatings on W substrate based on contact-mode AFM and EBSD. *Surf. Coat. Technol.* **2020**, *396*, 125966. [CrossRef]
25. Fu, T.; Cui, K.; Zhang, Y.; Wang, J.; Shen, F.; Yu, L.; Qie, J.; Zhang, X. Oxidation protection of tungsten alloys for nuclear fusion applications: A comprehensive review. *J. Alloys Compd.* **2021**, *884*, 161057. [CrossRef]
26. Tang, Y.; Guo, X. High temperature deformation behavior of an optimized Nb–Si based ultrahigh temperature alloy. *Scr. Mater.* **2016**, *116*, 16–20. [CrossRef]
27. Cui, K.K.; Fu, T.; Zhang, Y.; Wang, J.; Mao, H.B.; Tan, T.B. Microstructure and mechanical properties of $CaAl_{12}O_{19}$ rein-forced Al_2O_3-Cr_2O_3 composites. *J. Eur. Ceram. Soc.* **2021**, *41*, 7935–7945. [CrossRef]
28. Zhang, Y.; Fu, T.; Cui, K.; Shen, F.; Wang, J.; Yu, L.; Mao, H. Evolution of surface morphology, roughness and texture of tungsten disilicide coatings on tungsten substrate. *Vacuum* **2021**, *191*, 110297. [CrossRef]
29. Zhang, Y.Y.; Ni, W.J.; Li, Y.G. Effect of siliconizing temperature on microstructure and phase constitution of Mo-$MoSi_2$ functionally graded materials. *Ceram. Int.* **2018**, *44*, 11166–11171. [CrossRef]
30. Yu, L.; Zhang, Y.; Fu, T.; Wang, J.; Cui, K.; Shen, F. Rare earth elements enhanced the oxidation resistance of Mo-Si-based alloys for high temperature application: A review. *Coatings* **2021**, *11*, 1144. [CrossRef]
31. Jiang, W.; Li, M.; Sha, J.; Zhou, C. Microstructure and oxidation resistance of composition gradients Nb-Si based alloy thin film. *Mater. Des.* **2020**, *192*, 108687. [CrossRef]
32. Zhang, X.; Fu, T.; Cui, K.; Zhang, Y.; Shen, F.; Wang, J.; Yu, L.; Mao, H. The protection, challenge, and prospect of anti-oxidation coating on the surface of niobium alloy. *Coatings* **2021**, *11*, 742. [CrossRef]
33. Zhang, Y.Y.; Qie, J.M.; Cui, K.K.; Fu, T.; Fan, X.L.; Wang, J.; Zhang, X. Effect of hot dip silicon-plating temperature on microstructure characteristics of silicide coating on tungsten substrate. *Ceram. Int.* **2020**, *46*, 5223–5228. [CrossRef]
34. Zhang, Y.; Cui, K.; Fu, T.; Wang, J.; Qie, J.; Zhang, X. Synthesis WSi_2 coating on W substrate by HDS method with various deposition times. *Appl. Surf. Sci.* **2020**, *511*, 145551. [CrossRef]
35. Pu, D.; Pan, Y. Influence of high pressure on the structure, hardness and brittle-to-ductile transition of $NbSi_2$ ceramics. *Ceram. Int.* **2021**, *47*, 2311–2318. [CrossRef]
36. Zhang, Y.Y.; Zhao, Z.; Li, J.H.; Lei, J.; Cheng, X.K. Effect of hot-dip siliconizing time on phase composition and microstructure of Mo-$MoSi_2$ high temperature structural materials. *Ceram. Int.* **2019**, *45*, 5588–5593. [CrossRef]
37. Cui, K.K.; Zhang, Y.Y.; Fu, T.; Wang, J.; Zhang, X. Toughening mechanism of mullite substrate composites: A review. *Coatings* **2020**, *10*, 672. [CrossRef]

38. Geng, J.; Tsakiropoulos, P.; Shao, G.S. Oxidation of Nb-Si-Cr-Al in situ composites with Mo, Ti and Hf additions. *Mater. Sci. Eng. A* **2006**, *441*, 26–38. [CrossRef]
39. Geng, J.; Tsakiropoulos, P. A study of the microstructures and oxidation of Nb-Si-Cr-Al-Mo in situ composites alloyed with Ti, Hf and Sn. *Intermetallics* **2007**, *15*, 382–395. [CrossRef]
40. Vazquez, A.; Varma, S.K. High-temperature oxidation behavior of Nb-Si-Cr alloys with Hf additions. *J. Alloys Compd.* **2011**, *509*, 7027–7033. [CrossRef]
41. Zhang, S.; Guo, X. Alloying effects on the microstructure and properties of Nb–Si based ultrahigh temperature alloys. *Intermetallics* **2016**, *70*, 33–44. [CrossRef]
42. Zhang, S.; Guo, X.P. Microstructure, mechanical properties and oxidation resistance of Nb silicide based ultrahigh temperature alloys with Hf addition. *Mater. Sci. Eng. A* **2015**, *645*, 88–98. [CrossRef]
43. Cui, K.K.; Zhang, Y.Y.; Fu, T.; Hussain, S.; AlGarni, T.S.; Wang, J.; Zhang, X.; Ali, S. Effects of Cr_2O_3 content on microstructure and mechanical properties of Al_2O_3 substrate composites. *Coatings* **2021**, *11*, 234. [CrossRef]
44. Pan, Y. Structural Prediction and overall performances of $CrSi_2$ disilicides: DFT investigations. *ACS Sustain. Chem. Eng.* **2020**, *8*, 11024–11030. [CrossRef]
45. Chan, K.S. Cyclic oxidation response of multiphase niobium-based alloys. *Met. Mater. Trans. A* **2004**, *35*, 589–597. [CrossRef]
46. Wang, L.G.; Jia, L.N.; Cui, R.J.; Zheng, L.L.; Zhang, H. Microstructure, mechanical properties and oxidation resistance of Nb-22Ti-14Si-2Hf-2Al-xCr alloys. *Chin. J. Aeronaut.* **2012**, *25*, 292–296. [CrossRef]
47. Zelenitsas, K.; Tsakiropoulos, P. Study of the role of Al and Cr additions in the microstructureof Nb-Ti-Si in situ composites. *Intermetallics* **2005**, *13*, 1079–1095. [CrossRef]
48. Zelenitsas, K.; Tsakiropoulos, P. Effect of Al, Cr and Ta additions on the oxidation behaviour of Nb-Ti-Si in situ composites at 800 °C. *Mater. Sci. Eng. A* **2006**, *416*, 269–280. [CrossRef]
49. Esparza, N.; Rangel, V.; Gutierrez, A.; Arellano, B.; Varma, S.K. A comparison of the effect of Cr and Al additions on the oxidation behaviour of alloys from the Nb–Cr–Si system. *Mater. High Temp.* **2016**, *33*, 105–114. [CrossRef]
50. Zhang, S.; Guo, X. Effects of Cr and Hf additions on the microstructure and properties of Nb silicide based ultrahigh temperature alloys. *Mater. Sci. Eng. A* **2015**, *638*, 121–131. [CrossRef]
51. Pan, Y.; Pu, D.; Yu, E. Structural, electronic, mechanical and thermodynamic properties of Cr–Si binary silicides from first-principles investigations. *Vacuum* **2021**, *185*, 110024. [CrossRef]
52. Su, L.; Jia, L.; Jiang, K.; Zhang, H. The oxidation behavior of high Cr and Al containing Nb-Si-Ti-Hf-Al-Cr alloys at 1200 and 1250 °C. *Int. J. Refract. Met. Hard Mater.* **2017**, *69*, 131–137. [CrossRef]
53. Tsai, Y.L.; Wang, S.F.; Bor, H.Y.; Hsu, Y.F. Effects of Zr addition on the microstructure and mechanical behavior of a fine-grained nickel-based superalloy at elevated temperatures. *Mater. Sci. Eng. A* **2014**, *607*, 294–301. [CrossRef]
54. Luo, C.; Lu, N.; Zhu, C.L.; Li, H.Z.; Liu, X.Q. Effect of trace zirconium addition on high temperature mechanical properties of casting TiAl alloy. *Foundry* **2012**, *61*, 754–757.
55. Swadźba, R.; Swadźba, L.; Wiedermann, J.; Hetmańczyk, M.; Witala, B. Characterization of alumina scales grown on a 2nd generation single crystal Ni superalloy during isothermal oxidation at 1050, 1100 and 1150 °C. *Oxid. Met.* **2014**, *82*, 195–208. [CrossRef]
56. Hong, S.; Hwang, G.; Han, W.; Lee, K.; Kang, S. Effect of zirconium addition on cyclic oxidation behavior of platinum-modified aluminide coating on nickel-based superalloy. *Intermetallics* **2010**, *18*, 864–870. [CrossRef]
57. Zhang, S.-N.; Jia, L.-N.; Guo, Y.; Kong, B.; Zhang, F.-X.; Zhang, H. High-temperature oxidation behavior of Nb–Si-based alloy with separate vanadium, tantalum, tungsten and zirconium addition. *Rare Met.* **2021**, *40*, 607–615. [CrossRef]
58. Zhang, S.; Jia, L.; Guo, Y.; Kong, B.; Zhou, C.; Zhang, H. Improvement in the oxidation resistance of Nb-Si-Ti based alloys containing zirconium. *Corros. Sci.* **2020**, *163*, 108294. [CrossRef]
59. Qiao, Y.; Guo, X.; Zeng, Y. Study of the effects of Zr addition on the microstructure and properties of Nb-Ti-Si based ultrahigh temperature alloys. *Intermetallics* **2017**, *88*, 19–27. [CrossRef]
60. Ma, R.; Guo, X. Composite alloying effects of V and Zr on the microstructures and properties of multi-elemental Nb–Si based ultrahigh temperature alloys. *Mater. Sci. Eng. A* **2021**, *813*, 141175. [CrossRef]
61. Behrani, V.; Thom, A.J.; Kramer, M.J.; Akinc, M. Microstructure and oxidation behavior of Nb-Mo-Si-B alloys. *Intermetallics* **2006**, *14*, 24–32. [CrossRef]
62. Zhang, Y.Y.; Cui, K.K.; Fu, T.; Wang, J.; Shen, F.Q.; Zhang, X.; Yu, L.H. Formation of $MoSi_2$ and $Si/MoSi_2$ coatings on TZM (Mo-0.5Ti-0.1Zr-0.02C) alloy by hot dip silicon-plating method. *Ceram. Int.* **2021**, *47*, 23053–23065. [CrossRef]
63. Wu, M.L.; Jiang, L.W.; Qu, S.Y.; Guo, F.W.; Li, M.; Kang, Y.W.; Han, Y.F. Effect of trace Ce and B additions on the microstructure of Nb-3Si-22Ti alloys. *Prog. Nat. Sci. Mater. Int.* **2017**, *27*, 362–368. [CrossRef]
64. Zhang, Y.Y.; Li, Y.G.; Bai, C.G. Microstructure and oxidation behavior of $Si-MoSi_2$ functionally graded coating on Mo substrate. *Ceram. Int.* **2017**, *43*, 6250–6256. [CrossRef]
65. Zhang, S.; Guo, X. Effects of B addition on the microstructure and properties of Nb silicide based ultrahigh temperature alloys. *Intermetallics* **2015**, *57*, 83–92. [CrossRef]
66. Sun, Z.; Guo, X.; Guo, B. Effect of B and Ti on the directionally solidified microstructure of the Nb–Si alloys. *Int. J. Refract. Met. Hard Mater.* **2015**, *51*, 243–249. [CrossRef]
67. Thomas, K.S.; Varma, S.K. Oxidation response of three Nb-Cr-Mo-Si-B alloys in air. *Corros. Sci.* **2015**, *99*, 145–153. [CrossRef]

68. Su, L.F.; Jia, L.N.; Weng, J.F.; Hong, Z.; Zhou, C.G.; Zhang, H. Improvement in the oxidation resistance of Nb-Ti-Si-Cr-Al-Hf alloys containing alloyed Ge and B. *Corros. Sci.* **2014**, *88*, 460–465. [CrossRef]
69. Guo, Y.L.; Jia, L.N.; Kong, B.; Zhang, H.R.; Zhang, H. Simultaneous improvement in fracture toughness and oxidation re-sistance of Nb-Si based alloys by vanadium addition. *Mater. Sci. Eng. A* **2017**, *701*, 149–157. [CrossRef]
70. Kang, Y.W.; Qu, S.Y.; Song, J.X.; Huang, Q.; Han, Y.F. Microstructure and mechanical properties of Nb-Ti-Si-Al-Hf-xCr-yV multi-element in situ composite. *Mater. Sci. Eng. A* **2012**, *534*, 323–328. [CrossRef]
71. Zheng, J.S.; Hou, X.M.; Wang, X.B.; Meng, Y.; Zheng, X.; Zheng, L. Isothermal oxidation mechanism of Nb-Ti-V-Al-Zr alloy at 700–1200 °C Diffusion and interface reaction. *Corros. Sci.* **2015**, *96*, 186–195. [CrossRef]
72. Gang, F.; Klinski-Wetzel, K.V.; Wagner, J.N.; Heilmaieret, M. Influence of vanadium on the oxidation resistance of the intermetallic phase Nb_5Si_3. *Oxid. Met.* **2015**, *83*, 119–132. [CrossRef]
73. Xu, C.; Gao, W. Pilling-Bedworth ratio for oxidation of alloys. *Mater. Res. Innov.* **2000**, *3*, 231–235. [CrossRef]
74. Dasary, R.M.; Varma, S.K. Short-term oxidation response of Nb-15Re-15Si-10Cr-20Mo alloy. *J. Mater. Res. Technol.* **2014**, *3*, 25–34. [CrossRef]
75. Ma, R.; Guo, X. Effects of V addition on the microstructure and properties of multi-elemental Nb-Si based ultrahigh temperature alloys. *J. Alloys Compd.* **2020**, *845*, 156254. [CrossRef]
76. Murayama, Y.; Hanada, S. High temperature strength, fracture toughness and oxidation resistance of Nb-Si-Al-Ti multi-phase alloys. *Sci. Technol. Adv. Mater.* **2002**, *3*, 145–156. [CrossRef]
77. Chattopadhyay, K.; Balachandran, G.; Mitra, R.; Ray, K. Effect of Mo on microstructure and mechanical behaviour of as-cast Nbss–Nb5Si3 in situ composites. *Intermetallics* **2006**, *14*, 1452–1460. [CrossRef]
78. Ma, R.; Guo, X. Influence of molybdenum contents on the microstructure, mechanical properties and oxidation behavior of multi-elemental Nb–Si based ultrahigh temperature alloys. *Intermetallics* **2021**, *129*, 107053. [CrossRef]
79. Guo, Y.; Jia, L.; Zhang, H.; Zhang, F.; Zhang, H. Enhancing the oxidation resistance of Nb-Si based alloys by yttrium addition. *Intermetallics* **2018**, *101*, 165–172. [CrossRef]
80. Wang, Y.; Jia, L.; Sun, G.; Zhang, F.; Ye, C.; Zhang, H. Synchronous improvement in room-temperature fracture toughness and high-temperature oxidation resistance of Nb Si based alloys with Erbium addition. *Int. J. Refract. Met. Hard Mater.* **2021**, *94*, 105359. [CrossRef]
81. Liu, A.Q.; Sun, L.; Li, S.S.; Han, Y.F. Effect of cerium on micmstiuctms and high temperature oxidation resistance of an Nb-Si system in-situ composite. *J. Rare Earth* **2007**, *25*, 474–479. [CrossRef]
82. Guo, Y.; Jia, L.; Kong, B.; Zhang, F.; Liu, J.; Zhang, H. Improvement in the oxidation resistance of Nb-Si based alloy by selective laser melting. *Corros. Sci.* **2017**, *127*, 260–269. [CrossRef]
83. Guo, Y.L.; Jia, L.N.; Sun, S.B.; Kong, B.; Liu, J.H.; Zhang, H. Rapid fabrication of Nb-Si based alloy by selective laser melting: Microstructure, hardness and initial oxidation behavior. *Mater. Des.* **2016**, *109*, 37–46. [CrossRef]
84. Mathieu, S.; Knittel, S.; Berthod, P.; Vilasi, M. On the oxidation mechanism of niobium-base in situ composites. *Corros. Sci.* **2012**, *60*, 181–192. [CrossRef]
85. Alvarez, D.; Varma, S.K. Characterization of microstructures and oxidation behavior of Nb-20Si-20Cr-5Al alloy. *Corros. Sci.* **2011**, *53*, 2161–2167. [CrossRef]

Article

ZnO Nano-Flowers Assembled on Carbon Fiber Textile for High-Performance Supercapacitor's Electrode

Qasim Abbas [1], Muhammad Sufyan Javed [2,3,*], Awais Ahmad [4], Sajid Hussain Siyal [5,*], Idrees Asim [6,7], Rafael Luque [4,*], Munirah D. Albaqami [8] and Ammar Mohamed Tighezza [8]

1. Department of Intelligent Manufacturing, Yibin University, Yibin 644000, China; qsandhu@yibinu.edu.cn
2. School of Physical Science and Technology, Lanzhou University, Lanzhou 730000, China
3. Department of Physics, COMSATS University Islamabad, Lahore Campus, Islamabad 54000, Pakistan
4. Departamento de Quimica Organica, Universidad de Cordoba, Edificio Marie Curie (C-3), Ctra Nnal IV-A, Km 396, E14014 Cordoba, Spain; awaisahmed@gcuf.edu.pk
5. Metallurgy & Materials Engineering Department, Dawood University of Engineering and Technology, Karachi 74800, Pakistan
6. Key Laboratory of Materials Modification by Laser, Electron, and Ion-Beams, Dalian University of Technology, Dalian 116024, China; asim.idrees44@gmail.com
7. Department of Applied Sciences, National Textile University, Faisalabad 37610, Pakistan
8. Department of Chemistry, College of Science, King Saud University, Riyadh 11451, Saudi Arabia; muneerad@ksu.edu.sa (M.D.A.); ammar@ksu.edu.sa (A.M.T.)
* Correspondence: safisabri@gmail.com (M.S.J.); sajid.hussain@duet.edu.pk (S.H.S.); q62alsor@uco.es (R.L.)

Citation: Abbas, Q.; Javed, M.S.; Ahmad, A.; Siyal, S.H.; Asim, I.; Luque, R.; Albaqami, M.D.; Tighezza, A.M. ZnO Nano-Flowers Assembled on Carbon Fiber Textile for High-Performance Supercapacitor's Electrode. *Coatings* **2021**, *11*, 1337. https://doi.org/10.3390/coatings11111337

Academic Editor: Emerson Coy

Received: 3 October 2021
Accepted: 25 October 2021
Published: 30 October 2021

Publisher's Note: MDPI stays neutral with regard to jurisdictional claims in published maps and institutional affiliations.

Copyright: © 2021 by the authors. Licensee MDPI, Basel, Switzerland. This article is an open access article distributed under the terms and conditions of the Creative Commons Attribution (CC BY) license (https://creativecommons.org/licenses/by/4.0/).

Abstract: Herein, a crystalline nano-flowers structured zinc oxide (ZnO) was directly grown on carbon fiber textile (CFT) substrate via a simple hydrothermal process and fabricated with a binder-free electrode (denoted as ZnO@CFT) for supercapacitor (SC) utilization. The ZnO@CFT electrode revealed a 201 $F·g^{-1}$ specific capacitance at 1 $A·g^{-1}$ with admirable stability of >90% maintained after 3000 cycles at 10 $A·g^{-1}$. These impressive findings are responsible for the exceedingly open channels for well-organized and efficient diffusion of effective electrolytic conduction via ZnO and CFT. Consequently, accurate and consistent structural and morphological manufacturing engineering is well regarded when increasing electrode materials' effective surface area and intrinsic electrical conduction capability. The crystalline structure of ZnO nano-flowers could pave the way for low-cost supercapacitors.

Keywords: ZnO; nanoflowers; carbon-fiber-cloth; electrode; supercapacitors

1. Introduction

Recently, significant efforts have been made to develop electrochemically efficient-energy storing and converting devices such as batteries, supercapacitors (SCs), and fuel cells [1]. SCs have piqued the curiosity of the scientific community owing to their high energy density, better and longer cyclic life, commercial repairs, and the fact that they a higher energy density than conventional capacitors. The conductivity of the electrode material affects the performance of the SCs [2,3]. As a result, numerous nano-sized oxides are used to introduce a variety of SC electrodes containing more active sites, thereby increasing the electrode's electronic and ionic conductivity [4,5]. A number of studies have shown that transition metal oxides, such as RuO_2 [6,7], MnO_2 [8], Fe_3O_4 [9], and ZnO [10], in their respective oxidation states have been studied extensively for SC applications. Furthermore, these oxides have been demonstrated to self-assemble in a variety of nano-structured morphological features such as nanowires, nanosheets, nanorods, etc. The nanocomposite materials also considered as good for SCs such as Mn@Fe [11] and carbon coated metal oxides due to the hybrid properties of different materials [12,13]. Electrodes with all these morphological features may have considerably enhanced electrocatalytic activity for SCs due to fairly short diffusion paths, faster charge carrier transfer, and a greater number of electroactive sites [14,15].

Zn-based electrodes have been the subject of several studies as possible SC electrode material [16,17]. ZnO has the shortest energy band-gap of 3.37 eV and has recently attracted a lot of attention for its low cost and moderate electrochemical properties. For example, Chen et al. evidenced that nanorods of ZnO with MnO_2 coating had a good capacitance of 222 $F \cdot g^{-1}$ at 25 $mV \cdot s^{-1}$ with excellent cycling stability > 97.5% after 1200 cycles [18]. Gao et al. had testified that nanosheets of ZnO with graphene coating revealed a low 62.2 $F \cdot g^{-1}$ specific capacitance at 0.5 $A \cdot g^{-1}$; after 200 cycles, the capacitance decreased by only 5.1%, demonstrating good cycling stability [19]. Another study found that nanocomposites of ZnO/graphene which were synthesized had a 146 $F \cdot g^{-1}$ specific capacitance at 100 $mV \cdot s^{-1}$, showing long-term recycling performance [20]. Zhang et al. reported a greater 323.9 $F \cdot g^{-1}$ specific capacitance at 50 $mV \cdot s^{-1}$ for carbon nanotube-ZnO nanocomposites electrodes and showed that they exhibited good cyclic stability (83%) after 100 cycles [21]. Jayalakshmi et al. recounted a 21.7 $F \cdot g^{-1}$ specific capacitance at 50 $mV \cdot s^{-1}$ with better cyclic stability (of almost 100%) after 500 cycles for a composite containing ZnO and carbon [22]. It is evident that each of the aforementioned works was primarily concerned with ZnO-based composite materials for the electrode of SCs. However, since less importance is given to using ZnO as a pristine electrode material, there seems to be a deficiency in studies of ZnO applications for SCs. He et al. revealed that the ZnO-nanocones structure found by chemical fabrication exhibited 378.5 $F \cdot g^{-1}$ at 20 $mV \cdot s^{-1}$ with a cyclic stability of 65% completing 500 cycles and ZnO nanowires exhibited a capacitance of 191.5 $F \cdot g^{-1}$ at 20 $mV \cdot s^{-1}$ [23]. In another study, ZnO particles exhibited good specific capacitances prepared hydrothermally with nitrate (5.87 $F \cdot g^{-1}$), acetate (5.35 $F \cdot g^{-1}$), and chloride (4.14 $F \cdot g^{-1}$) specific capacitance at a 5 $mV \cdot s^{-1}$. Moreover, the electrodes' cyclic performances were tested for only 200 cycles [24]. According to the above findings, despite the fact that nanocomposites with zinc and ZnO pristine had also been observed to have similar capacitance values with other metal oxides although they are not very competitive [20,25]. Binders as well as additives could really assist in increasing the mechanical qualities of the SCs while also increasing electron and electrolyte ion resistance and lowering the device's power capability in commercial applications [26]. In another scenario, binder-free based electrodes could considerably improve conductivity and efficient ion diffusion routes while increasing electrode flexibility. Using a quick, easy, and cheap hydrothermal method, the electro-active electrodes can be directly grown with a range of electrically conductive bases such as nickel-foam, carbon steel foil, and carbon fabric [27–29]. Due to its superior conductivity, non-toxicity, and mechanical strength, carbon fiber textiles (CFT) are a particularly ideal base for direct creation of metallic oxide nano-structures for built-up SCs.

As a result, the growth of ZnO nano-structured via simple methods for the preparation is the key solution to these problems. To achieve this goal, nanostructures such as nano-wires, nano-rods, and nano-tubes could be directly grown on CFT or nickel-foam. Compared to powder electrodes coated using a standard liquid pasting technique, conductive surface-based electrodes offer good electrical interactions, assisting in the achievement of high electro-catalytic activity [30,31].

Herein, we reported the facile synthesis of ZnO-flower electrodes on CFT without the use of a binder which is suitable for use in solid-state SCs. Growing ZnO-flower on CFT with a post-annealing solvo-thermal procedure and a binder-free electrode is described in this paper as an easy but effective method. These qualities (such as having a low heating rate, and being cheap, fast, and easy) characterize the hydrothermal method. Studies on the ZnO-flower electrode's electrochemical properties revealed a high excellent specific capacitance of 201 $F \cdot g^{-1}$ at a current density of 1 and high stability of >90% capacitance maintained after 3000 cycles at 10 $A \cdot g^{-1}$. Growing ZnO-flower-structured using the method described in this paper could pave the way for low-cost SC electrodes that are free of the binder.

2. Experimental

2.1. The Preparation of ZnO Nano-Flowers on Carbon Fabric Cloth for a Binder-Free Electrode

All chemicals used for this work were bought from reputed chemicals suppliers named "Aladdin Chemicals suppliers and manufacturers", Shanghai, China. The carbon textile fiber (CFT), having a thickness of 0.2 mm with 1.8 g·cm^{-3} density, was purchased from a company known as "Shanghai Lishuo Composite Material Technology", Shanghai, China. The nano-flowers were directly grown on a carbon fiber textile (CFT) using the standard hydrothermal method. First, a solution mixture of 2 mmol of Zn(CH$_3$CO$_2$)$_2$, 4 mmol of NaOH, 2 mmol of cetrimonium bromide (CTAB) in 40 mL of deionized water. The prepared solution was then placed in an ultrasonicator for one hour to achieve a completely homogeneous mixture. Following that, this homogeneous ultrasonicated solution was then poured into a Teflon autoclave and filled to almost 70% capacity subsequently. A neat and clean CFT piece was put on the wall of the Teflon autoclave. The autoclave was perfectly sealed airtightly and placed inside an electronic oven at 100 °C for 10 h to initiate a simple hydrothermal process. After the hydrothermal process was completed, the CFT collected along with the ZnO ancestor was carefully washed several times with ethanol and deionized water and denoted as ZnO@CFT. The obtained ZnO@CFT material was ultra-sonicate continuously for 5 min to eliminate the remaining particles pated at the surface of CFT. The ZnO@CFT was then placed in a furnace set to 90 °C for complete dehydration. In conclusion, dehydrated ZnO@CFT was carbonized for 2 h at 350 °C to increase crystallinity. The experiment was repeated two times, and comparable results were achieved. The schematic representation of the fabrication of ZnO nano-flowers was given in Figure 1.

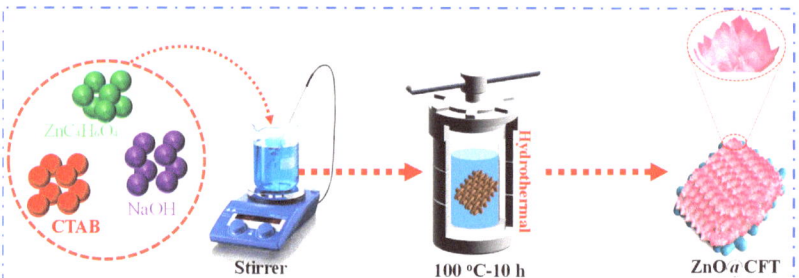

Figure 1. A schematic diagram shows the synthesis process used to develop ZnO nano-flowers directly on the surface of CFT.

2.2. ZnO@CFT Electrode Fabrication

The synthesized ZnO@CFT was cut into 1 × 1 cm^2 small pieces to use as a freestanding electrode in 1 M electrolyte of KOH for evaluating the electrochemical activity as an electrode in SCs. The active mass loading density of prepared ZnO@CFT for electrodes had been 1.26 mg·cm^2, calculated meticulously using only the mass difference between pristine CFT and annealed ZnO@CFT.

2.3. Materials Characterization

An as-prepared ZnO@CFT morphology study utilizes field emission scanning electron microscopy (FEI Nova 400). When it came to XRD powder crystal composition, the PAN-analytical X–pert diffractometer used Cu–K radiation to characterize the ZnO@CFT samples thoroughly. In order to carry out X-ray photoelectron spectroscopy (XPS), an X-ray source (Escalab (250Xi)) Al Ka (1486.5 eV) was utilized.

2.4. Measurements of Electrochemistry

The supercapacitive activity of the as-prepared ZnO@CFT electrode was evaluated using GCD (galvanostatic charge-discharge), CV (cyclic voltammetry), and EIS characterization (electrochemical impedance spectroscopy). The three-electrode configurations (for reference, Ag/AgCl, platinum sheet, and ZnO@CFT were used as reference, counter, and working electrode at room temperature) were verified in a 1M (mol·L^{-1}) KOH electrolyte using an electrochemical workstation (CHI660E, Chen Hua, Shanghai, China). Using Equation (1), the specific capacitance (C) of ZnO@CFT inside the three-electrode configuration was determined by calculating [27,30].

$$C = I\Delta t/mv \tag{1}$$

wherein C, I, m, and t represent specific capacitance (C·g^{-1}), discharge current (A), effective mass (g) of the as-prepared material, and discharge time (s), respectively.

3. Results and Discussion

Figure 1 illustrates a primary hydrothermal method used to directly develop the nanoflowers-like ZnO@CFT structure on the conductive CFT. In the presence of NaOH, Zn atoms react with O atoms, and 2 mmol of CTAB initiate a nano-flowers growing at 100 °C. FE-SEM was used to look at the surface morphology of prepared nanoparticles of ZnO, which can be seen in Figure 2. The ZnO nanostructure was found to be uniformly distributed over the conductive CFT, which is shown in Figure 2a,b. This surface is crucial to getting good electrochemical behavior from the system. An additional benefit of ZnO nanoflowers is their better working mobility, as shown in Figure 2c by the three-dimensionally consistent toughened structural properties with ununiform void space. As can be seen in Figure 2d, this high-energy storage host has a unique morphological characteristic. Using a nano-structure made of ZnO@CFT nano-flowers, a highly porous nanostructure can be created that increases surface area while also improving the conductivity of electrons and electrolyte ions.

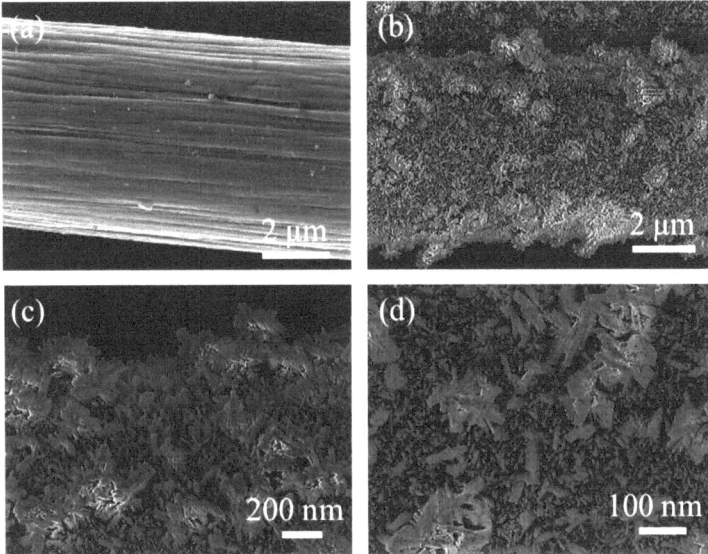

Figure 2. Morphological characteristics of ZnO@CFT electrode: (**a**) SEM image of pristine CFT and (**b**) a low-resolution FE-SEM image. (**c**) A high-resolution FE-SEM image and (**d**) a section from Figure 2c.

The XRD pattern of annealed ZnO is shown in Figure 3a, which explains its structural nature and purity. The ZnO nano-flowers diffraction peaks at 26.7°, 28.35°, 30.2°, 33.465°, 36.495°, 39.133°, 41.988°, 44.3°, 47.2°, 54.681°, 58.763°, and 62.8° could be freely indexed to the ZnO crystal planes (110), (112), (220), (222), (003), and (330), respectively. The diffraction pattern demonstrates the crystalline structure of the as-synthesized ZnO nano-flowers. There are no signs of Zn contamination, indicating that the nano-flowers are phase pure ZnO. The peak indicated with asterisk comes from CFT. Figure 3b depicts the three-dimensional crystallinity of a ZnO nano-flowers. The Zn and O atoms interconnected establish a bridge structure, which might assist quick ionic transport between the active material and the electrolyte [27,31]. ZnO is a slightly soft material with a hardness of almost 4.5 on the Mohs hardness scale. Its elastic properties are lower than those of comparable III-V semiconductor materials, such as GaN. The large specific heat, conductivity, low coefficient of thermal expansion, and high ZnO melting point are advantageous in energy storage applications.

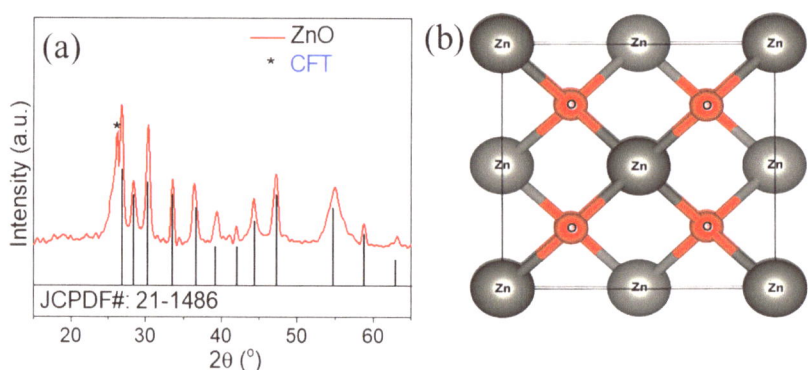

Figure 3. (a) XRD pattern of ZnO@CFT electrode and (b) crystal structure of ZnO@CF.

To better clarify these results, the oxidizing states and elemental composition of ZnO@CFT electrodes were also examined using XPS and fitted by Gaussian fitting process. The Zn 2p displayed two characteristic peaks, as shown in Figure 4a. The two diffraction peaks inside the fundamental Zn 2p spectral region at the binding affinity of 1021.6 and 1044.6 eV correlate to Zn $2p_{3/2}$ and Zn $2p_{1/2}$, respectively, with spin-orbit fracturing of 23 eV and develop a positive agreement with previous reports Zn^{2+} data [32]. The intensity peak at the least binding energy (530.8 eV) (Figure 4b) is associated with the development of metal-bonded reactive oxygen species, in which "bonded oxygen" relates to molecular oxygen formed a bond with Zn^{2+} (Zn-O) at a binding energy of 530.8 eV. The reaction kinetics redox potential is satisfied by the reduced oxygen atoms are bonded to Zn. The intensity peak (at 532.5 eV) is caused by the adsorption of loosely bonded oxygen atom "O" on the sample's surface from atmospheric humidity [33]. The XPS, as well as XRD outcomes, confirmed the ZnO nano-flowers formation on a CFT [34].

The specific surface morphology and porosity of the ZnO@CFT nano-flowers at 77 K were determined using isotherms of nitrogen adsorption-desorption. The amorphous sample takes a Brunauer–Emmett–Teller-specific surface area of approximately 89.65 $m^2 \cdot g^{-1}$. Accepting the IUPAC's classification system, the isotherms evidenced the hysteresis loop of type H1, indicating the presence of a highly porous channel [35], which can be seen in Figure 5a. The distribution of pore size, as measured by the BJH (Barrett–Joyner–Halenda) methodology from the desorption hysteresis loop, is displayed in Figure 5b. The pore size distribution of ZnO@CFT nano-flowers is largely centered at almost 6.5 nm, confirming the prepared sample's mesoporous structure. The large specific surface area improves the electrocatalytic performance and provides plenty of electro-active sites for an efficient supercapacitor [29].

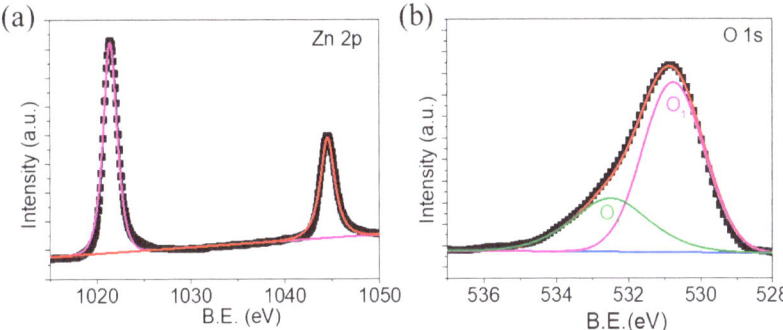

Figure 4. (**a**) Zn 2p and (**b**) O 1s core-level X-ray photoelectron spectra of ZnO nano-flowers.

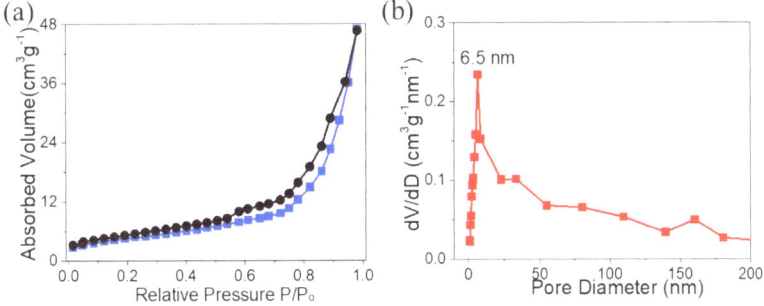

Figure 5. N$_2$ adsorption and desorption isotherms of ZnO nano-flowers (**a**) and pore-size distribution of ZnO nano-flowers (**b**).

The electronic properties such as CV and GCD of the ZnO@CFT electrode in the aqueous KOH electrolyte were investigated using a three-electrode cell system. Figure 6a depicts the characteristic of pristine CFT and CV curves of ZnO@CFT electrodes in a potential window of 0–0.8 V vs. Ag/AgCl at a 5 mV·s^{-1} scan rate. Figure 6b explains the ZnO nano-flowers electrode CV, in an aqueous KOH electrolyte, was measured at 5, 10, 20, 30, and 50 mV·s^{-1} scan rates in the enlarged potential window of 0 to +0.8 V vs. Ag/AgCl. Moreover, it was observed that increasing the scan rate value increases the intensity of the current, illustrating the ZnO nano-capacitance flower's behavior.

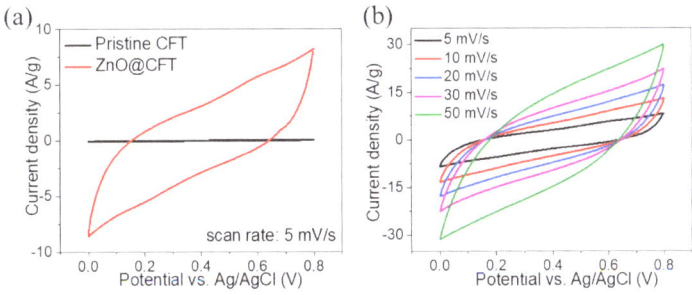

Figure 6. Electrochemical characterizations of ZnO@CFT electrode in an aqueous electrolyte of KOH: (**a**) ZnO@CFT and pristine CFT electrodes CV curves are compared (**b**) CV curves for ZnO@CFT electrodes.

The curves of GCD at current densities that vary from 1–10 $A·g^{-1}$ were also used to verify the charging/discharging characteristic features of ZnO nano-flowers, as shown in Figure 7a. The charge/discharge curve generated an excellent capacitive performance of 201.25 $F·g^{-1}$ at current density 1 $A·g^{-1}$ and 125 $F·g^{-1}$ at 10 $A·g^{-1}$ current density. A plot of specific capacitance against current density is shown in Figure 7b. Surprisingly, at a high discharge current density of 10 $A·g^{-1}$, the specific capacitance of the ZnO@CFT nano-flowers electrode is reduced to 125 $F·g^{-1}$, resulting in a good rate performance of 62%. At all current densities, all charge/discharge curves seem to be greatly symmetric, with such a Coulombic-efficiency exceeding 90%, indicating excellent reversible efficiency and agreeing very well CV evaluation. Specific capacitance values of the ZnO@CFT (measured to use discharging curves) were 201.25, 185, 150, and 140, 125 $F·g^{-1}$ at current densities of 1, 2, 5, 7, and 10 $A·g^{-1}$, respectively.

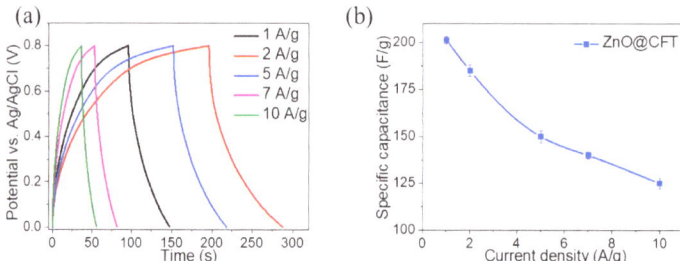

Figure 7. (**a**) Charge/discharge curves of ZnO@CFT electrodes measured against various current densities in the potential range of 0.0–0.8 V. (**b**) Specific capacitance of the ZnO@CFT electrode versus current density.

Electrochemical impedance spectroscopy (EIS) was used to investigate the electrochemical properties of the ZnO nano-flowers electrode surface. Typical Nyquist plots of ZnO@CFT nano-flowers electrodes are shown in Figure 8. Nyquist x-intercept curve represents the comparable series resistance (ESR) inside the region of high-frequency, which ultimately results from the source of impedance from multiple sources, including electrolyte, electrode material inductance, and resistance of interface [36]. The semicircle's diameter is proportional to the resistance of charge transmission on the surface of the electrode (R_{ct}) [37]. In the region of high-frequency, there is a semicircle, and in contrast, a region of low-frequency exhibited a quasi-vertical line and fitted with the electrical equivalent circuit. The inductance (Rs: 2.28 Ω) was determined by calculating the region's x-intercept of the high-frequency and the resistance of charge transmission (R_{ct}: 2.25 Ω), measured from the semicircle's diameter. The slope of the Nyquist plot keeps changing from lower to upper angles as frequency declines, exhibiting the charge-storage procedure. Additionally, EIS outcomes expose the combination of rapid electron transmission and extraordinary ion diffusion. One more key parameter for determining the possibility of Sc gadgets for industrial cases is their cycling stability.

Consequently, the cycling stability of the ZnO@CFT nano-flowers electrode was tested by deliberately running it consciously for 3000 cycles of charge/discharge at a 10 $A·g^{-1}$ elevated current density. The cycling attitude of the ZnO@CFT electrode was investigated for 3000 cycles to determine cyclic stability, as shown in Figure 9. Excluding a relatively small (under 10%) decline throughout specific capacitance after 3000 cycles, the lengthy cycling life of ZnO@CFT reveals no discernible fade. Such cyclic stability consequences for ZnO@CFT electrodes suggested that slight variations in the physical or chemical characteristics might take place during the charge/discharge cycling method. Subsequently, the ZnO@CFT electrode retains nearly 90.32% of its initial capacity after 3000 cycles, far better than earlier reported results on the same substance and other bimetal oxides. The findings suggest that the ZnO@CFT has a remarkable capacity to keep charge with superior stability and high efficiency during long cycling life inside the expanded potential window of 0.0–0.8 V.

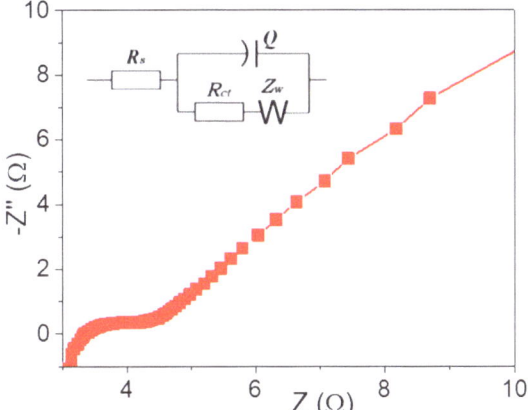

Figure 8. Nyquist plot of ZnO@CFT nano-flowers electrode.

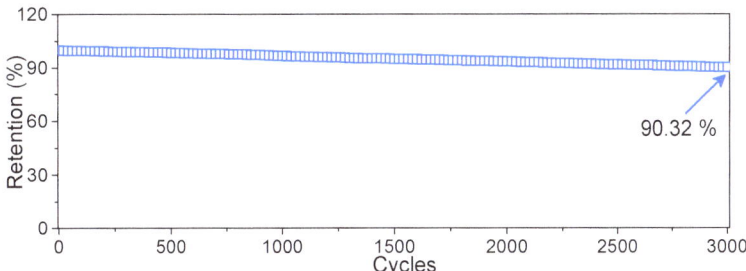

Figure 9. Capacitance retention vs. cycles number over 3000 cycles.

4. Conclusions

We designed a binder-free electrode of ZnO@CFT nano-flowers to improve the performance of SCs in the current study. A simple hydrothermal route is used to assemble ZnO nano-flowers on the surface of CFT directly. The as-prepared electrode ZnO@CFT demonstrates extraordinary electrochemical performance in a basic aqueous electrolyte, with a remarkable capacitance of 201 $F·g^{-1}$ at 1 $A·g^{-1}$ and the best retention at a great 10 $A·g^{-1}$ current density. More fascinatingly, after 3000 charge-discharge cycles, the ZnO@CFT electrode maintains excellent stability of 90.32%. Highly open electrolyte ion and electrons transmission channels in both ZnO and CFT are responsible for such great results. The increase of the area for active surface and improvement of the electrode material's underlying conductivity, accurate morphological characteristics, and surface manufacturing engineering are well regarded.

Author Contributions: Conceptualization, M.S.J. and Q.A.; methodology, A.A.; software, S.H.S.; validation, M.S.J., I.A. and R.L.; formal analysis, M.D.A.; investigation, R.L.; resources, A.M.T.; data curation, A.M.T.; writing—original draft preparation, A.A. and M.S.J.; writing—review and editing, R.L.; visualization, S.H.S.; supervision, R.L. All authors have read and agreed to the published version of the manuscript.

Funding: This work was funded by the Researchers Supporting Project Number (RSP-2021/267) King Saud University, Riyadh, Saudi Arabia.

Institutional Review Board Statement: Not applicable.

Informed Consent Statement: Not applicable.

Data Availability Statement: Not applicable.

Conflicts of Interest: The authors declare no conflict of interest.

References

1. Shaheen, I.; Ahmad, K.S.; Zequine, C.; Gupta, R.K.; Thomas, A.G.; Malik, M.A. Green synthesis of ZnO–Co$_3$O$_4$ nanocomposite using facile foliar fuel and investigation of its electro-chemical behaviour for supercapacitors. *New J. Chem.* **2020**, *44*, 18281–18292. [CrossRef]
2. Javed, M.S.; Khan, A.J.; Ahmad, A.; Siyal, S.H.; Akram, S.; Zhao, G.; Bahajjaj, A.A.A.; Ouladsmane, M.; Alfakeer, M. Design and fabrication of bimetallic oxide nanonest-like structure/carbon cloth composite electrode for supercapacitors. *Ceram. Int.* **2021**, *47*, 30747–30755. [CrossRef]
3. Syah, R.; Ahmad, A.; Davarpanah, A.; Elveny, M.; Ramdan, D.; Albaqami, M.D.; Ouladsmane, M. Incorporation of Bi$_2$O$_3$ residuals with metallic bi as high performance electrocatalyst toward hydrogen evolution reaction. *Catalysts* **2021**, *11*, 1099. [CrossRef]
4. Siyal, S.H.; Javed, M.S.; Ahmad, A.; Sajjad, M.; Batool, S.; Khan, A.J.; Akram, S.; Alothman, A.A.; Alshgari, R.A.; Najam, T. Free-standing 3D Co3O4@NF micro-flowers composed of porous ultra-long nanowires as an advanced cathode material for supercapacitor. *Curr. Appl. Phys.* **2021**, *31*, 221–227. [CrossRef]
5. Saleem, M.; Irfan, M.; Tabassum, S.; Albaqami, M.D.; Javed, M.S.; Hussain, S.; Pervaiz, M.; Ahmad, I.; Ahmad, A.; Zuber, M. Experimental and theoretical study of highly porous lignocellulose assisted metal oxide photoelectrodes for dye-sensitized solar cells. *Arab. J. Chem.* **2021**, *14*, 102937. [CrossRef]
6. Li, H.; Li, X.; Liang, J.; Chen, Y. Hydrous RuO 2 -decorated MXene coordinating with silver nanowire inks enabling fully printed micro-supercapacitors with extraordinary volumetric performance. *Adv. Energy Mater.* **2019**, *9*, 1803987. [CrossRef]
7. Asim, S.; Javed, M.S.; Hussain, S.; Rana, M.; Iram, F.; Lv, D.; Hashim, M.; Saleem, M.; Khalid, M.; Jawaria, R.; et al. RuO2 nanorods decorated CNTs grown carbon cloth as a free standing electrode for supercapacitor and lithium ion batteries. *Electrochim. Acta* **2019**, *326*, 135009. [CrossRef]
8. Javed, M.S.; Imran, M.; Assiri, M.A.; Hussain, I.; Hussain, S.; Siyal, S.H.; Saleem, M.; Shah, S.S.A. One-step synthesis of carbon incorporated 3D MnO2 nanorods as a highly efficient electrode material for pseudocapacitors. *Mater. Lett.* **2021**, *295*, 129838. [CrossRef]
9. Khan, A.J.; Khan, A.; Javed, M.S.; Arshad, M.; Asim, S.; Khalid, M.; Siyal, S.H.; Hussain, S.; Hanif, M.; Liu, Z. Surface assembly of Fe3O4 nanodiscs embedded in reduced graphene oxide as a high-performance negative electrode for supercapacitors. *Ceram. Int.* **2020**, *46*, 19499–19505. [CrossRef]
10. Samuel, E.; Londhe, P.U.; Joshi, B.; Kim, M.-W.; Kim, K.; Swihart, M.; Chaure, N.B.; Yoon, S.S. Electrosprayed graphene decorated with ZnO nanoparticles for supercapacitors. *J. Alloy. Compd.* **2018**, *741*, 781–791. [CrossRef]
11. Laureti, S.; Peddis, D.; Del Bianco, L.; Testa, A.; Varvaro, G.; Agostinelli, E.; Binns, C.; Baker, S.; Qureshi, M.; Fiorani, D. Exchange bias and magnetothermal properties in Fe@Mn nanocomposites. *J. Magn. Magn. Mater.* **2012**, *324*, 3503–3507. [CrossRef]
12. Golubeva, E.V.; Stepanova, E.A.; Balymov, K.G.; Volchkov, S.O.; Kurlyandskaya, G.V. Magnetic properties and the giant magnetoimpedance of amorphous co-based wires with a carbon coating. *Phys. Met. Met.* **2018**, *119*, 324–331. [CrossRef]
13. Spizzo, F.; Sgarbossa, P.; Sieni, E.; Semenzato, A.; Dughiero, F.; Forzan, M.; Bertani, R.; Del Bianco, L. Synthesis of ferrofluids made of iron oxide nanoflowers: Interplay between carrier fluid and magnetic properties. *Nanomaterials* **2017**, *7*, 373. [CrossRef] [PubMed]
14. Yang, Q.; Zhang, X.T.; Zhang, M.Y.; Gao, Y.; Gao, H.; Liu, X.C.; Lui, H.; Wong, K.W.; Lau, W.M. Rationally designed hierarchical MnO2-shell/ZnO-nanowire/carbon-fabric for high-performance supercapacitor electrodes. *J. Power Sources* **2014**, *272*, 654–660. [CrossRef]
15. Cao, F.; Pan, G.; Xia, X.; Tang, P.; Chen, H. Synthesis of hierarchical porous NiO nanotube arrays for supercapacitor application. *J. Power Sources* **2014**, *264*, 161–167. [CrossRef]
16. Javed, M.S.; Lei, H.; Li, J.; Wang, Z.; Mai, W. Construction of highly dispersed mesoporous bimetallic-sulfide nanoparticles locked in N-doped graphitic carbon nanosheets for high energy density hybrid flexible pseudocapacitors. *J. Mater. Chem. A* **2019**, *7*, 17435–17445. [CrossRef]
17. Javed, M.S.; Lei, H.; Shah, H.U.; Asim, S.; Raza, R.; Mai, W. Achieving high rate and high energy density in an all-solid-state flexible asymmetric pseudocapacitor through the synergistic design of binder-free 3D ZnCo2O4 nano polyhedra and 2D layered Ti$_3$C$_2$T$_x$-MXenes. *J. Mater. Chem. A* **2019**, *7*, 24543–24556. [CrossRef]
18. Chen, H.-C.; Lyu, Y.-R.; Fang, A.; Lee, G.-J.; Karuppasamy, L.; Wu, J.J.; Lin, C.-K.; Anandan, S.; Chen, C.-Y. The design of ZnO nanorod arrays coated with MnOx for high electrochemical stability of a pseudocapacitor electrode. *Nanomaterials* **2020**, *10*, 475. [CrossRef]
19. Wang, J.; Gao, Z.; Li, Z.; Wang, B.; Yan, Y.; Liu, Q.; Mann, T.; Zhang, M.; Jiang, Z. Green synthesis of graphene nanosheets/ZnO composites and electrochemical properties. *J. Solid State Chem.* **2011**, *184*, 1421–1427. [CrossRef]
20. Lu, T.; Pan, L.; Li, H.; Zhu, G.; Lv, T.; Liu, X.; Sun, Z.; Chen, T.; Chua, D.H. Microwave-assisted synthesis of graphene–ZnO nanocomposite for electrochemical supercapacitors. *J. Alloy. Compd.* **2011**, *509*, 5488–5492. [CrossRef]

21. Zhang, Y.; Li, H.; Pan, L.; Lu, T.; Sun, Z. Capacitive behavior of graphene–ZnO composite film for supercapacitors. *J. Electroanal. Chem.* **2009**, *634*, 68–71. [CrossRef]
22. Jayalakshmi, M.; Palaniappa, M.; Balasubramanian, K. Balasubramanian, single step solution combustion synthesis of ZnO/carbon composite and its electrochemical characterization for supercapacitor application. *Int. J. Electrochem. Sci.* **2008**, *3*, 96–103.
23. He, X.; Yoo, J.E.; Lee, M.H.; Bae, J. Morphology engineering of ZnO nanostructures for high performance supercapacitors: Enhanced electrochemistry of ZnO nanocones compared to ZnO nanowires. *Nanotechnology* **2017**, *28*, 245402. [CrossRef]
24. Alver, Ü.M.İ.T.; Tanrıverdi, A.; Akgül, Ö. Hydrothermal preparation of ZnO electrodes synthesized from different pre-cursors for electrochemical supercapacitors. *Synth. Met.* **2016**, *211*, 30–34. [CrossRef]
25. Luo, Q.; Xu, P.; Qiu, Y.; Cheng, Z.; Chang, X.; Fan, H. Synthesis of ZnO tetrapods for high-performance supercapacitor applications. *Mater. Lett.* **2017**, *198*, 192–195. [CrossRef]
26. Javed, M.S.; Lei, H.; Wang, Z.; Liu, B.-T.; Cai, X.; Mai, W. 2D V2O5 nanosheets as a binder-free high-energy cathode for ultrafast aqueous and flexible Zn-ion batteries. *Nano Energy* **2020**, *70*, 104573. [CrossRef]
27. Abbas, Y.; Yun, S.; Javed, M.S.; Chen, J.; Tahir, M.F.; Wang, Z.; Yang, C.; Arshad, A.; Hussain, S. Anchoring 2D NiMoO4 nano-plates on flexible carbon cloth as a binder-free electrode for efficient energy storage devices. *Ceram. Int.* **2020**, *46*, 4470–4476. [CrossRef]
28. Li, Y.H.; Li, Q.Y.; Wang, H.Q.; Huang, Y.G.; Zhang, X.H.; Wu, Q.; Gao, H.-Q.; Yang, J.H. Synthesis and electrochemical properties of nickel–manganese oxide on MWCNTs/CFP substrate as a super-capacitor electrode. *Appl. Energy* **2015**, *153*, 78–86. [CrossRef]
29. Javed, M.S.; Chen, J.; Chen, L.; Xi, Y.; Zhang, C.; Wan, B.; Hu, C. Flexible full-solid state supercapacitors based on zinc sulfide spheres growing on carbon textile with superior charge storage. *J. Mater. Chem. A* **2016**, *4*, 667–674. [CrossRef]
30. Gogotsi, Y.; Penner, R.M. Energy Storage in Nanomaterials–Capacitive, Pseudocapacitive, or Battery-Like? ACS Publications: Washington, DC, USA, 2018.
31. Javed, M.S.; Shaheen, N.; Hussain, S.; Li, J.; Shah, S.S.A.; Abbas, Y.; Ahmad, M.A.; Raza, R.; Mai, W. An ultra-high energy density flexible asymmetric supercapacitor based on hierarchical fabric decorated with 2D bimetallic oxide nanosheets and MOF-derived porous carbon polyhedra. *J. Mater. Chem. A* **2019**, *7*, 946–957. [CrossRef]
32. Lin, J.C.; Peng, K.C.; Liao, H.L.; Lee, S.L. Transparent conducting Sc-codoped AZO film prepared from ZnO: Al–Sc by RF-DC sputtering. *Thin Solid Film.* **2008**, *516*, 5349–5354. [CrossRef]
33. Chang, F.-M.; Brahma, S.; Huang, J.-H.; Wu, Z.-J.; Lo, K.-Y. Strong correlation between optical properties and mechanism in deficiency of normalized self-assembly ZnO nanorods. *Sci. Rep.* **2019**, *9*, 1–9. [CrossRef]
34. György, E.; del Pino, A.P.; Logofatu, C.; Duta, A.; Isac, L. Effect of nitrogen doping on wetting and photoactive properties of laser processed zinc oxide-graphene oxide nanocomposite layers. *J. Appl. Phys.* **2014**, *116*, 024906. [CrossRef]
35. Ravikovitch, P.I.; Neimark, A. Characterization of nanoporous materials from adsorption and desorption isotherms. *Colloids Surfaces A Physicochem. Eng. Asp.* **2001**, *187–188*, 11–21. [CrossRef]
36. Zhang, J.; Kong, L.-B.; Cai, J.-J.; Luo, Y.-C.; Kang, L. Nanoflake-like cobalt hydroxide/ordered mesoporous carbon composite for electrochemical capacitors. *J. Solid State Electrochem.* **2010**, *14*, 2065–2075. [CrossRef]
37. Shi, S.; Zhuang, X.; Cheng, B.; Wang, X. Solution blowing of ZnO nanoflake-encapsulated carbon nanofibers as electrodes for supercapacitors. *J. Mater. Chem. A* **2013**, *1*, 13779–13788. [CrossRef]

Review

Microstructure and Oxidation Behavior of Metal-Modified Mo-Si-B Alloys: A Review

Laihao Yu, Fuqiang Shen, Tao Fu, Yingyi Zhang *, Kunkun Cui, Jie Wang and Xu Zhang

School of Metallurgical Engineering, Anhui University of Technology, Maanshan 243002, China; aa1120407@126.com (L.Y.); sfq19556630201@126.com (F.S.); ahgydxtaofu@163.com (T.F.); 15613581810@163.com (K.C.); wangjiemaster0101@outlook.com (J.W.); zx13013111171@163.com (X.Z.)
* Correspondence: zhangyingyi@cqu.edu.cn

Abstract: With the rapid development of the nuclear industry and the aerospace field, it is urgent to develop structural materials that can work in ultra-high temperature environments to replace nickel-based alloys. Mo-Si-B alloys are considered to have the most potential for new ultra-high temperature structural material and are favored by researchers. However, the medium-low temperature oxidizability of Mo-Si-B alloys limits their further application. Therefore, this study carried out extensive research and pointed out that alloying is an effective way to solve this problem. This work provided a comprehensive review for the microstructure and oxidation resistance of low silicon and high silicon Mo-Si-B alloys. Moreover, the influence of metallic elements on the microstructure, phase compositions, oxidation kinetics and behavior of Mo-Si-B alloys were also studied systematically. Finally, the modification mechanism of metallic elements was summarized in order to obtain Mo-Si-B alloys with superior oxidation performance.

Keywords: Mo-Si-B alloys; alloying and modification; microstructure; oxidation resistance; mechanism

Citation: Yu, L.; Shen, F.; Fu, T.; Zhang, Y.; Cui, K.; Wang, J.; Zhang, X. Microstructure and Oxidation Behavior of Metal-Modified Mo-Si-B Alloys: A Review. *Coatings* **2021**, *11*, 1256. https://doi.org/10.3390/coatings11101256

Academic Editor: Hideyuki Murakami

Received: 22 September 2021
Accepted: 12 October 2021
Published: 15 October 2021

Publisher's Note: MDPI stays neutral with regard to jurisdictional claims in published maps and institutional affiliations.

Copyright: © 2021 by the authors. Licensee MDPI, Basel, Switzerland. This article is an open access article distributed under the terms and conditions of the Creative Commons Attribution (CC BY) license (https://creativecommons.org/licenses/by/4.0/).

1. Introduction

As the most commonly used blade material for aircraft engines, the nickel-based superalloy can operate normally at temperatures below 1150 °C. However, the lower melting point limits its further use. To improve the working efficiency and reduce the fuel consumption of heat engines, the exploration research of new structural materials is promoted. The Mo-Si-B superalloy is considered to be the candidate material with the most potential to replace the nickel-based superalloy because of its outstanding high temperature mechanical properties and excellent oxidation performance [1–8].

In recent decades, Mo-Si-B superalloy has been developed rapidly as the most attractive ultra-high temperature material; however, there are still a large number of problems in the practical industrial application. For example, poor oxidation resistance at medium-low temperatures is one of the non-negligible problems [9–12]. This is because Mo-Si-B superalloys usually contain multiple phases, which exhibit different oxidation behaviors at moderate temperatures. For example, in the low silicon Mo-Si-B alloys (i.e., Si content less than 30 at.%) that contain Mo_{ss}-Mo_5SiB_2-Mo_3Si phases, Mo_{ss} can improve the fracture toughness of the alloy, but it is easily oxidized and forms volatile molybdenum trioxide in environment temperatures over 350 °C, resulting in significant weight loss of the alloy; Mo_3Si metal compound contains a high content of molybdenum, which leads to poor oxidation resistance [13–17]. For high silicon Mo-Si-B alloys (i.e., Si content more than 30 at.%) containing $MoSi_2$ phase, "pesting oxidation" is a very tricky problem [18–25]. Therefore, the exploration to further improve the antioxidation ability of Mo-Si-B superalloy at moderate temperature has not been interrupted.

Many studies showed that the alloying of metallic elements such as Zr, Al, Fe, and Cr into Mo-Si-B ternary alloys can enhance antioxidation properties by refining the microstructure or forming the duplex layer, etc. In this work, the oxidation behaviors of low silicon and high silicon Mo-Si-B ternary alloys were reviewed, and studied their oxidation dynamic, microstructure, and oxidation resistance in detail. In addition, this work gave a special and systematic review of the oxidation performance of metal-modified Mo-Si-B alloys, and also analyzed the effects of different metallic elements on the microstructure and oxidation behavior of the Mo-Si-B alloys. Finally, the modification mechanism of the metallic elements was summarized and analyzed to ameliorate the antioxidation performance of the alloy.

2. Mo-Si-B Ternary Alloys

The addition of silicon element in the Mo matrix can distinctly enhance the antioxidation capability of the Mo metal at high temperature environment [26,27]. However, the antioxidant performance of the Mo-Si binary alloys is extremely bad at the intermediate temperature (400–800 °C), which often shows "pesting behavior" or accelerated oxidation. The researchers found that a B element can significantly enhance the antioxidation properties of Mo-Si binary alloys [28–31]. Therefore, it is very imperative to discuss the microstructure and oxidation behavior of the Mo-Si-B ternary alloys.

2.1. Low Silicon Content of the Mo-Si-B Ternary Alloys

In Mo-Si-B ternary alloys, the silicon and boron elements can form protective borosilicate scale when oxidized on the substrate surface and reduce the oxidation rate of the alloy. Furthermore, the added boron can facilitate the flow of the SiO_2 layer on the substrate surface to fill the exterior defects such as cracks and pores, thus enhancing the antioxidation properties of molybdenum metal [32–34]. The microstructure and oxidation performance of Mo-Si-B ternary alloys are closely related to the content of silicon. When the silicon content is low, the research of Mo-Si-B alloys is mostly concentrated at the region of Mo_{ss}-Mo_5SiB_2-Mo_3Si [35–37]. Wang et al. [38] prepared the Mo-12.5Si-25B (at.%) alloy by arc melting from powder materials of Mo, Si and B, the corresponding microstructure is represented in Figure 1a. EDS and XRD results showed that Mo-12.5Si-25B sample was composed of Mo_{ss} and Mo_3Si implanted into the Mo_5SiB_2 matrix. Figure 2a exhibits the oxidation dynamics of this sample at 1200 °C. It is seen that the oxidation dynamics of the alloy consists of two stages: transient oxidation and steady-state oxidation. The mass loss of the alloy is rapid during transient oxidation. This is because the formed borosilicate scale was thin and intermittent at the initial stage of oxidation, which made it difficult to inhibit the rapid evaporation of MoO_3. In addition, the gaps provided a channel for the oxygen diffusion into the matrix, which accelerated the oxidation of molybdenum and silicon and formed a Mo-Si-O oxide region at the interface of substrate and scale, as displayed in Figure 1b. As can be observed from Figure 1c, a thicker and continuous oxide film formed on the surface of the alloy when it was oxidized at 1200 °C for 2 h. XRD analysis proved that the scale consisted of silica and dispersed molybdenum dioxide. In addition, an inner layer of Mo_{ss} and diffuse silica was found below the oxide scale. After 2 h of exposure, the oxidation of alloy reached a steady-state stage. However, due to the existence of holes in the alloy (Figure 1c), the oxidation dynamics was still characterized by mass loss, but the loss rate was evidently reduced (Figure 2a). After 100 h of oxidation, a dense and thick oxide scale formed on the surface of the substrate, which was composed of three layers. The outermost layer consisted of silica and a small amount of dispersed molybdenum dioxide, and the inner layer was mainly composed of Mo_{ss} and diffuse SiO_2 particles. Combined with EDS and XRD analysis results, there is also a MoO_2 layer between the outermost layer and inner layer. Moreover, it can be seen from the local enlarged diagram that the Mo_3Si phase in the interface between the matrix and the inner layer was preferentially oxidized and generated molybdenum and SiO_2, as depicted in Figure 1d. Compared with the Mo-12.5Si-25B alloy, the Mo-14Si-28B (at.%) alloy fabricated by arc melting method showed superior antioxidation properties during cyclic oxidation at 1200 °C for 100 h [39,40]. This is because the

Mo-14Si-28B alloy takes only about 2 h to achieve the steady-state oxidation and forms a continuous protective oxide film, which is much earlier than Mo-12.5Si-25B alloy.

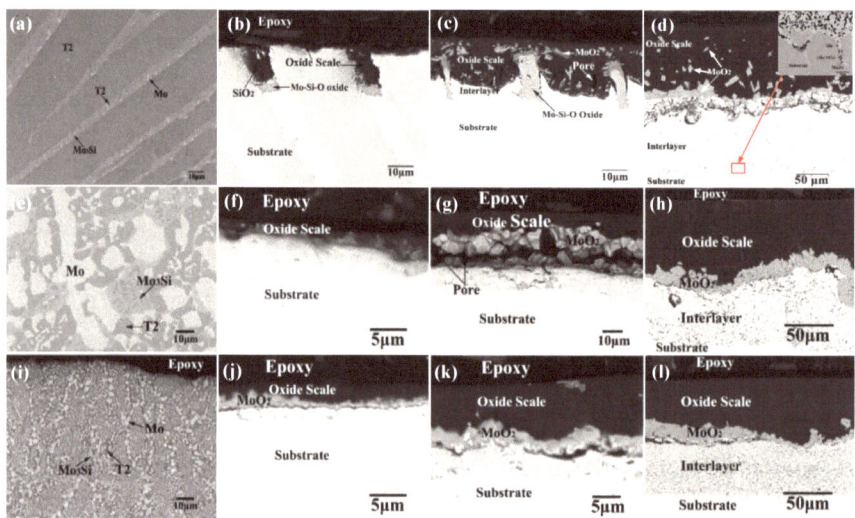

Figure 1. BSE micrographs of arc-melted Mo–12.5Si–25B (**a**), annealed (**e**) and laser-remelted Mo–10Si–14B (**i**) samples; Cross-sectional SEM images of the three samples oxidized at 1200 °C for different time: 10 min (**b**,**f**,**j**), 2 h (**c**,**g**,**k**) and 100 h (**d**,**h**,**l**). (**a**–**l**) reproduced with permission [38] and [41], respectively. Copyright 2007 Elsevier.

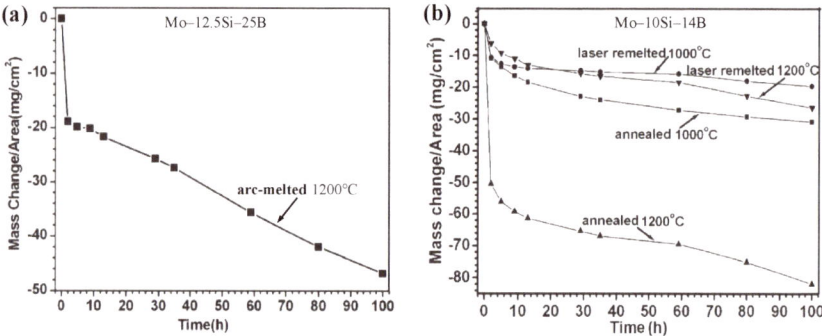

Figure 2. Mass change curves: (**a**) arc-melted alloy cyclic oxidation at 1200 °C, (**b**) laser-remelted and annealed alloys oxidized at 1000 °C and 1200 °C. (**a**,**b**) reproduced with permission [38,41], respectively. Copyright 2007 Elsevier.

In addition, Wang et al. [41] also prepared the Mo-10Si-14B (at.%) alloy through arc melting and studied its oxidation behavior after annealing and laser remelting, respectively. The microstructure of the alloy surface after annealing and laser remelting is shown in Figure 1e,i. The results of EDS and XRD analysis revealed that both alloy surfaces were composed of Mo_{ss}, T2 and Mo_3Si three phases, but the alloy after laser remelting had a more refined and evenly distributed three phase structure. The cyclic oxidation kinetics of the two alloys at 1000 °C and 1200 °C is presented in Figure 2b. It was found that the mass loss per unit time of the laser remelted was smaller than that of the annealed alloy, and this difference was more obvious at 1200 °C. Therefore, it can be inferred that laser remelting can effectively improve the antioxidation ability of the Mo-10Si-14B sample. To further study their antioxidation mechanism, the cross-sectional SEM images of two treated alloys after oxidation of 100 h at 1200 °C was analyzed. It can be seen that the annealed alloy formed a thin discontinuity scale consisting of silicon dioxide and dispersed molybdenum dioxide when oxidized for 10 min. After oxidation for 2 h, the scale thickened with many holes. When the oxidation time is 100 h, a relatively continuous scale formed. Moreover, a MoO_2 middle layer and an interlayer composed of Mo_{ss} and SiO_2 also appeared below the scale, as being presented in Figure 1f–h. By comparison, the laser remelted alloy could form a continuous and uniform SiO_2 scale after oxidation for 10 min, and the scale became obviously thicker after oxidation for 2 h. While oxidized for 100 h, the SiO_2 scale became denser, as presented in Figure 1j–l. Therefore, it can be inferred that the better antioxidation performance of laser-remelted alloy is attributed to refinement of grain size, which makes the alloy form protective continuous borosilicate scale earlier during oxidation. The studies of Rioult [42] and Choi et al. [43] also confirmed that grain refinement will enhance the antioxidant function of the alloy.

It is reported that boron plays a positive role in improving oxidation protection of the low silicon Mo-Si-B ternary alloys. Because boron can decrease the viscosity of silica film, thus prompting it to flow and cover the surface of the alloy substrate. Li et al. [44] fabricated the 0.9 wt.% La_2O_3-doped Mo-12Si-xB (x = 5, 8.5, 17 at.%) samples through mechanical alloying and hot pressing sintering. Figure 3a–c display the microstructures of the three alloys after hot pressure. It can be seen that all the alloy surfaces have fine structure and uniform composition. EDS analysis results of Figure 3a–c show that the bright area is α-Mo phase, the gray area is Mo_3Si phase, and the dark gray area is T2 phase, which is consistent with the result of XRD analysis (Figure 4a). Furthermore, the area of bright or gray decreases, while the dark gray area increases significantly, indicating that the volume fractions of α-Mo phase and Mo_3Si phase gradually reduce with the raising boron content, while the volume fraction of T2 phase gradually increases. However, the boron content does hardly affect the granularity of each phase, as presented in Figure 4b,c. Figure 3d–f is the TEM bright field micrographs of three alloys. It can be observed that the presence of small La_2O_3 particles is inside the grain or on the grain boundary. However, the diffraction peak of the La_2O_3 phase is not observed in the XRD pattern (Figure 4a), because its content is too small to be detected. In addition, La_2O_3 particles act as "pinning". On the one hand, it can inhibit grain growth, to achieve grain refining effect; on the other hand, it can serve as a barrier phase to inhibit the diffusion of molybdenum ions and the volatilization of MoO_3. For the above conclusions, more systematic studies were reported by Zhang [45], Majumda [46,47], and Burk et al. [48,49]. To further study the effect of boron content on the antioxidation behavior of Mo-Si-B alloys, Li et al. [44] also carried out isothermal oxidation experiment of the three samples at 1000 °C and obtained the oxidation dynamics through taking notes the change of mass loss with time during oxidation process, as shown in Figure 4d. It can be found that with the increase of the boron content, the mass loss per unit time of the alloy decreased gradually in the oxidation process, and this phenomenon was the most obvious during the transient oxidation period. Moreover, the Mo-12Si-5B sample had the most weight loss after oxidation for 30 h, showing relatively poor oxidation resistance, which indicated that the increase of boron content played a beneficial role in enhancing the oxidation properties of Mo-Si-B alloys. Figure 5a–d depicts the surface SEM images of Mo-12Si-xB samples when oxidized at 1200 °C for 30 h. On the whole, borosilicate scales were formed on the surface of all Mo-12Si-xB samples, and (Mo, La)-oxide particles with relatively dispersed distribution were observed in this scale. However, after oxidizing Mo-12Si-5B alloy, the borosilicate scale was rough, and numerous cavities, cracks and coarse silica-rich regions appeared on the surface. This is attributed to the comparatively low boron concentration in the alloy that makes it difficult to form a continuous stable borosilicate scale, as shown in Figure 5a,b. As the boron content increased, the borosilicate scale gradually became smooth, continuous and dense, and the cavities, cracks, and coarse silica-rich regions on the alloy surface gradually disappeared, as depicted in Figure 5c,d. However, when the content of boron reaches a certain degree, the continuous increase of boron has little effect on the oxidation rate of the alloy (Figure 4d). Meantime, too much boron will also lead to the shedding of silica scale due to too low adhesion, which results in poor oxidation resistance. Therefore, increasing the boron content appropriately is of great significance to enhance the antioxidation action of Mo-Si-B alloys.

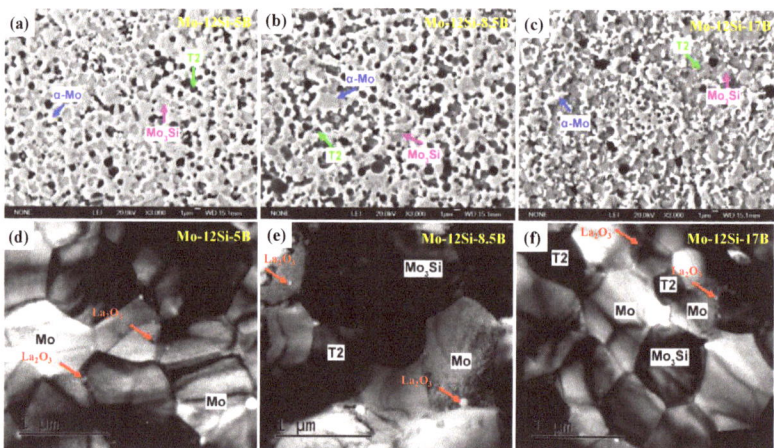

Figure 3. Typical SEM (**a–c**) and TEM bright field images (**d–f**) present the micromorphology of three different alloys. Reproduced with permission [44]. Copyright 2019 Elsevier.

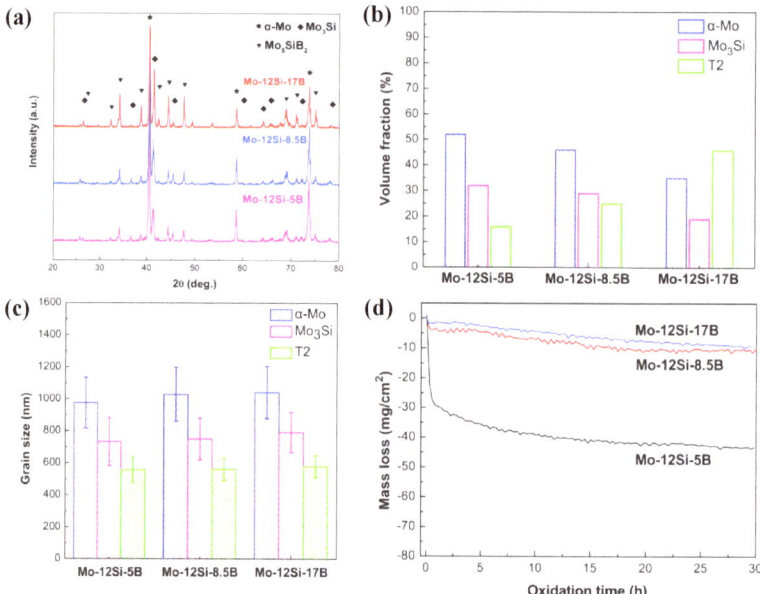

Figure 4. (**a**) XRD patterns of different alloys; (**b**) The volume fraction of different phases in three alloys; (**c**) Average grain size of each phase; (**d**) Oxidation behaviors of various alloys at 1000 °C. Reproduced with permission [44]. Copyright 2019 Elsevier.

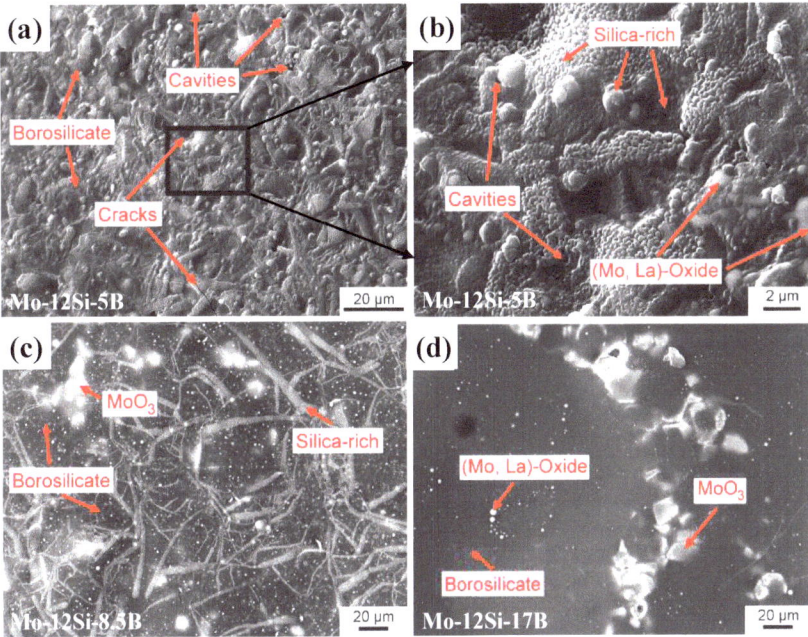

Figure 5. Surface SEM images of various alloys during oxidation 30 h at 1000 °C. (**a**) Mo-12Si-5B alloy with magnified image (**b**), (**c**) Mo-12Si-8.5B alloy, (**d**) Mo-12Si-17B alloy. Reproduced with permission [44]. Copyright 2019 Elsevier.

2.2. High Silicon Content of the Mo-Si-B Ternary Alloys

In contrast with the low silicon Mo-Si-B alloys, the high silicon Mo-Si-B alloys tend to possess superior antioxidation capacity owing to the high silicon concentration, which is easier to form borosilicate scale covering the substrate surface. For going a step further study the antioxidation performance of high silicon Mo-Si-B alloys, Wen et al. [50] prepared two kinds of Mo-62Si-5B (at.%) alloys with different particle sizes through spark plasma sintering (SPS), using two alloy powders obtained by mechanical disruption (MD) and ball milling (BM) as raw materials. They were also recorded as SPS-MD alloy and SPS-BM alloy, respectively. The typical BSE micrographs of the sintered Mo-Si-B alloys are displayed in Figure 6a,d. XRD analysis results show that the two alloys are composed of MoB phase and $MoSi_2$ phase. However, there are two additional Mo_5Si_3 phase and SiO_2 phase in SPS-BM alloy, which are unevenly dispersed on $MoSi_2$ matrix, resulting in a fine and dense microstructure. While a large number of holes and cracks appeared on the surface of SPS-MD alloy (Figure 6a), which may lead to relatively poor oxidation resistance. In addition, Wen et al. also investigated the antioxidation function of the two alloys at 1250 °C and 1350 °C respectively, which further confirmed this inference. After oxidation 200 h at 1250 °C or 1350 °C, the oxidation rate of SPS-BM alloy was slower than SPS-MD alloy. In addition, the difference of oxidation rate between the two alloys was more obvious at 1250 °C, which revealed that the SPS-BM alloy had better antioxidant effect, as shown in Figure 6b,c,e,f and Figure 7a,b, which give the cross-sectional images of both alloys when oxidized at 1250 °C and 1350 °C for 30 h, respectively. They point out that with the increase of oxidation temperature, the dimensions of the oxide film becomes thicker, and the change of the oxide layer thickness of SPS-BM is more pronounced. A similar conclusion is also confirmed in Figure 7c,d. Moreover, Figure 7c,d also shows that the oxide layer thickness of SPS-MD alloy is thicker than that of SPS-BM alloy at the same temperature, and the difference of the oxide layer thickness is more obvious at 1250 °C, which is because the oxidation resistance of SPS-MD alloy is relatively poor. This is consistent with the results of the oxidative dynamics (Figure 7a,b). Furthermore, a lot of cracks and holes were found in the cross-sectional morphology of SPS-MD alloy during oxidation for 30 h, which provided shortcuts for the internal diffusion of oxygen, as shown in Figure 6b,c. No cracks and holes were observed in the cross section of SPS-BM alloy after oxidation, but there were silica particles on the interface between borosilicate scale and substrate. These particles connected borosilicate scale and substrate and played the role of "mechanical locking". Moreover, it promoted the rapid formation of consecutive and protective B_2O_3-SiO_2 scale on the alloy surface, thus enhancing the antioxidation properties of SPS-BM alloy, as shown in Figure 6e,f. This result was also observed in the study of Pan et al. [51].

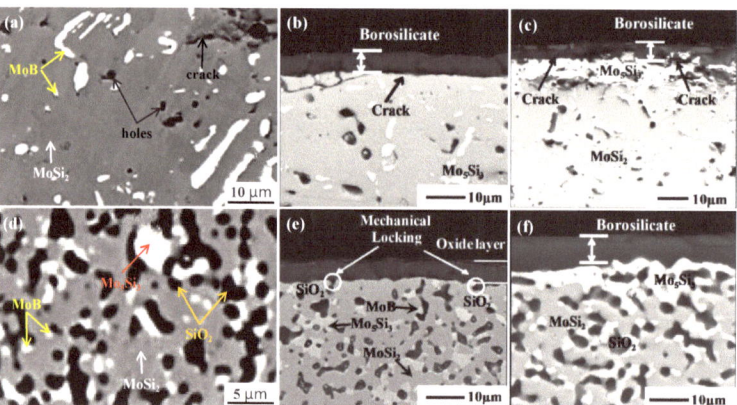

Figure 6. BSE micrographs: SPS-MD (**a**), SPS-BM (**d**); Cross-sectional micrographs of SPS-MD and SPS-BM oxidized at different temperatures for 30 h: 1250 °C (**b,e**); 1350 °C (**c,f**). Reproduced with permission [50]. Copyright 2017 Elsevier.

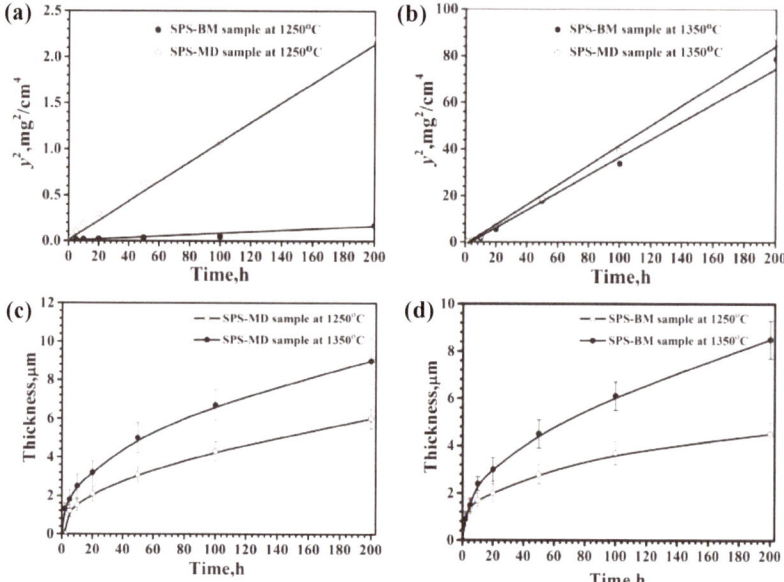

Figure 7. (**a,b**) Square of mass change (y^2) plots at different temperatures; (**c,d**) Oxide layer thickness change curve at different temperatures. Reproduced with permission [50]. Copyright 2017 Elsevier.

3. Metallic Elements Modified Mo-Si-B Alloys

Although there are significant differences in the oxidation behavior between low silicon and high silicon Mo-Si-B ternary alloys, the antioxidation mechanism of both types of alloys is similar, i.e., the continuous dense B_2O_3-SiO_2 film is formed on the substrate surface to passivate alloy, so as to strengthen the antioxidation properties of the alloy. However, for Mo-Si-B ternary alloys containing the Mo_{ss} phase, although the Mo_{ss} phase improves the fracture toughness of the alloy, it is likely to be oxidized to volatile molybdenum trioxide even at 350 °C. Furthermore, with regard to Mo-Si-B ternary alloys containing $MoSi_2$ phase, the phenomenon of "pesting behavior" or accelerated oxidation will occur when oxidized at 800 °C [51,52]. Therefore, the improvement of antioxidation function of Mo-Si-B ternary alloys is going to be extensively studied in the following.

3.1. Single Metallic Elements Modified Mo-Si-B Alloys

Adding metallic elements such as Zr, Al, and Cr to the Mo-Si-B ternary alloys will refine their microstructure to shorten the period of transient oxidation, and improve the stability of borosilicate scale in the steady oxidation stage. Therefore, it is important to systematically study the Mo-Si-B alloys modified by metal elements.

3.1.1. Zr Element Modified Mo-Si-B Alloys

Zr is a kind of active metal element which can refine grain with great application prospect. It cannot only improve the mechanical properties of the alloy, such as strength and ductility, but shorten the formation period of stable continuous B_2O_3-SiO_2 layer, thus improving the antioxidation properties of alloy. Therefore, it is favored by the majority of researchers. Wang et al. [53] synthesized Mo-12Si-8.5B (at.%, referred to as MSB) and 1.0 wt.% ZrB_2-doped MSB (referred to as MSBZ) by mechanical alloying. Figure 8a,e shows the microstructures of MSB and MSBZ alloys. On the whole, the surface composition of the two alloys is uniform and the structure is fine, and they are composed of three regions with different brightness. EDS analysis derived that the bright, gray and dark gray regions in Figure 8a,e were α-Mo, Mo_3Si and Mo_5SiB_2, respectively. Moreover, Zr_{ss} particles dispersed in the grain interior or boundary of MSBZ alloy. In addition, compared with MSB alloy, MSBZ alloy microstructure has a finer grain size, as shown in Figure 8i,j. The oxidation kinetics of MSB and MSBZ alloys at 900 °C is displayed in Figure 9a. It can be seen that the net mass variations of MSBZ and MSB after oxidation for 30 h were about +10 mg/cm^2 and −180 mg/cm^2 respectively, indicating that adding ZrB_2 could effectively prevent the weight loss of MSB alloy. It is due to the formation of flat and dense B_2O_3-SiO_2 film on the surface of MSBZ alloy during the oxidation. Furthermore, lots of white particles were scattered in this scale, as shown in Figure 8f,g. According to the analysis results of EDS (Figure 8k) and XRD (Figure 9b), these white particles were composed of $ZrSiO_4$ or ZrO_2. On the one hand, these particles can improve the stability of protective oxide film on the alloy surface. On the other hand, they can be used as diffusion barriers to prevent the inward spread of oxygen or the outward volatilization of MoO_3, thus improving the oxidation resistance of the alloy. Similar conclusions were reported by Pan et al. [51]. In contrast, the B_2O_3-SiO_2 film formed on the MSB's surface during oxidation was uneven, which might be due to the formation of Molybdenum trioxide bubbles at the borosilicate scale. In addition, the MSB's surface also appeared pores, which might be due to the ruptured volatilization of the MoO_3 bubbles, as shown in Figure 8b,c. The pores provided a fast path for oxygen inward diffusion to accelerate the oxidation of the alloy. Therefore, the alloy with ZrB_2 has a better oxidation resistance compared to the MSB alloy. Furthermore, Wang et al. [54] also studied the antioxidation of MSB and MSBZ alloys at 1300 °C, which further confirmed this conclusion. Figure 9c presents the oxidation kinetics of the MSB and MSBZ alloys at 1300 °C. Generally speaking, the weight variation trend of the two alloys was similar, namely transient oxidation stage and steady oxidation stage. For MSB alloy, the weight loss was serious in the transient stage, and it took a long time (2.7 h) to achieve steady-state. However, the weight variation of the MSBZ alloy in the transient period was little, and it merely needed three hundred seconds to achieve steady-state. It indicated that MSBZ alloy had a strong oxidation resistance, which was consistent with the results at 900 °C. From the SEM images (Figure 8d,h,l), the smooth and continuous borosilicate scale generated on the surface of MSBZ sample when oxidized for 30 h, and many white particles were embedded in the scale. XRD analysis results (Figure 9d) show that the bright granules were $ZrSiO_4$, t-ZrO_2 and m-ZrO_2, which were different from the composition of white particles observed at 900 °C (Figure 9b). By comparison, after 30 h of oxidation, many holes and cracks appeared on the surface of MSB alloy, which led to its poor oxidation resistance. This result further demonstrated that the addition of ZrB_2 had a positive role in improving the antioxidation of MSB alloy.

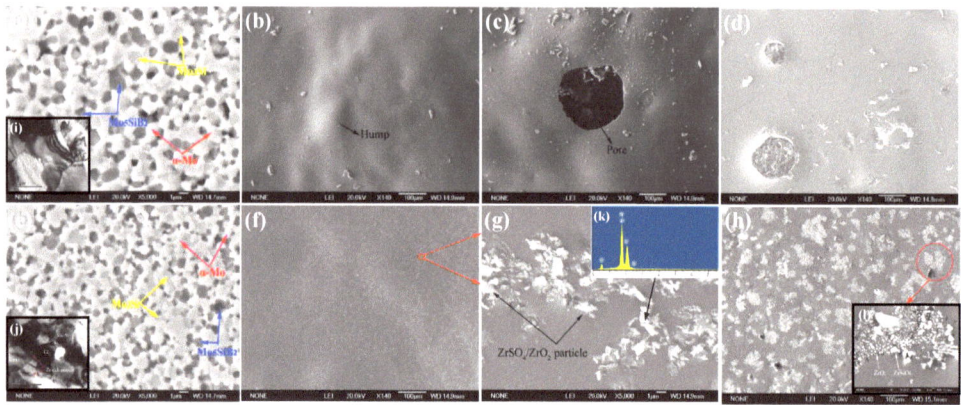

Figure 8. SEM micrographs: (**a**) MSB, (**e**) MSBZ with (**i**,**j**) TEM micrographs; Surfaces of SEM micrographs of the MSB alloy and MSBZ alloy oxidized for 30 h at: (**b**,**c**,**f**,**g**) 900 °C and (**d**,**h**) 1300 °C; (**k**) EDS pattern of the bright granules in (**g**); (**l**) local magnification image. (**a**–**c**,**e**–**g**,**k**,**d**,**h**–**j**,**l**) reproduced with permission [53,54], respectively. Copyright 2017 Elsevier.

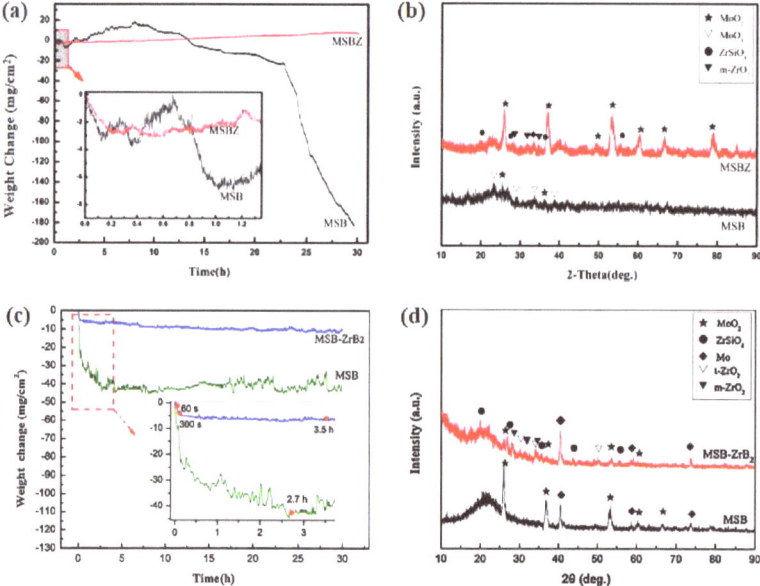

Figure 9. The different alloys after oxidation 30 h at (**a**,**b**) 900 °C and (**c**,**d**) 1300 °C: (**a**,**c**) mass change curve, (**b**,**d**) XRD analysis results. (**a**–**d**) reproduced with permission [53,54], respectively. Copyright 2017 Elsevier.

In addition, Burk et al. [48,55,56] also systematically researched the effect of adding 1 at.% Zr on the antioxidation capacity of Mo-9Si-8B (at.%) alloy at the medium-high temperature. It was found that the addition of Zr can significantly reduce oxidation rate of Mo-9Si-8B alloy when the temperature is below 1200 °C. In particular, the oxidation speed of Mo-9Si-8B-1Zr (at.%) alloy was nearly three orders of magnitude slower than Mo-9Si-8B sample, which revealed that the addition of Zr was instrumental in improving antioxidation ability of Mo-9Si-8B composite, as depicted in Table 1. However, this beneficial effect of Zr gradually disappeared when the temperature exceeded 1200 °C. Figure 10a displayed the variation curve of the unit area mass with time of Mo-9Si-8B-(1Zr) samples during oxidation at 1300 °C. It could be observed that the oxidation trend of Mo-9Si-8B sample was stable and its weight loss was small. However, the mass change curve of 1 at.% Zr-doped Mo-9Si-8B sample showed a rapid decreasing trend after oxidation for 18 min, which suggested that the antioxidation of Mo-9Si-8B sample had deteriorated after adding Zr. The reason was that when the temperature reached 1300 °C, the phase transformation of ZrO_2 in the SiO_2 scale occurred, which led to the volume change of ZrO_2 particles (expansion or shrinkage), thus destroying the integrity of the SiO_2 scale and leaving holes on the alloy surface, providing channels for the volatilization of MoO_3 and the inward spread of O_2, as shown in Figure 10c,d. In contrast, the surface of the Mo-9Si-8B alloy was overspread with a complete SiO_2 scale. Even after oxidation for 72 h, this scale still maintained a continuous and dense structure, as shown in Figure 10b. Therefore, the Mo-9Si-8B composite exhibited superior oxidation resistance at 1300 °C, which was very different from the conclusion of Wang et al. [54]. Kumar et al. [57] argued that this can be caused by the divergences in the B/Si ratio and the microstructural length scales, which made Zr show different effects on the antioxidation of Mo-Si-B alloys.

Table 1. The oxidation rate constants (kg^2/m^4s) of Mo–9Si–8B–(1Zr) at 1000–1200 °C. Reproduced with permission [55]. Copyright 2009 Springer Nature.

Temperature (°C)	Oxidation Rate Constants (kg^2/m^4s)	
	Mo-9Si-8B	Mo-9Si-8B-1Zr
1000	1.0×10^{-9}	2.78×10^{-12}
1100	5.85×10^{-11}	9.0×10^{-12}
1150	-	1.22×10^{-10}
1200	1.2×10^{-8}	3.0×10^{-8}

3.1.2. Ti Element Modified Mo-Si-B Alloys

The effect of Ti on the properties of Mo-Si-B alloys has been widely studied. On the one hand, the addition of Ti can make Mo-Si-B alloys have lower density and superior creep resistance. On the other hand, Ti can also act as a stabilizer for Mo_5SiB_2 and Mo_5Si_3 phases. Schliephake [58] and Azim et al. [59] further explored the impact of adding Ti on the oxidation resistance of the Mo-9Si-8B (at.%) alloy based on the Burk [48] research. By comparing the isothermal oxidation weight change curves of Mo-9Si-8B-(29Ti) (at.%) composites at 1100 °C to 1300 °C, it was found that the weight loss of both alloys increased with the rise of oxidation temperature. In addition, the mass wastage of Mo-9Si-8B-29Ti composite was more serious than Mo-9Si-8B composite, indicating that adding Ti did not enhance the antioxidation of Mo-9Si-8B composite, as depicted in Figure 11. To further analyzed the causes of this result, the surface and cross-sectional microstructure of the 29 at.% Ti-doped Mo-9Si-8B composite before and after oxidation were deeply studied. Unlike the microstructural composition of the Mo-9Si-8B alloy, the Mo-9Si-8B-29Ti alloy was mainly composed of Mo_{ss}, Mo_5SiB_2 and $(Ti, Mo)_5Si_3$, as being displayed in Figure 12a. After oxidation 100 h at 1100 °C, the rutile film formed on the surface of Mo-9Si-8B-29Ti composite, and lots of unevenly dispersed silica was embedded in the scale. In addition, the existence of microcracks was also observed from the scale enlarged diagram. With the increase of oxidation temperature, these microcracks grew and expanded, thus forming many gaps between or within the ruite particles. It may be related to the expansion of large oxides (TiO$_2$) and the release of internal stress, as shown in Figure 12b,c,f. As previously mentioned, the excellent antioxidation of Mo-9Si-8B composite at temperatures above 1100 °C was caused by the formation of continuous and dense oxide scale (Figure 10b). By contrast, the oxide layer structure of 29 at.% Ti-doped Mo-9Si-8B composite was completely different. When the Ti-doped sample was oxidized at 1100 °C for 100 h, a thick and porous oxide layer formed. The outermost layer was rutile layer, which was mostly composed of TiO$_2$ and little embedded borosilicate. However, the rutile scale had no protective effect, thus it could not serve as barriers against the inward spread of oxygen. Directly below the rutile layer was the TiO$_2$-borosicate duplex layer. However, there were many pores in the duplex layer, which provided a fast route for oxygen diffusion into the substrate and accelerated the alloy oxidation, as described in Figure 12d,g,h. Moreover, With the rise of oxidation temperature, the oxide layer of Mo-9Si-8B-29Ti alloy became thicker and more porous. The author believed that it was attributed to the accelerated growth of TiO$_2$ and the mismatched thermal expansion coefficients between TiO$_2$ and SiO$_2$, as presented in Figure 12e. Therefore, the addition of Ti to the Mo-9Si-8B alloy destroyed the integrity of the borosilicate scale, resulting in its poor oxidative resistance. Schliephake et al. [60] also reached the same conclusion.

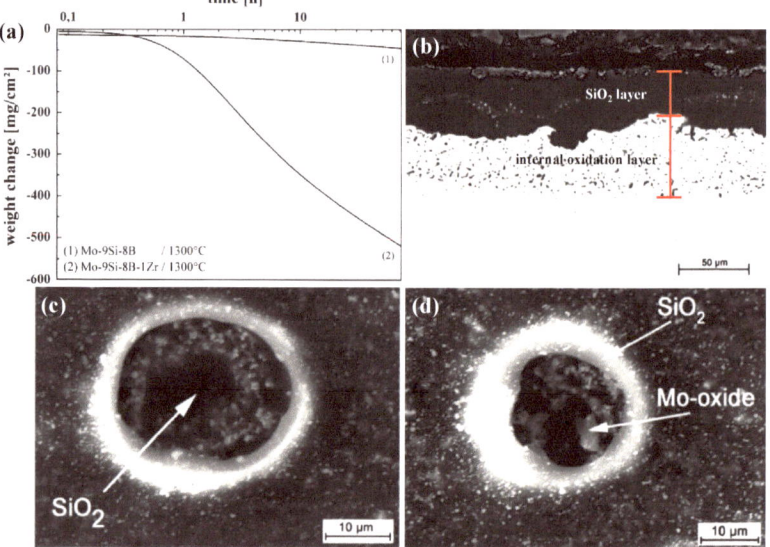

Figure 10. (a) Weight change curve of both samples at 1300 °C; (b) Cross-sectional SE image of Mo–9Si–8B sample after oxidation 72 h at 1300 °C; (c,d) SE images of Mo–9Si–8B–Zr sample surface after oxidation 15 min at 1300 °C. Reproduced with permission [55]. Copyright 2009 Springer Nature.

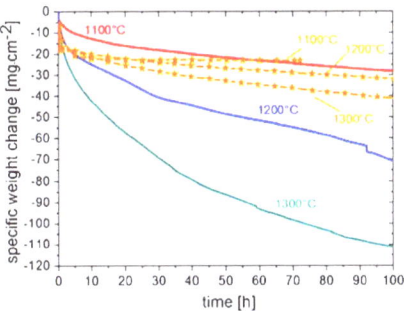

Figure 11. Specific weight change vs. time for the Ti-doped (continuous curves) and Ti-free (discontinuous curves) Mo-9Si-8B composites oxidized at various temperatures. Reproduced with permission [58]. Copyright 2013 Springer Nature.

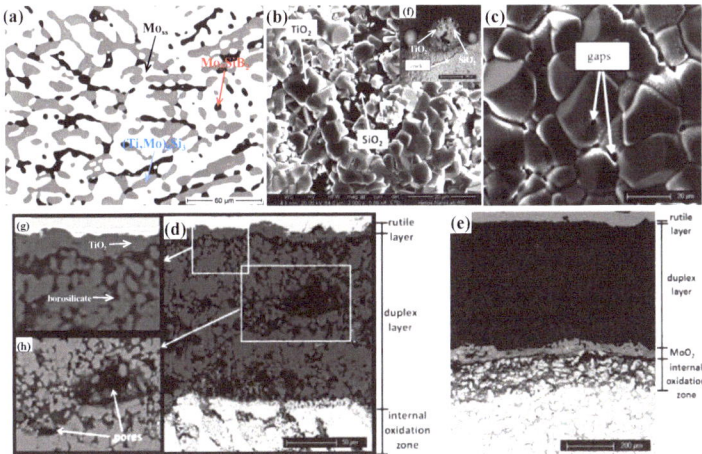

Figure 12. SEM (BSE) images of Mo-9Si-8B-29Ti composite: (**a**) microstructure, (**b**,**c**,**f**) surface and (**d**,**e**,**g**,**h**) cross section after oxidation 100 h at (**b**,**d**,**f**–**h**) 1100 °C, (**c**) 1200 °C and (**e**) 1300 °C. (**a**–**h**) reproduced with permission [58,59], respectively. Copyright 2013 Springer Nature.

3.1.3. Al Element Modified Mo-Si-B Alloys

Al element modified Mo-Si-B alloys can improve the microstructure and form a continuous protective $Al_2(MoO_4)_3$ scale on the outer surface of borosilicate film, therefore improving their antioxidation capacity. On the other hand, the added Al competes with Si, B and Mo for oxidation and brings about the formation of protective B_2O_3-SiO_2 film more slowly, which makes the oxidation resistance of the alloy worse [61–63]. This characteristic of Al makes it show different effects on antioxidation of the Mo-Si-B alloys. Paswan et al. [64–66] carried out isothermal and non-isothermal oxidation tests about Mo-14Si-10B (at.%) alloy and 7.3 at.%. Al-doped Mo-11.2Si-8.1B (at.%) alloy fabricated by reaction hot pressing, and recorded the mass variations of both alloys after oxidation at 400–1300 °C for 24 h respectively, as shown in Table 2. It was seen that the net mass change value of the Mo-11.2Si-8.1B-7.3Al alloy was more than the Mo-14Si-10B alloy at each temperature, which revealed that the addition of the Al reduced the oxidative resistance. In the study of the alloy microstructure, it was found that the two alloys presented multi-phase microstructures, and WDS analysis concluded that the white, gray, dark gray and black regions were Mo, Mo_3Si, T2 and SiO_2 phases, respectively, as being shown in Figure 13a,c. The SiO_2 might be produced by the alloy powder during hot pressing. Moreover, there was an additional Al_2O_3 phase in the black region of the Mo-11.2Si-8.1B-7.3Al alloy microstructure. These Al_2O_3 particles hindered protective borosilicate scale coverage of the alloy surface, providing a pathway for inward diffusion of oxygen. Moreover, with the increase of oxidation temperature, aluminum borate or mullite might also be precipitated in the borosilicate scale on the surface of Mo-11.2Si-8.1B-7.3Al alloy [67], which further inhibited the passivation of the alloy. Figure 13b,d depicted the surface morphology of Mo-14Si-10B alloy and Mo-11.2Si-8.1B-7.3Al alloy after isothermal oxidation at 700 °C for 24 h, respectively. It served to show the surface oxide layer of Mo-11.2Si-8.1B-7.3Al alloy was rougher and existed a large number of holes, which might be related to thermal chock and mismatch of the thermal expansion coefficient of various oxides in the oxide layer. Therefore, the addition of Al was detrimental to the oxidation resistance of the alloy. However, Yamauchi et al. [68] drew a different conclusion in the study of antioxidation performance of the Mo-Si-B alloy. They observed that adding Al could promote the quick formation of protective Al-Si-O oxide layers on the surface of Mo-Si-B sample between 800 °C and 1300 °C, thus effectively ameliorating its oxidative resistance. The difference in these research results may be caused by the different preparation methods and composition of the alloys.

Table 2. Mass loss of both alloys after oxidation at different temperatures for 24 h. Reproduced with permission [64,65]. Copyright 2006 Elsevier and 2007 Elsevier.

S. No.	T (°C)	Mass Loss (mg/cm^2)	
		Mo-14Si-10B	Mo-11.2Si-8.1B-7.3Al
1	500	+0.6	−0.7
2	600	+1.2	−4
3	700	−92.43	−369.45
4	800	−36.97	−677.43
5	900	−18.83	−621.54
6	1150	−1.11	−131.44
7	1300	−112.81	−192.37

3.1.4. W Element Modified Mo-Si-B Alloys

In the Mo-Si-B ternary alloys, the presence of Mo_3Si can neither enhance the oxidation resistance nor improve the fracture toughness. Therefore, the removal of the Mo_3Si phase has a positive effect on improving the properties of the alloy. Ray et al. [69] found that adding tungsten to the Mo-Si metal compound would destabilize the Mo_3Si structure. It is worth mentioning that adding tungsten to Mo-Si-B alloys can also remove the brittle Mo_3Si phase and help to form the desired Mo_{ss}, Mo_5SiB_2 and Mo_5Si_3 phases, which will improve the stability, fracture toughness and oxidation resistance of the alloy. Based on this property of tungsten, Karahan et al. [70] further discussed the influences of tungsten addition on microstructure and antioxidation performance of the Mo-Si-B alloys. They prepared Mo-15Si-15B (at.%) and Mo-15W-15Si-15B (at.%) alloys using drop-casting, which were abbreviated as W0 and W15, respectively. The microstructures of the two alloys obtained by casting are shown in Figure 14a,b. It can be observed that there were three phases in the microstructure of W0 alloy, namely metal rich phase in the white region, A15 phase in the gray region and T2 phase in the black region, respectively. In contrast, there were metal rich phase (white region), T2 phase ((Mo, W)$_5$SiB$_2$, gray region) and T1 phase ((Mo, W)$_5$Si$_3$, black region) in the microstructure of W15 alloy. However, the presence of the A15 phase was not observed, which might be because adding tungsten destroyed stability of A15 and made it form the T1 phase. Figure 15a depicted the mass change curve per unit area of alloys with different W content oxidized at 1200 °C. It can be found that compared with the W0 alloy, the mass loss of the alloy with the addition of W was more serious, and the mass loss of the W15 alloy was the most serious, which indicated that the addition of tungsten was harmful to the antioxidation of the alloy. It was because at 1200 °C, the radioactive WO_3 generated by oxidation was unevenly dispersed in the borosilicate scale and destroyed the composition of the alloy [71–73], resulting in B_2O_3-SiO_2 film discontinuity, as being depicted in Figure 14c. The discontinuous scale could not serve as barriers to the inward spread of O_2, so the antioxidation of W15 alloy became worse. However, as the oxidation temperature was further elevated, WO_3 gradually volatilized (Figure 15b). It was noteworthy that when the temperature exceeded 1400 °C, WO_3 in the borosilicate scale was almost exhausted by volatilization, and the oxide film became dense (Figure 14d). Therefore, the addition of W might lead to poor antioxidation performance of Mo-Si-B alloys, but it would not cause significant deterioration. Moreover, it also suggested that the pretreatment of Mo-Si-B-W composites at higher temperatures (>1400 °C) can effectively improve the oxidation protection effect.

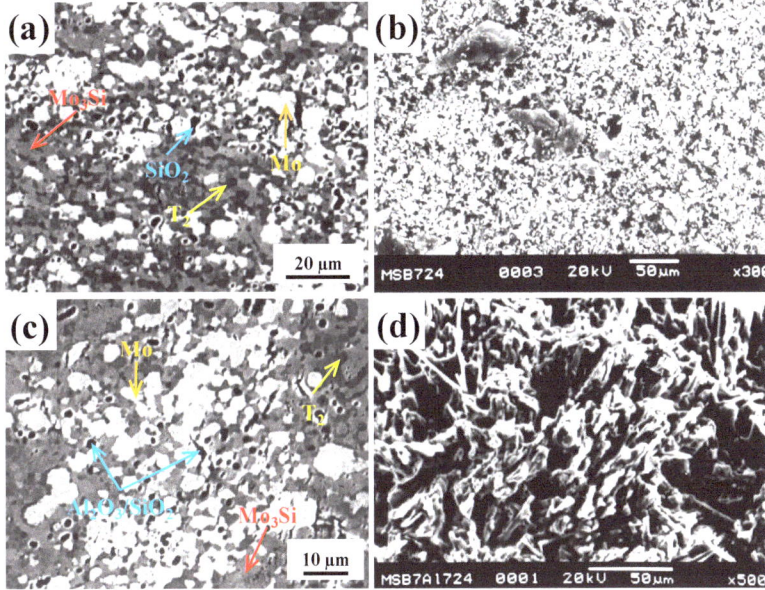

Figure 13. BSE images of the Mo-14Si-10B alloy (**a**) and Mo-11.2Si-8.1B-7.3Al alloy (**c**) before the oxidation; Surface SE images of Mo-14Si-10B alloy (**b**) and Mo-11.2Si-8.1B-7.3Al alloy (**d**) after oxidation at 700 °C for 24 h. (**a**–**d**) reproduced with permission [64,65], respectively. Copyright 2006 Elsevier and 2007 Elsevier.

3.1.5. Other Single Metallic Elements Modified Mo-Si-B Alloys

Similar to tungsten action, adding niobium to Mo-Si compounds can also destroy the stability of A15 structure [74]. In addition, niobium has a low density and is easy to dissolve in Mo, thus reducing the density of molybdenum-based alloys and making it exhibit excellent high temperature strength and fracture toughness. In recent years, researchers have carried out lots of experiments about improving the properties of Mo-Si-B alloys by niobium. Yang et al. [75] produced Nb-free and 26 at.% Nb-doped Mo-12Si-10B (at.%) alloys through mechanical alloying and hot pressing, abbreviated to 0 Nb and 26 Nb alloys, respectively. They found that the mechanical behaviors such as fracture toughness, compactibility and compression strength of 0 Nb alloy were significantly improved after adding Nb. However, the oxidation resistance of the alloy became worse. Figure 16a gave the oxidation kinetic functions of the 0 Nb and 26 Nb alloys at 1300 °C. It was seen that the 26 Nb alloy lost more mass after oxidation for the same time. Furthermore, it was also observed from the macroscopic image that the 26 Nb alloy was severely destroyed after oxidation for 5 h. In contrast, the 0 Nb alloy still retained the original shape, which further indicated that adding Nb could reduce the antioxidation of the alloy. It may be associated with non-protective porous oxides formed on the 26 Nb alloy's surface. In addition, XRD and EDX records suggested that these oxides were Nb_2O_5 (Figure 16b). Furthermore, the dispersion of Nb_2O_5 on the surface of the alloy caused the silica scale to become loose and porous, which provided a fast channel for the diffusion of oxygen into the substrate, as being shown in Figure 17a. Behrani et al. [76] also reported a similar conclusion.

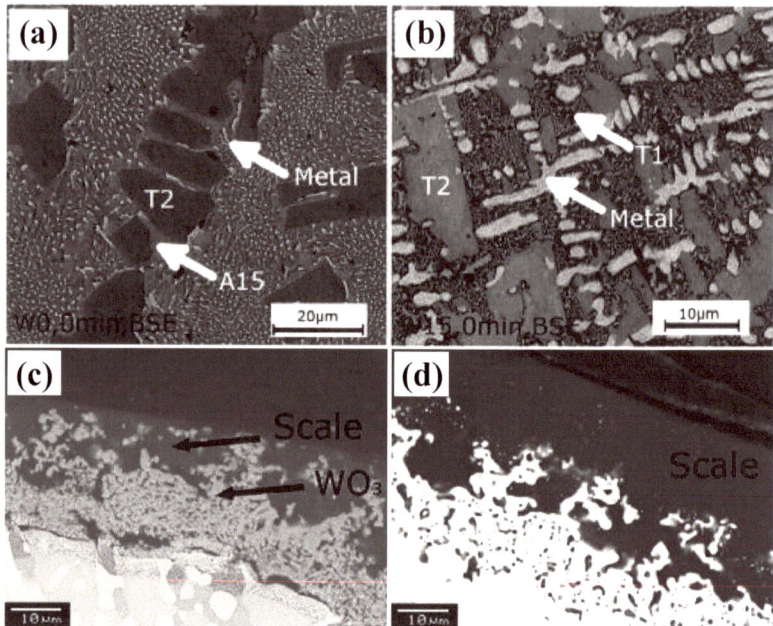

Figure 14. BSE images of the W0 (**a**) and W15 (**b**) surface prior to oxidation; Cross-sectional images of W15 after oxidation at 1200 °C (**c**) and 1500 °C (**d**) for 2 h. Reproduced with permission [70]. Copyright 2017 Elsevier.

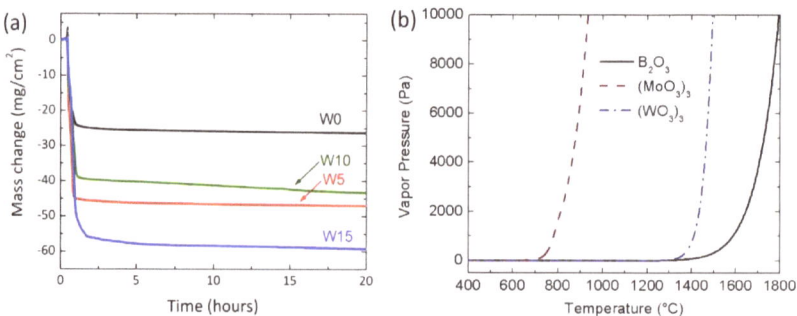

Figure 15. (**a**) Weight change curve of all samples at 1200 °C; (**b**) Variation of vapor pressure with temperatures. Reproduced with permission [70]. Copyright 2017 Elsevier.

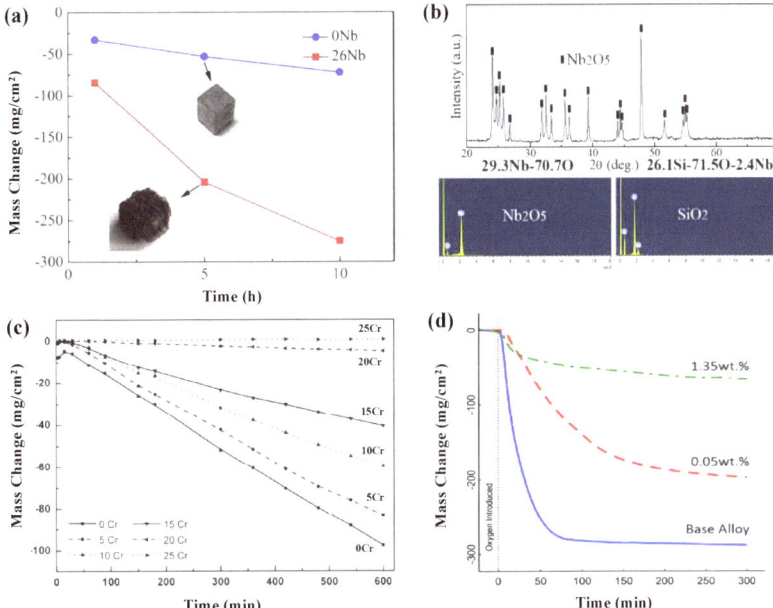

Figure 16. Mass change curves: (**a**) 0 Nb and 26 Nb after oxidation at 1300 °C with macroscopic images of samples oxidized for 5 h, (**c**) Mo-9Si-8B samples with different amounts of Cr oxidized at 750 °C, (**d**) Mo–2Si–1B samples with different amounts of Fe oxidized at 1100 °C; (**b**) XRD and EDX results of 26 Nb oxidized at 1300 °C for 5 h. (**a**–**d**) reproduced with permission [75,77,78], respectively. Copyright 2019 Elsevier, 2011 Springer Nature and 2012 Elsevier.

Compared with Nb, traditional metallic elements (like Cr, Fe) have a positive impact on the oxidative protection of Mo-Si-B ternary alloys, so they are favored by many researchers. The oxidation behavior of Cr modified M-Si-B alloy was relatively systematically explored by Burk et al. [49,77]. Figure 16c shows the mass change curve of Mo-9Si-8B (at.%) alloy containing different amounts of chromium when oxidized at 750 °C. It can be seen that the addition of Cr reduces the mass wastage of Mo-9Si-8B sample, and with the increase of Cr content, its weight loss gradually decreases. When the Cr content reached 25%, the alloy experienced little weight loss, showing excellent oxidative resistance. This was because the relatively high content of chromium made the alloy surface form a stable continuous Cr-oxide scale (consisting mainly of $Cr_2(MoO_4)_3$) that acted as a diffusion barrier of oxygen to passivate the alloy and prevent further oxidation, as being shown in Figure 17b. Moreover, Sossaman et al. [78] researched the antioxidation performance of Mo-2Si-1B (wt.%) composite at 1100 °C. In addition, they found that adding Fe can observably reduce the mass loss of the Mo-2Si-1B composite during transient oxidation process, especially after adding 1.35 wt.% Fe, the mass loss of the alloy even reduced by 75%, as displayed in Figure 16d. It was due to the fact that Fe reduced the stickiness of borosilicate film, thus facilitating the flow of oxide film and making it quickly cover the alloy surface to form a continuous scale (Figure 17d). In contrast, the Mo-2Si-1B alloy failed to form continuous borosilicate film during transient oxidation process (Figure 17c), so MoO_3 volatilized rapidly in the early oxidation stage, resulting in severe mass loss. In addition, a large number of $Fe_2(MoO_4)_3$ particles attached to the surface of the Mo-2Si-1B-1.35Fe (wt.%) alloy, which were filled in the pores of borosilicate scale and prevented the diffusion of oxygen. Therefore, the addition of Fe can significantly enhance the antioxidation properties of the alloy, and the same results were reported by Woodard et al. [79].

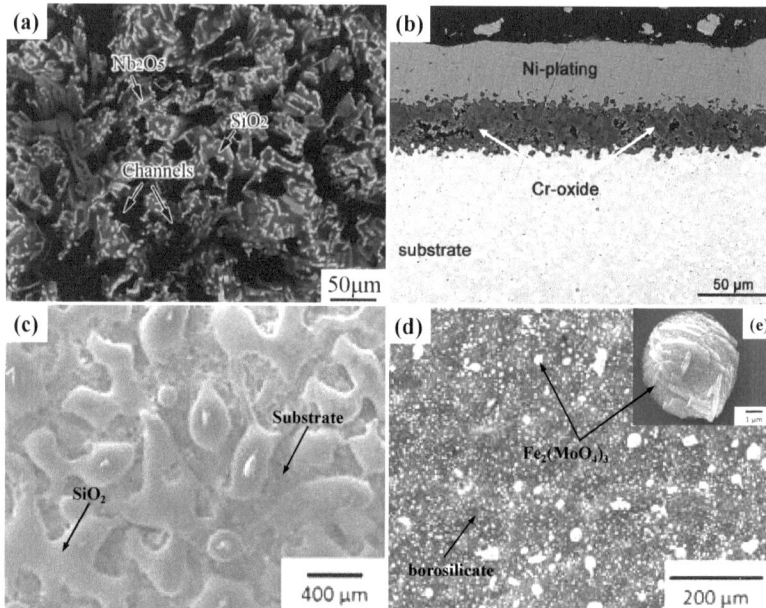

Figure 17. Surface SEM-BSE images of 26 Nb oxidized at 1300 °C for 5 h (**a**); Cross section of Mo-9Si-8B-25Cr after oxidation 10 h at 750 °C (**b**); SEM images of both samples after oxidation 1 min at 1100 °C: Mo–2Si–1B (**c**) and 1.35 wt.% Fe-doped Mo–2Si–1B (**d**) with a high magnification image (**e**). (**a**–**d**) reproduced with permission [75,77,78], respectively. Copyright 2019 Elsevier, 2011 Springer Nature and 2012 Elsevier.

3.2. Multiple Metallic Elements Co-Modified Mo-Si-B Alloys

In contrast with single metallic elements modified Mo-Si-B alloys, the modification mechanism of multiple metallic elements tends to be more complex. It has been previously described that thick and porous TiO2-borosilicate duplex scale was generated on the surface of Mo-Si-B composites after adding Ti, which led to rapid oxidation of the alloys. To further enhance the antioxidant capacity of Mo-Ti-Si-B composites, Zhao et al. [80] again added the fifth metallic element, such as Al or Cr, and synthesized the 35Mo-35Ti-20Si-10B (at.%), 35Mo-35Ti-15Si-10B-5Al (at.%) and 32.5Mo-32.5Ti-20Si-10B-5Cr (at.%) alloys, which were referred to as BASE, AL and CR, respectively. The microstructures of three samples are shown in Figure 18a–c. It was observed that the microstructures of the three alloys were composed of Mo_{ss} (white region), Mo_5SiB_2 (gray area) and Ti_5Si_3 (dark gray region). However, the addition of Al or Cr affects the morphology or volume fraction of individual phase. For example, adding Al refines the structure of the Mo_{ss} and Ti_5Si_3 phases (Figure 18b), and the addition of Cr reduces the volume fraction of the Mo_{ss} phase (Figure 18c). Therefore, BASE, AL and CR alloys can exhibit different oxidation behaviors at the same temperature. Figure 19a,b gives the change of unit area weight with oxidation time of the three alloys at 700 °C and 1100 °C, respectively. Overall, the weight loss of AL and CR alloys was lower than the BASE alloy at the both oxidation temperatures, which indicated that adding Al or Cr can improve the antioxidation function of the BASE sample. This was related to AL or CR alloy with finer Mo_{ss} and Ti_5Si_3 structures and lower Mo_{ss} volume fraction. It was worth noting that CR alloy had better antioxidant than AL and BASE alloys at 1100 °C (Figure 19b). For a deeper understanding of the enhancement mechanism of adding aluminum or chromium to the antioxidation of Mo-Ti-Si-B alloys, the surface and section of the postoxidation alloys were studied. Figure 18d,e are the surface SEM micrographs of AL and CR alloys during oxidation for 30 min at 700 °C, respectively.

It can be found that the surface borosilicate scales of AL and CR alloys had low viscosity and were easy to flow, which made the AL and CR alloys form the TiO$_2$-borosilicate duplex scale covering the substrate surface earlier than the BASE alloy. Therefore, the antioxidation of the AL and CR alloys were better than that of the BASE alloy at 700 °C. Compared to the cross-sections of three samples oxidized at 1100 °C for 24 h (Figure 18f–h), it can be seen that lots of holes and cracks appeared in the duplex scale of BASE and AL alloys, which provided a way for inward diffusion of oxygen. In contrast, continuous protection films are generated on the surface of CR alloy, and no holes appeared in the duplex scale. Therefore, the CR alloy possessed optimum oxidative resistance at 1100 °C. It was worth mentioning that there was no continuous Cr-oxide scale formed on the surface of CR sample during oxidation (Figure 18i). Similarly, there was no continuous Al-oxide scale on the surface of AL sample. This may be caused by the small content of Al or Cr in the samples. Therefore, it can be envisaged that to further enhance the antioxidation of alloys can be realized by increasing the concentration of Al or Cr.

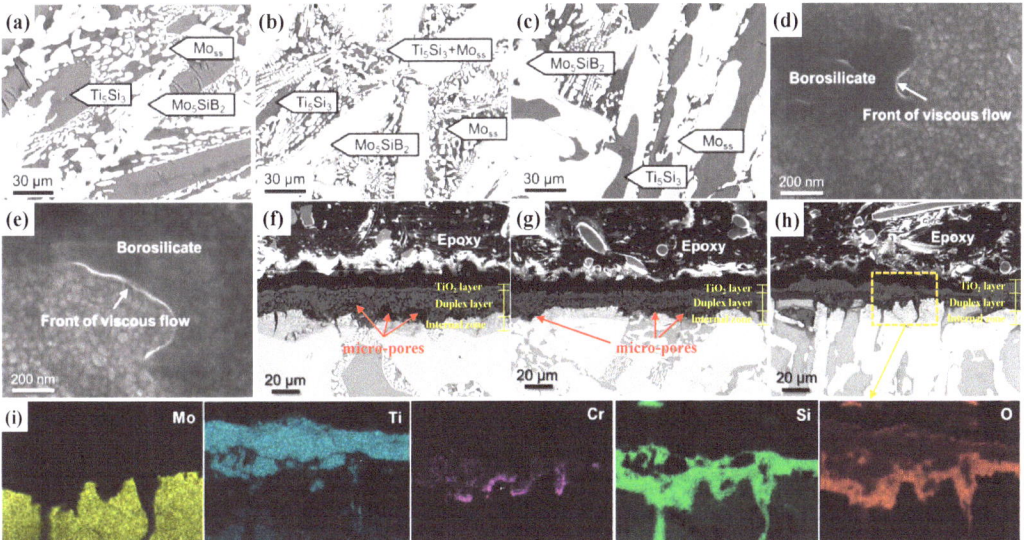

Figure 18. BSE images: (**a**) BASE, (**b**) AL and (**c**) CR; Surface SE images of different samples after oxidation 30 min at 700 °C: (**d**) AL and (**e**) CR; Cross-sectional BSE images of different samples after oxidation 24 h at 1100 °C: (**f**) BASE, (**g**) AL and (**h**) CR; (**i**) Elemental mappings of the micro-zone. Reproduced with permission [80]. Copyright 2020 Elsevier.

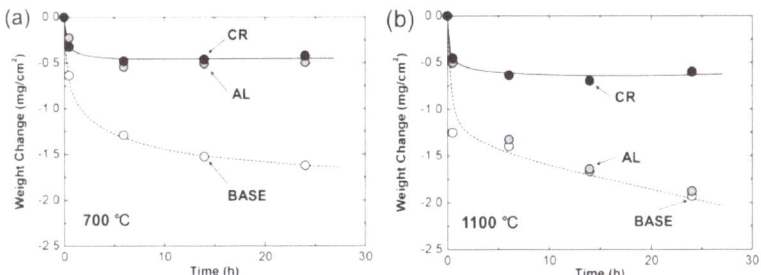

Figure 19. Weight change curves of different samples: (**a**) oxidized at 700 °C; (**b**) oxidized at 1100 °C. Reproduced with permission [80]. Copyright 2020 Elsevier.

3.3. Modification Mechanism of Metal Elements

According to the research on the microstructure and antioxidation performance of Mo-Si-B alloys containing metal elements, the modification mechanism of metal elements is obtained, as shown in Figure 20. The study reveals that the modification of the metal elements to the Mo-Si-B alloys is commonly achieved by changing the transient oxidation stage behavior of the alloy. As is seen from Figure 20b, the addition of Zr or Al cannot only refine the structure of the alloy, accelerate the formation of the borosilicate scale, but also play a "pinning" effect to inhibit the volatilization of molybdenum trioxide and internal diffusion of oxygen, thus improving the oxidation resistance of the alloy. Furthermore, the addition of an appropriate amount of Al can also promote the formation of dense $Al_2(MoO_4)_3$ or Al_2O_3 scale on the alloy surface, to passivate the alloy and prevent further oxidation. However, Zr or Al will compete with silicon and boron for oxidation, thus delaying the appearance of protective B_2O_3-SiO_2 film. Moreover, the oxidation of Zr or Al can form the large volume of oxide particles and distribute in the borosilicate scale. These particles will undergo phase transition at high temperature environments, leading to volume expansion or stress release. As a result, the oxide layer cracked and peeled off, and the oxidation resistance of the alloy deteriorated. Therefore, the addition of Zr or Al will show different modification effects on the antioxidation performance of the Mo-Si-B alloys, which may be related to preparation process, microstructure or B/Si ratio. Adding proper amount of Cr can form continuous protective $Cr_2(MoO_4)_3$ or Cr_2O_3 scale on the outer surface of B_2O_3-SiO_2 film, which can act as a diffusion barrier of oxygen, thus significantly enhancing the oxidative resistance of the alloy. The addition of a small amount of Fe can enhance the fluidity of borosilicate scale, and the $Fe_2(MoO_4)_3$ particles generated by oxidation fill in the gaps of the borosilicate scale and act as blocking phases. On the one hand, they prevent the internal diffusion of oxygen, on the other hand, they limit the volatilization of MoO_3, which have a positive effect on improving the antioxidation properties of the alloy. It can be observed from Figure 20c that the addition of Ti leads to the formation of thick and porous rutile-borosilicate duplex film on the surface of the alloy. In addition, the pores in the scale provide channels for the volatilization of molybdenum trioxide and inward spread of O_2, thus reducing the oxidative resistance of the alloy. The addition of Nb or W can destabilize the Mo_3Si structure, thus improving mechanical properties such as high temperature compressive strength and fracture toughness of the alloy. However, Nb oxidizes to form coarse and porous Nb_2O_5 particles, which attach to the borosilicate scale and provide a fast route for the inward diffusion of O_2, resulting in the deterioration of the oxidation behavior of alloy. Moreover, the WO_3 particles formed by W oxidation are also embedded in the borosilicate scale, which destroys the integrity of the borosilicate scale and adversely affects the antioxidation of the alloy. However, as the oxidation temperature rises, the WO_3 will gradually volatilize, so W will not seriously damage the oxidation resistance of the alloy. This also reveals that the pretreatment of the alloy at higher temperatures can effectively improve its service life.

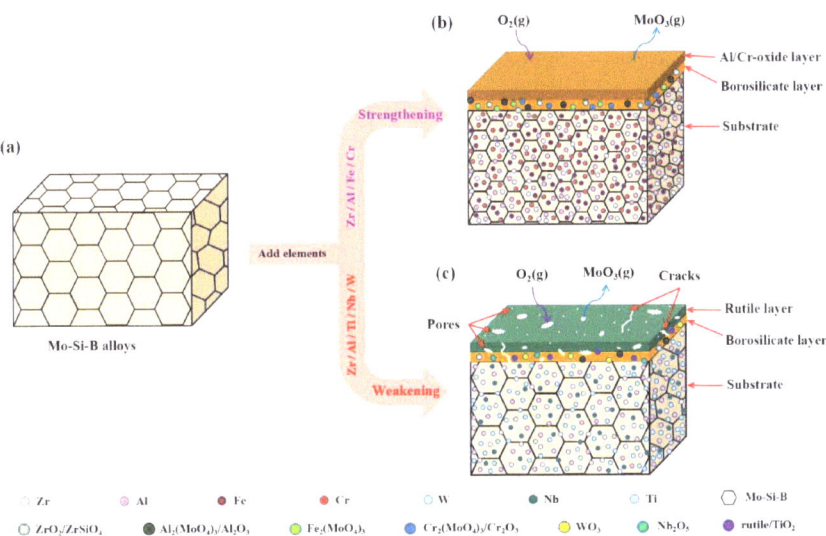

Figure 20. The schematic diagram of modification mechanism of metal elements. (**a**) Mo-Si-B alloys before oxidation (**b**) Strengthening (**c**) Weakening.

4. Conclusions

This work systematically studies the microstructure and oxidation behavior of Mo-Si-B ternary alloys. It is found that the antioxidation properties of Mo-Si-B ternary alloys are related to the processing method, grain siz,e and the content of silicon and boron elements. In the low silicon Mo-Si-B alloys, due to adding silicon and boron, the alloys form protective B_2O_3-SiO_2 films on the surface during oxidation process, so they have good oxidation resistance. Moreover, refining the microstructure of the alloy and moderately increasing the content of boron in the alloy can further enhance the antioxidation capacity of the low silicon Mo-Si-B alloys. By contrast, high silicon Mo-Si-B alloys are more prone to form continuous borosilicate films on the surface because of higher silicon content. Therefore, they exhibit better oxidation resistance than low silicon Mo-Si-B alloys.

Furthermore, the influences of metallic elements on the oxidation behavior of Mo-Si-B ternary alloys were also studied. It is shown that adding metallic elements can refine the grains and improve the microstructure of the Mo-Si-B alloys. Among them, adding elements such as Cr and Fe can significantly improve the antioxidation of Mo-Si-B alloys. In contrast, adding elements such as Ti, W and Nb will reduce the oxidation resistance of Mo-Si-B alloys. The modification effect of multiple metallic elements is usually more significant than single-metal modifications. It is worth noting that if the composition, preparation process or microstructure of the alloy are different, they may exhibit different oxidation behavior even if the same amount of Zr or Al elements is added. To further enhance the antioxidant effect of the alloy, the concentration of metal elements (such as Cr, Al) can be increased appropriately or the alloy can be pretreated at a higher temperature.

Author Contributions: The manuscript was written through the contributions of all authors. Y.Z.: conceptualization, investigation and supervision. Y.Z. and L.Y.: writing—original draft and image processing. L.Y., K.C. and F.S.: validation, resources, investigation, writing—review & editing. J.W., T.F. and X.Z.: visualization, writing—review & editing. All authors have read and agreed to the published version of the manuscript.

Funding: This work was supported by the Anhui Province Science Foundation for Excellent Young Scholars (2108085Y19) and National Natural Science Foundation of China (No. 51604049).

Institutional Review Board Statement: Not applicable.

Informed Consent Statement: Not applicable.

Data Availability Statement: Not applicable.

Acknowledgments: This work was supported by the Anhui Province Science Foundation for Excellent Young Scholars (2108085Y19) and National Natural Science Foundation of China (No. 51604049).

Conflicts of Interest: The authors declare no competing financial interest. The authors declare no conflict of interest.

References

1. Yoshimi, K.; Nakatani, S.; Suda, T.; Hanada, S.; Habazaki, H. Oxidation behavior of Mo5SiB2-based alloy at elevated tempera-tures. *Intermetallics* **2002**, *10*, 407–414. [CrossRef]
2. Liu, L.; Shi, C.; Zhang, C.; Voyles, P.; Fournelle, J.; Perepezko, J. Microstructure, microhardness and oxidation behavior of Mo-Si-B alloys in the Moss+Mo2B+Mo5SiB2 three phase region. *Intermetallics* **2020**, *116*, 106618. [CrossRef]
3. Zhang, Y.; Cui, K.; Gao, Q.; Hussain, S.; Lv, Y. Investigation of morphology and texture properties of WSi2 coatings on W substrate based on contact-mode AFM and EBSD. *Surf. Coat. Technol.* **2020**, *396*, 125966. [CrossRef]
4. Zhang, X.; Fu, T.; Cui, K.; Zhang, Y.; Shen, F.; Wang, J.; Yu, L.; Mao, H. The Protection, Challenge, and Prospect of Anti-Oxidation Coating on the Surface of Niobium Alloy. *Coatings* **2021**, *11*, 742. [CrossRef]
5. Cui, K.; Fu, T.; Zhang, Y.; Wang, J.; Mao, H.; Tan, T. Microstructure and mechanical properties of CaAl12O19 reinforced Al2O3-Cr2O3 composites. *J. Eur. Ceram. Soc.* **2021**, *41*, 7935–7945. [CrossRef]
6. Jung, J.; Zhou, N.X.; Jian, L. Effects of sintering aids on the densification of Mo–Si–B alloys. *J. Mater. Sci.* **2012**, *47*, 8308–8319. [CrossRef]
7. Wang, J.; Zhang, Y.; Cui, K.; Fu, T.; Gao, J.; Hussain, S.; AlGarni, T.S. Pyrometallurgical recovery of zinc and valuable metals from electric arc furnace dust—A review. *J. Clean. Prod.* **2021**, *298*, 126788. [CrossRef]
8. Kramer, M.; Thom, A.; Degirmen, O.; Behrani, V.; Akinc, M. Oxidation Behavior of Mo-Si-B Alloys in Wet Air. *Mater. Sci. Eng. A* **2004**, *371*, 335–342. [CrossRef]
9. Fu, T.; Cui, K.; Zhang, Y.; Wang, J.; Shen, F.; Yu, L.; Qie, J.; Zhang, X. Oxidation protection of tungsten alloys for nuclear fusion applications: A comprehensive review. *J. Alloys Compd.* **2021**, *884*, 161057. [CrossRef]
10. Obert, S.; Kauffmann, A.; Seils, S.; Boll, T.; Weiss, S.K.; Chen, H.; Anton, R.; Heilmaier, M. Microstructural and chemical consti-tution of the oxide scale formed on a pesting-resistant Mo-Si-Ti alloy. *Corros. Sci.* **2021**, *178*, 109081. [CrossRef]
11. Zhang, Y.Y.; Cui, K.K.; Fu, T.; Wang, J.; Shen, F.Q.; Zhang, X.; Yu, L.H. Formation of MoSi2 and Si/MoSi2 coatings on TZM (Mo-0.5Ti-0.1Zr-0.02C) alloy by hot dip silicon-plating method. *Ceram. Int.* **2021**, *47*, 23053–23065. [CrossRef]
12. Zhang, Y.Y.; Li, Y.G.; Bai, C.G. Microstructure and oxidation behavior of Si-MoSi2 functionally graded coating on Mo substrate. *Ceram. Int.* **2017**, *43*, 6250–6256. [CrossRef]
13. Wirkus, C.D.; Wilder, D.R. High-Temperature Oxidation of Molybdenum Disilicide. *J. Am. Ceram. Soc.* **1966**, *49*, 173–177. [CrossRef]
14. Fu, T.; Cui, K.; Zhang, Y.; Wang, J.; Zhang, X.; Shen, F.; Yu, L.; Mao, H. Microstructure and Oxidation Behavior of Anti-Oxidation Coatings on Mo-Based Alloys through HAPC Process: A Review. *Coatings* **2021**, *11*, 883. [CrossRef]
15. Thom, A.J.; Summers, E.; Akinc, M. Oxidation behavior of extruded Mo5Si3Bx–MoSi2–MoB intermetallics from 600–1600 °C. *Intermetallics* **2002**, *10*, 555–570. [CrossRef]
16. Majumdar, S.; Schliephake, D.; Gorr, B.; Christ, H.-J.; Heilmaier, M. Effect of Yttrium Alloying on Intermediate to High-Temperature Oxidation Behavior of Mo-Si-B Alloys. *Met. Mater. Trans. A* **2013**, *44*, 2243–2257. [CrossRef]
17. Majumdar, S. A study on microstructure development and oxidation phenomenon of arc consolidated Mo-Nb-Si-(Y) alloys. *Int. J. Refract. Met. Hard Mater.* **2018**, *78*, 76–84. [CrossRef]
18. Halvarsson, M.; Jonsson, T.; Ingemarsson, L.; Sundberg, J.M.; Svensson, E.; Johansson, L.G. Microstructural investigation of the initial oxidation at 1450 °C and 1500 °C of a Mo(Si,Al)2-based composite. *Mater. High Temp.* **2009**, *26*, 137–143. [CrossRef]
19. Zhang, Y.; Hussain, S.; Cui, K.; Fu, T.; Wang, J.; Javed, M.S.; Lv, Y.; Aslam, B. Microstructure and Mechanical Properties of MoSi2 Coating Deposited on Mo Substrate by Hot Dipping Processes. *J. Nanoelectron. Optoelectron.* **2019**, *14*, 1680–1685. [CrossRef]
20. Pan, Y.; Pu, D.; Yu, E. Structural, electronic, mechanical and thermodynamic properties of Cr–Si binary silicides from first-principles investigations. *Vacuum* **2021**, *185*, 110024. [CrossRef]
21. Seidenschnur, P. The intermediate and high-temperature oxidation behaviour of Mo(Si1-xAlx)2 intermetallic alloys. *Intermetallics* **1997**, *5*, 69–81.
22. Sharif, A. Effects of Re- and Al-alloying on mechanical properties and high-temperature oxidation of MoSi2. *J. Alloys Compd.* **2012**, *518*, 22–26. [CrossRef]
23. Ingemarsson, L.; Halvarsson, M.; Hellström, K.; Jonsson, T.; Sundberg, M.; Johansson, L.G.; Svensson, J.E. Oxidation behavior at 300–1000 °C of a (Mo,W)Si2-based composite containing boride. *Intermetallics* **2010**, *18*, 77–86. [CrossRef]
24. Samadzadeh, M.; Oprea, C.; Sharif, H.K.; Troczynski, T. Comparative studies of the oxidation of MoSi 2 based materials: High-temperature oxidation (1000–1600 °C). *Int. J. Refract. Met. Hard Mater.* **2017**, *69*, 31–39. [CrossRef]

25. Zhang, L.T.; Zhu, O.; Zhang, F.; Shan, A.D.; Wu, J.S. High-temperature oxidation of Mo-rich (Mo1-xNbx)Si2 pseudo-binary compounds. *Scr. Mater.* **2007**, *57*, 305–308. [CrossRef]
26. Zhang, Y.Y.; Zhao, J.; Li, J.H.; Lei, J.; Cheng, X. KEffect of hot-dip siliconizing time on phase composition and microstructure of Mo-MoSi2 high temperature structural materials. *Ceram. Int.* **2019**, *45*, 5588–5593. [CrossRef]
27. Sharif, A.A. High-temperature oxidation of MoSi2. *J. Mater. Sci.* **2010**, *45*, 865–870. [CrossRef]
28. Hellström, K.; Persson, P.; Ström, E. Oxidation behaviors and microstructural alterations of a Mo(Si,Al)2-based composite after heating at 1580°C either in a furnace (ex-situ) or via alternating current (in-situ). *J. Eur. Ceram. Soc.* **2015**, *35*, 513–523. [CrossRef]
29. Ingemarsson, L.; Halvarsson, M.; Engkvist, J.; Jonsson, T.; Hellström, K.; Johansson, L.G.; Svensson, J.E. Oxidation behavior of a Mo(Si,Al)2-based composite at 300–1000 °C. *Intermetallics* **2010**, *18*, 633–640. [CrossRef]
30. Yu, L.; Zhang, Y.; Fu, T.; Wang, J.; Cui, K.; Shen, F. Rare Earth Elements Enhanced the Oxidation Resistance of Mo-Si-Based Alloys for High Temperature Application: A Review. *Coatings* **2021**, *11*, 1144. [CrossRef]
31. Ingemarsson, L.; Hellström, K.; Canovic, S.; Jonsson, T.; Halvarsson, M.; Johansson, L.G.; Svensson, J.E. Oxidation behavior of a Mo(Si,Al)2 composite at 900–1600 °C in dry air. *J. Mater. Sci.* **2013**, *48*, 1511–1523. [CrossRef]
32. Roy, B.; Das, J.; Mitra, R. Transient stage oxidation behavior of Mo76Si14B10 alloy at 1150 °C. *Corros. Sci.* **2012**, *68*, 231–237. [CrossRef]
33. Supatarawanich, V.; Johnson, D.R.; Liu, C.T. Oxidation behavior of multiphase Mo–Si–B alloys. *Intermetallics* **2004**, *12*, 721–725. [CrossRef]
34. Supatarawanich, V.; Johnson, D.R.; Liu, C.T. Effects of microstructure on the oxidation behavior of multiphase Mo–Si–B alloys. *Mater. Sci. Eng. A* **2003**, *344*, 328–339. [CrossRef]
35. Becker, J.; Fichtner, D.; Schmigalla, S.; Schultze, S.; Heinze, C.; Küsters, Y.; Hasemann, G.; Schmelzer, J.; Krüger, M. Oxidation response of additively manufactured eutectic Mo-Si-B alloys. *IOP Conf. Ser. Mater. Sci. Eng.* **2020**, *882*, 12002. [CrossRef]
36. Mendiratta, M.G.; Parthasarathy, T.A.; Dimiduk, D.M. Oxidation behavior of αMo–Mo3Si–Mo5SiB2 (T2) three phase system. *Intermetallics* **2002**, *10*, 225–232. [CrossRef]
37. Parthasarathy, T.; Mendiratta, M.; Dimiduk, D. Oxidation mechanisms in Mo-reinforced Mo5SiB2(T2)–Mo3Si alloys. *Acta Mater.* **2002**, *50*, 1857–1868. [CrossRef]
38. Wang, F.; Shan, A.D.; Dong, X.P.; Wu, J.S. Oxidation behavior of Mo–12.5Si–25B alloy at high temperature. *J. Alloys Compd.* **2008**, *459*, 362–368. [CrossRef]
39. Wang, F.; Shan, A.-D.; Dong, X.-P.; Wu, J.-S. Oxidation behavior of multiphase Mo5SiB2 (T2)-based alloys at high temperatures. *Trans. Nonferrous Met. Soc.* **2007**, *17*, 1242–1247. [CrossRef]
40. Wang, F.; Shan, A.D.; Dong, X.P.; Wu, J.S. Microstructure and oxidation behavior of directionally solidifified Mo–Mo5SiB2 (T2)–Mo3Si alloys. *J. Alloys Compd.* **2008**, *462*, 436–441. [CrossRef]
41. Wang, F.; Shan, A.D.; Dong, X.P.; Wu, J.S. Microstructure and oxidation resistance of laser-remelted Mo-Si-B alloy. *Scr. Mater.* **2007**, *56*, 737–740. [CrossRef]
42. Rioult, F.; Imhoff, S.; Sakidja, R.; Perepezko, J. Transient oxidation of Mo–Si–B alloys: Effect of the microstructure size scale. *Acta Mater.* **2009**, *57*, 4600–4613. [CrossRef]
43. Choi, W.J.; Park, C.W.; Park, J.H.; Kim, Y.D.; Byun, J.M. Volume and size effects of intermetallic compounds on the high-temperature oxidation behavior of Mo-Si-B alloys. *Int. J. Refract. Met. Hard Mater.* **2019**, *81*, 94–99. [CrossRef]
44. Li, R.; Li, B.; Chen, X.; Wang, J.; Wang, T.; Gong, Y.; Ren, Z.; Zhang, G. Variation of phase composition of Mo-Si-B alloys induced by boron and their mechanical properties and oxidation resistance. *Mater. Sci. Eng. A* **2019**, *749*, 196–209. [CrossRef]
45. Zhang, G.J.; Kou, H.; Dang, Q.; Liu, G.; Sun, J. Microstructure and oxidation resistance behavior of lanthanum oxide-doped Mo-12Si-8.5B alloys. *Int. J. Refract. Met. Hard Mater.* **2021**, *30*, 6–11.
46. Majumdar, S.; Gorr, B.; Christ, H.J.; Schliephake, D.; Heilmaier, M. Oxidation mechanisms of lanthanum-alloyed Mo–Si–B. *Corros. Sci.* **2014**, *88*, 360–371. [CrossRef]
47. Majumdar, S.; Burk, S.; Schliephake, D.; Krüger, M.; Christ, H.-J.; Heilmaier, M. A Study on Effect of Reactive and Rare Earth Element Additions on the Oxidation Behavior of Mo–Si–B System. *Oxid. Met.* **2013**, *80*, 219–230. [CrossRef]
48. Burk, S.; Gorr, B.; Trindade, V.B.; Christ, H.J. High temperature oxidation of mechanically alloyed Mo-Si-B alloys. *Br. Corros. J.* **2009**, *44*, 168–175. [CrossRef]
49. Burk, S.; Christ, H.J. High-Temperature Oxidation Performance of Mo-Si-B Alloys: Current Results, Developments and Opportunities. *Adv. Mater. Res.* **2011**, *278*, 587–592. [CrossRef]
50. Wen, S.H.; Zhou, C.G.; Sha, J.B. Improvement of oxidation resistance of a Mo-62Si-5B (at.%) alloy at 1250 °C and 1350 °C via an in situ pre-formed SiO2 fabricated by spark plasma sintering. *Corros. Sci.* **2017**, *127*, 175–185. [CrossRef]
51. Pan, K.; Yang, Y.; Wei, S.; Wu, H.; Dong, Z.; Wu, Y.; Wang, S.; Zhang, L.; Lin, J.; Mao, X. Oxidation behavior of Mo-Si-B alloys at medium-to-high temperatures. *J. Mater. Sci. Technol.* **2021**, *60*, 113–127. [CrossRef]
52. Wang, J.; Li, B.; Li, R.; Chen, X.; Zhang, G. Bimodal α-Mo grain structure inducing excellent oxidation resistance in Mo-12Si-8.5B alloy at 1100 °C. *Int. J. Refract. Met. Hard Mater.* **2021**, *98*, 105533. [CrossRef]
53. Wang, J.; Li, B.; Li, R.; Chen, X.; Wang, T.; Zhang, G.J. Unprecedented oxidation resistance at 900 °C of Mo–Si–B composite with addition of ZrB2. *Ceram. Int.* **2020**, *46*, 14632–14639. [CrossRef]
54. Wang, J.; Li, B.; Shuai, R.; Wang, T.; Zhang, G. Enhanced oxidation resistance of Mo-12Si-8.5B alloys with ZrB2 addition at 1300 °C. *J. Mater. Sci. Technol.* **2018**, *34*, 635–642. [CrossRef]

55. Burk, S.; Gorr, B.; Trindade, V.B.; Christ, H.-J. Effect of Zr Addition on the High-Temperature Oxidation Behaviour of Mo–Si–B Alloys. *Oxid. Met.* **2009**, *73*, 163–181. [CrossRef]
56. Burk, S.; Gorr, B.; Christ, H.-J. High temperature oxidation of Mo–Si–B alloys: Effect of low and very low oxygen partial pressures. *Acta Mater.* **2010**, *58*, 6154–6165. [CrossRef]
57. Kumar, N.K.; Das, J.; Mitra, R. Effect of Zr Addition on Microstructure, Hardness and Oxidation Behavior of Arc-Melted and Spark Plasma Sintered Multiphase Mo-Si-B Alloys. *Met. Mater. Trans. A* **2019**, *50*, 2041–2060. [CrossRef]
58. Schliephake, D.; Azim, M.; Von Klinski-Wetzel, K.; Gorr, B.; Christ, H.-J.; Bei, H.; George, E.; Heilmaier, M. High-Temperature Creep and Oxidation Behavior of Mo-Si-B Alloys with High Ti Contents. *Met. Mater. Trans. A* **2013**, *45*, 1102–1111. [CrossRef]
59. Azim, M.; Burk, S.; Gorr, B.; Christ, H.J.; Schliephake, D.; Heilmaier, M.; Bornemann, R.; Bolívar, P.H. Effect of Ti (macro-) alloying on the high-temperature oxidation behavior of ternary Mo-Si-B alloys at 820–1300 °C. *Oxid. Met.* **2013**, *80*, 231–242. [CrossRef]
60. Schliephake, D.; Gombola, C.; Kauffmann, A.; Heilmaier, M.; Perepezko, J.H. Enhanced oxidation resistance of Mo–Si–B–Ti alloys by pack cementation. *Oxid. Met.* **2017**, *88*, 267–277. [CrossRef]
61. Das, J.; Mitra, R.; Roy, S.K. Oxidation behaviour of Mo-Si-B-(Al, Ce) ultrafine-eutectic dendrite composites in the temperature range of 500–700 °C. *Intermetallics* **2011**, *19*, 1–8. [CrossRef]
62. Das, J.; Roy, B.; Kumar, N.K.; Mitra, R. High temperature oxidation response of Al/Ce doped Mo-Si-B composites. *Intermetallics* **2017**, *83*, 101–109. [CrossRef]
63. Cui, K.K.; Zhang, Y.Y.; Fu, T.; Hussain, S.; Al Garni, T.S.; Wang, J.; Zhang, X.; Ali, S. Effects of Cr2O3 content on microstructure and mechanical properties of Al2O3 matrix composites. *Coatings* **2021**, *11*, 234. [CrossRef]
64. Paswan, S.; Mitra, R.; Roy, S.K. Isothermal oxidation behaviour of Mo-Si-B and Mo-Si-B-Al alloys in the temperature range of 400–800 °C. *Mater. Sci. Eng. A* **2006**, *424*, 251–265. [CrossRef]
65. Paswan, S.; Mitra, R.; Roy, K.S. Oxidation behaviour of the Mo-Si-B and Mo-Si-B-Al alloys in the temperature range of 700–1300 °C. *Intermetallics* **2007**, *15*, 1217–1227. [CrossRef]
66. Paswan, S.; Mitra, R.; Roy, S. Nonisothermal and Cyclic Oxidation Behavior of Mo-Si-B and Mo-Si-B-Al Alloys. *Met. Mater. Trans. A* **2009**, *40*, 2644–2658. [CrossRef]
67. Cui, K.; Zhang, Y.; Fu, T.; Wang, J.; Zhang, X. Toughening Mechanism of Mullite Matrix Composites: A Review. *Coatings* **2020**, *10*, 672. [CrossRef]
68. Yamauchi, A.; Yoshimi, K.; Murakami, Y.; Kurokawa, K.; Hanada, S. Oxidation Behavior of Mo-Si-B In Situ Composites. *Solid State Phenom.* **2007**, *127*, 215–220. [CrossRef]
69. Ray, P.; Ye, Y.; Akinc, M.; Kramer, M. Effect of Nb and W substitutions on the stability of the A15 Mo3Si phase. *J. Alloys Compd.* **2012**, *537*, 65–70. [CrossRef]
70. Karahan, T.; Ouyang, G.; Ray, P.; Kramer, M.; Akinc, M. Oxidation mechanism of W substituted Mo-Si-B alloys. *Intermetallics* **2017**, *87*, 38–44. [CrossRef]
71. Zhang, Y.Y.; Qie, J.M.; Cui, K.K.; Fu, T.; Fan, X.L.; Wang, J.; Zhang, X. Effect of hot dip silicon-plating temperature on microstructure characteristics of silicide coating on tungsten substrate. *Ceram. Int.* **2020**, *46*, 5223–5228. [CrossRef]
72. Zhang, Y.; Cui, K.; Fu, T.; Wang, J.; Qie, J.; Zhang, X. Synthesis WSi2 coating on W substrate by HDS method with various deposition times. *Appl. Surf. Sci.* **2020**, *511*, 145551. [CrossRef]
73. Zhang, Y.; Fu, T.; Cui, K.; Shen, F.; Wang, J.; Yu, L.; Mao, H. Evolution of surface morphology, roughness and texture of tungsten disilicide coatings on tungsten substrate. *Vacuum* **2021**, *191*, 110297. [CrossRef]
74. Ray, P.K.; Akinc, M.; Kramer, M.J. Applications of an extended Miedema's model for ternary alloys. *J. Alloys Compd.* **2010**, *489*, 357–361. [CrossRef]
75. Yang, T.; Guo, X.P. Comparative studies on densification behavior, microstructure, mechanical properties and oxidation re-sistance of Mo-12Si-10B and Mo3Si-free Mo-26Nb-12Si-10B alloys. *Int. J. Refract. Met. Hard Mater.* **2019**, *84*, 104993. [CrossRef]
76. Behrani, V.; Thom, A.J.; Kramer, M.J.; Akinc, M. Microstructure and oxidation behavior of Nb-Mo-Si-B alloys. *Intermetallics* **2006**, *14*, 24–32. [CrossRef]
77. Burk, S.; Gorr, B.; Krüger, M.; Heilmaier, M.; Christ, H.-J. Oxidation behavior of Mo-Si-B-(X) alloys: Macro- and microalloying (X = Cr, Zr, La2O3). *JOM J. Miner. Met. Mater. Soc.* **2011**, *63*, 32–36. [CrossRef]
78. Sossaman, T.; Sakidja, R.; Perepezko, J. Influence of minor Fe addition on the oxidation performance of Mo–Si–B alloys. *Scr. Mater.* **2012**, *67*, 891–894. [CrossRef]
79. Woodard, S.R.; Raban, R.; Myers, J.F.; Berczik, D.M. Oxidation Resistant Molybdenum. U.S. Patent 6652674-B1, 25 November 2003.
80. Zhao, M.; Xu, B.; Shao, Y.; Liang, J.; Wu, S.; Yan, Y. Oxidation behavior of Moss–Ti5Si3–T2 composites at intermediate and high temperatures. *Intermetallics* **2020**, *118*, 106702. [CrossRef]

Article

Physical Experiment and Numerical Simulation on Thermal Effect of Aerogel Material for Steel Ladle Insulation Layer

Limin Zhang [1,2], Liguang Zhu [1,3,*], Caijun Zhang [1,2], Zhiqiang Wang [4], Pengcheng Xiao [1,2] and Zenxun Liu [1,2]

1. Hebei Provincial High-Quality Steel Continuous Casting Engineering Technology Research Center, North China University of Science and Technology, Tangshan 063000, China; Lichenxiao@ncst.edu.cn (L.Z.); zhangcaijun@126.com (C.Z.); xiaopc@ncst.edu.cn (P.X.); liuzenxun@heuu.edu.cn (Z.L.)
2. College of Metallurgy and Energy, North China University of Science and Technology, Tangshan 063210, China
3. School of Science and Engineering, Hebei University of Science and Technology, Shijiazhuang 050018, China
4. Social Science, University of California, Irvine, CA 92612, USA; zhiqiaw3@uci.edu
* Correspondence: zhuliguang@ncst.edu.cn

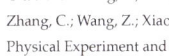

Citation: Zhang, L.; Zhu, L.; Zhang, C.; Wang, Z.; Xiao, P.; Liu, Z. Physical Experiment and Numerical Simulation on Thermal Effect of Aerogel Material for Steel Ladle Insulation Layer. *Coatings* **2021**, *11*, 1205. https://doi.org/10.3390/coatings11101205

Academic Editor: Paolo Castaldo

Received: 14 August 2021
Accepted: 26 September 2021
Published: 1 October 2021

Publisher's Note: MDPI stays neutral with regard to jurisdictional claims in published maps and institutional affiliations.

Copyright: © 2021 by the authors. Licensee MDPI, Basel, Switzerland. This article is an open access article distributed under the terms and conditions of the Creative Commons Attribution (CC BY) license (https://creativecommons.org/licenses/by/4.0/).

Abstract: The selection of lining material for a steel ladle is important to heat preservation of molten steel. Aerogel insulation materials have very low thermal conductivity, however, they are rarely used in steel ladles. In this paper, the application of a new silica aerogel material on the steel ladle insulation layer is tested, and a new calculation method is designed to study its insulation effect. In other words, the ladle wall temperature is obtained by finite element model (FEM) and experiments, then the heat emission from the ladle wall is calculated by the Boltzmann mathematical model according to the ladle wall temperature, and the temperature loss of molten steel is calculated inversely according to the heat emission of ladle wall. Compared with the original steel ladle (comparison ladle), the application effect is analyzed. Due to the stable heat storage of the ladle wall after refining, the validity of the models are verified in ladle furnace (LF) process. The results show that the new calculation method is feasible, and the relevant parameter settings in the FEM and Boltzmann mathematical model are correct. Finally, after using the new aerogel insulation material, the temperature of molten steel is reduced by 16.67 °C, and the production cost is reduced by CNY 5.15/ton of steel.

Keywords: aerogel insulation material; steel ladle; finite element model; Boltzmann mathematical model; temperature; cost

1. Introduction

In the steelmaking process, the steel ladle is an extremely important piece of equipment. It has multi-functional uses, including temporary smelting, molten steel heat preservation and transportation [1,2]. In the context of the current global advocacy of energy conservation and emission reduction, the steel ladle is particularly important for the thermal insulation effect of molten steel. When the heat preservation effect of the steel ladle is poor [3–5]: (1). The steel ladle dissipates more heat and the temperature of molten steel drops quickly. To increase the temperature of the molten steel, higher power consumption is necessary, which will also lead to longer power supply time and more deoxidizer consumption. The corrosion of the refractory lining is accelerated, thus the steel ladle life is reduced and the cost is significantly increased; (2). The extension of the steelmaking time has led to lower steelmaking efficiency, insufficient production capacity and increased cost; (3). The surface temperature of the steel shell is increased, which increases the thermal fatigue effect of the steel shell and significantly reduces the service life of the steel shell; (4). Rapid heat dissipation of the steel ladle leads to increased temperature fluctuations of the molten steel, which has a very adverse effect on stable continuous casting and improving the yield of continuous casting slabs. In short, the steel ladle heat preservation effect is good, and it will have a positive effect on all aspects of production.

The steel ladle insulation layer plays a vital role in the temperature preservation of molten steel. There are many types of steel ladle insulation materials, mainly including asbestos panel, ordinary ceramic fiber panel, calcium silicate panel, clay brick and other materials [6–8]. At present, 10 mm or 20 mm thick asbestos sheets are mainly used for steel ladle insulating layers. For example, South Africa Iscor ltd and Indian Pat Industries Company uses 10 mm thickness asbestos sheets; China Hegang Iron and Steel Group, China Shandong Iron and Steel Group, and other enterprises use 20 mm thickness asbestos sheets [9,10]. However, these materials have a poor insulation effect, and the service life is short [11,12]. In recent years, the silica aerogel composite insulation panel (SACIP) has attracted great interest in thermal insulation applications due to its ultralow thermal conductivity [13,14] (see Figure 1), and is applied to the high temperature industry, gradually replacing the products with poor thermal insulation performance, high energy consumption and those that are harmful to the human body and the environment, such as glass fiber, rock, wool, and aluminum silicate fiber [15,16]. However, at present, the application of aerogel insulation materials in the metallurgical industry is relatively rare, and there are few researchers studying its application, especially calculating the influence of steel ladle insulation layer heat transfer on molten steel temperature. For example, in the period 2018–2020, Jia Changjiang, Sun Ye and others [17–19] measured the application of 10mm thickness Mk type aerogel insulation panel in 80-ton of steel ladle insulating layer in Hebei Iron and Steel Group and Tianjin Steel Plant, and achieved good results. However, they didn't provide a calculation method for the thermal insulation of the aerogel composite insulation panel. It is only a general conclusion based on the actual measurement results. Moreover, the aerogel material and the thickness of the insulating layer were also quite different from the research in this paper, and the insulation and energy saving effects were not considered from the whole steelmaking process. In the course of the research, Romao, I and others [20] established a 2D heat transfer model of the ladle wall and obtained a better effect. However, they did not include the effect of free surface radiation. Taddeo [21] compared and analyzed the temperature drop rate of liquid steel between a traditional ladle and the one with an insulation layer in a steel plant. The results indicated that the speed of temperature drop of the latter was lower, that is to say, the heat preservation effect of the ladle with insulating lining structure was better, and 10.5% electric energy was saved. Others [22] used composite reflective insulation panels in the insulation layer, and used lightweight high-strength castables in the insulation layer, which significantly improved the performance of insulation of steel ladle, but the radiative heat and considered only convective heat transfer wasn't addressed. Other researchers [23] have undertaken research of ladle heat preservation performance, molten steel temperature, ladle wall temperature and so on, but as an integrated approach, especially the research on the interaction of the three, hasn't been adequately addressed. Therefore, based on the test ladle, in this paper, the thermal insulation performance of SACIP is studied comprehensively, that being the relationship between the temperature of molten steel, the temperature of the ladle wall and SACIP insulation efficiency, as well as the effect of using the test ladle on the cost compared with the original ladle (comparison ladle) were studied.

In this work, the surface temperature of the outer steel shell (OSS) and the temperature of the molten steel were measured with an infrared temperature measuring thermometer and a thermocouple temperature measuring thermometer, respectively. Then in the analysis process, the 2D heat transfer finite element model (FEM) of the ladle wall was established and analyzed by SOLIDWORKS 19.0 and ANSYS APDL software to study the temperature change of the ladle wall, then the temperature loss of molten steel was calculated by Boltzmann mathematical equation through the heat dissipation of the ladle wall. According to the measured temperature of molten steel and the ladle wall, the accuracy of the 2D heat transfer model and Boltzmann mathematical equation were verified. In the analysis process, the method used to measure the temperature of the molten steel and the temperature of the ladle wall, 2D heat transfer model of grid points, heat transfer units, the boundary temperature parameter, and the selection and calculation methods of some parameters such as the Rayleigh number in Boltzmann mathematical model were discussed extensively. In addition, the test ladle was compared with the original ladle in terms of ladle wall temperature, molten steel temperature, steelmaking cost, temperature drop rate, etc. Finally, some reasonable suggestions for a building model were presented.

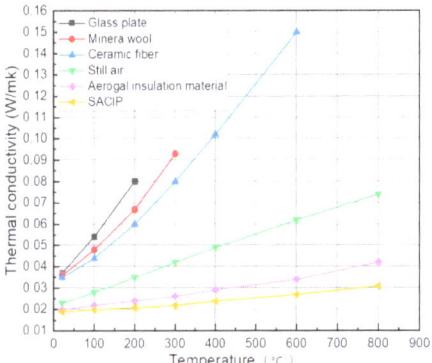

Figure 1. Comparison of thermal conductivity of different insulation materials.

2. Material and Methods

2.1. Silica Aerogel Composite Insulation Panel

2.1.1. Composition and Morphology

SACIP was purchased from Beijing Zhongheng New Material Refractory Co., Ltd., Beijing, China, it is a kind of high strength viscoelastic microporous insulation material and it is composed of nanoscale silicon powder, metal aluminum foil, and fiber cloth. Its surface is metal aluminum foil (the detailed structure is shown in Figure 2). The size is 10 × 400 × 600 mm (thickness × width × length). The finished product is shown in Figure 3.

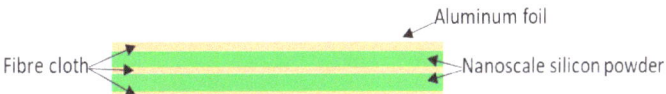

Figure 2. SACIP structure diagram.

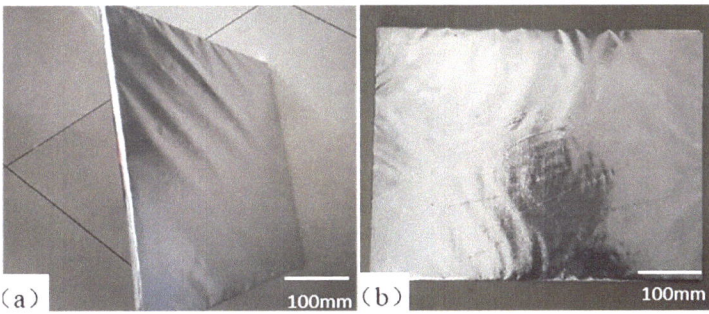

Figure 3. (a) Side of SACIP and (b) front of SACIP.

2.1.2. Main Performance Parameters

The SACIP and OCRIP were sent to the National Refractory Quality Supervision and Inspection Center of Luoyang, China for testing. The test results are shown in Table 1.

Table 1. Test data.

Material	Physical Parameters		
	Density (kg/m^3)	Refractoriness (°C)	Thermal Conductivity (W/mK) {Hot Surface Temperature: 800 (°C)}
OCRIP	536	1560	0.073
SACIP	149	1760	0.031

Table 1 shows that the SACIP has high refractoriness, low bulk density and low thermal conductivity. The thermal conductivity is only 0.031 W/mK at 800 °C, which is 0.042 W/mK lower than the OCRIP.

2.1.3. Method of Use

The steel ladle is composed of two parts: steel shell and refractory materials. The refractory materials are built against the inner wall of the steel shell in order, which are inner (working) layer, back (safety) layer, and insulating layer. This study took 130-ton of slab steel ladle as the research object. Between the back layer and steel shell, avoiding the steel nails on the steel shell, were pasted two layers of 20 mm thickness SACIP with castable. The profile structure of the test ladle is shown in Figure 4.

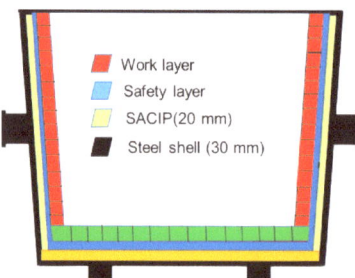

Figure 4. The profile structure of the test steel ladle.

The on-site construction of SACIP is shown in Figure 5.

Figure 5. (**a**) on-site construction operation and (**b**) construction completed.

2.2. Surface Temperature of Outer Steel Shell

2.2.1. Numerical simulation of Temperature Field of Steel Ladle Wall

(1). Establishment of Model

Without affecting the calculation results, in order to facilitate the establishment of the model, the following assumptions were proposed [24,25]: (1). The steel ladle trunnels, bottom breathable bricks and some driving devices structural reinforcements valves have little effect on the temperature field in the FEM of steel ladle wall, so these parts are omitted in the model; (2). The contact thermal resistance between the steel ladle insulating layer and the steel shell is negligible; (3). Due to the small slope angle of the steel ladle, the steel ladle can be regarded as a cylinder after removing the mechanical parts for loading and unloading around the steel ladle; (4). The temperature of molten steel is the same everywhere in the steel ladle; (5). Because the steel ladle has axially symmetrical geometry, a part of the steel ladle wall is taken for the FEM. The thermal conductivity equation is shown in Equation (1).

$$\lambda \left(\frac{\partial^2 T}{\partial r^2} + \frac{\partial^2 T}{\partial z^2} \right) = 0 \qquad (1)$$

where λ is the thermal conductivity of the material, W/mK; T is the temperature, K; r is the normal distance, m; z is axial distance, m.

(2). Meshing

Figure 6 is a 2D model of the ladle wall which was designed by SOLIDWORKS 19.0, and imported into APDL of ANSYS 2020 R2 for processing. Because it's a solid heat transfer model, plane 77 was chosen specifically for the heat transfer model which is a 2-dimensional 8-node thermal unit entity (each node has only one degree of freedom– temperature, and each element has a consistent temperature shape function), suitable for 2D steady-state or transient thermal analysis [26]. Therefore, the plane 77 was used in the ladle wall model to improve calculation accuracy. The steel ladle wall model was divided into 4450 units, and the total number of nodes is 11,236. The type of unit is quadrilateral. The advantage of this quadrilateral is that on the one hand it can reduce the amount of calculation. However, on the other hand, it doesn't cause the mesh to be too distorted, resulting in inaccurate results.

By comparing temperature difference of ladle wall with the counterparts calculated in the numerical model which contains a double quantity of mesh elements (including the test ladle wall model and the comparison ladle model), the difference in temperature difference remains constant, further refining of the mesh did not affect the results significantly, and the numerical computation independence on mesh is confirmed. Considering this result, the steel ladle wall model was divided into 4450 units, and the total number of nodes is 11,236. The type of unit is quadrilateral.

The temperature condition and convective heat transfer coefficient were used at the inner wall of thermal insulation material and the surface of OSS, the density, specific heat capacity, and thermal conductivity of materials were set, respectively. The calculation process adopted the stepped; the solution time was automatically set, and the steady-state solution was adopted.

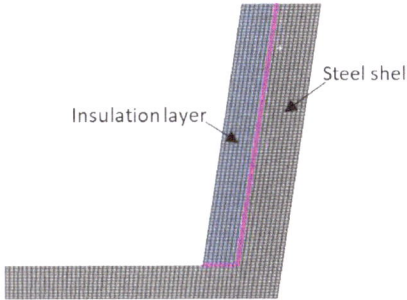

Figure 6. Mesh division of steel ladle wall.

(3). Initial and Boundary Conditions

According to steel ladle heat transfer form, in the process of heat transfer to the steel ladle wall, the inner wall of the steel ladle was set as the first type of boundary condition, as shown in Equation (2).

$$T|_{r=0} = T_\omega \tag{2}$$

where r is the radius of steel ladle wall, m; and T_ω is the temperature of the inner wall, °C.

The steel shell was set as the third type of boundary condition. As shown in Equation (3).

$$-K\frac{dT}{dr}\bigg|r = h(T|r - T_f) \tag{3}$$

In the Equation (3), k is thermal conductivity along the normal direction, W/mK; T_f is the ambient temperature, °C, and h is the comprehensive convective heat transfer coefficient of air, W/m²K. T is the OSS temperature, °C.

Under the working condition of steel ladle containing molten steel, the working layer of the steel ladle directly contacts the high temperature molten steel of 1570–1630 °C. According to related references [27,28], the temperature of the inner wall of the insulating layer is between 700–900 °C, take the average value 800 °C (T_ω) in this paper according to references [27,28], the OSS is in an external environment of 30 °C (T_f), and the heat transfer between OSS and the surrounding environment is convective heat exchange and radiation heat exchange. Generally, when the temperature exceeds 300 °C, the radiation heat transfer is strong, when the temperature is below 300 °C, the radiation heat transfer phenomenon is not obvious [29]. In the process of steel ladle turnaround, the surface temperatures of OSS are mainly 200–300 °C, the radiation heat transfer has little effect during the entire heat transfer process. Therefore, convective heat transfer and radiation heat transfer are converted into comprehensive convective heat transfer coefficient [30], so h is 8.1 W/mK. The physical parameters of the insulating layer are derived from the test data of Table 1.

2.2.2. The Surface Temperature of Outer Steel Shell Measurement

Many factors affect the temperature of the steel shell, especially the steel ladle used for converter steelmaking. The steel ladle is affected by its own conditions and external factors (such as the state of steel ladle, vacancy time, the tapping temperature of the molten steel, the type of alloy material, the added weight, and argon blowing stirring time). The temperature of the steel shell changes greatly, which is not easy to accurately compare. When the steel ladle is out-station of the LF, the molten steel temperature is uniform, sufficient and stable heat storage in each layer of the steel ladle wall, and the temperature fluctuation is small. After comprehensive thinking, the surface temperature data of the OSS after soft blowing at the out-station of the LF were collected.

Meanwhile, in order to reduce the influence of other production conditions on the measurement of surface temperature of OSS, it is stipulated that the steel ladle, steel type, temperature of molten steel, measuring temperature time and measuring temperature position are all measured under the same conditions.

In addition, during the use of steel ladle in a mid-maintenance period (the time from the steel ladle to the repair of the inner lining of the steel ladle is called a mid-maintenance period), the condition of steel ladle will gradually deteriorate due to the erosion of molten steel and other reasons, the heat storage of steel ladle wall is slightly different in a medium repair period, so in order to improve the measurement accuracy, the furnace age(the time interval between steel ladle passing through the same process each time is called one furnace age) was divided into early stage (1–50) and later stage(51–100). Details are as follows:

(1) Steel ladle: steel ladle during the first mid-maintenance period.
(2) Steel type: SPHC (The first S is the abbreviation of Steel, P is the abbreviation of Plate, H is the abbreviation of hot Heat, and C is the abbreviation of Commercial, which generally means hot-rolled Steel Plate and Steel belt).
(3) Temperature of molten steel: 1580 ± 2 °C.

(4) Temperature measuring tools: infrared temperature measuring thermometer.
(5) Measuring temperature time: after soft blowing at the out-station of the LF.
(6) Measuring temperature position: the same position of the steel ladle.

The temperature measuring tool uses a handheld infrared temperature measuring thermometer (Smart Sensor AS882) (Dongguan Wanchuang Electronic Products Co., Ltd., Dongguan, China). The temperature measuring range is from −18 °C to 1650 °C. The infrared temperature measuring thermometer and its working principle are shown in Figure 7.

Figure 8 shows on-site temperature measurement method. That is, within the effective distance, the infrared probe is aligned with the 0.01 m^2 circle area marked of the surface OSS which is 1000 mm away from the upper stiffener and 800 mm from the right stiffener, and the temperature key is pressed for about 2 s to display the measured temperature. Measuring three times, then take the average of the three measured temperature values. A total of 70 steel ladles were measured (there are 36 test furnaces and 34 comparison furnaces).

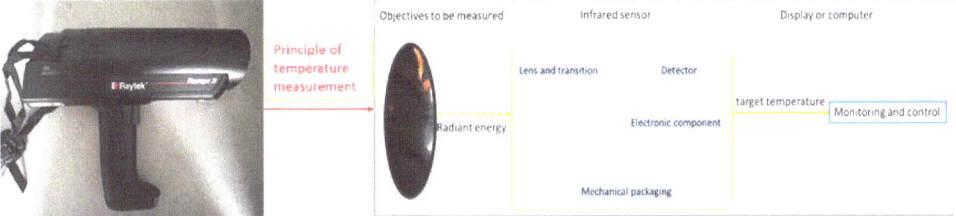

Figure 7. Infrared temperature measuring thermometer and working principle.

Figure 8. On-site temperature measurement.

2.3. Mathematical Model of Temperature Loss of Molten Steel

2.3.1. Heat Transfer Model

There are three basic ways of heat transfer: heat conduction, heat convection and heat radiation [31]. The heat transfer of high-temperature molten steel from the inside to the outside of the steel ladle is also divided into three parts. The first part is that the high-temperature molten steel heat enters the steel ladle wall through thermal conduction, among them, part of the heat is stored in the steel ladle wall, and the other part of the heat is radiated and convective through the OSS; the second part is the heat dissipation through the bottom of the steel ladle, the principle is the same as the first part; the third part is the heat dissipation of molten steel indirectly through the slag surface, which also includes three basic methods of heat conduction, heat convection and heat radiation at the same time. The proportion of heat dissipation in these three parts is shown in Figure 9 [32]. So the outward heat dissipation of the steel ladle becomes the calculation of the outward heat dissipation from the steel ladle wall.

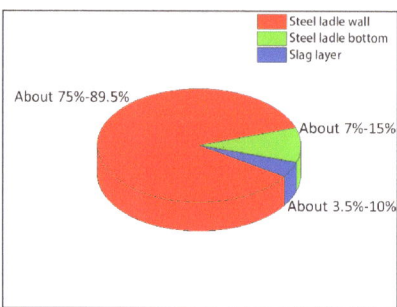

Figure 9. Steel ladle heat dissipation distribution ratio.

According to the steel ladle heat transfer model, think of the steel ladle as a heat source with a constant surface temperature that dissipates heat to the outside. The difference of heat dissipation at different surface temperatures is only investigated here. In other words, the steel ladle and the molten steel are considered as a whole object. Since the steel ladle filled with molten steel has much more heat than its surface heat dissipation, the internal temperature of the steel ladle is much higher than the ambient temperature, and there is no forced cooling medium on the outer surface of the steel ladle. Therefore, the steel ladle can be used as a cylinder with a constant surface temperature. The amount of heat dissipation has nothing to do with the temperature of the molten steel, only related to factors such as steel ladle surface temperature and environmental conditions. Therefore, the formula can be used to calculate the change of molten steel temperature [33].

Heat loss of the steel ladle heat transfer is Equation (4).

$$\varphi = \varphi_1 + \varphi_2 \tag{4}$$

where φ_1 is the heat flow of thermal radiation of OSS, W; φ_2 is the heat flow of thermal convection of the OSS, W.

The steel shell's radiant heat flow can be described as follows.

$$\varphi_1 = \varepsilon A \sigma \left(T_1^4 - T_2^4 \right) \tag{5}$$

where ε is the emissivity of steel shell; A is the OSS surface area, m^2; σ is the Boltzmann constant (5.67×10^{-8} W/m^2 K^4); T_1 is the surface temperature of OSS, K; T_2 is the ambient temperature, K.

φ_2 can be regarded as the convective heat transfer of a vertical cylinder, which is applicable to the convective heat transfer Equation (6).

$$\varphi_2 = A h \Delta T \tag{6}$$

where h is convective heat transfer coefficient the surface of OSS, W/m^{2}k; A is the heat transfer surface area of OSS, m^2; ΔT is the difference between the surface of OSS and the surrounding environment, K. h can be estimated as (7).

$$h = \frac{Nu \lambda}{l} \tag{7}$$

where Nu is Nusselt Number, λ is the thermal conductivity of air, W/mK; l is the height of the OSS, m. Nu can be estimated as (8).

$$Nu = C(GrPr)^n \tag{8}$$

where Gr is the Grashof Number, Pr is the Prandtl Number, C, n is the constant. Gr can be estimated as (9).

$$Gr = \left(\frac{g\alpha\Delta T H^3}{v^2}\right) \quad (9)$$

where g is the gravitational acceleration, m/s^2; α is the volume expansion coefficient of air (the air in this paper is an ideal gas), the value is 3.676×10^{-3} [34]; ΔT is the difference between the surface of OSS and the surrounding environment, K; H is the height of steel ladle, m; v is the kinematic viscosity of air, m^2/s.

2.3.2. Related Parameters of Model

According to the surface properties of different objects "Table of Emissivity of Various Surfaces" [35], the value of the steel shell ε is 0.80. According to Table 2, A is 44.71 m^2.

Table 2. Steel ladle related parameters.

Parameters	Value
D_{Ladle}	3.56 m
H	4.0 m
π	Constant

$T_{qualitative\ temperature}$ as the qualitative temperature of air, and its value is half the sum of ambient temperature and surface temperature of OSS. The values of v, λ, and Pr are shown in Table 3.

Table 3. Physical parameters of air (303 K).

Temperature $T_{qualitative\ temperature}$ (+273 K)	Thermal Conductivity λ ($\times 10^{-2}$ W/mK)	Kinematic Viscosity v ($\times 10^{-6}$ m^2/s)	Prandtl Number Pr
130	3.42	26.63	0.6850
135	3.45	27.21	0.6846
140	3.49	27.80	0.6840
145	3.53	28.38	0.6834
150	3.57	28.95	0.6830
155	3.60	29.56	0.6824
160	3.64	30.09	0.6820
165	3.67	30.66	0.6817
170	3.71	31.31	0.6815
175	3.74	31.96	0.6810

The value of C and n can be determined by the product of $GrPr$ (see Table 4). When the minimum and maximum surface temperatures of the OSS are taken into $GrPr$, the value range of $GrPr$ is shown in Formula (11). According to Formula (11) and Table 4, C is 0.135 and n is 1/3.

$$\frac{9.8 \times 3.676 \times 10^{-3} \times 289 \times 4^3}{(31.9 \times 10^{-6})^2} \leq GrPr \leq \frac{9.8 \times 3.676 \times 10^{-3} \times 203 \times 4^3}{(26.63 \times 10^{-6})^2} \quad (10)$$

$$6.54 \times 10^{11} \leq GrPr \leq 6.59 \times 10^{11} \quad (11)$$

Table 4. The relationship of $GrPr$ to C and n.

Rayleigh Number (the Product of $GrPr$)	C	n
1×10^{-3}–5×10^2	1.18	1/8
5×10^2–2×10^7	0.54	1/4
2×10^7–1×10^{13}	0.135	1/3

2.3.3. Verification of the Model

Temperature data of molten steel are collected respectively: (1). Measure molten steel temperature data immediately after soft blowing of LF; (2). The temperature data of molten steel with an interval of 15 min from the last measurement. According to the temperature difference, the actual temperature drop rate of molten steel is calculated. Then, the results of the mathematical model are verified.

3. Results and Discussion

3.1. Surface Temperature of Outer Steel Shell

3.1.1. The Simulated Results

According to the simulation results, the steady-state temperature field distribution cloud map of the two different thermal insulation materials on the ladle wall is obtained (see Figures 10 and 11). After the heat flux passes through the insulation layer, the following conclusions were obtained (see Table 5):

(1). The temperature of the comparative ladle's OSS was 309 °C, and the temperature at the interface OCRIP and steel shell was 417 °C. It can be calculated that the thermal resistance rate of the SACIP is 47.9%.

(2). The temperature of the test ladle OSS was 242 °C, and the temperature at the interface between SACIP and steel shell was 304 °C (see Figure 8). It can be calculated that the thermal resistance rate of the SACIP is 62.0%.

It can be seen that the thermal resistance of the SACIP was 14.1% higher than the OCRIP, and the surface temperature of OSS was 67 °C lower. The difference of the temperature of OSS between comparison ladle and test ladle is similar to that reported by Luo B [36], where they established that convective mechanisms were dominant in the heat transfer of temperature.

Table 5. Simulation results of steel ladle wall.

Items	Temperature (°C)		Thermal Resistance
	OSS	Interface	
Comparison ladle wall	309	417	47.9%
Test ladle wall	242	304	62.0%
Difference value (the comparison ladle wall minus the test ladle wall)	67	113	14.1%

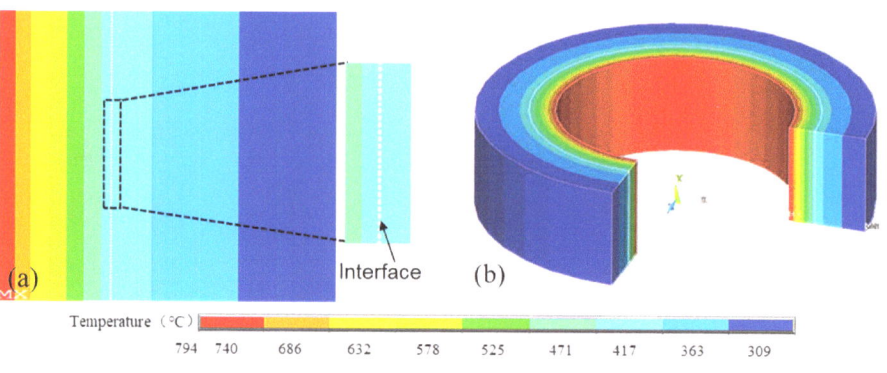

Figure 10. (**a**) Comparison ladle wall slice and (**b**) expand 3D temperature field distribution cloud map.

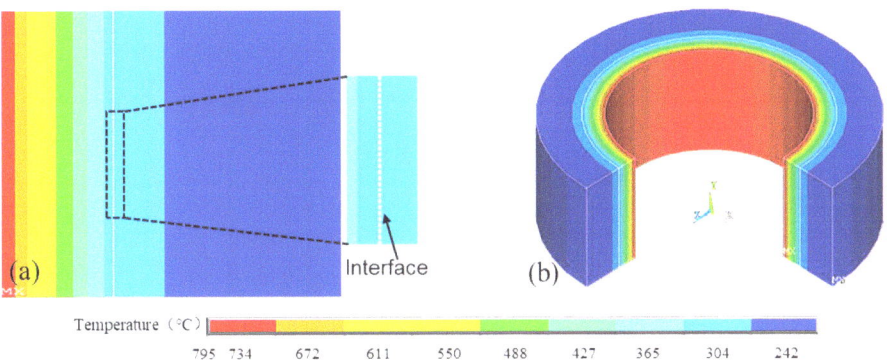

Figure 11. (**a**) Test ladle wall slice and (**b**) expand 3D temperature field distribution cloud map.

3.1.2. Measuring Result

Figure 12 shows the surface temperature of the OSS measured with an infrared temperature measuring thermometer. It can be seen that the surface average temperature of the test ladle's OSS fluctuates between 233–260 °C, and the surface average temperature of the comparison ladle's OSS fluctuates between 309–319 °C. In addition, the surface average temperature of the test ladle's OSS is 59–73 °C lower than the comparison ladle when the steel ladle furnace age is 1–100.

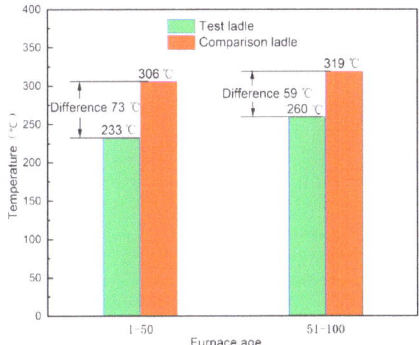

Figure 12. The surface average temperature of OSS measured.

3.1.3. Simulation Result Verification

Table 6 shows that the measurement results of the surface temperature OSS of test ladle and comparison ladle, and their average values range from 233–260 °C and 306–319 °C, respectively. The simulation results of test ladle and comparison ladle are 242 °C and 309 °C, respectively. The simulation results of the OSS surface temperature are all within the range of the measured temperature fluctuations, which verifies the validity of the finite element analysis of the ladle wall. This shows that the initial boundary temperature of the finite element was a reasonable choice at 800 °C, the selection of Plane 77 element is correct, also shows that the grid segmentation method, step solution setting is correct. However, Farrera-Buenrostro, J. E and others [37] pointed out that the temperature of the inner wall boundary of the insulation layer varies with factors such as the thermal conductivity of the insulation material and the state of molten steel. Therefore, this is the deficiency of the 2D heat transfer model. However, in this paper, on the basis of a large number of previous simulation results [38,39], the selected boundary temperature and model assumptions are feasible, and the experimental results also prove that the selected boundary conditions are within a reasonable range.

The results of measurement and simulation show that the insulation effect of the test ladle is better than for the comparison ladle; that is to say, SACIP has a better insulation effect than OCRIP. However, from the actual measurement results, the difference between the temperature of the test ladle wall in the early stage and the later stage is 27 °C, and the difference between the temperature of the comparison ladle wall is 13 °C, indicating that the temperature of the test ladle wall is greatly changed, and the specific reasons are still not clear. In subsequent tests, a large number of ladle wall temperatures should be measured to reduce the impact of less temperature data on the data. Meanwhile, more stringent temperature measurement measures should be established.

Table 6. Surface temperature of OSS.

Steel Ladle Condition	Surface Temperature of OSS (°C)		
	Measurement Result (Average Value)		Simulation Result
	Early Stage (1–50 Furnace Age)	Later Stage (51–100 Furnace Age)	
Test ladle	233	260	242
Comparison ladle	306	319	309

3.2. Temperature Loss of Molten Steel

3.2.1. Results of Mathematical Model Calculation

The heat flux of the comparison ladle and the test ladle $\varphi_{c\,ladle}$ and $\varphi_{t\,ladle}$ can be calculated by Formulas (4)–(9). According to (12), the temperature loss of molten steel can be calculated.

$$\Delta T = \frac{Q}{mC_p} = \frac{\varphi t}{mC_p} \quad (12)$$

where: ΔT is the temperature difference of molten steel, °C; Q is the total heat transferred by OSS, J; m is the quality of molten steel, kg; C_P is the specific heat capacity of molten steel, J/kg °C; t is the time of a furnace age under working conditions, s.

The drop rate of molten steel temperature can be calculated by Formula (13).

$$\Delta V = \frac{\Delta T}{t} \quad (13)$$

where: ΔV is the drop rate of steel molten temperature, °C/min; ΔT is the temperature difference of molten steel, °C; t is the time of a furnace age under working conditions, s.

Related parameter values are shown in Table 7.

Table 7. Related parameter values for Formulas (12) and (13).

Parameter	Value
φ	0.717×10^6 W
m	1.30×10^5 kg
t	5400 s
C_P (1600 °C)	0.837×10^3 J/kg °C

To sum up, in a furnace age, the test ladle saves 16.67 °C of molten steel temperature loss than the comparison ladle. The calculated results of the mathematical model are very close to the 18.8 °C drop of steel water temperature measured in Reference [40], the main reason for the difference of 2.13 °C is that the thermal conductivity of the insulation layer is 0.042 W/mK, ladle walls transfer heat outward quickly. The molten steel drop rate of the test ladle is 0.18 °C/min lower than the comparison ladle (see Appendix A for calculation process of temperature loss of molten steel).

3.2.2. Measuring Result

Table 8 shows the actual measured drop rate of steel molten temperature.

Table 8. Actual molten steel temperature drop rate.

Steel Ladle Condition	A [a] (Furnace Age)	N [b] (Furnace)	T [c] (°C)	T [d] (°C)	T [e] (min)	V [f] (°C/min)
Comparison ladle	Early stage	24	1585.4	1577.1	15.4	0.54
Test ladle	(1–50)	23	1583.1	1576.6	15.7	0.41
Comparison ladle	later stage	18	1584.9	1577.2	14.8	0.52
Test ladle	(51–100)	20	1585.2	1579.2	15.1	0.40

[a] The stage of steel ladle; [b] Total number of steel ladle furnaces collected; [c] Average temperature of molten steel after soft blowing; [d] Average temperature of molten steel on continuous caster platform; [e] Temperature measurement interval. [f] Temperature drop rate of molten steel.

It can be seen from Table 8, when the steel ladle furnace age is 1–50, the temperature drop rates of comparison ladle and test ladle are 0.54 °C/min and 0.41 °C/min, respectively, the difference temperature drop rates of is 0.13 °C /min. when the steel ladle furnace age is 51–100; the temperature drop rates of comparison ladle and test ladle are 0.52 °C/min and 0.40 °C/min, respectively, the difference temperature drop rates of is 0.12 °C /min. Therefore, when the steel ladle furnace age is 1–100, the temperature drop rate of the test ladle can be 0.12–0.13 °C/min lower than comparison ladle.

According to the calculating model, the molten steel drop rate of the test ladle is 0.18 °C/min lower than the comparison ladle in a furnace age. The difference of molten steel temperature drop rate between test ladle and comparison ladle is 0.05–0.06 °C/min lower than obtained by calculation model. The reason is that the thermal insulation effect of the test ladle is better than comparison ladle, the heat loss of the ladle wall is small, the molten steel temperature is high, and the heat loss of the ladle slag layer and bottom is greater than the comparison ladle. However, in the mathematical calculation model, the heat loss of the slag layer and the ladle bottom is not considered, only the heat loss of ladle wall is considered [41].

3.3. Cost Comparison

In the production, the test ladle adopts the measure that the refining temperature of the LF is lowered by 16.67 °C than the comparison ladle. The relevant smelting parameters of the 130-ton LF are shown in Table 9 (CNY: Chinese Yuan).

It can be seen from Table 9, When the LF out-station molten steel temperature of the test ladle is 16.67 °C lower than the comparison ladle, compared with the comparison ladle, the total cost of electricity and electrode saved by the test ladle is CNY 6.01/ton.

When the insulating layer is removed, the steel ladle furnace age is 600 furnaces, the steel ladle volume is 130 tons on average, the price of the SACIP is CNY 1200/m^2, the steel ladle insulating layer is double layer, and the usable area is 89.4 m^2, Therefore, it is concluded that the cost of using the SACIP in the test ladle is CNY 1.38/ton.

According to the factory's annual production cost data, the cost of using the OCRIP for the comparison ladle is CNY 0.52/ton. Therefore, the cost of the insulating layer of the test ladle exceeds the cost of the comparison ladle by CNY 0.86/ton. The cost of the test ladle for saving electrodes and electricity is CNY 6.01/ton, so the total cost savings of the test ladle is CNY 5.15/ton compared to the comparison ladle.

Table 9. The LF related parameters.

Each Ton of Molten Steel Rises by 1 °C		Reference Electricity Price (CNY/kwh)	Reference Electrode Price (CNY/kg)
Electricity Consumption (°C/kwh)	Electrode Consumption (kg/ton)		
0.69	0.0034	0.475	9.72

4. Conclusions

To study the application effect of steel ladle insulation materials, a new research method is designed in this paper. The main conclusions can be drawn as follows:

1. The actual measurement of the surface OSS temperature shows that the surface OSS average temperature of the test ladle is 59–73 °C lower than the comparison OSS. According to the simulation results of the steel ladle wall, the surface OSS temperature of the test ladle is 67 °C lower than that the comparison OSS. The simulation results are within the range of the average temperature of the actual measured surface OSS, which proves the accuracy of the FEM and the selection of related parameters.
2. According to the calculation model of molten steel temperature loss, compared with the comparison ladle, the test ladle in a furnace age, can save the temperature loss of molten steel 16.67 °C, and reduce the temperature drop rate of molten steel by 0.18 °C/min. In addition, when the steel ladle furnace age is 1–100, the temperature drop rate of the test ladle can be 0.12–0.13 °C/min lower than comparison ladle. Although the difference of molten steel temperature drop rate between test ladle and comparison ladle is 0.05–0.06 °C/min lower than obtained by calculation model, the Boltzmann mathematical model is still valid in evaluating the effect of the insulation layer and relevant parameters and solutions are correct.
3. In the LF, by reducing the temperature of molten steel in the test ladle by 16.67 °C, the cost of a test ladle can be reduced by CNY 5.15/ton compared with the comparison ladle. Using silica aerogel composite insulation panels on a steel ladle plays a very positive role in reducing production costs and energy consumption.
4. The new method designed in this paper to calculate the thermal insulation effect of steel ladle insulation layers is feasible. The ladle wall temperature is obtained by finite element model and experiment, then the heat emission from the ladle wall is calculated by the Boltzmann equation according to ladle wall temperature, and the temperature loss of molten steel is calculated inversely according to the heat emission.
5. In a furnace age, the surface temperature of OSS of the test ladle measured fluctuated by a larger amount than did the comparison ladle. The specific reason for this is still unclear, and further study is required.

Author Contributions: L.Z. (Limin Zhang): Writing—original draft, Writing—review and editing, Conduct experiment, Data, Graphics; L.Z. (Liguang Zhu): Project administration, Methodolog, Review, Funding, Goals and Aims; C.Z.: Contacting with the plant, Formal Analysis; P.X.: Contacting with the plant; Z.W.: Assist in translation, Formal Analysis; Z.L.: Visualization, review. All authors have read and agreed to the published version of the manuscript.

Funding: This work was funded by the Nature Science Foundations of Hebei Grant Nos. CXZZBS2020130, E2020209005, National Natural Science Foundation of China (51904107), Tangshan Talent Subsidy project(A202010004).

Institutional Review Board Statement: Not applicable.

Informed Consent Statement: Not applicable.

Data Availability Statement: No new data were created or analyzed in this study. Data sharing is not applicable to this article.

Conflicts of Interest: There is no interest conflict with others.

Nomenclature

Abbreviation	Description	Unit
C_P	Special heat capacity	J/kgk
T	Temperature	°C
t	Time	s
r	Ladle wall radius	m
z	Ladle wall thickness	m
k	Thermal conductivity	W/mK
h	convective heat transfer coefficient	W/m²K
A	Area	m²
Nu	Nusselt Number	dimensionless
D	Diameter	m
H	Height	m
Gr	Grashof Number	dimensionless
Pr	Prandtl Number	dimensionless
C	Constant determined by experiment	dimensionless
n	Constant determined by experiment	dimensionless
g	Gravitational acceleration	m/s²
Q	Heat	J
m	Quality	kg
Greek Symbols		
ρ	Density	kg/m³
λ	Thermal conductivity	W/mK
φ	Heat flow	W
ε	Emissivity	dimensionless
σ	Boltzmann constant	W/m² K⁴
α	Volume expansion coefficient	dimensionless
ν	Kinematic viscosity	m²/s

Appendix A. Mathematical Model Calculation Process of Temperature Loss of Molten Steel

The steel ladle furnace age is 1–50, suppose: the radiation heat dissipation of the test OSS is φ_{t1}, W; the convective heat transfer is φ_{t2}, W; the radiation heat dissipation of the comparative OSS is φ_{c1}, W; the convective heat transfer is φ_{c2}, W; $\varphi_{C\ ladle(1-50)}$ is the sum of φ_{c1} and φ_{c2}, W; $\varphi_{T\ ladle(1-50)}$ is the sum of φ_{t1} and φ_{t2}, W.

According to the Formula (A1):

$$\varphi_1 = \varepsilon A \sigma \left(T_1^4 - T_2^4 \right) \tag{A1}$$

Parameter value in the Formula (A1): $\varepsilon = 0.8$; $A = 44.71$ m²; $\sigma = 5.67 \times 10^{-8}$ W/m² K⁴; the values of T_1 and T_2 are shown in Table A1.

Calculated:

$$\varphi_{t1} = 0.8 \times 44.71 \times 5.67 \times 10^{-8} \times \left\{ (233+273.15)^4 - (30+273.15)^4 \right\} = 0.116 \times 10^6 \text{ W}$$

$$\varphi_{t2} = 0.8 \times 44.71 \times 5.67 \times 10^{-8} \times \left\{ (260+273.15)^4 - (30+273.15)^4 \right\} = 1.018 \times 10^6 \text{ W}$$

$$\varphi_{c1} = 0.8 \times 44.71 \times 5.67 \times 10^{-8} \times \left\{ (306+273.15)^4 - (30+273.15)^4 \right\} = 0.211 \times 10^6 \text{ W}$$

$$\varphi_{c2} = 0.8 \times 44.71 \times 5.67 \times 10^{-8} \times \left\{ (319+273.15)^4 - (30+273.15)^4 \right\} = 1.246 \times 10^6 \text{ W}$$

$$\varphi_{C\ ladle(1-50)} - \varphi_{T\ ladle(1-50)} = (1.246+0.211) \times 10^6 - (0.116+1.018) \times 10^6 = 0.323 \times 10^6 \text{ W}$$

Table A1. Surface temperature of OSS after the LF out-station.

Steel Ladle Condition	Surface Temperature of OSS (+273 K)		
	Measurement Result		Simulation Result
	Early Stage (1–50 Furnace Age)	Later Stage (51–100 Furnace Age)	
Test ladle	233	260	242
Comparison ladle	306	319	309

When the steel ladle furnace age is 51–100, suppose: the heat dissipation of the test OSS is φ_{t3}, W; the convective heat transfer is φ_{t4}, W; The radiation heat dissipation of the comparative shell is φ_{c3}, W; the convective heat transfer is φ_{c4}, W; $\varphi_{C\,ladle(51-100)}$ is the sum of φ_{c3} and φ_{c4}, W; $\varphi_{T\,ladle(51-100)}$ is the sum of φ_{t3} and φ_{t4}, W.

According to the Formulas (A2)–(A5):

$$\varphi_2 = Ah\Delta T \tag{A2}$$

$$h = \frac{Nu\lambda}{l} \tag{A3}$$

$$Nu = C(GrPr)^n \tag{A4}$$

$$Gr = \left(\frac{g\alpha\Delta TH^3}{v^2}\right) \tag{A5}$$

Parameter value in the Formulas (A2)–(A5) are shown in Section 2.3 of the paper. Calculated:

$$\varphi_{t3} = 44.71 \times \frac{0.135 \times (6.59 \times 10^{11})^{\frac{1}{3}} \times 3.42 \times 10^{-2}}{4} \times (233 - 30) = 0.09 \times 10^6 \text{ W};$$

$$\varphi_{t4} = 44.71 \times \frac{0.135 \times (6.58 \times 10^{11})^{\frac{1}{3}} \times 3.53 \times 10^{-2}}{4} \times (260 - 30) = 0.11 \times 10^6 \text{ W};$$

$$\varphi_{c3} = 44.71 \times \frac{0.135 \times (6.55 \times 10^{11})^{\frac{1}{3}} \times 3.71 \times 10^{-2}}{4} \times (306 - 30) = 0.13 \times 10^6 \text{ W};$$

$$\varphi_{t3} = 44.71 \times \frac{0.135 \times (6.54 \times 10^{11})^{\frac{1}{3}} \times 3.74 \times 10^{-2}}{4} \times (319 - 30) = 0.14 \times 10^6 \text{ W}.$$

$$\varphi_{C\,ladle(50-100)} - \varphi_{T\,ladle(50-100)} = 0.07 \times 10^6 \text{ W}.$$

φ_{Total} is the difference between the heat dissipation of the comparative steel ladle and the test steel ladle. Therefore, the mid-repair period of each steel ladle age (the steel ladle furnace age is 1–100, the steel ladle turnaround once). The value of φ is as follows.

$$\varphi = (0.323 + 0.07) \times 10^6 \text{ W} = 0.33 \times 10^6 \text{ W}.$$

According to the heat dissipation Formula (A6).

$$\Delta T = \frac{Q}{mC_p} = \frac{\varphi_{Total} t}{mC_P} \tag{A6}$$

where: ΔT is the decreasing temperature of molten steel due to the heat dissipation by the steel ladle shell, °C; Q is the heat dissipation of the OSS, J; m is the quality of molten steel, kg; C_P is the specific heat capacity of molten steel, J/kg °C; t is the steel ladle turnaround once time under working conditions, s.

Related parameters of Formula (A6) are shown in Table A2.

Table A2. Related parameter values for Formula (A6).

Parameter	Value
φ	0.717×10^6 W
m	1.30×10^5 kg
t	5400 s
C_P (1600 °C)	0.837×10^3 J/kg·°C

$$\Delta T = \frac{Q}{mC_p} = \frac{\varphi_{\text{Total}} t}{mC_P} = \frac{0.33 \times 10^6 \times 5400}{1.3 \times 10^5 \times 0.837 \times 10^3} = 16.37 \,°\text{C}$$

The drop rate of molten steel temperature can be calculated by Formula (A7).

$$\Delta V = \frac{\Delta T}{t} \tag{A7}$$

where: ΔV is the drop rate of steel water temperature, °C/min; ΔT is the temperature difference of molten steel, 16.37 °C; t is the steel ladle turnaround once time under working conditions, 5400 s.

To sum up,

$$\Delta V = \frac{\Delta T}{t} = \frac{16.37}{5400} = 0.18 \,°\text{C/min}$$

References

1. Glaser, M.; Gornerup, M.; Du, S.C. Fluid flow and heat transfer in the ladle during teeming. *Steel Res. Int.* **2011**, *82*, 827–835. [CrossRef]
2. Wang, Y.; Ai, X.G.; Liu, F.; Zhong, W.M.; Huang, R.H. Physical simulation of symmetric alternating bottom blowing mixing behavior in double orifice ladle. *China Metall.* **2017**, *27*, 18–21.
3. Glaser, B.; GorRnerup, M.; Du, S.C. Thermal modelling of the ladle preheating process. *Steel Res. Int.* **2011**, *82*, 1425–1434. [CrossRef]
4. Wang, S.S.; Guo, X.C.; Jiang, W.H.; Li, Q.; Zhang, F. Numerical simulation and field practice of flow field in 100 t bottom blown argon ladle. *China Metall.* **2018**, *28*, 51–55.
5. Liu, Z.Z.; Guo, H.Z. Development and present situation concering heat transfer research of ladle. *Iron Steel Res.* **2007**, *2*, 59–63.
6. Dong, P.L. Physical simulation of 210 t ladle bottom blowing process optimization. *Iron Steel* **2016**, *51*, 41–45.
7. Zhang, Y.Y.; Cui, K.K.; Fu, T.; Wang, J.; Shen, F.Q.; Zhang, X.; Yu, L.H. Formation of $MoSi_2$ and $Si/MoSi_2$ coatings on TZM (Mo-0.5Ti-0.1Zr-0.02C) alloy by hot dip silicon-plating method. *Ceram. Int.* **2021**, *47*, 23053–23065. [CrossRef]
8. Wohrmeyer, C.; Gao, S.M.; Ping, Z.F.; Parr, C. Corrosion mechanism of MgO-CMA-C ladle brick with high service life. *Steel Res. Int.* **2019**, *91*, 1900436. [CrossRef]
9. Chang, W.J. Influence Analysis and Life Prediction of New Structure Ladle with Super-Insulation and Lightweight. Master's Thesis, Wuhan University of Science and Technology, Wuhan, China, 1 June 2018.
10. He, W.X. Development on Longevity and High Efficient Thermal Insulation Technique of Ladle with Thin Layer and Its Application. Master's Thesis, Chongqing University, Chongqing, China, 1 June 2007.
11. Onyeaju, M.C.; Osarolube, E.; Chukwuocha, E.O.; Ekuma, C.E.; Omasheye, G. Comparison of the thermal properties of asbestos and polyvinylchloride (PVC) ceiling sheets. *Mater. Sci. Appl.* **2012**, *3*, 240–244. [CrossRef]
12. Khalaf, F.M. Using crushed clay brick as coarse aggregate in concrete. *J. Mater. Civ. Eng.* **2006**, *18*, 518–526. [CrossRef]
13. Hall, C.; Wilson, M.A.; Hoff, W.D. Kinetics of long-term moisture expansion in fired-clay brick. *J. Am. Ceram. Soc.* **2011**, *94*, 3651–3654. [CrossRef]
14. Zhang, Y.Y.; Fu, T.; Cui, K.K.; Shen, F.Q.; Wang, J.; Yu, L.H.; Mao, H.B. Evolution of surface morphology, roughness and texture of tungsten disilicide coatings on tungsten substrate. *Vacuum* **2021**, *191*, 110297. [CrossRef]
15. Cui, K.; Zhang, Y.; Fu, T.; Hussain, S.; Saad Algarni, T.; Wang, J.; Zhang, X.; Ali, S. Effects of Cr_2O_3 content on microstructure and mechanical properties of Al_2O_3 matrix composites. *Coatings* **2021**, *11*, 234. [CrossRef]
16. Gruber, D.; Harmuth, H. Thermomechanical behavior of steel ladle linings and the influence of insulations. *Steel Res. Int.* **2014**, *84*, 512–518. [CrossRef]
17. Jia, C.J.; Song, D.J.; Sun, J.B. Research on the application of new type thermal insulation material for casting ladle. *Foundry Equip. Technol.* **2018**, *1*, 42–44.
18. Sun, Y.; Zhao, S.H.; Li, Z.S. Development of aerogel insulation board and its thermal insulation effect in ladle. *Tianjin Sci. Technol.* **2018**, *8*, 59–61.
19. Lv, W.; Mao, Z.; Yuan, P.; Jia, M. Pruned bagging aggregated hybrid prediction models for forecasting the steel temperature in ladle furnace. *Steel Res. Int.* **2014**, *85*, 405–414. [CrossRef]

20. Romao, I.; Nduagu, E.; Fagerlund, J.; Gando-Ferreira, L.M.; Zevenhoven, R. CO_2 fixation using magnesium silicate minerals. Part 2: Energy efficiency and integration with iron-and steelmaking. *Energy* **2012**, *41*, 203–211. [CrossRef]
21. Taddeo, L.; Gascoin, N.; Fedioun, I.; Chetehouna, K.; Lamoot, L.; Fau, G. Dimensioning of automated regenerative cooling: Setting of high-end experiment. *Aerospace Sci. Tech.* **2015**, *43*, 350–359. [CrossRef]
22. Cui, Z.; Zhang, J.; Wu, D. Hybrid many-objective particle swarm optimization algorithm for green coal production problem. *Inf. Sci.* **2020**, *518*, 256–271. [CrossRef]
23. Hu, J.; Sun, Y.; Li, G.; Jiang, G.; Tao, B. Probability analysis for grasp planning facing the field of medical robotics. *Measurement* **2019**, *141*, 227–234. [CrossRef]
24. Wu, X.H.; Meng, X.W.; Chen, L.; Zheng, C.L. Numerical simulation of ladle temperature field under the working conditions of ladle baking and steel holding. *Electron. Qual.* **2018**, *376*, 84–86.
25. Li, H.C.; Huang, S.M.; Chen, X.Y. Simulation research of ladle temperature field and stress field based on ANSYS. *Equip. Manuf. Technol.* **2007**, *40*, 35–37.
26. Warzecha, M.; Jowsa, J.; Warzecha, P.; Pfeifer, H. Numerical and experimental investigations of steel mixing time in a 130 t ladle. *Steel Res. Int.* **2008**, *79*, 852–857. [CrossRef]
27. Chen, Y.C.; Bao, Y.; Wang, M.; Zhao, L.; Peng, Z. A mathematical model for the dynamic desulfurization process of ultra-low-sulfur steel in the LF refining process. *Metall. Res. Technol.* **2014**, *111*, 37–43. [CrossRef]
28. Wu, Y.F.; He, D.F.; Xu, A.J.; Tian, N.Y. Numerical simulation and optimization of temperature field in the baking ladle. *J. Iron Steel Res.* **2012**, *21*, 21–24.
29. Li, Q. Study on Heat Transfer Model of Ladle in Steelmaking and Continuous Casting Process. Master's Thesis, Northeastern University, Shenyang, China, 7 July 2010.
30. Wu, T.; Chen, M.; Zhang, L.; Xu, X.; Liu, Y.; Yan, J.; Wang, W.; Gao, J. Three-dimensional graphene-based aerogels prepared by a self-assembly process and its excellent catalytic and absorbing performance. *J. Mater. Chem. A* **2013**, *1*, 7612–7621. [CrossRef]
31. Li, D.Y. Physical and mathematical model of heat transfer in plasma ladle furnace. *Iron Steel* **1997**, *3*, 16–18.
32. Li, Q.; Wei, D.Y.; Yang, C. Calculation and analysis of steady-state heat transfer of 150-Ton ladle wall. *Metall. Ser.* **2015**, *1*, 8–11.
33. Solorio-Diaz, G.; Davila-Morales, R.; Jose, B.S.; Vergara-Hernández, H.J.; Ramos-Banderas, A.; Galvan, S.R. Numerical modelling of dissipation phenomena in a new ladle shroud for fluidynamic control and its effect on inclusions removal in a slab tundish. *Steel Res. Int.* **2014**, *85*, 863–874. [CrossRef]
34. Zhou, Y.; Dong, Y.C.; Wang, H.C.; Wang, S.J.; Liu, Y.B. Mathematical simulation of flow phenomena in CAS-OB refining ladle. *J. Iron Steel Res.* **2003**, *10*, 8–12.
35. Singham, J.R. Tables of emissivity of surfaces. *Int. J. Heat Mass Transf.* **1962**, *5*, 67–76. [CrossRef]
36. Luo, B.; Li, G.; Kong, J.; Jiang, G.; Sun, Y.; Liu, Z.; Chang, W.; Liu, H. Simulation analysis of temperature field and its influence factors of the new structure ladle. *Appl. Math. Inf. Sci.* **2017**, *11*, 589–599. [CrossRef]
37. Farrera-Buenrostro, J.E.; Hernández-Bocanegra, C.A.; Ramos-Banderas, J.A.; Torres-Alonso, E.; López-Granados, N.M.; Ramírez-Argáez, M.A. Analysis of temperature losses of the liquid steel in a ladle furnace during desulfurization stage. *Trans. Indian Inst. Met.* **2019**, *11*, 22–23. [CrossRef]
38. Lu, B.; Meng, X.; Zhu, M. Numerical analysis for the heat transfer behavior of steel ladle as the thermoelectric waste-heat source—ScienceDirect. *Catal. Today* **2018**, *318*, 180–190. [CrossRef]
39. Prabhakaran, S.; Maboob, S.A. Numerical analysis and experimental investigation on behavior of cold-formed steel castellated beam with diamond castellation. *Int. J. Steel Struct.* **2021**, *21*, 1082–1091. [CrossRef]
40. Xia, J.L.; Ahokainen, T. Transient flow and heat transfer in a steelmaking ladle during the holding period. *Met. Mater. Trans. B* **2001**, *32*, 733–741. [CrossRef]
41. Urióstegui-Hernández, A.; Garnica-González, P.; Ángel Ramos-Banderas, J.; Hernández-Bocanegra, C.A.; Solorio-Díaz, G. Multiphasic study of fluid-dynamics and the thermal behavior of a steel ladle during bottom gas injection using the eulerian model. *Metals* **2021**, *11*, 1082. [CrossRef]

Rare Earth Elements Enhanced the Oxidation Resistance of Mo-Si-Based Alloys for High Temperature Application: A Review

Laihao Yu, Yingyi Zhang *, Tao Fu, Jie Wang, Kunkun Cui and Fuqiang Shen

School of Metallurgical Engineering, Anhui University of Technology, Maanshan 243002, China; aa1120407@126.com (L.Y.); ahgydxtaofu@163.com (T.F.); wangjiemaster0101@outlook.com (J.W.); 15613581810@163.com (K.C.); sfq19556630201@126.com (F.S.)
* Correspondence: zhangyingyi@cqu.edu.cn; Tel.: +86-17375076451

Abstract: Traditional refractory materials such as nickel-based superalloys have been gradually unable to meet the performance requirements of advanced materials. The Mo-Si-based alloy, as a new type of high temperature structural material, has entered the vision of researchers due to its charming high temperature performance characteristics. However, its easy oxidation and even "pesting oxidation" at medium temperatures limit its further applications. In order to solve this problem, researchers have conducted large numbers of experiments and made breakthrough achievements. Based on these research results, the effects of rare earth elements like La, Hf, Ce and Y on the microstructure and oxidation behavior of Mo-Si-based alloys were systematically reviewed in the current work. Meanwhile, this paper also provided an analysis about the strengthening mechanism of rare earth elements on the oxidation behavior for Mo-Si-based alloys after discussing the oxidation process. It is shown that adding rare earth elements, on the one hand, can optimize the microstructure of the alloy, thus promoting the rapid formation of protective SiO_2 scale. On the other hand, it can act as a diffusion barrier by producing stable rare earth oxides or additional protective films, which significantly enhances the oxidation resistance of the alloy. Furthermore, the research focus about the oxidation protection of Mo-Si-based alloys in the future was prospected to expand the application field.

Keywords: Mo-Si-based alloys; alloying; rare earth elements; oxidation behavior; mechanism

1. Introduction

As the world population increases, the problem of global resource shortage has become increasingly prominent. It is well known that in addition to waste recycling, improving energy utilization and exploring new energy are effective methods to solve resource problems, and are also the main trend of future scientific and technological development [1,2]. Nowadays, the development of new energy is overwhelming, which urges scientists to study the properties of materials on a smaller and smaller scale, so that it is easier to develop new materials [3,4]. For example, Ghidelli and Ast et al. [5–7] systematically studied the mechanical properties of ceramics at the submicrometre scale such as fracture toughness, and improved the recently developed pillar splitting method that can provide reliable and simple ways to measure fracture toughness over a broad range of material properties. As a new high temperature structural material, the Mo-Si-based alloy is expected to replace the nickel-based alloy and play an important role in turbine engine and industrial furnace components [8,9].

A large number of studies have pointed out that Mo-Si-based alloys have outstanding high temperature performance characteristics, such as moderate density, strong electrical and thermal conductivity, ultra-high melting point, high thermal impact resistance, etc., which have been widely used in various industries [10,11]. However, these alloys also have some inherent defects that limit their generalization as structural materials [12–14]. For example, $MoSi_2$-based alloys exhibit low room temperature fracture toughness, poor high temperature creep resistance, and catastrophic oxidation at 400 to 800 °C. Although Mo_5Si_3-based alloys have relatively strong creep resistance, they generally present an accelerated oxidation or "pesting oxidation" phenomenon below 1000 °C, which is a thorny problem [15–19]. To overcome these shortcomings, relevant research work has not stopped since the end of the 20th century [20–25]. Fortunately, people finally succeeded in improving the properties of alloy through material designs [26–29], preparation techniques [30–34] or surface modification methods [35–38]. Among them, doping second phases such as W, Nb, ZrO_2, La_2O_3, Al_2O_3, and Cr_2O_3 in material designs to strengthen the performance of substrate is generally regarded as an important measure [39]. In the case of preparation technique, Ghidelli et al. [40] used a pulsed laser deposition technique to prepare a class of new nano-structured $Zr_{50}Cu_{50}$ (at.%) metallic glass films with excellent and tunable mechanical properties. In the case of surface modification methods, Besozzi et al. [41] studied the thermomechanical properties of different systems of amorphous tungsten–oxygen and tungsten–oxide coatings, which was of great significance to the design and construction of devices.

There is no doubt that in some refractory metal materials like niobium-based, molybdenum-based, and tungsten-based materials, adding active elements may significantly improve the material properties [42–44]. For example, the addition of Si and B to molybdenum-based materials can produce protective borosilicate scales on the substrate surface, which helps to reduce the oxidation rate. Moreover, the presence of B can also facilitate the SiO_2 scale flow by reducing the scale viscosity, thereby filling the pores and cracks that may occur on the alloy surface. Rare earth or its oxides, as a kind of active element, have received particular attentions and widespread applications [45–48]. In recent years, researchers have made great breakthroughs in studying rare earth elements to enhance the mechanical properties of metal materials. However, so far, there are few reports on the oxidation behavior of Mo-Si-based alloy doped with rare earth elements. Therefore, this paper comprehensively and systematically reviewed the actions of rare earth elements such as La, Hf, Ce and Y on the antioxidant properties of Mo-Si-based alloys, especially the Mo-Si-B system, and summarized the relevant strengthening mechanisms.

2. Effects of Rare Earth Elements on Oxidation Behavior

It's well known that the increase of oxide layer thickness is primarily caused by the internal diffusion of O_2 [49–52], and studies have shown that alloying with active elements can effectively reduce the internal diffusion rate of O_2 [53–55]. For example, appropriate adding rare earth elements, on the one hand, can separate oxygen atoms at the metal-oxide interfaces and oxide scale grain boundaries and react with O_2, thus hindering the diffusion of O_2 [56–59]. On the other hand, it can optimize the oxide scale microstructure and improve the scale adhesion [60–63].

2.1. Effects of Rare Earth Elements on Mo-Si-B Alloys

Among various Mo-Si-based alloys, Mo-Si-B alloys are the most widely studied [64–70]. Therefore, we first discuss the effect of rare earth elements on the oxidation properties of Mo-Si-B alloys.

2.1.1. Effects of La Element Addition

Compared with pure Mo-Si alloys, adding La element can significantly optimize the microstructure of these alloys by the means of reducing the grain size and making the intermetallic particles disperse more evenly, thus improving the fracture toughness, bending and compressive strength significantly [71–74]. Based on the existing studies of Mo-12Si-8.5B alloy (at.%) [75–79], we further analyzed the actions of doping La or La_2O_3 s phase on the oxidation behavior.

Zhang et al. [80] prepared Mo-12Si-8.5B (at.%, abbreviated as MSB) samples added with different contents of La_2O_3 through arc-melted and spark plasma sintered methods, and the specific contents were presented in Table 1. Figure 1a gives the XRD patterns of MSB + xLa_2O_3 samples (x = 0, 0.3, 0.6, 1.2 wt.%). It can be seen that all the samples consist of Mo_5SiB_2, Mo_3Si and α-Mo three phases, which is consistent with the phase diagram of isothermal Mo-Si-B composites (Figure 1b) [76]. At the same time, it also reveals that even if La_2O_3 is added, it will not affect the phase composition of samples. Figure 1c–f are micrographs of the four samples prepared, where the white regions are α-Mo phase, and the black regions are Mo_5SiB_2/Mo_3Si phases dispersed in the α-Mo matrix. It can also be found from micrographs that the grain size of α-Mo and Mo_5SiB_2/Mo_3Si will be reduced after adding La_2O_3, in which the α-Mo size change is more pronounced, whereas the decrease of each phase size is not sensitive to the La_2O_3 mass fraction. Moreover, the distribution of Mo_5SiB_2/Mo_3Si phases is more uniform after doping La_2O_3. This is because parts of La_2O_3 can be used as nucleation sites, which leads to the increase of nucleation density. On the other hand, La_2O_3 particles play a "pinning" role on the α-Mo boundary to inhibit its grains growth. The results in Table 1 further reveal the effect of La_2O_3 on grain size.

Table 1. The La_2O_3 mass fractions and grain sizes of various samples studied. Reproduced with permission [80]. Copyright 2011 Elsevier.

Materials	MSB	MSB + 0.3	MSB + 0.6	MSB + 1.2
Mass fraction of La_2O_3 (wt.%)	0	0.3	0.6	1.2
Grain sizes of α-Mo (μm)	19.78	10.88	9.56	9.46
Grain sizes of Mo_3Si/Mo_5SiB_2 (μm)	3.04	2.46	2.55	2.17

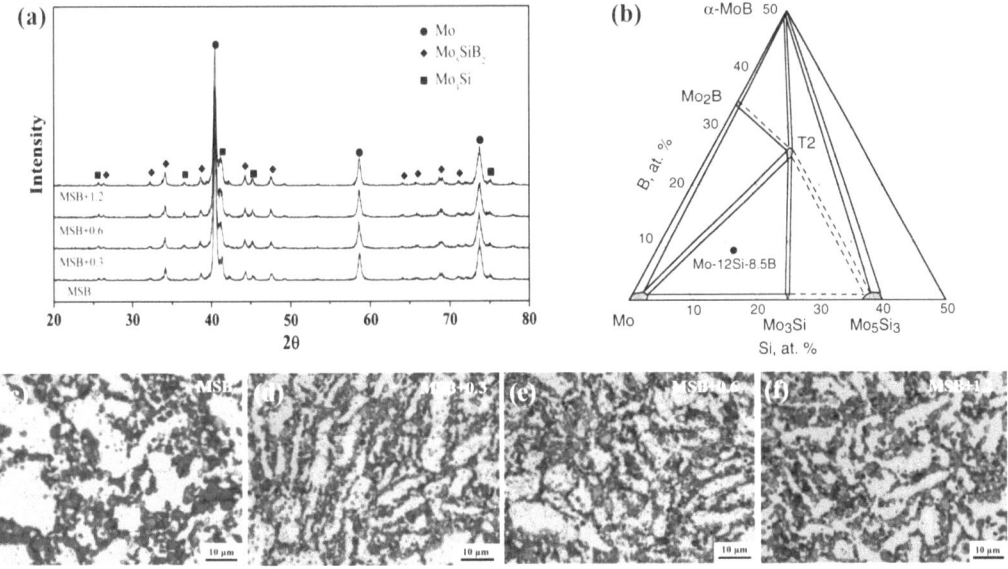

Figure 1. XRD patterns (**a**) and microstructure images (**c**–**f**) of MSB + xLa_2O_3 samples (x = 0, 0.3, 0.6, 1.2 wt.%) before oxidation; schematic section of the 1600 °C isothermal ternary Mo-Si-B phase diagram (**b**). (**a**,**c**–**f**) and (**b**) reproduced with permission from [80] and [76], respectively. Copyright 2011 Elsevier and 2001 Elsevier.

Weight variation curves for different mass fraction La_2O_3-doped MSB samples after oxidation at 800 °C are shown in Figure 2a. The results point out that the mass loss of alloy can be significantly reduced after adding La_2O_3, where MSB + 0.6 sample exhibits the least mass loss during transient oxidation stage. It is due to the fact that La$_2$O$_3$-doped MSB samples present a finer grain size, which makes it faster to form a protective borosilicate scale to prevent further volatilization of MoO$_3$. The role of grain refinement has also been reported elsewhere [81–87]. Again, La$_2$O$_3$ in the alloy can decrease the grain boundary transport rate, leading to the reduction of weight loss rate. To determine the impact of La$_2$O$_3$ on the oxidation behavior, Zhang et al. [80] further studied the cross-sectional structure of MSB + 0.6 sample oxidized at 800 °C. It has been noticed from Figure 2d that the cross section of this sample is composed of oxidation scale, interlayer and substrate. Among them, the top layer is the dense B$_2$O$_3$-SiO$_2$ scale and the interlayer comprised with MoO$_2$, which is confirmed by XRD analysis and the content of each element (i.e., Mo, Si, B, O). Furthermore, compared with the MSB sample, the intensity and peaks of B$_2$O$_3$ and SiO$_2$ of MSB + 0.6 sample are raised visibly (Figure 2b), indicating that the antioxidant capacity of the MSB + 0.6 sample is enhanced. Burk [88,89] and Jéhanno [90] et al. also reported similar results.

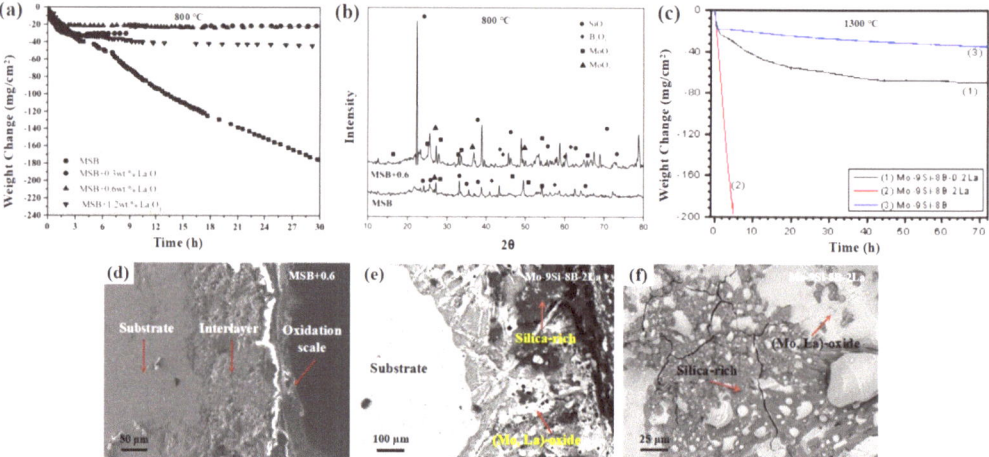

Figure 2. Weight change curves of various samples oxidized at different temperatures (**a,c**); surface and cross-sectional SEM/BSE images of various samples oxidized at different conditions: 800 °C (**d**) with corresponding surface XRD analysis results (**b**), 1200 °C for 2 h (**e**), and 1300 °C for 23 h (**f**). (**a,b,d**) and (**c,e,f**) reproduced with permissions from [80] and [60], respectively. Copyright 2011 Elsevier and 2014 Elsevier.

In addition, Majumdar et al. [59,60] also investigated the oxidizability of the Mo-9Si-8B (at.%) sample doped with 2 at.% La at 750–1400 °C. It is established that the La-doped sample exhibits relatively good antioxidant capability below 1000 °C, which is the result of stable lanthanum oxides like 3La$_2$O$_3$·MoO$_3$, La$_2$O$_3$ and La$_2$O$_3$·3MoO$_3$ produced at the oxidation scale to inhibit the formation and volatilization of MoO$_3$. However, when the temperature exceeded 1000 °C, the addition of La might adversely affect the sample oxidation properties. Figure 2c displays the weight change curves of La-doped and undoped samples at 1300 °C, and it can be seen that the weight loss of La-doped sample is significantly higher than that of undoped sample. This is because adding La makes the sample cross section present a loose and porous oxide layer structure when oxidized at 1200 °C (Figure 2e). Meanwhile, as oxidation temperature rises to 1300 °C, a large number of cracks and holes are observed on the sample surface (Figure 2f), which provides a pathway for O$_2$ internal diffusion and MoO$_3$ volatilization. In contrast, even if undoped sample is oxidized at 1300 °C for 72 h, a continuous and compact oxidation scale can be still observed in its cross section [88]. Thus, undoped sample has better antioxidant properties in high-temperature environments.

2.1.2. Effects of Hf Element Addition

Extensive experiments have reported that adding Hf/HfB$_2$ to Mo-Si-B composites can clearly improve their performance features, such as high temperature strength, high temperature stability, creep resistance, fracture toughness, etc. [91–95]. Potanin et al. [96] discussed the oxidation behavior of MoB-HfB$_2$-MoSi$_2$ composites at 1200 °C in detail. The composition of each alloy is illustrated in Table 2, where the difference between X34$_2$ and X34$_1$ samples is that the former presents a two-level structure (TLS), while the latter presents a single-level structure (SLS). The microstructures of the studied samples are depicted in Figure 3a–c, and on the whole, the three samples all contain MoB and MoSi$_2$ phases. The difference is that X34$_1$ and X34$_2$ samples also have additional HfSiO$_4$ and HfB$_2$ phases and their grain sizes are finer than that of the X0 sample. Figure 4 gives the oxidation kinetics curves of the three samples, and it can been observed that the weight increases of X34$_1$ and X34$_2$ samples are more obvious because adding HfB$_2$ can make the samples generate HfSiO$_4$ (6.97 g/cm^3) and HfO$_2$ (9.68 g/cm^3), whose specific weights are greater than SiO$_2$ (2.36 g/cm^3) [97]. By the way, the weight gain of the X34$_1$ sample is smaller than that of the X34$_2$ sample. This is because HfB$_2$ and MoB grains in the X34$_2$ sample present finer size and more uniform distribution than the X34$_1$ sample (Figure 3b,c), which causes the more intense oxidation of X34$_2$ sample.

Table 2. The elemental composition of various samples. Reproduced with permission [96]. Copyright 2019 Elsevier.

Samples	Composition (at.%)			
	Mo	Hf	Si	B
X0	35.0	–	60.0	5.0
X34$_1$ X34$_2$	23.2	11.3	39.7	25.8

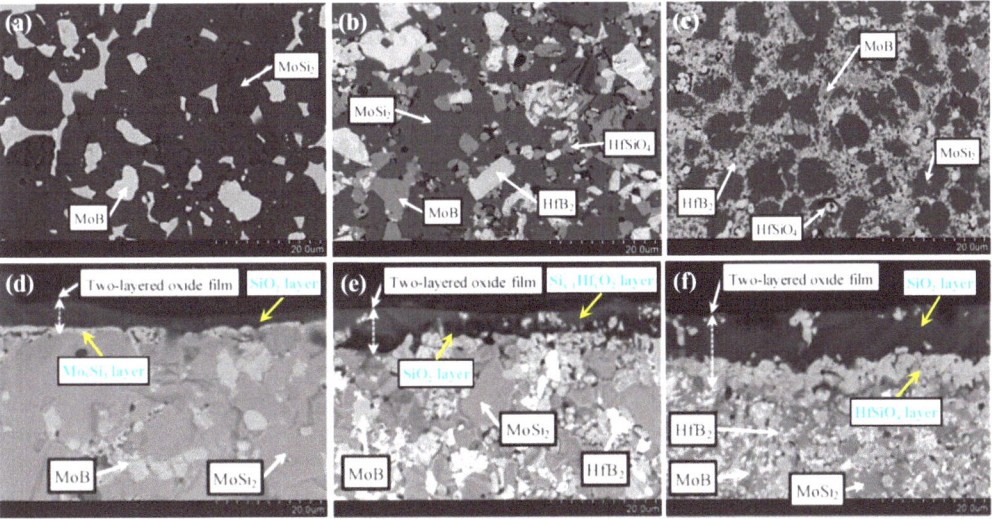

Figure 3. Microstructure of various samples: (**a**) X0, (**b**) X34$_1$ and (**c**) X34$_2$; cross-sectional SEM images of samples after oxidation at 1200 °C for 30 h: (**d**) X0, (**e**) X34$_1$ and (**f**) X34$_2$. Reproduced with permission [96]. Copyright 2019 Elsevier.

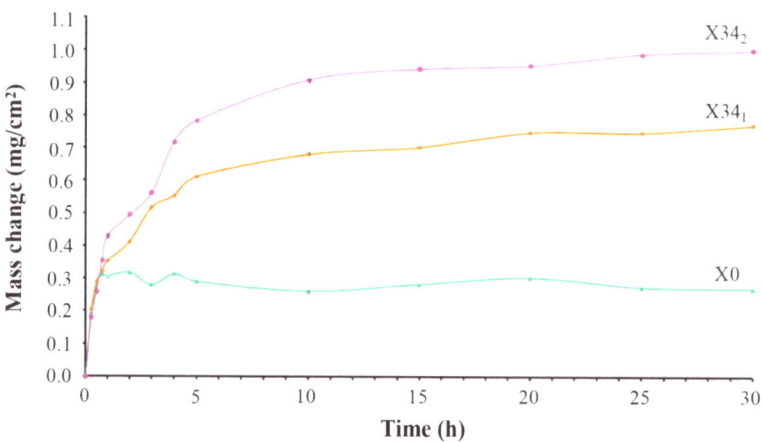

Figure 4. Kinetic curves of the samples oxidized at 1200 °C. Reproduced with permission [96]. Copyright 2019 Elsevier.

Figure 3d–f shows the cross-sectional structure of X0, X34$_1$ and X34$_2$ samples after oxidation at 1200 °C for 30 h. It has been noticed that the X34$_1$ and X34$_2$ samples do produce Si$_{x-1}$Hf$_x$O$_2$/HfSiO$_4$ phases during the oxidation process, which is consistent with the above analysis results. Meanwhile, two-layered oxide films are formed on the surface of all samples, whereas there are great differences in the oxide-film composition and structure. In other words, the X0 sample oxide scale containing the SiO$_2$ layer (top layer) and Mo$_5$Si$_3$ layer (bottom layer), the X34$_1$ sample oxide scale consists of a Si$_{x-1}$Hf$_x$O$_2$-doped amorphous SiO$_2$ layer (upper layer) and a crystalline α-SiO$_2$ layer (lower layer), while the X34$_2$ sample oxide scale is comprised of a crystalline α-SiO$_2$ layer (outermost layer) and a HfSiO$_4$ layer (interlayer). Figure 5 gives the TEM images of X34$_1$ sample after oxidation. It can be seen that two SiO$_2$-based layers have been formed on the sample surface: crystalline lower layer and amorphous upper layer, which is close to the unoxidized MoSi$_2$ ceramic. At the same time, the EDS analysis shows that the upper layer contains 2.4 ± 0.7 at.% Hf in addition to O and Si. It is speculated that the amorphization of SiO$_2$ may be related to the dissolution of Hf. As Hf and Si are equivalent elements, Hf can be incorporated into SiO$_2$ lattice to form Si$_{x-1}$Hf$_x$O$_2$ solid solution and suppress the transition of SiO$_2$ from amorphous to the crystalline state upon heating. Meanwhile, the Si–O–Si bond is stretched owing to the higher atomic radius of Hf (0.167 nm) than Si (0.132 nm). Furthermore, the amorphous phenomenon of SiO$_2$ may also be associated with the oxidation and cooling rate of X34$_1$ sample. It is worth noting that the X34$_2$ sample oxide scale exhibits denser structure is caused by the existence of high wetting angle makes SiO$_2$-B$_2$O$_3$ melt can shrink HfSiO$_4$ particles together, resulting in the formation of smooth and compact oxide films [98].

To summarize, addition of HfB$_2$ to X0 sample can produce HfSiO$_4$/Si$_{x-1}$Hf$_x$O$_2$ particles dispersed in oxide film, and even form an HfSiO$_4$ interlayer. It has been proven that the HfSiO$_4$ and ZrSiO$_4$ particles have similar effects. On the one hand, they can promote the healing of cracks and holes in the borosilicate scale [99,100], on the other hand, they can both act as barriers and HfSiO$_4$ particles can also increase the crystallization temperature of amorphous scale [101]. Therefore, adding HfB$_2$ can enhance the alloy antioxidant effects. The research results of Sciti et al. also confirmed this conclusion [97].

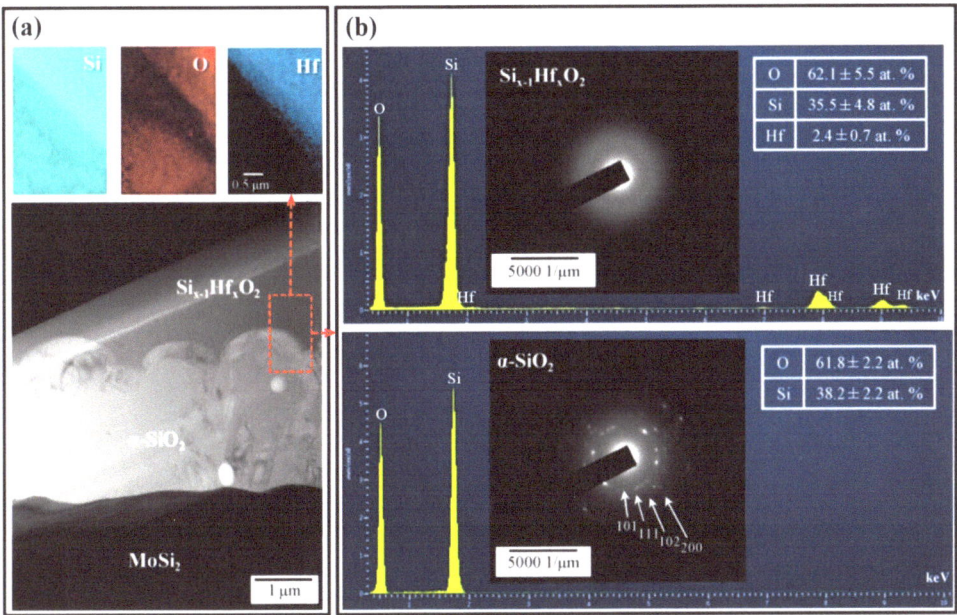

Figure 5. (**a**) TEM micrograph of the oxidized X34$_1$ sample; (**b**) EDS and diffraction patterns of the oxide scale. Reproduced with permission [96]. Copyright 2019 Elsevier.

2.1.3. Effects of Ce Element Addition

Actions of Ce on oxidation performance of Mo-Si-B alloys have been investigated deeply. Das et al. [102,103] performed isothermal oxidation experiments about Mo-11.18Si-8.09B-7.29Al-0.16Ce, Mo-11.2Si-8.1B-7.3Al, Mo-13.98Si-9.98B-0.16Ce and Mo-14Si-10B alloys (at.%, abbreviated as MSBACe, MSBA, MSBCe and MSB$_1$, respectively) synthesized by arc melting. The experiment results have suggested that doping a small amount of Ce has little effect on oxidation kinetics of the alloy at 500 and 700 °C (Figure 6a), while the presence of Ce presents a palpable effect on the alloy oxidation behavior at 900–1300 °C. Figure 6b–d provide the mass variation curves of MSB$_1$ and MSBCe oxidized at 900–1300 °C, respectively. It can be seen that the mass loss of MSB$_1$ increases after the addition of Ce (Figure 6c,d). Even so, MSBCe exhibits shorter transient oxidation periods as compared to MSB$_1$, and its steady-state stage curve almost presents a horizontal trend, revealing that MSBCe is more effectively protected than MSB$_1$.

At the same time, the microstructural morphology of two oxidized alloys is shown in Figure 7. During oxidation at 900 °C, one-layered oxide film (i.e., Mo-oxide film) and flowing glassy phase are observed on the surface of both alloys (Figure 7a,d). The difference is that the glassy phase in MSBCe flows faster due to the presence of Ce, hence MSBCe presents a smaller mass variation at 900 °C (Figure 6b). However, after oxidation at 1100 °C for 24 h, the oxide-layer structure of both alloys has changed distinctly, namely SiO$_2$ layer is formed on top of the Mo-oxide layer (Figure 7b,e). When the oxidation temperature reaches 1300 °C, B$_2$O$_3$ has begun to evaporate from the oxide scale, resulting in increased viscosity and weak fluidity of the scale, which leads to the deterioration of alloy antioxidation ability [104–108]. Nevertheless, the addition of Ce increases the volatilization temperature of B$_2$O$_3$, leaving the fluidity of oxide scale largely unchanged. Therefore, the flow traces of glassy scale can still be observed on the MSBCe surface even at 1300 °C; in contrast, the MSB$_1$ surface has little flow traces (Figure 7c,f). There is no doubt that scale flow can heal pores and cracks on the alloy surface, thus Ce addition has a positive effect on the oxidation protection of MSB$_1$.

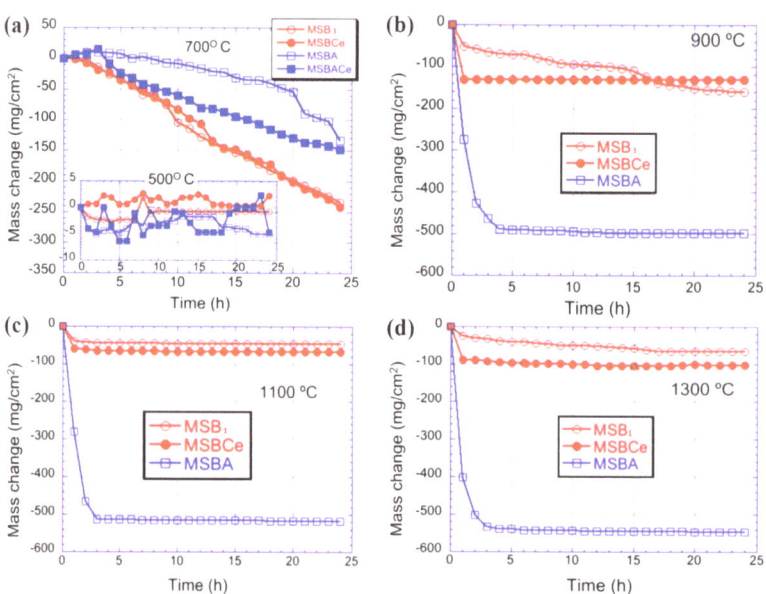

Figure 6. Mass loss curves of various alloys at different temperatures. (**a**) and (**b**–**d**) reproduced with permission from [102] and [103], respectively. Copyright 2010 Elsevier and 2016 Elsevier.

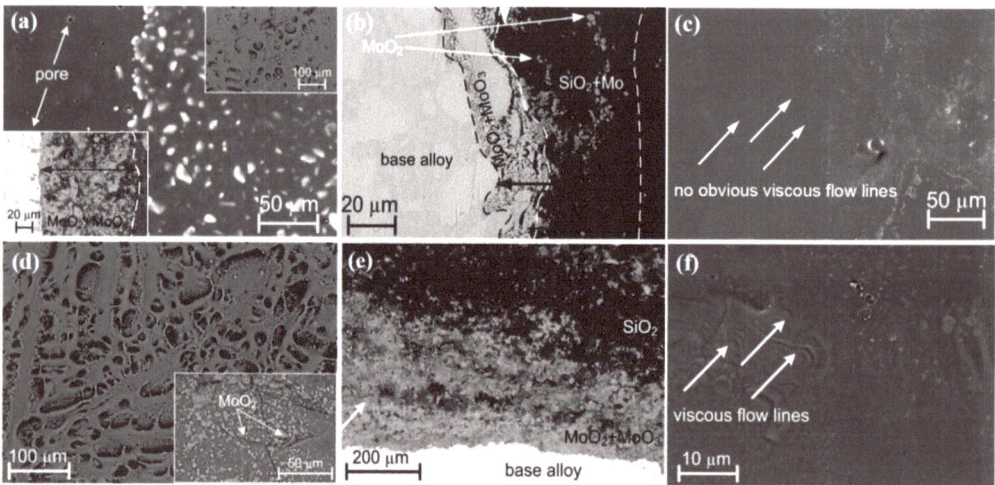

Figure 7. Surface and cross-sectional SEM (BSE) images of MSB$_1$ (**a**–**c**) and MSBCe (**d**–**f**) alloys oxidized at different temperatures for 24 h: 900 °C (**a,d**), 1100 °C (**b,e**) and 1300 °C (**c,f**). Reproduced with permission [103]. Copyright 2016 Elsevier.

In addition, Das et al. [109] also researched the oxidation reaction of Mo-Si-B-Al-Ce alloys at 1100 °C. It is well known that adding Al to Mo-Si-B systems may lead to the failure of alloy oxidation protection owing to the formation of mullite [110–114], which is also verified by the mass loss curves of Al-doped alloy in Figure 6b–d. Das [109] noted that adding Ce can further inhibit the malignant oxidation of Mo-Si-B-Al systems at 1100 °C, since the presence of Ce hindered the generation of mullite and promoted the formation of dense protective Al-oxide films on the alloy surface, thereby improving the antioxidant capacity.

2.1.4. Effects of Y Element Addition

It has been sure that adding a Y element can significantly prolong the service life of Mo-based alloys due to the fact that Y will exhibit higher oxygen affinity than Mo [115,116]. Moreover, Y can also inhibit ion diffusion in grain boundaries and decrease the oxide-scale growth rate [117–119]. The presence of Y improves the adhesion between oxide film and substrate, thus improving the alloyed oxidation resistance [120,121]. Therefore, researchers try to add the Y element to Mo-Si-B alloys to obtain materials with better performance. After comparing the variation of Y-doped and Y-free Mo-9Si-8B (at.%) samples in the oxidation behavior at 650–1400 °C, Majumdar et al. [122–124] found that all samples presented a trend of transient mass increase followed by continuous rapid decrease at 650 °C (Figure 8a). Among all the samples, the 2 at.% Y-doped sample had the minimum mass loss at 750–1000 °C (Figure 8b–e), and the 0.2 at.% Y-doped sample presented the lowest mass loss at 1100 °C (Figure 8f). This phenomenon is attributed to the stable Y_6MoO_{12} and $Y_5Mo_2O_{12}$ oxides produced in the initial period of oxidation at 750–1000 °C. These stable yttrium-molybdate oxides were formed rapidly, which inhibited the generation and vaporization of MoO_3, leading to a decrease in the mass loss of 2 at.% Y-doped sample. In addition, these yttrium-molybdate oxides often exist in the form of fine particles, which can accelerate the nucleation and growth of outer protective SiO_2 films.

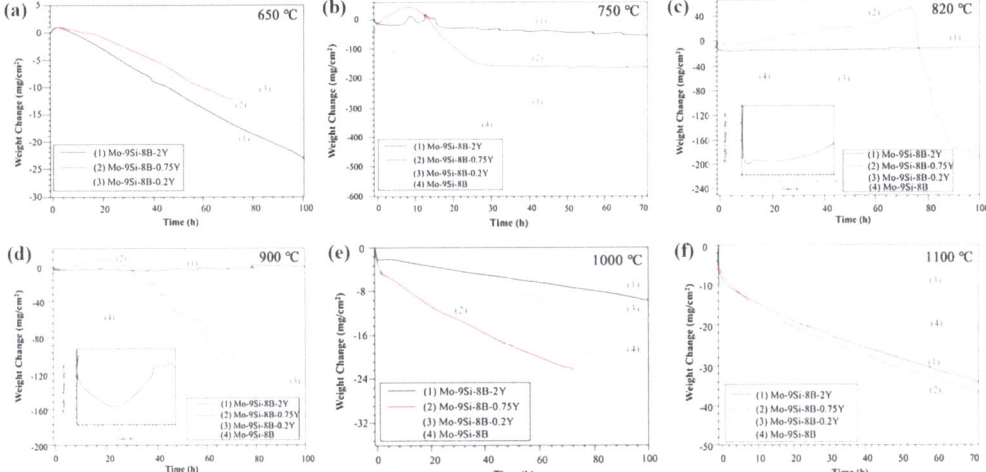

Figure 8. Weight change curves of Y-doped and undoped samples oxidized at (**a**) 650 °C; (**b**) 750 °C; (**c**) 820 °C; (**d**) 900 °C; (**e**) 1000 °C; (**f**) 1100 °C, respectively. Reproduced with permission [122]. Copyright 2014 Elsevier.

In order to further clarify the role of Y doping on the sample antioxidation behavior, the microstructure of oxidized samples are analyzed in detail [58,122]. Figure 9a,b give the cross-sectional micrographs of 2 at.% and 0.75 at.% Y-doped samples oxidized at 900 °C for 24 h, respectively. It can be seen that 0.75 at.% Y-doped sample has a thicker inner MoO_2 scale than that of 2 at.% Y-doped sample, which indicates that properly increasing the concentration of Y can inhibit the formation of MoO_2 to some extent. Furthermore, microcracks are also observed on the outer SiO_2 scale of 0.75 at.% Y-doped sample due to the quite high growth velocity (about 7.6 μm·h^{-1}) of inner MoO_2 scale, which produces such a large tensile stress that outer SiO_2 scale breaks (Figure 9b,d). It is worth noting that the thickness of inner MoO_2 scale for 0.75 at.% Y-doped sample is significantly thinner after oxidation at 1100 °C for 72 h (Figure 9c). As SiO_2 will present viscous flow to cover holes and cracks on the alloy surface during the temperature exceeds 965 °C [125,126], which prevents further oxidation of the substrate. Figure 9e shows the thickness changes of SiO_2 and MoO_2 layers for 0.2 at.% Y-dpoed sample at 1100 and 1200 °C, which further supports the above analysis results. When the oxidation temperature is higher than 1200 °C, a thin yttrium-silicate ($Y_2Si_2O_7$) scale is observed on the outer surface of Y-doped samples (Figure 10a–d), and the thickness of yttrium-silicate film gradually increases as the oxidation temperature raises (Figure 10c,e) [123]. It has been proven that the outer yttrium-silicate film is conducive to the alloy oxidation protection through preventing SiO_2 from forming volatile silicon hydroxide in humid conditions above 1200 °C [127–129]. Similar studies have been reported by Gorr et al. [130].

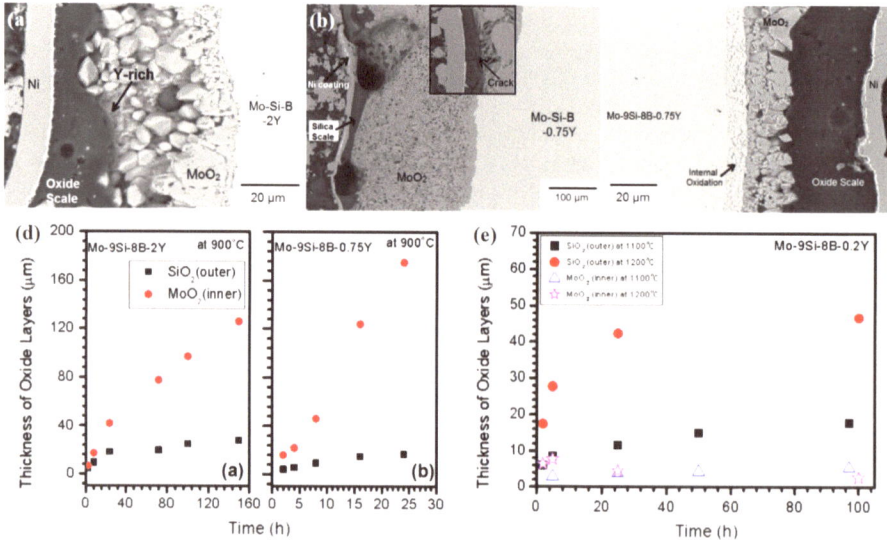

Figure 9. Cross-sectional BSE images of 2 at.% and 0.75 at.% Y-doped samples oxidized at differernt conditions: (a,b) 900 °C for 24 h, (c) 1100 °C for 72 h; (d,e) Changes of MoO_2 laye and SiO_2 layer thickness in Y-doped samples oxidized at different temperatures. Reproduced with permission [122]. Copyright 2014 Elsevier.

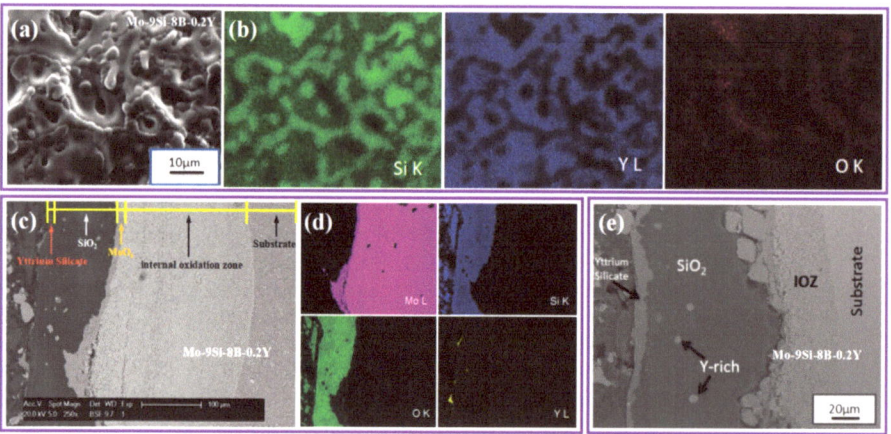

Figure 10. Surface SE (a) and cross-sectional BSE (c,e) images of 0.2 at.% Y-doped samples oxidized at different conditions: 1300 °C for 100 h (a,c) and 1400 °C for 2 h (e); (b,d) are the element mappings of (a,c), respectively. Reproduced with permission [123]. Copyright 2013 Springer Nature.

There is no doubt that alloying with Zr has a great impact on the antioxidant ability of Mo-Si-B materials. This is because the addition of Zr may produce polymorphic ZrO_2

or monomorphic ZrSiO$_4$, which mainly depends on the oxidation temperature. Among them, the ZrSiO$_4$ can act as an obstacle phase, which is beneficial to improve the alloy oxidation behavior; whereas the ZrO$_2$ will expand in volume at high temperatures (>1200 °C), which destroys the integrity of the SiO$_2$ scale so that it loses the protective effect [131,132]. Therefore, inhibiting the formation of ZrO$_2$ phase is essential to improve the alloy oxidation resistance.

Based on the fact that yttria suppresses the zirconia phase transition [133,134], Yang et al. [135] designed and fabricated Mo-12Si-10B-1Zr-0.3Y, Mo-12Si-10B-1Zr and Mo-12Si-10B samples (at.%, abbreviated as 1Zr-0.3Y, 1Zr-0Y and 0Zr-0Y, respectively). Figure 11a shows the mass variation of the three samples at 1250 °C, it can be seen that adding 1 at.% Zr to the 0Zr-0Y sample will lead to continuous and sharp mass loss. As the 0Zr-0Y sample has formed dense protective SiO$_2$ films during the oxidation, it avoids the sample sustained mass loss (Figure 12a), whereas the addition of Zr causes the SiO$_2$ scale to become loose and porous due to the formation of ZrO$_2$, and the porous structure provides channels for O$_2$ inward diffusion, thus accelerating the sample oxidation (Figure 12b). It is encouraging that further adding 0.3 at.% Y can effectively prevent the rapid mass loss of the 1Zr-0Y sample. As can be seen from Figure 12c, ZrSiO$_4$ rather than ZrO$_2$ appears on the sample surface after the addition of Y, thus eliminating the adverse effect of Zr doping. Meanwhile, the Y-Mo-rich oxide is also observed around the ZrSiO$_4$ phase. EDS analysis shows that the Y/Mo atomic ratio of this oxide is about 1/2, revealing that the oxide may be Y$_2$Mo$_4$O$_{15}$. Again, the XPS spectra also presents that the oxide has nearly the same Mo 3d and Y 3d bonding energies as Y$_2$Mo$_4$O$_{15}$ gauged through You et al. [136], which further verifies the above inference (Figure 11b,c). Furthermore, the 1Zr-0.3Y sample surface also forms a uniformly dense outer Y$_2$Si$_2$O$_7$ scale with the increase of oxidation time, which provides a better protection effect than 0Zr-0Y sample (Figure 12d). It has been observed from the cross-section enlarged Figure 12e,f that Y diffuses outward with the metastable Y$_2$Mo$_4$O$_{15}$ as the carrier and produces Y$_2$Si$_2$O$_7$ after a series of reactions at the top of SiO$_2$ scale, which will be accumulated and compressed to form the outer Y$_2$Si$_2$O$_7$ layer. Therefore, 1Zr-0.3Y sample presents the best antioxidant performance among the three samples.

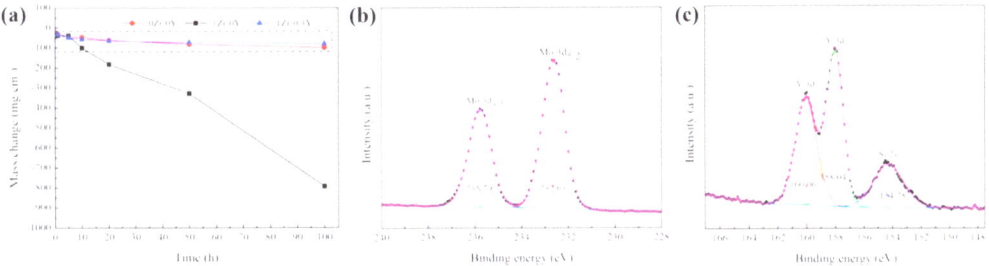

Figure 11. Mass change curves of three studied samples at 1250 °C (**a**); XPS analysis of Mo (**b**) and Y (**c**) characterized the 1Zr-0.3Y sample surface oxidized for 1 h. Reproduced with permission [135]. Copyright 2020 Elsevier.

2.2. Effects of Rare Earth Elements on Other Mo-Si Alloys

Previous studies have pointed out that adding Nb to the Mo-Si-based materials can play a satisfactory effect in improving mechanical properties due to damaging the stability of the Mo$_3$Si phase [137–139]. However, the presence of Nb will lead to catastrophic oxidation of the material [140–142]. Inspired by the above study that adding Y can enhance the antioxidant properties of Zr-doped Mo-Si-B alloys, we further discussed the role of adding Y on the oxidation behavior of Nb-doped Mo-Si alloys.

Majumdar [143] used the non-consumable arc-melted method to prepare the undoped and 0.5Y-doped Mo-26Nb-19Si samples (at.%), which are simply referred to as Alloy1 and Alloy2, respectively. The microstructures of both samples are shown in Figure 13a,e. It can be found that they are both composed of dark and bright areas. According to XRD analysis

(Figure 14a) and EBSD mappings (Figure 13b–d,f–h), the dark and bright areas are (Mo, Nb)$_5$Si$_3$ and (Mo, Nb)$_{ss}$ phases, respectively. Moreover, Y$_2$O$_3$ particles are also observed on the Alloy2 grain boundaries. These particles can suppress the elongated grain growths, which results in the difference of microstructure morphology between the two samples. Meanwhile, Majumdar [143] also studied the oxidation process of Alloy2 at 1000 and 1300 °C. It is established that the sample exhibits continuous linear mass loss when oxidized at 1000 °C. When the oxidation temperature increases to 1300 °C, the sample is oxidized more vigorously and loses its antioxidant capacity within 2 h of oxidation, as shown in Figure 14b. Figure 15 shows the cross-section and surface micrographs of the oxidized sample. It can be discovered that the Alloy2 surface has formed a thick oxide layer after oxidation at 1000 °C for 24 h (Figure 15a). As can be seen from Figure 15d, the oxide layer is mainly composed of MoO$_2$, Nb$_2$O$_5$ and SiO$_2$, wherein Nb$_2$O$_5$ can act as a channel for O$_2$ internal diffusion due to the lack of protective action, which leads to rapid oxidation of the sample. In addition, the sample surface oxide film, which consists of Y$_2$O$_3$, Nb$_2$O$_5$ and SiO$_2$, appears numerous cracks and holes during oxidation at 1300 °C for 2 h (Figure 15b,c), resulting in the loss of protection from oxidation. Therefore, adding Y to Mo-Si-Nb alloys cannot overcome the oxidizing problem.

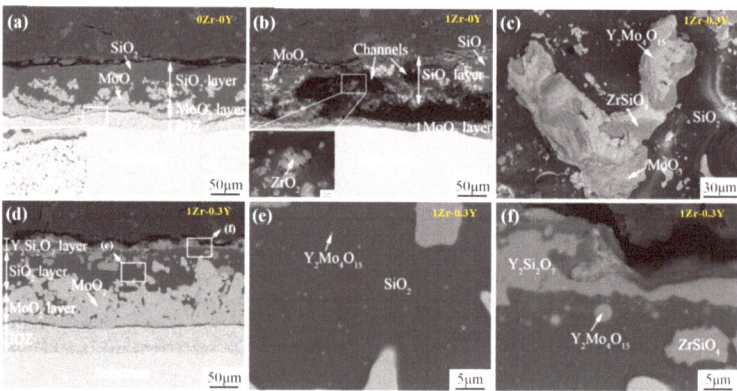

Figure 12. Cross-sectional and surface BSE images of three studied samples oxidized at 1250 °C for different time: (**a**) 20 h, (**b**) 5 h, (**c**) 1 h, (**d–f**) 50 h. Reproduced with permission [135]. Copyright 2020 Elsevier.

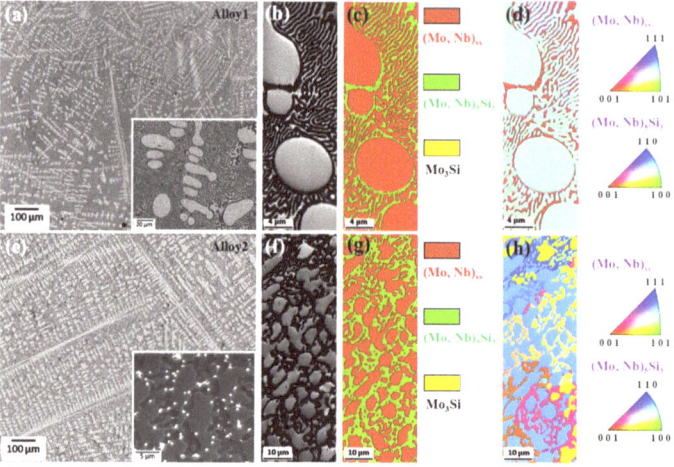

Figure 13. Microstructure morphologies of Alloy1 (**a**) and Alloy2 (**e**); EBSD mappings of Alloy1 and Alloy2: band contrast (**b,f**), phase (**c,g**), IPF (**d,h**). Reproduced with permission [143]. Copyright 2018 Elsevier.

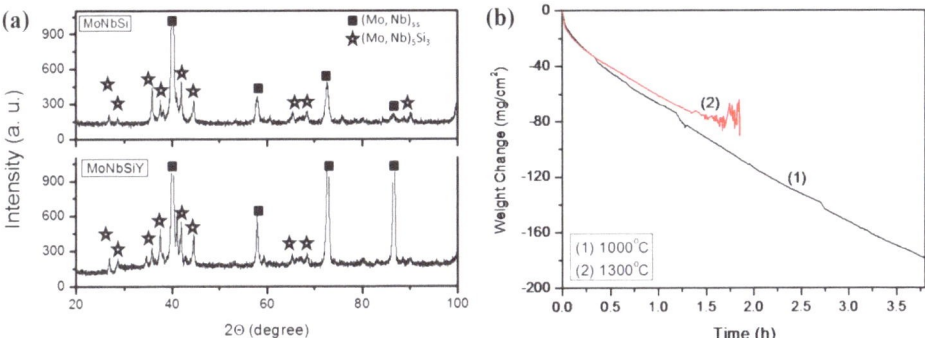

Figure 14. (a) XRD patterns of both samples before the oxidation; (b) weight change curves of Alloy2 oxidized at 1000 and 1300 °C, respectively. Reproduced with permission [143]. Copyright 2018 Elsevier.

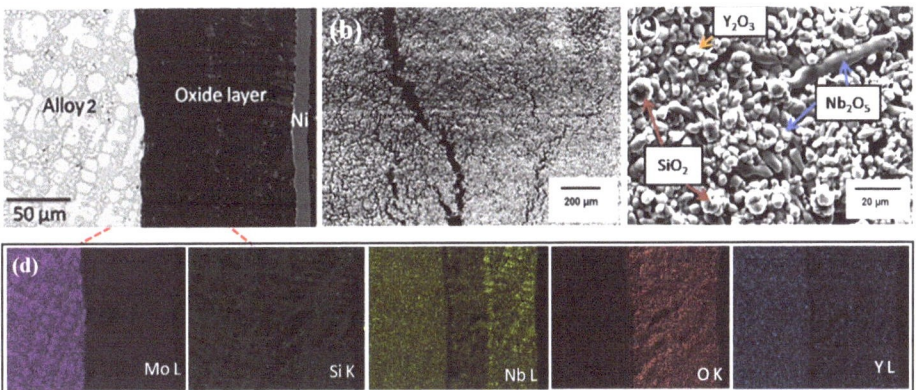

Figure 15. Cross-section BSE micrograph of Alloy2 (a) with the corresponding EDS mappings (d) oxidized at 1000 °C for 24 h; surface SEM/SE images of low (b) and high magnifications (c) for Alloy2 oxidized at 1300 °C for 2 h. Reproduced with permission [143]. Copyright 2018 Elsevier.

3. Strengthening Mechanism of Rare Earth Elements

According to the above research, it can be determined that the improvement of oxidation behavior of Mo-Si-based alloys by rare earth elements is mainly achieved through the following three ways. First, optimizing the alloy microstructure by refining grains or uniformly distributing phases, which is conducive to the rapid formation of protective oxide films on the alloy surface [80,96,103,144]. Second, producing stable rare earth oxides, these oxides are dispersed in scale and act as obstacle phases or diffusion barriers, which is conducive to suppressing the MoO_3 volatilization and O_2 inward diffusion [59,122,123]. Third, forming an additional rare earth oxide layer, thus further improving the antioxidant capacity [96,135].

However, it is disappointing that sometimes the addition of rare earth elements may even lead to the deterioration of alloy oxidation behavior in high temperature environments. For example, adding La to the Mo-Si-B system above 1100 °C has leaded to its accelerated oxidation attribute to the formation of large amounts of cracks and holes [60]. Therefore, the challenges ahead remain severe.

Figure 16 shows a schematic diagram of the oxidation process for rare earth element doped and undoped Mo-Si-based alloys at medium-high temperatures, which is helpful to further understand the strengthening mechanism of rare earth elements. It can be seen that the alloy with finer grain size can be prepared after adding rare earth elements like La, which will affect the oxidation behavior to some extent. Overall, the oxidation process of the two kinds of alloys can be divided into two stages: initial and stable oxidation stages. During the initial oxidation stage, a discontinuous SiO_2 scale is formed on the surface of alloy without rare earth doping, which cannot effectively isolate oxygen. As a result, the alloy is oxidized violently and forms a Mo-oxide (MoO_2 and MoO_3) layer below the SiO_2 scale. Among them, MoO_3 is highly volatile, which leads to a severe mass loss of the alloy and leaves some holes and cavities on the surface [145]. Fortunately, SiO_2 gradually increases and flows to heal these holes and cavities as the oxidation time increases, thus facilitating the formation of continuous SiO_2 scale [146]. During the stable oxidation stage, the complete scale can provide sufficient protection for the substrate due to the effective restriction of O_2 internal diffusion, resulting in the reduction of oxygen pressure inside the alloy. Obviously, low oxygen partial pressure inhibits the continuous generation of MoO_2, and the original MoO_2 will continue to oxidize to produce MoO_3 and slowly volatilize so that the Mo-oxide interlayer becomes thinner [147]. Meanwhile, the substrate below the MoO_2 layer has been oxidized selectively, leading to the emergence of an internal oxidation zone [122], as shown in Figure 16a. In contrast, the alloy doped with rare earth can generate rare earth oxides such as La_2O_3, Y_6MoO_{12}, $Y_5Mo_2O_{12}$, etc., in the initial oxidation stage. These stable oxides, on the one hand, promote the formation of continuous SiO_2 scale. On the other hand, they fill holes in the scale to eliminate the shortcut of O_2 inward diffusion and MoO_3 volatilization, so that the alloy can enter the stable oxidation stage faster. In addition, a double-layer protective oxide film (i.e., $Y_2Si_2O_7$-SiO_2 or SiO_2-$HfSiO_4$ duplex scales) is formed on the alloy surface during the stable oxidation stage, providing more effective protection against oxidation, as shown in Figure 16b.

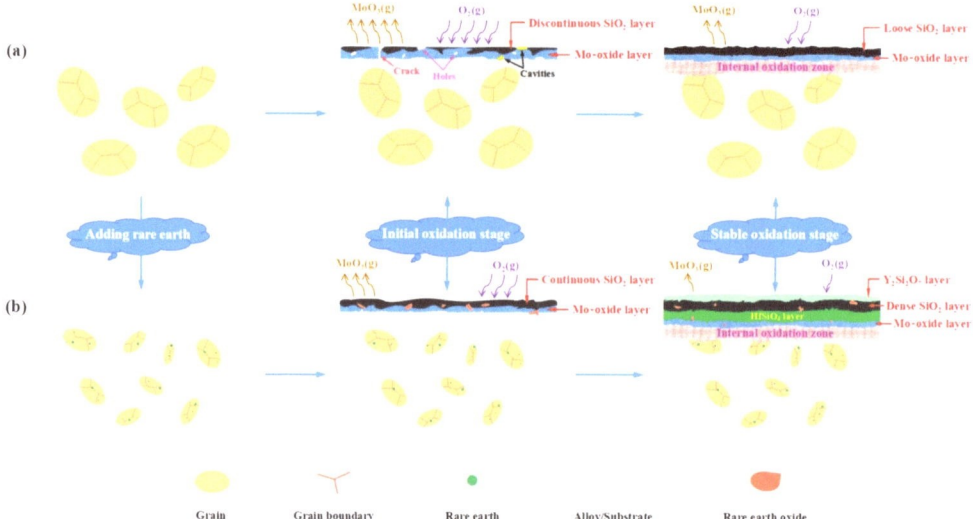

Figure 16. Schematic diagram of the oxidation process for Mo-Si-based alloy at medium-high temperature: (**a**) without rare earth elements, (**b**) doped with rare earth elements.

4. Conclusions and Outlook

This paper reviewed the role of rare earth elements on the oxidation behavior of Mo-Si-based alloys, and summarized the strengthening mechanism of various rare earth elements. Adding La to Mo-Si-B alloys can make grains become finer and produce stable La-containing oxides with a "pinning" effect, which significantly enhances the oxidation resistance below 1000 °C. Mo-Si-B alloys doped with Hf can generate $HfSiO_4$ particles that promote the healing of holes and cracks in oxide scales; besides, it is likely that a $HfSiO_4$ inner layer will be formed to inhibit the MoO_3 volatilization and O_2 inward diffusion.

Alloying with Ce can shorten the transient oxidation period of Mo-Si-B alloys, meanwhile it also raises the volatilization temperature of B_2O_3 in oxide films, which is conducive to maintaining the viscosity and integrity of borosilicate scale. Moreover, adding Ce to the Mo-Si-B-Al system can also hinder the formation of mullite and promote the emergence of protective Al-oxide scale, which can provide more effective protection against oxidation. Adding Y to Mo-Si-B alloys will produce stable Y_6MoO_{12} and $Y_5Mo_2O_{12}$ oxides or create a $Y_2Si_2O_7$ outer layer that can act as diffusion barriers. Furthermore, adding Y to the Mo-Si-B-Zr system also suppresses ZrO_2 formation, thus eliminating the adverse effect of Zr doping on oxidation behavior.

However, it is noteworthy that adding rare earth elements do not always improve the antioxidation ability of Mo-Si-based alloys. For example, adding Y to the Mo-Si-Nb system cannot prevent its catastrophic oxidation; adding La also leads to accelerated oxidation of Mo-Si-B alloys above 1000 °C. Therefore, further research for other oxidation protection methods is necessary. Some research schemes worth exploring in the future are listed below. Before using, preoxidation treatment at an appropriate temperature can obtain protective silica scales on the alloy surface, thus effectively inhibiting the inward diffusion of O_2 and obviously extending the service life of alloy. Processing of preceramic polymers is a very promising method owing to its simple operation and low cost. Establishing a relevant numerical simulation or mathematical model to quantitatively study the relationship between oxidation behavior and microstructure. Combined with the emerging coating technology to design a suitable silicide-based coating such as $MoSi_2$ ceramic coating for the Mo-Si-based system.

Author Contributions: The manuscript was written through the contributions of all authors. Y.Z.: conceptualization, investigation, and supervision. Y.Z. and L.Y.: writing—original draft and image processing. L.Y., T.F., K.C. and F.S.: validation, resources, investigation, writing—review and editing. Y.Z., L.Y. and J.W.: visualization, Writing—review and editing. All authors have read and agreed to the published version of the manuscript.

Funding: This research was funded by the National Natural Science Foundation of China, Grant No. 51604049.

Institutional Review Board Statement: Not applicable.

Informed Consent Statement: Not applicable.

Data Availability Statement: Not applicable.

Acknowledgments: This work was supported by the National Natural Science Foundation of China (No. 51604049).

Conflicts of Interest: The authors declare no conflict of interest.

Notes: The authors declare no competing financial interest.

References

1. Wang, J.; Zhang, Y.Y.; Cui, K.K.; Fu, T.; Gao, J.J.; Hussain, S.; AlGarni, T.S. Pyrometallurgical recovery of zinc and valuable metals from electric arc furnace dust—A review. *J. Clean. Prod.* **2021**, *298*, 126788. [CrossRef]
2. Azim, M.A.; Burk, S.; Gorr, B.; Christ, H.J.; Schliephake, D.; Heilmaier, M.; Bornemann, R.; Bolívar, P.H. Effect of Ti (Macro-)alloying on the high-temperature oxidation behavior of ternary Mo-Si-B alloys at 820–1300 °C. *Oxid. Met.* **2013**, *80*, 231–242. [CrossRef]
3. Gianola, D.S.; Sedlmayr, A.; Mönig, R.; Volkert, C.A.; Major, R.C.; Cyrankowski, E.; Asif, S.A.S.; Warren, O.L.; Kraft, O. In situ nanomechanical testing in focused ion beam and scanning electron microscopes. *Rev. Sci. Instrum.* **2011**, *82*, 063901. [CrossRef] [PubMed]
4. Kiener, D.; Motz, C.; Dehm, G.; Pippan, R. Overview on established and novel FIB based miniaturized mechanical testing using in-situ SEM. *Int. J. Mater. Res.* **2009**, *100*, 1074–1087. [CrossRef]
5. Ghidelli, M.; Sebastiani, M.; Johanns, K.E.; Pharr, G.M. Effects of indenter angle on micro-scale fracture toughness measurement by pillar splitting. *J. Am. Ceram. Soc.* **2017**, *100*, 5731–5738. [CrossRef]
6. Ast, J.; Ghidelli, M.; Durst, K.; Göken, M.; Sebastiani, M.; Korsunsky, A.M. A review of experimental approaches to fracture toughness evaluation at the micro-scale. *Mater. Des.* **2019**, *173*, 107762. [CrossRef]

7. Ast, J.; Schwiedrzik, J.J.; Wehrs, J.; Frey, D.; Polyakov, M.N.; Michler, J.; Maeder, X. The brittle-ductile transition of tungsten single crystals at the micro-scale. *Mater. Des.* **2018**, *152*, 168–180. [CrossRef]
8. Lemberg, J.A.; Ritchie, R.O. Mo-Si-B alloys for ultrahigh-temperature structural applications. *Adv. Mater.* **2012**, *24*, 3445–3480. [CrossRef] [PubMed]
9. Perepezko, J.H. The hotter the engine, the better. *Science* **2009**, *326*, 1068–1069. [CrossRef]
10. Becker, J.; Fichtner, D.; Schmigalla, S.; Schultze, S.; Heinze, C.; Küsters, Y.; Hasemann, G.; Schmelzer, J.; Krüger, M. Oxidation response of additively manufactured eutectic Mo-Si-B alloys. *IOP Conf. Ser. Mater. Sci. Eng.* **2020**, *882*, 012002. [CrossRef]
11. Ingemarsson, L.; Hellström, K.; Čanović, S.; Jonsson, T.; Halvarsson, M.; Johansson, L.G.; Svensson, J.E. Oxidation behavior of a Mo(Si,Al)$_2$ composite at 900–1600 °C in dry air. *J. Mater. Sci.* **2013**, *48*, 1511–1523. [CrossRef]
12. Petrovic, J.J. Mechanical behavior of MoSi$_2$ and MoSi$_2$ composites. *Mater. Sci. Eng. A* **1995**, *192–193*, 31–37. [CrossRef]
13. Schneibel, J.H.; Sekhar, J.A. Microstructure and properties of MoSi$_2$-MoB and MoSi$_2$-Mo$_5$Si$_3$ molybdenum silicides. *Mater. Sci. Eng. A* **2003**, *340*, 204–211. [CrossRef]
14. Zhao, M.; Xu, B.Y.; Shao, Y.M.; Liang, J.F.; Wu, S.S.; Yan, Y.W. Oxidation behavior of Mo$_{ss}$-Ti$_5$Si$_3$-T$_2$ composites at intermediate and high temperatures. *Intermetallics* **2020**, *118*, 106702. [CrossRef]
15. Vasudévan, A.K.; Petrovic, J.J. A comparative overview of molybdenum disilicide composites. *Mater. Sci. Eng. A* **1992**, *155*, 1–17. [CrossRef]
16. Zhang, H.; Wang, D.Z.; Chen, S.P.; Liu, X.Y. Toughening of MoSi$_2$ doped by La$_2$O$_3$ particles. *Mater. Sci. Eng. A* **2003**, *345*, 118–121. [CrossRef]
17. SWen, H.; Zhou, C.G.; Sha, J.B. Improvement of oxidation resistance of a Mo-62Si-5B (at.%) alloy at 1250 °C and 1350 °C via an in situ pre-formed SiO$_2$ fabricated by spark plasma sintering. *Corros. Sci.* **2017**, *127*, 175–185.
18. Zhang, Y.Y.; Zhao, J.; Li, J.H.; Lei, J.; Cheng, X.K. Effect of hot-dip siliconizing time on phase composition and microstructure of Mo-MoSi$_2$ high temperature structural materials. *Ceram. Int.* **2019**, *45*, 5588–5593. [CrossRef]
19. Ito, K.; Hayashi, T.; Yokobayashi, M.; Numakura, H. Evolution kinetics and microstructure of MoSi$_2$ and Mo$_5$Si$_3$ surface layers on two-phase Mo-9Si-18B alloy during pack-cementation and high-temperature oxidation. *Intermetallics* **2004**, *12*, 407–415. [CrossRef]
20. Pint, B.A. On the formation of interfacial and internal voids in a-Al$_2$O$_3$ scales. *Oxid. Met.* **1997**, *48*, 303–328. [CrossRef]
21. Lohfeld, S. Oxidation behaviour of particle reinforced MoSi$_2$ composites at temperatures up to 1700 °C. Part II: Initial screening of the oxidation behaviour of MoSi$_2$ composites. *Mater. Corros.* **2005**, *56*, 149–158. [CrossRef]
22. Lohfeld, S.; Schütze, M.; Böhm, A.; Güther, V. Oxidation behaviour of particle reinforced MoSi$_2$ composites at temperatures up to 1700 °C. Part III: Oxidation behaviour of optimised MoSi$_2$ composites. *Mater. Corros.* **2005**, *56*, 250–258. [CrossRef]
23. Murty, B.S.; Ping, D.H.; Hono, K.; Inoue, A. Direct evidence for oxygen stabilization of icosahedral phase during crystallization of Zr$_{65}$Cu$_{27.5}$Al$_{7.5}$ metallic glass. *Appl. Phys. Lett.* **2000**, *76*, 55–57. [CrossRef]
24. Zhang, Y.Y.; Li, Y.G.; Bai, C.G. Microstructure and oxidation behavior of Si-MoSi$_2$ functionally graded coating on Mo substrate. *Ceram. Int.* **2017**, *43*, 6250–6256. [CrossRef]
25. Zhang, Y.Y.; Qie, J.M.; Cui, K.K.; Fu, T.; Fan, X.L.; Wang, J.; Zhang, X. Effect of hot dip silicon-plating temperature on microstructure characteristics of silicide coating on tungsten substrate. *Ceram. Int.* **2020**, *46*, 5223–5228. [CrossRef]
26. Petrovic, J.J. Toughening strategies for MoSi$_2$-based high temperature structural silicides. *Intermetallics* **2000**, *8*, 1175–1182. [CrossRef]
27. Sossaman, T.; Sakidja, R.; Perepezko, J.H. Influence of minor Fe addition on the oxidation performance of Mo-Si-B alloys. *Scr. Mater.* **2012**, *67*, 891–894. [CrossRef]
28. Burk, S.; Gorr, B.; Krüger, M.; Heilmaier, M. Oxidation behavior of Mo-Si-B-(X) alloys: Macro- and microalloying (X = Cr, Zr, La$_2$O$_3$), JOM: The journal of the Minerals. *Met. Mater. Soc.* **2011**, *63*, 32–36. [CrossRef]
29. Mandal, P.; Thom, A.J.; Kramer, M.J.; Behrani, V.; Akinc, M. Oxidation behavior of Mo-Si-B alloys in wet air. *Mater. Sci. Eng. A* **2004**, *371*, 335–342. [CrossRef]
30. Jeng, Y.L.; Lavernia, E. Processing of molybdenum disilicide. *J. Mater. Sci.* **1994**, *29*, 2557–2571. [CrossRef]
31. Morris, D.G.; Leboeuf, M.; Morris, M.A. Hardness and toughness of MoSi$_2$ and MoSi$_2$-SiC composite prepared by reactive sintering of powders. *Mater. Sci. Eng. A* **1998**, *251*, 262–268. [CrossRef]
32. Chrysanthou, A.; Jenkins, R.C.; Whiting, M.J.; Tsakiropoulos, P. A study of the combustion synthesis of MoSi$_2$ and MoSi$_2$-matrix composites. *J. Mater. Sci.* **1996**, *31*, 4221–4226. [CrossRef]
33. Ignat, S.; Sallamand, P.; Nichici, A.; Vannes, A.B.; Grevey, D.; Cicală, E. MoSi$_2$ laser cladding-a comparison between two experimental procedures: Mo-Si online combination and direct use of MoSi$_2$. *Opt. Laser Technol.* **2001**, *33*, 461–469. [CrossRef]
34. Fu, T.; Cui, K.K.; Zhang, Y.Y.; Wang, J.; Zhang, X.; Shen, F.Q.; Yu, L.H.; Mao, H.B. Microstructure and oxidation behavior of antioxidation coatings on Mo-based alloys through HAPC process: A review. *Coatings* **2021**, *11*, 883. [CrossRef]
35. Wang, F.; Shan, A.D.; Dong, X.P.; Wu, J.S. Microstructure and oxidation resistance of laser-remelted Mo-Si-B alloy. *Scr. Mater.* **2007**, *56*, 737–740. [CrossRef]
36. Petitbon, A.; Boquet, L.; Delsart, D. Laser surface sealing and strengthening of zirconia coatings. *Surf. Coat. Technol.* **1991**, *49*, 57–61. [CrossRef]
37. Zhang, Y.Y.; Cui, K.K.; Fu, T.; Wang, J.; Qie, J.M.; Zhang, X. Synthesis WSi$_2$ coating on W substrate by HDS method with various deposition times. *Appl. Surf. Sci.* **2020**, *511*, 145551. [CrossRef]

38. Zhang, Y.Y.; Fu, T.; Cui, K.K.; Shen, F.Q.; Wang, J.; Yu, L.H.; Mao, H.B. Evolution of surface morphology, roughness and texture of tungsten disilicide coatings on tungsten substrate. *Vacuum* **2021**, *191*, 110297. [CrossRef]
39. Cui, K.K.; Fu, T.; Zhang, Y.Y.; Wang, J.; Mao, H.B.; Tan, T.B. Microstructure and mechanical properties of $CaAl_{12}O_{19}$ reinforced Al_2O_3-Cr_2O_3 composites. *J. Eur. Ceram. Soc.* **2021**. [CrossRef]
40. Ghidelli, M.; Orekhov, A.; Bassi, A.L.; Terraneo, G.; Djemia, P.; Abadias, G.; Nord, M.; Béché, A.; Gauquelin, N.; Verbeeck, J.; et al. Novel class of nanostructured metallic glass films with superior and tunable mechanical properties. *Acta Mater.* **2021**, *213*, 116955. [CrossRef]
41. Besozzi, E.; Dellasega, D.; Russo, V.; Conti, C.; Passoni, M.; Beghi, M.G. Thermomechanical properties of amorphous metallic tungsten-oxygen and tungsten-oxide coatings. *Mater. Des.* **2019**, *165*, 107565. [CrossRef]
42. Hou, P.Y.; Stinger, J. The effect of reactive element additions on the selective oxidation, growth and adhesion of chromia scales. *Mater. Sci. Eng. A* **1995**, *202*, 1–10. [CrossRef]
43. Krüger, M.; Franz, S.; Saage, H.; Heilmaier, M.; Schneibel, J.H.; Jéhanno, P.; Böning, M.; Kestler, H. Mechanically alloyed Mo-Si-B alloys with a continuous α-Mo matrix and improved mechanical properties. *Intermetallics* **2008**, *16*, 933–941. [CrossRef]
44. Christensen, R.J.; Tolpygo, V.K.; Clarke, D.R. The influence of the reactive element yttrium on the stress in alumina scales formed by oxidation. *Acta Mater.* **1997**, *45*, 1761–1766. [CrossRef]
45. Sharif, A.A. Effects of Re- and Al-alloying on mechanical properties and high-temperature oxidation of $MoSi_2$. *J. Alloys Compd.* **2012**, *518*, 22–26. [CrossRef]
46. Yang, Y.; Bei, H.; Tiley, J.; George, E.P. Re effects on phase stability and mechanical properties of Mo_{SS}+Mo_3Si+Mo_5SiB_2 alloys. *J. Alloys Compd.* **2013**, *556*, 32–38. [CrossRef]
47. Jin, H.M.; Zhang, L.N. Rare earth effects on adhesion of Cr_2O_3 oxide scale formed on surface of Co–40Cr alloy. *J. Rare Earths* **2001**, *19*, 34–49.
48. Qi, H.; Lees, D.G.; He, Y. Effect of surface-applied rare earth containing thin oxide film on high-temperature oxidation of Fe25Cr. *Corros. Sci. Prot. Technol.* **1999**, *11*, 200–201.
49. Zhang, Y.Y.; Cui, K.K.; Gao, Q.J.; Hussain, S.; Lv, Y. Investigation of morphology and texture properties of WSi_2 coatings on W substrate based on contact-mode AFM and EBSD. *Surf. Coat. Technol.* **2020**, *396*, 125966. [CrossRef]
50. Zhang, X.; Fu, T.; Cui, K.K.; Zhang, Y.Y.; Shen, F.Q.; Wang, J.; Yu, L.H.; Mao, H.B. The protection, challenge, and prospect of anti-oxidation coating on the surface of niobium alloy. *Coatings* **2021**, *11*, 742. [CrossRef]
51. Zhang, Y.Y.; Cui, K.K.; Fu, T.; Wang, J.; Shen, F.Q.; Zhang, X.; Yu, L.H. Formation of $MoSi_2$ and $Si/MoSi_2$ coatings on TZM (Mo-0.5Ti-0.1Zr-0.02C) alloy by hot dip silicon-plating method. *Ceram. Int.* **2021**, *47*, 23053–23065. [CrossRef]
52. Zhang, Y.Y.; Hussain, S.; Cui, K.K.; Fu, T.; Wang, J.; Javed, M.S.; Lv, Y.; Aslam, B. Microstructure and mechanical properties of $MoSi_2$ coating deposited on Mo substrate by hot dipping processes. *Nanoelectron. Optoelectron.* **2019**, *14*, 1680–1685. [CrossRef]
53. Pint, B.A. The oxidation behavior of oxide-dispersed β-NiAl: I. short-term performance at 1200 °C. *Oxid. Met.* **1998**, *49*, 531–559. [CrossRef]
54. Hiramatsu, N.; Stott, F.H. The effect of lanthanum on the scales developed on thin foils of Fe20Cr5Al at very high temperatures. *Oxid. Met.* **1999**, *51*, 479–494. [CrossRef]
55. Fu, T.; Cui, K.K.; Zhang, Y.Y.; Wang, J.; Shen, F.Q.; Yu, L.H.; Qie, J.M.; Zhang, X. Oxidation protection of tungsten alloys for nuclear fusion applications: A comprehensive review. *J. Alloys Compd.* **2021**, *884*, 161057. [CrossRef]
56. Pint, B.A. Experimental observations in support of the dynamic-segregation theory to explain the reactive element effect. *Oxid. Met.* **1996**, *45*, 1–37. [CrossRef]
57. Nijdam, T.J.; Sloof, W.G. Effect of reactive element oxide inclusions on the growth kinetics of protective oxide scales. *Acta Mater.* **2007**, *55*, 5980–5987. [CrossRef]
58. Pan, K.; Yang, Y.P.; Wei, S.Z.; Wu, H.H.; Dong, Z.L.; Wu, Y.; Wang, S.Z.; Zhang, L.Q.; Lin, J.P.; Mao, X.P. Oxidation behavior of Mo-Si-B alloys at medium-to-high temperatures. *J. Mater. Sci. Technol.* **2021**, *60*, 113–127. [CrossRef]
59. Majumdar, S.; Burk, S.; Schliephake, D.; Krüger, M.; Christ, H.J.; Heilmaier, M. A study on effect of reactive and rare earth element additions on the oxidation behavior of Mo-Si-B system. *Oxid. Met.* **2013**, *80*, 219–230. [CrossRef]
60. Majumdar, S.; Gorr, B.; Christ, H.J.; Schliephake, D.; Heilmaier, M. Oxidation mechanisms of lanthanum-alloyed Mo-Si-B. *Corros. Sci.* **2014**, *88*, 360–371. [CrossRef]
61. Whittle, D.P.; Stringer, J. Improvement in properties: Additives in oxidation resistance. *Philos. Trans. R. Soc. Lond.* **1980**, *295*, 309–329.
62. Bennett, I.J.; Sloof, W.G. Modelling the influence of reactive elements on the work of adhesion between a thermally grown oxide and a bond coat alloy. *Mater. Corros.* **2015**, *57*, 223–229. [CrossRef]
63. Naumenko, D.; Gleeson, B.; Wessel, E.; Singheiser, L.; Quadakkers, W.J. Correlation between the microstructure, growth mechanism, and growth kinetics of alumina scales on a FeCrAlY alloy. *Metall. Mater. Trans. A* **2007**, *38*, 2974–2983. [CrossRef]
64. Thom, A.J.; Summers, E.; Akinc, M. Oxidation behavior of extruded $Mo_5Si_3B_x$–$MoSi_2$–MoB intermetallics from 600°–1600 °C. *Intermetallics* **2002**, *10*, 555–570. [CrossRef]
65. Wang, F.; Shan, A.D.; Dong, X.P.; Wu, J.S. Oxidation behavior of Mo-12.5Si-25B alloy at high temperature. *J. Alloys Compd.* **2008**, *459*, 362–368. [CrossRef]
66. Yoshimi, K.; Nakatani, S.; Suda, T.; Hanada, S.; Habazaki, H. Oxidation behavior of Mo_5SiB_2-based alloy at elevated temperatures. *Intermetallics* **2002**, *10*, 407–414. [CrossRef]

67. Liu, L.; Shi, C.; Zhang, C.; Voyles, P.M.; Fournelle, J.H.; Perepezko, J.H. Microstructure, microhardness and oxidation behavior of Mo-Si-B alloys in the $Mo_{ss}+Mo_2B+Mo_5SiB_2$ three phase region. *Intermetallics* **2020**, *116*, 106618. [CrossRef]
68. Zhang, G.J.; He, W.; Li, B.; Zha, Y.; Sun, J. Effect of Si/B ratio on the microstructure and mechanical properties of lanthanum oxide-doped Mo-Si-B alloys. *J. Alloys Compd.* **2013**, *577*, 217–221. [CrossRef]
69. Kumar, N.K.; Das, J.; Mitra, R. Efect of moist air and minor Zr addition on oxidation behavior of arc-melted multiphase Mo-Si-B alloys in the temperature range of 1000 °C–1300 °C. *Oxid. Met.* **2020**, *93*, 483–513. [CrossRef]
70. Roy, B.; Khushboo; Das, J.; Mitra, R.; Roy, S.K. Effect of oxygen partial pressure on the cyclic oxidation behavior of $Mo_{76}Si_{14}B_{10}$. *Metall. Mater. Trans. A* **2013**, *44*, 2910–2913. [CrossRef]
71. Yang, X.Q.; Tan, H.; Lin, N.; Li, Z.X.; He, Y.H. Effects of the lanthanum content on the microstructure and properties of the molybdenum alloy. *Int. J. Refract. Met. Hard Mater.* **2016**, *61*, 179–184. [CrossRef]
72. Zhang, G.J.; Dang, Q.; Kou, H.; Wang, R.H.; Liu, G.; Sun, J. Microstructure and mechanical properties of lanthanum oxide-doped Mo-12Si-8.5B (at.%) alloys. *J. Alloys Compd.* **2013**, *577*, 493–498. [CrossRef]
73. Zhang, G.J.; Zha, Y.; Li, B.; He, W.; Sun, J. Effects of lanthanum oxide content on mechanical properties of mechanical alloying Mo-12Si-8.5B (at.%) alloys. *Int. J. Refract. Met. Hard Mater.* **2013**, *41*, 585–589. [CrossRef]
74. Yan, J.H.; Zhang, H.A.; Tang, S.W.; Xu, J.G. Room temperature mechanical properties and high temperature oxidation behavior of $MoSi_2$ matrix composite reinforced by adding La_2O_3 and Mo_5Si_3. *Mater. Charact.* **2009**, *60*, 447–450. [CrossRef]
75. Choe, H.; Chen, D.; Schneibel, J.H.; Ritchie, R.O. Ambient to high temperature fracture toughness and fatigue-crack propagation behavior in a Mo-12Si-8.5B (at.%) intermetallic. *Intermetallics* **2001**, *9*, 319–329. [CrossRef]
76. Schneibel, J.H.; Kramer, M.J.; Unal, O.; Wright, R.N. Processing and mechanical properties of a molybdenum silicide with the composition Mo-12Si-8.5B (at.%). *Intermetallics* **2001**, *9*, 25–31. [CrossRef]
77. Wang, J.; Li, B.; Li, R.; Chen, X.; Zhang, G.J. Bimodal α-Mo grain structure inducing excellent oxidation resistance in Mo-12Si-8.5B alloy at 1100 °C. *Int. J. Refract. Met. Hard Mater.* **2021**, *98*, 105533. [CrossRef]
78. Wang, J.; Li, B.; Ren, S.; Li, R.; Wang, T.; Zhang, G.J. Enhanced oxidation resistance of Mo-12Si-8.5B alloys with ZrB_2 addition at 1300 °C. *J. Mater. Sci. Technol.* **2018**, *34*, 635–642. [CrossRef]
79. Wang, J.; Li, B.; Li, R.; Chen, X.; Wang, T.; Zhang, G.J. Unprecedented oxidation resistance at 900 °C of Mo-Si-B composite with addition of ZrB_2. *Ceram. Int.* **2020**, *46*, 14632–14639. [CrossRef]
80. Zhang, G.J.; Kou, H.; Dang, Q.; Liu, G.; Sun, J. Microstructure and oxidation resistance behavior of lanthanum oxide-doped Mo-12Si-8.5B alloys. *Int. J. Refract. Met. Hard Mater.* **2012**, *30*, 6–11.
81. Fang, W.; Shan, A.D.; Dong, X.P.; Wu, J.S. Microstructure and oxidation behavior of directionally solidified Mo-Mo_5SiB_2 (T2)-Mo_3Si alloys. *J. Alloys Compd.* **2008**, *462*, 436–441.
82. Fang, W.; Shan, A.D.; Dong, X.P.; Wu, J.S. Oxidation behavior of multiphase Mo_5SiB_2 (T2)-based alloys at high temperatures. *Trans. Nonferrous Met. Soc. China* **2007**, *17*, 1242–1247.
83. Choi, W.J.; Park, C.W.; Park, J.H.; Kim, Y.D.; Byun, J.M. Volume and size effects of intermetallic compounds on the high-temperature oxidation behavior of Mo-Si-B alloys. *Int. J. Refract. Met. Hard Mater.* **2019**, *81*, 94–99. [CrossRef]
84. Schliephake, D.; Kauffmann, A.; Cong, X.; Gombola, C.; Azim, M.; Gorr, B.; Christ, H.J.; Heilmaier, M. Constitution, oxidation and creep of eutectic and eutectoid Mo-Si-Ti alloys. *Intermetallics* **2019**, *104*, 133–142. [CrossRef]
85. Supatarawanich, V.; Johnson, D.R.; Liu, C.T. Effects of microstructure on the oxidation behavior of multiphase Mo-Si-B alloys. *Mater. Sci. Eng. A* **2003**, *344*, 328–339. [CrossRef]
86. Yamauchi, A.; Yoshimi, K.; Murakami, Y.; Kurokawa, K.; Hanada, S. Oxidation behavior of Mo-Si-B in-situ composites. *Solid State Phenom.* **2007**, *127*, 215–220. [CrossRef]
87. Rosales, I.; Martinez, H.; Bahena, D.; Ruiz, J.A.; Guardian, R.; Colin, J. Oxidation performance of Mo_3Si with Al additions. *Corros. Sci.* **2009**, *51*, 534–538. [CrossRef]
88. Burk, S.; Gorr, B.; Trindade, V.B.; Krupp, U. Temperature oxidation of mechanically alloyed Mo-Si-B alloys. *Br. Corros. J.* **2009**, *44*, 168–175. [CrossRef]
89. Burk, S.; Christ, H.J. High-temperature oxidation performance of Mo-Si-B alloys: Current results. developments and opportunities. *Adv. Mater. Res.* **2011**, *278*, 587–592. [CrossRef]
90. Jéhanno, P.; Böning, M.; Kestler, H.; Heilmaier, M.; Saage, H.; Krüger, M. Molybdenum alloys for high temperature applications in air. *Powder Metall.* **2008**, *51*, 99–102. [CrossRef]
91. Mattuck, B. High-temperature oxidation III. Zirconium and hafnium diborides. *J. Electrochem. Soc.* **1966**, *113*, 908–914. [CrossRef]
92. Cook, J.; Khan, A.; Lee, E.; Mahapatra, R. Oxidation of $MoSi_2$-based composites. *Mater. Sci. Eng. A* **1992**, *155*, 183–198. [CrossRef]
93. Yang, Y.; Bei, H.; Chen, S.L.; George, E.P.; Tiley, J.; Chang, Y.A. Effects of Ti, Zr, and Hf on the phase stability of Mo_ss + Mo_3Si + Mo_5SiB_2 alloys at 1600 °C. *Acta Mater.* **2010**, *58*, 541–548. [CrossRef]
94. Vorotilo, S.; Potanin, A.Y.; Pogozhev, Y.S.; Levashov, E.A.; Kochetov, N.A.; Kovalev, D.Y. Self-propagating high-temperature synthesis of advanced ceramics $MoSi_2$-HfB_2-MoB. *Ceram. Int.* **2019**, *45*, 96–107. [CrossRef]
95. Sciti, D.; Silvestroni, L.; Bellosi, A. Fabrication and properties of HfB_2-$MoSi_2$ composites produced by hot pressing and spark plasma sintering. *J. Mater. Res.* **2006**, *21*, 1460–1466. [CrossRef]
96. Potanin, A.Y.; Vorotilo, S.; Pogozhev, Y.S.; Rupasov, S.I.; Loginov, P.A.; Shvyndina, N.V.; Sviridova, T.A.; Levashov, E.A. High-temperature oxidation and plasma torch testing of $MoSi_2$-HfB_2-MoB ceramics with single-level and two-level structure. *Corros. Sci.* **2019**, *158*, 108074. [CrossRef]

97. Sciti, D.; Balbo, A.; Bellosi, A. Oxidation behaviour of a pressureless sintered HfB$_2$-MoSi$_2$ composite. *J. Eur. Ceram. Soc.* **2009**, *29*, 1809–1815. [CrossRef]
98. Parthasarathy, T.A.; Rapp, R.A.; Opeka, M.; Kerans, R.J. A model for the oxidation of ZrB$_2$, HfB$_2$ and TiB$_2$. *Acta Mater.* **2007**, *55*, 5999–6010. [CrossRef]
99. Kumar, N.K.; Das, J.; Mitra, R. Effect of Zr addition on microstructure, hardness and oxidation behavior of arc-melted and spark plasma sintered multiphase Mo-Si-B alloys. *Metall. Mater. Trans. A* **2019**, *50*, 2041–2060. [CrossRef]
100. Yu, Y.; Luo, R.; Xiang, Q.; Zhang, Y.; Wang, T.Y. Anti-oxidation properties of a BN/SiC/Si$_3$N$_4$-ZrO$_2$-SiO$_2$ multilayer coating for carbon/carbon composites. *Surf. Coat. Technol.* **2015**, *277*, 7–14. [CrossRef]
101. Choi, J.H.; Mao, Y.; Chang, J.P. Development of hafnium based high- materials—A review. *Mater. Sci. Eng. R* **2011**, *72*, 97–136. [CrossRef]
102. Das, J.; Mitra, R.; Roy, S.K. Oxidation behaviour of Mo-Si-B-(Al,Ce) ultrafine-eutectic dendrite composites in the temperature range of 500–700 °C. *Intermetallics* **2011**, *19*, 1–8. [CrossRef]
103. Das, J.; Roy, B.; Kumar, N.K.; Mitra, R. High temperature oxidation response of Al/Ce doped Mo-Si-B composites. *Intermetallics* **2017**, *83*, 101–109. [CrossRef]
104. Dimiduk, D.M.; Perepezko, J.H. Mo-Si-B alloys: Developing a revolutionary turbine-engine material. *Mrs Bull.* **2003**, *28*, 639–645. [CrossRef]
105. Perepezko, H.J.; Sakidja, R.; Kumar, K.S. Mo-Si-B alloys for ultrahigh temperature applications. *Adv. Struct. Mater.* **2006**, 437–473. [CrossRef]
106. Mendiratta, M.G.; Parthasarathy, T.A.; Dimiduk, D.M. Oxidation behavior of αMo-Mo$_3$Si-Mo$_5$SiB$_2$ (T2) three phase system. *Intermetallics* **2002**, *10*, 225–232. [CrossRef]
107. Cofer, C.G.; Economy, J. Oxidative and hydrolytic stability of boron nitride—A new approach to improving the oxidation resistance of carbonaceous structures. *Carbon* **1995**, *33*, 389–395. [CrossRef]
108. Karahan, T.; Ouyang, G.; Ray, P.K.; Kramer, M.J. Oxidation mechanism of W substituted Mo-Si-B alloys. *Intermetallics* **2017**, *87*, 38–44. [CrossRef]
109. Das, J.; Mitra, R.; Roy, S.K. Effect of Ce addition on the oxidation behaviour of Mo-Si-B-Al ultrafine composites at 1100 °C. *Scr. Mater.* **2011**, *64*, 486–489. [CrossRef]
110. Paswan, S.; Mitra, R.; Roy, S.K. Isothermal oxidation behaviour of Mo-Si-B and Mo-Si-B-Al alloys in the temperature range of 400–800 °C. *Mater. Sci. Eng. A* **2006**, *424*, 251–265. [CrossRef]
111. Paswan, S.; Mitra, R.; Roy, S.K. Oxidation behaviour of the Mo-Si-B and Mo-Si-B-Al alloys in the temperature range of 700–1300 °C. *Intermetallics* **2007**, *15*, 1217–1227. [CrossRef]
112. Cui, K.K.; Zhang, Y.Y.; Fu, T.; Hussain, S.; AlGarni, T.S.; Wang, J.; Zhang, X.; Ali, S. Effects of Cr$_2$O$_3$ content on microstructure and mechanical properties of Al$_2$O$_3$ matrix composites. *Coatings* **2021**, *11*, 234. [CrossRef]
113. Cui, K.K.; Zhang, Y.Y.; Fu, T.; Wang, J.; Zhang, X. Toughening mechanism of mullite matrix composites: A review. *Coatings* **2020**, *10*, 672. [CrossRef]
114. Paswan, S.; Mitra, R.; Roy, S.K. Nonisothermal and cyclic oxidation behavior of Mo-Si-B and Mo-Si-B-Al alloys. *Metall. Mater. Trans. A* **2009**, *40*, 2644. [CrossRef]
115. Yanagihara, K.; Maruyama, T.; Nagata, K. Effect of third elements on the pesting suppression of Mo-Si-X intermetallics (X = Al, Ta, Ti, Zr and Y). *Intermetallics* **1996**, *4*, S133–S139. [CrossRef]
116. Liu, D.G.; Zheng, L.; Luo, L.M.; Zan, X.; Song, J.P.; Xu, Q.; Zhu, X.Y.; Wu, Y.C. An overview of oxidation-resistant tungsten alloys for nuclear fusion. *J. Alloys Compd.* **2018**, *765*, 299–312. [CrossRef]
117. Liu, Z.Y.; Gao, W.; He, Y.D. Modeling of oxidation kinetics of Y-doped Fe-Cr-Al alloys. *Oxid. Met.* **2000**, *53*, 341–350. [CrossRef]
118. Fujikawa, H.; Morimoto, T.; Nishiyama, Y.; Newcomb, S.B. The effects of small additions of yttrium on the high-temperature oxidation resistance of a Si-containing austenitic stainless steel. *Oxid. Met.* **2003**, *59*, 23–40. [CrossRef]
119. Telu, S.; Mitra, R.; Pabi, S.K. Effect of Y$_2$O$_3$ addition on oxidation behavior of W-Cr alloys. *Metall. Mater. Trans. A* **2015**, *46*, 5909–5919. [CrossRef]
120. Pérez, F.J.; Cristóbal, M.J.; Hierro, M.P.; Arnau, G.; Botella, J. Corrosion protection of low-nickel austenitic stainless steel by yttrium and erbium-ion implantation against isothermal oxidation. *Oxid. Met.* **2000**, *54*, 87–101. [CrossRef]
121. Wu, Y.; Umakoshi, Y.; Li, X.W.; Narita, T. Isothermal oxidation behavior of Ti-50Al alloy with Y additions at 800 and 900 °C. *Oxid. Met.* **2006**, *66*, 321–348. [CrossRef]
122. Majumdar, S.; Dönges, B.; Gorr, B.; Christ, H.J.; Schliephake, D.; Heilmaier, M. Mechanisms of oxide scale formation on yttrium-alloyed Mo-Si-B containing fine-grained microstructure. *Corros. Sci.* **2015**, *90*, 76–88. [CrossRef]
123. Majumdar, S.; Schliephake, D.; Gorr, B.; Christ, H.J.; Heilmaier, M. Effect of yttrium alloying on intermediate to high-temperature oxidation behavior of Mo-Si-B alloys. *Metall. Mater. Trans. A* **2013**, *44*, 2243–2257. [CrossRef]
124. Majumdar, S.; Kumar, A.; Schliephake, D.; Christ, H.J.; Jiang, X.; Heilmaier, M. Microstructural and micro-mechanical properties of Mo-Si-B alloyed with Y and La. *Mater. Sci. Eng. A* **2013**, *573*, 257–263. [CrossRef]
125. EerNisse, E.P. Stress in thermal SiO$_2$ during growth. *Appl. Phys. Lett.* **1979**, *35*, 8–10. [CrossRef]
126. Irene, E.A. Silicon oxidation studies: A revised model for thermal oxidation. *J. Appl. Phys.* **1983**, *54*, 5416–5420. [CrossRef]
127. Opila, E.J. Variation of the oxidation rate of SiC in water vapor. *J. Am. Ceram. Soc.* **1999**, *82*, 625–636. [CrossRef]

128. EOpila, J.; Smialek, J.L.; Robinson, R.C.; Fox, D.S.; Jacobson, N.S. SiC recession caused by SiO_2 scale volatility under combustion conditions: II, thermodynamics and gaseous-diffusion model. *J. Am. Ceram. Soc.* **2004**, *82*, 1826–1834. [CrossRef]
129. Schneider, J.; Biswas, K.; Rixecker, G.; Aldinger, F. Microstructural changes in liquid-phase-sintered silicon carbide during creep in an oxidizing environment. *J. Am. Ceram. Soc.* **2004**, *86*, 501–507. [CrossRef]
130. Gorr, B.; Wang, L.; Burk, S.; Azim, M.; Majumdar, S.; Christ, H.J.; Mukherji, D.; Rösler, J.; Schliephake, D.; Heilmaier, M. High-temperature oxidation behavior of Mo-Si-B-based and Co-Re-Cr-based alloys. *Intermetallics* **2014**, *48*, 34–43. [CrossRef]
131. Burk, S.; Gorr, B.; Trindade, V.B.; Christ, H.J. Effect of Zr addition on the high-temperature oxidation behaviour of Mo-Si-B alloys. *Oxid. Met.* **2010**, *73*, 163–181. [CrossRef]
132. Burk, S.; Gorr, B.; Christ, H.J. High temperature oxidation of Mo-Si-B alloys: Effect of low and very low oxygen partial pressures. *Acta Mater.* **2010**, *58*, 6154–6165. [CrossRef]
133. Asadikiya, M.; Foroughi, P.; Zhong, Y. Re-evaluation of the thermodynamic equilibria on the zirconia-rich side of the ZrO_2-$YO_{1.5}$ system. *Calphad* **2018**, *61*, 264–274. [CrossRef]
134. Du, C.S.; Yuan, Q.M.; Yang, Z.F. Lowering the synthesis temperature of zircon powder by yttria addition. *J. Mater. Sci. Lett.* **1999**, *18*, 965–966. [CrossRef]
135. Yang, T.; Guo, X.P. Oxidation behavior of Zr-Y alloyed Mo-Si-B based alloys. *Int. J. Refract. Met. Hard Mater.* **2020**, *88*, 105200. [CrossRef]
136. You, J.H.; Wang, R.C.; Liu, C.B.; Shi, X.J.; Han, F.; Guo, R.; Liu, X.W. Facile synthesis and highly efficient selective adsorption properties of $Y_2Mo_4O_{15}$ for methylene blue: Kinetics, thermodynamics and mechanical analyses. *J. Rare Earths* **2018**, *36*, 844–850. [CrossRef]
137. Takata, N.; Sekido, N.; Takeyama, M.; Perepezko, J.H.; Follett-Figueroa, M.; Zhang, C. Solidification of $Bcc/T_1/T_2$ three-phase microstructure in Mo-Nb-Si-B alloys. *Intermetallics* **2016**, *172*, 1–8. [CrossRef]
138. Ray, P.K.; Akinc, M.; Kramer, M.J. Applications of an extended Miedema's model for ternary alloys. *J. Alloys Compd.* **2010**, *489*, 357–361. [CrossRef]
139. Byun, J.M.; Bang, S.R.; Kim, S.H.; Choi, W.J.; Kim, Y.D. Mechanical properties of Mo-Nb-Si-B quaternary alloy fabricated by powder metallurgical method. *Int. J. Refract. Met. Hard Mater.* **2017**, *65*, 14–18. [CrossRef]
140. Yang, T.; Guo, X. Comparative studies on densification behavior, microstructure, mechanical properties and oxidation resistance of Mo-12Si-10B and Mo_3Si-free Mo-26Nb-12Si-10B alloys. *Int. J. Refract. Met. Hard Mater.* **2019**, *84*, 104993. [CrossRef]
141. Behrani, V.; Thom, A.J.; Kramer, M.J.; Akinc, M. Microstructure and oxidation behavior of Nb-Mo-Si-B alloys. *Intermetallics* **2006**, *14*, 24–32. [CrossRef]
142. Behrani, V.; Thom, A.J.; Kramer, M.J.; Akinc, M. Chlorination treatment to improve the oxidation resistance of Nb-Mo-Si-B alloys. *Metall. Mater. Trans. A* **2005**, *36*, 609–615. [CrossRef]
143. Majumdar, S. A study on microstructure development and oxidation phenomenon of arc consolidated Mo-Nb-Si-(Y) alloys. *Int. J. Refract. Met. Hard Mater.* **2019**, *78*, 76–84. [CrossRef]
144. Udoeva, L.Y.; Chumarev, V.M.; Larionov, A.V.; Zhidovinova, S.V.; Tyushnyakov, S.N. Influence of rare earth elements on the structural-phase state of Mo–Si–X (X = Sc, Y, Nd) in situ composites. *Inorg. Mater. Appl. Res.* **2018**, *9*, 257–263. [CrossRef]
145. Roy, B.; Das, J.; Mitra, R. Transient stage oxidation behavior of $Mo_{76}Si_{14}B_{10}$ alloy at 1150 °C. *Corros. Sci.* **2013**, *68*, 231–237. [CrossRef]
146. Rioult, F.A.; Imhoff, S.D.; Sakidia, R.; Perepezko, J.H. Transient oxidation of Mo-Si-B alloys: Effect of the microstructure size scale. *Acta Mater.* **2009**, *57*, 4600–4613. [CrossRef]
147. Parthasarathy, T.A.; Mendiratta, M.G.; Dimiduk, D.M. Oxidation mechanisms in Mo-reinforced Mo_5SiB_2 (T2)-Mo_3Si alloys. *Acta Mater.* **2002**, *50*, 1857–1868. [CrossRef]

Article

Ecofriendly Ultrasonic Rust Removal: An Empirical Optimization Based on Response Surface Methodology

Lijie Zhang [1], Bing He [1,2,3], Shengnan Wang [2,3], Guangcun Wang [2,3] and Xiaoming Yuan [1,*]

[1] Hebei Provincial Key Laboratory of Heavy Machinery Fluid Power Transmission and Control, Yanshan University, Qinhuangdao 066004, China; zhangljys@126.com (L.Z.); hebing5280@163.com (B.H.)
[2] Jiangsu XCMG Construction Machinery Research Institute Co., Ltd., Xuzhou 221004, China; tcyy225577@163.com (S.W.); wanggc632@163.com (G.W.)
[3] State Key Laboratory of Intelligent Manufacturing of Advanced Construction Machinery, Xuzhou Construction Machinery Group, Xuzhou 221004, China
* Correspondence: yuanxiaoming@ysu.edu.cn

Citation: Zhang, L.; He, B.; Wang, S.; Wang, G.; Yuan, X. Ecofriendly Ultrasonic Rust Removal: An Empirical Optimization Based on Response Surface Methodology. *Coatings* **2021**, *11*, 1127. https://doi.org/10.3390/coatings11091127

Academic Editors: Awais Ahmad, Shahid Hussain and Yingyi Zhang

Received: 26 August 2021
Accepted: 7 September 2021
Published: 16 September 2021

Publisher's Note: MDPI stays neutral with regard to jurisdictional claims in published maps and institutional affiliations.

Copyright: © 2021 by the authors. Licensee MDPI, Basel, Switzerland. This article is an open access article distributed under the terms and conditions of the Creative Commons Attribution (CC BY) license (https://creativecommons.org/licenses/by/4.0/).

Abstract: This study shows that the hard-to-remove rust layer on the guide sleeve surface of a used cylinder can be removed using a specially developed, environmentally friendly formula for cleaning rust. Furthermore, we studied the rust removal technology that is based on ultrasonic cavitation and chemical etching. The surface morphology and structural components of the rust layer were observed using an electron microscope and an X-ray powder diffractometer. These tools were used to explore the mechanism of combined rust removal. Using response surface methodology (RSM) and central composite design (CCD), with the rust removal rate as our index of evaluation, data were analyzed to establish a response surface model that can determine the effect of cleaning temperature and ultrasonic power interaction on the rate of rust removal. Results showed that the main components of the rust layer on a 45 steel guide sleeve were α-FeOOH, γ-FeOOH, and Fe_3O_4. The rust was unevenly distributed with a loose structure, which was easily corroded by chemical reagents and peeled off under ultrasonic cavitation. With the increase in the cleaning temperature, the chemical reaction effect was intensified, and the cleaning ability was enhanced. With the increase in ultrasonic power, the cavitation effect was aggravated, the ultrasonic agitation was enhanced, and the rust removal rate was improved. According to response surface analysis and the application scope of the rust remover, we determined that the optimal cleaning temperature is 55 °C, and that the optimal ultrasonic power is 2880 W. The descaling rate under these parameters is 0.15 $g \cdot min^{-1} \cdot m^{-2}$.

Keywords: construction machinery; ultrasonic rust removal; response surface methodology; process parameters optimization; derusting rate

1. Introduction

There are a lot of pollutants such as rust on the surface of waste construction machinery parts. It is necessary to remove the pollutants on the surface of these parts to ensure their quality when they are inspected, repaired, and assembled [1,2]. At present, single- or multiple-combined processes, such as high-pressure water jet cleaning, steam cleaning, shot blasting, high-temperature roasting, chemical cleaning, and manual polishing, are usually used for treating pollutants [3–6]. However, steam cleaning and high-temperature roasting have high costs, while shot blasting easily damages the substrate. Manual polishing is low in efficiency and high in labor costs, and the cleaning effect is not obvious. We have independently developed a derusting test machine with an ultrasonic generator and a heating device to solve the problem of cleaning rust layers on surfaces. A special, environmentally friendly formula, with citric acid as the main body, has been developed and a process of ultrasonic rust removal has been used to remove the rust layer on the surface of the sample parts. Under the action of the sound field, the cavitation bubbles in the cleaning tank rapidly expand and contract, resulting in local high temperature and high pressure. This has the effect of destroying the rust layer on the sample's surface. At the same time, the strong ultrasonic oscillation continuously agitates the cleaning liquid [7,8], accelerates the chemical interaction between the rust remover and the rust layer [9,10], and continuously dissolves and penetrates the dirt. At present, there have been many achievements in the field of ultrasonic cleaning and environmentally friendly rust removal. Zhang Baocai et al. [11] used ultrasonic compounding of molten salt cleaning technology to remove thick paint on the surface of remanufactured end caps. They combined the technology of chemical paint removal and ultrasonic cavitation, and explored the impact of cleaning temperature and ultrasonic power on the composite cleaning cycle. Wang Jian et al. [12] used the potential tracking method to study the dynamic changes in the ultrasonic pickling process of steel and the removal mechanism of the oxide layer. They found that the introduction of ultrasonic waves in the pickling process produced ultrasonic cavitation, which accelerated the reaction and greatly improved the rust removal rate. Lin Jinzhu [13] analyzed the physical and chemical properties of citric acid and the mechanism of rust removal. He pointed out the necessity and importance of using environmentally friendly acid to remove rust, laying a foundation for the establishment of a rust removal program.

A large number of single-factor tests have proved that cleaning fluid temperature and ultrasonic power have a direct effect on the rust removal rate. However, the joint effect of the two factors on the rust removal process is rarely studied. Response surface methodology (RSM) can be used to study the effects of one or more factors by facilitating the design of a reasonable test scheme, while the optimal conditions or results in the experimental design are found by analyzing the response surface or contours [14,15]. Therefore, this experiment intends to explore the combined effect of temperature and ultrasonic power through the response surface method. Response surface methodology mainly includes central composite design (CCD) and Box–Behnken experimental design (BBD). Of these two, the most widely used is central composite design [16,17]. In the CCD test design, the test points are composed of cube points, center points, and axial points [18,19], which are sequential, efficient, and flexible [20,21]. There are many practical applications of response surface methodologies in process parameter optimization. Yan Dongping et al. [22] used the central composite design to study the effect of process parameters in the milling experiment on the cutting force of the titanium alloy TC21. Yuan Julong et al. [23] optimized the polishing process of YG8 cemented carbide inserts via a response surface methodology so as to quickly determine the best process parameters of YG8 rake face polishing. Wang Qun et al. [24] used a response surface methodology to explore the effects of potassium ferrate dosage in flocculant and water pH on the water turbidity and UV254 removal rate, and optimized process parameters by establishing a secondary response model. Therefore, not only can the RSM establish a continuous mathematical model, but it can also show the interaction between factors, which is often used in process parameter optimization.

In this paper, the rust removal rate test was designed by combining RSM and CCD. A regression equation and a response surface model were established to study the effect of cleaning temperature and ultrasonic power on the rust removal rate. The optimal parameters of the rust removal process were found, and the cleaning technology was optimized, which provided theoretical support for the application of ultrasonic rust derusting technology in the remanufacturing cleaning field.

2. Experimental Procedure

2.1. Test Samples

A hydraulic cylinder guide sleeve with rust on its surface was used as the test sample. The guide sleeve was made of 45 steel and cut into 35 mm × 25 mm × 10 mm blocks for physical and chemical analysis of the rust layer. The corrosion morphology and the cross-section of the rust layer were observed using a Fei Inspection S50 scanning electron microscope (Thermo Fisher Scientific, Waltham, MA, USA), and the elements in the corrosion layer were analyzed using an Oxford X-act spectrometer. The structure and composition of iron oxide in the rust layer were analyzed using a D8 ADVANCE X-ray diffractometer (Bruker, Karlsruhe, Germany). X-ray diffraction (XRD) measurements were carried out using Cu targets and Kα radiation at 40 kV. The rust layer on the surface of the sample was scraped off, and the scraped sample was fully ground in an agate mortar. This was followed by sample preparation, and then test and result analysis.

2.2. Test Design

The sample for the process parameter optimization test was a uniformly rusted iron sheet with a size of 40 mm × 40 mm × 1 mm. The derusting formula has a main body of 30‰ citric acid, with 6–10‰ sodium dodecylbenzene sulfonate, 6–10‰ JFC (fatty alcohol-polyoxyethylene ether), and 6–10‰ benzotriazole added to perform a 45 s rust removal test. At room temperature, the formula can clean 80% of the rust layer in 45 s. If the cleaning time is extended, the derusting end point cannot be accurately determined, and the rust removal effect of each group cannot be compared. The sample was weighed using FA2004 precision electronic (Yoke Instrument, Shanghai, China) balance with a measurement accuracy of 0.1 mg. We used a self-made derusting test machine (Xuzhou Construction Machinery Group, Xuzhou, China) with an ultrasonic generator and a heating device. The ultrasonic power and cleaning temperature can be adjusted.

The lowest test temperature was the local annual average temperature. The maximum test temperature was related to the material of the derusting test machine. The tank of the machine is made of corrosion-resistant polypropylene material, and the temperature resistance of its bonding part is about 80 °C Since the machine is oriented to industrial applications and has a high frequency of use, the reliability and safety of the machine are particularly important. We chose 60 °C as the maximum heating parameter. The lowest value of power was 0, which is the value where ultrasound was not applied. The maximum power was the maximum value of the ultrasonic generator. Thus, the power regulation range was 0–2880 W, and the cleaning temperature range was 20–60 °C.

3. Results and Discussion

3.1. Response Surface Methodology

Figure 1 shows the scanning electron microscope (SEM) spectra of the rust layer surface and the cross-section, respectively. As shown in Figure 1a, the rust layer on the surface of the sample is uneven, with a maximum cross-section thickness of 66.7 μm and a minimum of 16.7 μm. There are a large number of irregular particles randomly distributed on the surface of the sample. The size of the particles is uneven, and they are connected to form a layer structure on the surface of the sample. The main components of 45 steel are Fe, C, Si, and Mn. The electron diffraction spectroscopy (EDS) results indicate that the elements of the rust layer are in accordance with 45 steel, with the exception of the O content. This shows that the main component of the rust layer is an iron oxide compound.

Iron oxides mainly include (α-β-γ-)Fe_2O_3, (α-β-γ-δ-)FeOOH, and Fe_3O_4, which have different valence, crystal form, and structure. Therefore, in order to determine the components of the rust layer, XRD was performed on the rust layer of the sample. The results show that the main components of the rust products are α-FeOOH, γ-FeOOH, and Fe_3O_4.

According to the SEM and XRD data, the main components of the rust layer are α-FeOOH, γ-FeOOH, and Fe_3O_4, and the rusted surface is loose and unevenly distributed. Therefore, the pickling agent can be continuously introduced through the act of ultrasonic pickling so that the rust layer can be peeled off from the sample surface. The mechanism of ultrasonic pickling and rust removal is shown in Figure 2.

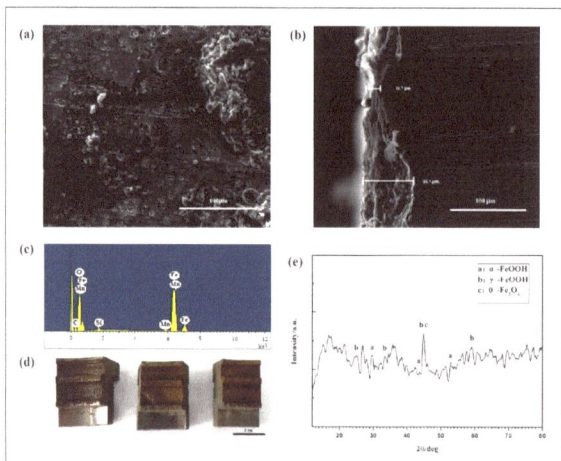

Figure 1. The surface rust characterization of the corroded specimen: (**a**) SEM picture of rust layer surface. (**b**) SEM picture of rust layer section. (**c**) EDS diagram of rust layer surface. (**d**) Corroded specimen of test. (**e**) XRD spectrum of corrosion layer.

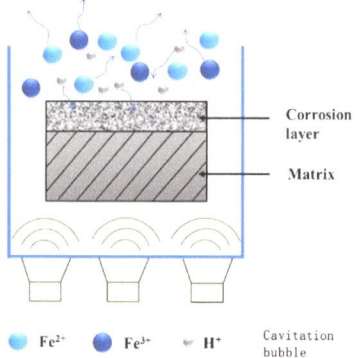

Figure 2. Schematic diagram of ultrasonic descaling process.

The environmentally friendly rust remover is mainly composed of an organic acid. It not only etches the rust on the surface of the sample but also chemically reacts with Fe^{2+} and Fe^{3+} in the rust layer. Moreover, organic acids will chelate with iron oxides to form stable complexes, which sheds rust from the surface of the sample.

On the basis of chemical etching, the bond between the surface oxide layer and the substrate is destroyed by the strong oscillation from the ultrasonic waves. Because of the ultrasonic cavitation, the cavitation bubbles in the liquid shrink or expand rapidly, resulting in a huge pressure to peel off the loose rust layer on the surface of the sample. In addition, the reacted pickling solution near the rust layer leaves the sample surface through ultrasonic stirring, and the unreacted pickling solution is replenished in time to ensure the efficient and continuous derusting reaction.

3.2. Central Composite Design

3.2.1. Central Compound Test Design

In this rust-cleaning process parameter optimization experiment, the central composite design is used to optimize the cleaning temperature (X) and ultrasonic power (Y). Taking the rust removal rate as the response value (Z), a two-level full factor test is established.

Four cubic points, five central points, and four axis points are selected to generate 13 sets of experiments.

The factors and levels are shown in Table 1.

Table 1. Test factor level table.

Factor	Code	Variable Level				
		−1	−0.7	0	0.7	1
Cleaning temperature (/°C)	X	20	26	40	54	60
Ultrasonic power (/W)	Y	0	432	1440	2448	2880

3.2.2. Rust Removal Test

1. According to the generated parameters, the cleaning temperature and ultrasonic power are adjusted. There are 13 groups of test parameters. The test factors and levels are shown in Table 2.
2. According to the order of each experimental group, the rusted iron sheets are numbered and cleaned with absolute alcohol. After drying, the rusted iron sheets are weighed and marked as m_0. The weighed iron sheets are then placed into the ultrasonic rust removal tank and cleaned at the specified temperature and ultrasonic frequency for 45 s. The descaling sheets are cleaned with anhydrous ethanol and weighed with electronic balance, which is marked as m_1.
3. According to Formula (1), the rust removal rate of a rusted iron sheet under various process parameters is calculated.

Table 2. Center composite test design and the rust removal test results.

Number	X	Y	Cleaning Temperature/(°C)	Ultrasonic Power/(W)	Rust Removal Rate/(g·min^{-1}·m^{-2})
1	0	0	40	1440	0.087
2	0	1	40	2880	0.119
3	1	0	60	1440	0.158
4	0.7	−0.7	54	432	0.095
5	0	0	40	1440	0.083
6	0	0	40	1440	0.098
7	−0.7	−0.7	26	432	0.044
8	0	0	40	1440	0.105
9	0	−1	40	0	0.058
10	−0.7	0.7	26	2448	0.090
11	0	0	40	1440	0.089
12	0.7	0.7	54	2448	0.153
13	−1	0	20	1440	0.069

$$V = \frac{m_0 - m_1}{S \times t} \quad (1)$$

In the formula, v represents the rust removal rate (g·min^{-1}·m^{-2}); m_0 and m_1 represent the mass of the rusted sample before and after 45 s rust removal (g), respectively; S represents the sample area (m^2); and t represents the rust removal time (min).

In this experiment, the rust removal rate is used to describe the cleaning effect. The higher the rust removal rate within 45 s, the better the rust removal effect.

The results of the rust removal test and the center composite test design are shown in Table 2.

3.2.3. Regression Analysis

According to the results of central composite test, a quadratic regression model is established to describe the rust removal rate under different cleaning temperature and ultrasonic power conditions. The corresponding cleaning temperature and ultrasonic power are coded and analyzed. Table 3 shows the estimated regression coefficients, with coded units for analysis. X represents cleaning temperature, and Y represents ultrasonic power.

Table 3. Estimated regression coefficient of rust removal rate under various cleaning parameters.

Term	Coefficient	Standard Error of Coefficient	T-Value (abs.)	p-Value (abs.)
Constant	0.0923	0.0036	25.330	0.000
X	0.0300	0.0029	10.422	0.000
Y	0.0223	0.0029	7.739	0.000
X* X	0.0098	0.0031	3.179	0.016
Y* Y	−0.0048	0.0031	−1.540	0.167
X* Y	0.0031	0.0041	0.767	0.468

Through analysis, it was found that the p-values corresponding to the main effect of X and Y are (X = 0.000) and (Y = 0.000), which are less than the 0.05 level of significance. The original hypothesis is rejected and the regression is significant. The p-value of X* X is (X* X = 0.016), which is less than 0.05, and the impact is significant. The p-values of Y* Y and X* Y are (Y* Y = 0.167) and (X* Y = 0.468), greater than 0.05. The impact is not significant, so it should be removed from the quadratic regression model. Y* Y and X* Y were removed, and the adjusted coefficients were redistributed, as shown in Table 4.

Table 4. Estimated regression coefficient of rust removal rate under various cleaning process parameters of corroded iron sheet (removal of insignificant items).

Term	Coefficient	Standard Error of Coefficient	T-Value (abs.)	p-Value (abs.)
Constant	0.0890	0.0031	28.745	0.000
X	0.0300	0.0030	9.906	0.000
Y	0.0223	0.0030	7.356	0.000
X* X	0.0104	0.0032	3.240	0.010

Table 4 shows the results of the model analysis after removing the insignificant items. The p-values of each group are all less than 0.05, and the model is generally valid.

The analysis of variance is shown in Table 5. The two determination coefficients are (R-Sq = 94.76) and (R-Sq (adjusted) = 93.01%). The difference between them is small and close to 1, which means that the regression is high. The cleaning cycle model is applicable.

Table 5. Variance analysis of rust removal rate under various cleaning process parameters (R-Sq = 94.76%, R-Sq (forecast) = 89.30%, R-Sq (adjustment) = 93.01%).

Source	Freedom	Seq SS	Adj SS	Adj MS	F-Value (abs.)	p-Value (abs.)
Regression	3	0.011967	0.011967	0.003989	54.25	0.000
Linear	2	0.011195	0.011195	0.005597	76.12	0.000
X	1	0.007216	0.007216	0.007216	98.14	0.000
Y	1	0.003979	0.003979	0.003979	54.11	0.000
Square	1	0.000772	0.000772	0.000772	10.50	0.010
X* X	1	0.000772	0.000772	0.000772	10.50	0.010
Error	9	0.000662	0.000662	0.000074		
Misfit	5	0.000352	0.000352	0.000070	0.91	0.553
Pure error	4	0.000310	0.000310	0.000078		
Total	12	0.012629				

Table 6 shows the estimated regression coefficient of the rust removal rate under various cleaning process parameters of a rusted iron sheet (using uncoded unit data).

Table 6. Regression coefficient of the rust removal rate under various cleaning process parameters of a rusted iron sheet (using uncoded unit data).

Term	Coefficient
Constant	0.0560859
X	−0.00205385
Y	0.0000219022
X* X	0.0000522192

According to the calculation results after coefficient correction, the modified regression equation can be obtained. The influencing factors in the equation include the first-order effect and second-order effect of cleaning temperature and the second-order effect of ultrasonic power.

3.2.4. Residual Analysis

Residual refers to the difference between the actual observed value and the regression estimated value. Residual analysis is used to analyze the reliability, periodicity, or other disturbances of the data through the information provided by the residual and further characterizes the adaptability of the model equation. The main purpose of residual diagnosis is to diagnose whether the model fits well with the data based on the status of the residuals. The residual diagram of rust removal rate is shown in Figure 3.

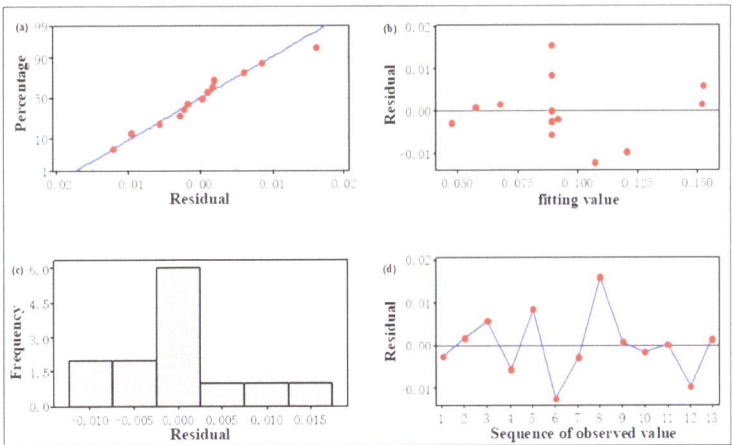

Figure 3. Residual diagram of rust removal rate under various cleaning parameters.

As is shown in Figure 3, the residuals of different process parameters conform to normal distribution from the normal probability diagram of residuals. According to the scatter plot with the predicted value of the response variable fitting to the horizontal axis of the residual, the residuals maintain equal variances and have no obvious regularity. There is no 'funnel shape' or 'trumpet shape'. According to the residuals for the scatter plot with the order of observations as the horizontal axis, each point randomly fluctuates up and down the horizontal axis. In summary, the residuals under different process parameters are not abnormal.

3.2.5. Establishment of Regression Equation

After model adjustment, variance analysis, and residual diagnosis, the regression equation of rust removal rate under various cleaning parameters of the corroded iron sheet is obtained, as shown in Equation (2).

$$RT = 0.0560859 - 0.00205385x + 0.0000219022y + 0.0000522192x^2 + \varepsilon \quad (2)$$

Among them, RT represents the rust removal rate, x represents the cleaning temperature, y represents ultrasonic power, and ε represents error.

3.2.6. Response Surface

The response surface of rust removal rate is shown in Figure 4.

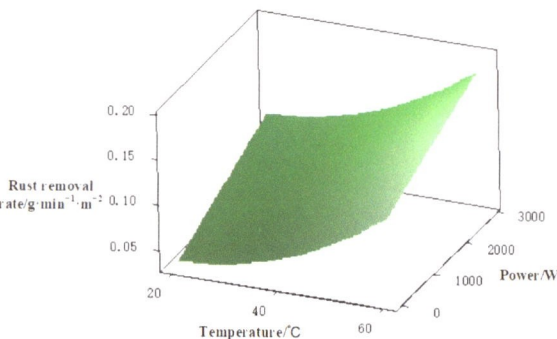

Figure 4. Evolution of rust removal rate with respect to cleaning temperature and ultrasonic power under various cleaning parameters.

The light green area in the upper right corner of Figure 4 shows a good rust removal rate, which is greater than 0.175 g·min^{-1}·m^{-2}. It can be seen from Figure 5 that the higher the cleaning temperature, the higher the rust removal rate, and the greater the ultrasonic power, the higher the rust removal rate, which is consistent with the phenomenon observed in the test. This behavior can be explained by considering the aqueous ultrasonic bath type, i.e., the acidic solution. The acid reacts with the rust layer in the rust removal tank, and the hydrogen bubbles thus generated have a peeling effect on the rust layer [25,26]. When the temperature increases, the chemical reaction rate increases, resulting in a faster rust removal rate. The vibration of the ultrasonic vibrator in the rust removal tank generates ultrasonic waves, and the ultrasonic cavitation effect appears in the liquid. When the sound energy reaches the critical point, the cavitation nucleus breaks down and produces a huge impact force, which causes the dirt on the liquid surface to decompose and peel off [27]. At the same time, due to the effect of micro-acoustic flow and local high-pressure impact, the ultrasonic high-speed stirring rust removal solution accelerates the chemical reaction on the surface of the sample, which accelerates the dissolution rate of the dirt [28]. When the ultrasonic power in the sound field increases, the cavitation effect and ultrasonic oscillation effect are intensified. The reaction between the rust remover and the rust layer is accelerated, and the rust removal rate increases.

When the cleaning temperature is at a low level, the response surface increases greatly with the increase in ultrasonic power. When the cleaning temperature is high, the increase in the response surface is small with the increase in ultrasonic power. Therefore, according to the regression equation, when the cleaning temperature is 20 °C, the rust removal efficiency is 0.036 g·min^{-1}·m^{-2} without ultrasonic power. When ultrasonic power is applied, the rust removal efficiency is 0.099 g·min^{-1}·m^{-2}, which is 2.76 times greater than the former. When the ultrasonic power is low, the increase in the response surface is larger. When the ultrasonic power is high, the increase in the response surface is small with the increase in cleaning temperature. Therefore, according to the regression equation, when the ultrasonic power is 0, the rust removal efficiency is 0.036 g·min^{-1}·m^{-2} without heating. After heating, the rust removal efficiency is 0.12 g·min^{-1}·m^{-2}, which is 3.37 times greater than the former. Thus, in the experimental range, the influence of temperature on the rust removal rate of 45 steel is greater than that of ultrasonic power.

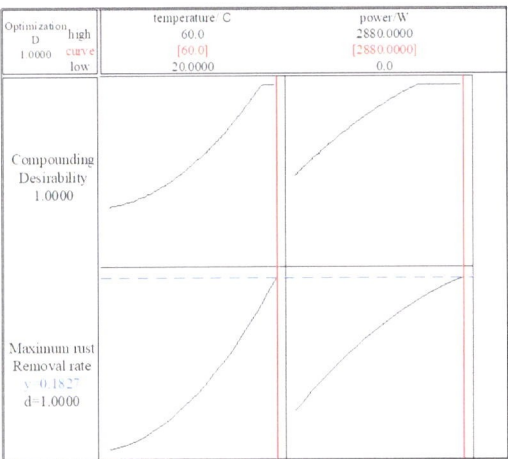

Figure 5. Optimization of rust removal rate under various cleaning parameters.

3.2.7. Response Optimization

The response variable optimizer is used to obtain the minimization optimization result of the cleaning cycle of the rusty iron sheet, as shown in Figure 5.

As shown in Figure 5, the optimal value of the rust removal rate is reached when the cleaning temperature is 60 °C and the ultrasonic power is 2880 W. At this point, the rust removal rate is 0.1827 g·min^{-1}·m^{-2}, and the desirability (d) is 1.0000.

3.2.8. Result Verification

With the increase in temperature, the speed of chemical reaction and rust removal increases. However, in practical applications, the increase in temperature will lead to the corrosion of the substrate material by the acidic rust remover, which causes hydrogen embrittlement. Results show that the pickling temperature is generally controlled within 60~70 °C. Therefore, considering the rust removal rate and the application scope of rust remover, we can reduce the temperature appropriately and choose 55 °C as the optimal rust-cleaning temperature.

The cleaning temperature should therefore be set to 55 °C and the ultrasonic power to 2880 W. The experiment was designed to ultrasonically clean the sample cylinder guide sleeve for 45 s. Figure 6 shows the micromorphology of the sample before and after cleaning. The results show that most of the rust on the sample surface was removed after cleaning. The rust removal rate at this time is calculated to be 0.15 g·min^{-1}·m^{-2}.

Figure 6. SEM surface morphology of the cylinder guide sleeve: (**a**) before rust cleaning; (**b**) after rust cleaning.

4. Conclusions

1. This paper takes the rust of a cylinder guide sleeve as an example of how to optimize rust removal efficiency and use the environmentally friendly citric acid as an alternative to traditional cleaning chemicals for rust removal. Under the action of H^+ and ultrasonic cavitation impact, the rust layer reacts and peels off.
2. The regression equation and response surface model of rust removal rate were obtained by using a central composite test method. The higher the cleaning temperature and the ultrasonic power, the higher the rust removal rate. Considering the rust removal rate and the application scope of rust remover, we chose 55 °C as the optimal rust-cleaning temperature.
3. The optimal process parameters of ultrasonic rust removal have been determined. The cleaning temperature is 55 °C, the ultrasonic power is 2880 W, and the descaling rate under the optimal parameters is 0.15 $g \cdot min^{-1} \cdot m^{-2}$.

Author Contributions: Conceptualization, L.Z. and B.H.; methodology, B.H.; investigation, S.W. and G.W.; writing—original draft preparation, B.H.; writing—review and editing, L.Z. and B.H.; supervision, X.Y. and G.W. All authors have read and agreed to the published version of the manuscript.

Funding: This research was funded in part by the National Key Research and Development Program 2019YFB2005302 and 2019YFB2005301.

Institutional Review Board Statement: Not applicable.

Informed Consent Statement: Not applicable.

Data Availability Statement: Not applicable.

Acknowledgments: Not applicable.

Conflicts of Interest: The authors declare no conflict of interest.

References

1. Xu, B. Remanufacturing of construction machinery and its key technology. *Constr. Mach. Equip.* **2009**, *40*, 1–7.
2. Liu, S.W. Research on Cleaning Technology of Remanufacturing Blank. Ph.D. Thesis, Department Mechanical Design and Theory Engineering, Shanghai Jiaotong University, Shanghai, China, 2010.
3. Kiyak, M.; Altan, M.; Altan, E. Prediction of chip flow angle in orthogonal turning of mild steel by neural network approach. *Int. J. Adv. Manuf. Technol.* **2006**, *33*, 251–259. [CrossRef]
4. Zhao, H.H. *Study on Preparation and Properties of Water-Based Rust Conversion Agent*; Tianjin University of Science and Technology: Tianjin, China, 2017.
5. Niemczewski, B. Influence of concentration of substances used in ultrasonic cleaning in alkaline solutions on cavitation intensity. *Ultrason. Sonochem.* **2009**, *16*, 402–407. [CrossRef] [PubMed]
6. Bi, C.X.; Yang, W.; Meng, M.L. Research on ultrasonic derusting technology for ferrous metal parts of armored equipment. *Equip. Manuf. Technol.* **2009**, *1*, 105–107.
7. Qin, S.S. Research and Application of Ultrasonic Cleaning for Remanufacturing. Ph.D. Thesis, Department Industry Engineering, Shandong University, Jinan, China, 2012.
8. Zhang, M.C. *Study on Influence Factors of Ultrasonic Cleaning and Cavitation Field*; Shaanxi Normal University: Xian, China, 2013.
9. Cao, Y.Y.; Lin, S.Y. Study on the Interaction Between Two Ultrasonic Cavities. *J. Shaanxi Norm. Univ.* **2010**, *38*, 46–50.
10. Collazo, A.; Novoa, X.R.; Pérez, C.; Puga, B. The corrosion protection mechanism of rust converters: An electrochemical impedance spectroscopy study. *Electrochim. Acta* **2010**, *55*, 6156–6162. [CrossRef]
11. Zhang, C.; Sun, Y.H.; Jia, X.J.; Li, F.Y.; Yang, M.B.; Xin, B.L.; Wang, X.; Wang, G.C. Research and process optimization of paint removal technology based on molten salt ultrasonic composite. *Surf. Technol.* **2018**, *47*, 280–287.
12. Wang, J.; Fu, M.; Ding, P.D.; Zhang, M.H. Study on ultrasonic pickling process of iron and steel. *Surf. Technol.* **2004**, *33*, 30–32.
13. Lin, J.Z. Application analysis of citric acid in chemical cleaning. *Guizhou Chem. Ind.* **2011**, *36*, 33–35.
14. Shao, M.W.; Wang, J.; Qiao, X.; Li, X.G.; Zhao, H. Friction sensitivity theory of solid propellant based on response surface central composite design. *Chin. J. Energetic Mater.* **2019**, *26*, 509–515.
15. Wang, G.; Wang, X.; Song, X.; Zheng, X. Comparison of BBD and CCD in optimization of preparation conditions of mercaptoacetylated chitosan by Response Surface Methodology. *J. Environ. Eng.* **2018**, *12*, 2502–2511.
16. Wang, Y.F.; Wang, C.G. Theory and application of response surface methodology. *J. Minzu Univ. China* **2005**, *14*, 236–240.
17. Li, J.F. *Optimization of Extraction Process of Low Grade Ionic Rare Earth by Response Surface Methodology*; Jiangxi University of Science and Technology: Ganzhou, China, 2018.
18. Sun, C.; Ning, J.; Song, Z.X.; Xie, P.; Tang, Z.S. Optimization of extraction process of total flavonoids from seabuckthorn pomace by central composite design response surface methodology. *Mod. Chin. Med.* **2018**, *20*, 74–82.
19. Zi, J.Y.; Zhang, S.P.; Li, Y.Q.; Zhou, W. Optimization of diesel engine performance parameters based on response surface design. *Small Intern. Combust. Engine Veh. Technol.* **2020**, *49*, 28–34.
20. Zhang, Z.H.; Zhen, H.; Guo, W. A comparative study of three types of central composite design in response surface methodology. *J. Shenyang Aerosp. Univ.* **2007**, *24*, 87–91.
21. Liu, W.; Zhang, Y.; Cheng, T.C.; Wang, H.Y. Optimization design of exhaust manifold response surface based on face center combination. *Mech. Strength.* **2021**, *43*, 137–144.
22. Yan, D.P.; Jiang, B. Optimization of milling parameters for titanium alloy TC21 based on response surface methodology. *Tool Eng.* **2016**, *50*, 18–22.
23. Yuan, J.L.; Mao, M.J.; Li, M.; Liu, S.; Wu, F. Optimization of chemical mechanical polishing process parameters for YG8 cemented carbide blade based on response surface methodology. *China Mech. Eng.* **2018**, *29*, 2290–2297.
24. Wang, Q.; Chai, B.; Lu, X.H.; Bai, X.F.; Liu, L.J. Optimization of coagulation effect by potassium ferrate pre oxidation by response surface methodology. *Technol. Water Treat.* **2017**, *43*, 42–46.
25. Bi, C.X.; Yang, W.; Lu, M.L. Research on ultrasonic rust removal technology for ferrous metal parts of armored equipment. *Equip. Manuf. Technol.* **2009**, *30*, 105–107.
26. Bi, J.C.; Zhang, X.D.; Xu, F. Neutral Rust Remover. China Patent 103,938,217, 2 July 2014.
27. Sun, Y.; Huang, S.Y.; Mao, Y.L.; Zhu, L.H. Crushing effect of near wall ultrasonic cavitating micro jet on fine particles. *China Mech. Eng.* **2019**, *30*, 2953–2960.
28. Huang, S.Y. *Study on the Crushing Effect of Ultrasonic Cavitation Near the Wall in Rotating Flow Field on Fine Particles*; Zhejiang University of Technology: Hangzhou, China, 2019.

Article

Fracture Toughness Analysis of Aluminum (Al) Foil and Its Adhesion with Low-Density Polyethylene (LPDE) in the Packing Industry

Umer Sharif [1,2,*], Beibei Sun [1,*], Md Shafiqul Islam [2], Kashif Majeed [2], Dauda Sh. Ibrahim [1], Orelaja Oluseyi Adewale [1], Naseem Akhtar [3], Zaki Ismail Zaki [4] and Zeinhom M. El-Bahy [5]

1. School of Mechanical Engineering, Southeast University, Nanjing 211189, China; sidauda.mct@buk.edu.ng (D.S.I.); 233179946@seu.edu.cn (O.O.A.)
2. Department of Mechanical Engineering, Blekinge Institute of Technology, 37179 Karlskrona, Sweden; shafiqul.islam@bth.se (M.S.I.); kama10@bth.se (K.M.)
3. Department of Chemistry, The Government Sadiq College Women University, Bahawalpur 63100, Pakistan; drnaseem@gscwu.edu.pk
4. Department of Chemistry, College of Science, Taif University, P.O. Box 11099, Taif 21944, Saudi Arabia; zakimohamed@tu.edu.sa
5. Department of Chemistry, Faculty of Science, Al-Azhar University, Nasr City, Cairo 11884, Egypt; zeinelbahy@azhar.edu.eg
* Correspondence: 233179945@seu.edu.cn (U.S.); bbsun@seu.edu.cn (B.S.)

Citation: Sharif, U.; Sun, B.; Islam, M.S.; Majeed, K.; Ibrahim, D.S.; Adewale, O.O.; Akhtar, N.; Zaki, Z.I.; El-Bahy, Z.M. Fracture Toughness Analysis of Aluminum (Al) Foil and Its Adhesion with Low-Density Polyethylene (LPDE) in the Packing Industry. *Coatings* **2021**, *11*, 1079. https://doi.org/10.3390/coatings11091079

Academic Editor: Yingyi Zhang

Received: 31 July 2021
Accepted: 2 September 2021
Published: 6 September 2021

Publisher's Note: MDPI stays neutral with regard to jurisdictional claims in published maps and institutional affiliations.

Copyright: © 2021 by the authors. Licensee MDPI, Basel, Switzerland. This article is an open access article distributed under the terms and conditions of the Creative Commons Attribution (CC BY) license (https://creativecommons.org/licenses/by/4.0/).

Abstract: Liquid food packages consist of various polymers films, which are bonded together with Aluminum foil (Al-foil) using adhesion or by direct heat. The main aim of this research was to define important material properties such as fracture toughness and some FE-simulation material model parameters such as damage initiation, damage evolution, and the adhesion between Al-foil and low-density polyethylene (LDPE) film. This investigation is based on both physical experiments and FE simulations in ABAQUS with and without initial cracks of different lengths for comparison purposes. The final FE model in ABAQUS was used to compare the numerical input parameters in an extensive study with the ambition to investigate the materials' parameters in cases with or without adhesion between laminates. Finally, the relation between the theoretical and experimental results for Al-foil using linear elastic fracture mechanics and modified strip yield model were shown, and the fracture toughness was calculated for two different thicknesses of Al-foil.

Keywords: fracture toughness; aluminum foil; LDPE film; laminate; FE model

1. Introduction

Due to the continuous development of packaging industries, it has become possible to buy well-stored food distributed from far away locations [1]. An important step is to understand the mechanical and fracture properties of the packaging material. Packaging materials generally comprise various layers of paperboard, polymers and aluminum foil (Al-foil) [2]. Al-foil and the low-density polyethylene (LDPE) film are studied in this paper. Paperboard together with these materials is broadly utilized as a material structure in aseptic food packages. The Al-foil is utilized as a proficient barrier against oxygen and light in food packages [3,4].

The final packaging material is exposed to diverse loading conditions during its lifetime: forming, folding, filling, distribution, storage, handling, and eventually opening, wasting and recycling by the consumer [5]. Al-foil is incapable of withstanding high local strains like polymer film and the paper layers. Pre-cracks in the Al-foil can subsequently spread into the polymer and the paper layers. Therefore, it is important to recognize the individual and combined fracture behaviors of the Al-foil as well as its adhesion with LDPE layers and their roles as contributors towards the laminated structure when creating an opening device mechanism [6,7]. In order to predict the impairment evolution within the laminate, the fracture performance of the Al-foil and the LDPE are first studied distinctly. Numerous studies of the fracture performance of specific packaging material layers, for example, paperboard, are presented in [8–10]. To present the foundation for the study of Al-foil and LDPE, pre-center cracked panels were exposed to in-plane tensile mode I loading for further examination in this study. The bonds amongst the distinctive layers are allocated similar strengths as the induced traction forces generated when the distinctive material layers contract due to stress localization in the tensile tests, therefore leading to separation of the two material layers locally. De-lamination and the level of adhesion are interesting topics [11–16], which must be reviewed in the future. As pointed out in [17], mechanical modeling of polymer materials is still in a relatively premature phase when compared with the advances in metallic materials. A cell model has been developed and applied to study the effect of voids on matrix yielding and localized plastic deformation [18,19]. The modified Dugdale model approach was developed to investigate the localized plastic deformation of the single layer in addition to the laminate. The fracture behavior of the Al-foil and LDPE laminate has also been reviewed in previous work in [11,12,20–22]. A common opinion is the huge diversity of failure mechanisms in laminates of distinctive structures. Diverse loading happens at the time during transportation that triggers the opening phenomena of food packages [23]. Linear elastic fracture mechanics (LEFM), Elastic Plastic Fracture mechanics (EPFM) and modified strip yield model theories were used in the investigation of outcomes whilst loading components, response and fracture of material [24].

The purpose of this work was to present the results of the mechanical and fracture behaviors of the Al-foil and its adhesion with LDPE. These results will be used as input to virtual material modeling for calculation of the different package properties. The aim of the discussed approaches should be utilized to create a capable and potential tool for decision making and package improvement with practical and analytical simulation models. This article has three parts: a theoretical part, description of the fracture toughness methods and established practices. It explains important fracture mechanics parameters such as stress intensity factor K, J-integral, from the essential conception and meaning to experimental assessment and test approaches. In Section 3, the material properties are discussed. Experimental setup and experimental results are also presented in this section. Section 4 describes the mechanical behavior of the pre-cracked material and laminates. Analytical models were used to examine the experimental data in this section. The study will hopefully lead to further understanding of each of the laminate parts and how they contribute to crack development in the laminate. The differences in material behavior between adhesion and cohesion will also be investigated to gain further understanding.

2. Theoretical Model

2.1. Fracture Mechanics

Some theoretical models of fracture mechanics from the literature were used to study the breakage behavior of different materials. There are three different basic modes of loading on a crack tip [25,26]. All the three modes have different stress intensity factors at the crack, which is required to propagate the crack [27,28]. The stress intensity factors describing different modes of fracture, K_I, K_{II} and K_{III}, are given by Equations (1)–(3), respectively, while σ_y is the stress caused by normal force, τ_{xy} and τ_{yz} are the shear forces along xy and yz planes, respectively.

$$k_I = \lim_{x \to 0} \sigma_y(x,0)\sqrt{2\pi x} \quad (1)$$

$$k_{II} = \lim_{x \to 0} \tau_{xy}(x,0)\sqrt{2\pi x} \quad (2)$$

$$k_{III} = \lim_{x \to 0} \tau_{yz}(x,0)\sqrt{2\pi x} \quad (3)$$

The materials' physical properties often depend on their direction or plane. In the case of Al and LDPE, both materials are assumed to be isotropic because their properties remain the same in all directions. The laminate in this work consists of Al-foil/Adhesive/LDPE. LDPE is, in reality, not linear, but under small strains it can be approximated by a linear elastic material. It can also be considered to be isotropic. The adhesive layer is assumed to have the same mechanical properties as LDPE [29].

By the theory of elasticity, the stress field near the crack tip is a function of the geometry of the specimen location and loading. LEFM explains the stresses surrounding a crack tip, which are given as follows [30]:

$$\sigma_{ij} = \frac{K}{\sqrt{2\pi r}} f_{ij}(\theta) \, as \, r \to 0 \tag{4}$$

Here, "r" is the distance from the crack tip, "θ" is the angle of the crack plane and "K" is the stress intensity factor, where "i" and "j" are indices values 1 and 2 referring to the Cartesian axes X_1 and X_2, and are known as angular functions [30]. The stress intensity factor "K" becomes equal to the fracture toughness at the start of crack growth. The LEFM limiting stress can be mathematically expressed as:

$$\sigma_c = \frac{K_c}{\sqrt{\pi a \times \emptyset}} \tag{5}$$

Here, the above notation is described as: a = half of crack length, σ = stress at crack growth (obtained from experiments), and K_c = fracture toughness. When inelastic deformation n is small compared to the crack size, the LEFM theory is valid, which is called small scale yielding (SSY) [31].

2.2. Model of Modified Strip Yield

The model of strip yield for a crack, using two elastic solutions given below, defines the elastic plastic behavior, crack under remote tension and crack with closure stress at the tip.

The discontinuous displacement segment was assumed to model the strip yield plastic zone. The initial crack length including the discontinuous displacement segment length is 2(a + l), where "l" is the length at each end of the discontinuity, which are under equal stress σ_b, and 2a is the crack length and length "l" is the plastic zone represented by nonlinear behavior of materials shown in Figure 1 [32,33]. For calculating the yielding, in thin steel sheets, Dugdale suggested the strip yield model, and later this model was used for a variety of materials. The strip yield model is applicable to polymer materials.

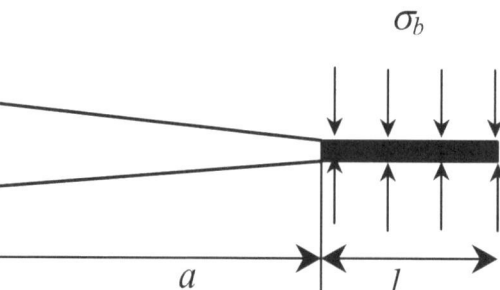

Figure 1. The strip yield model.

The relation between applied stress and crack length can be derived, as written below

$$\frac{\sigma_c}{\sigma_b} = \frac{2}{\pi} arc \, \cos \left[\exp\left(-\frac{\pi K_c^2}{8a\sigma_b^2} \right) \right] \tag{6}$$

Equation (6) is derived for infinite plates. To obtain reasonable results for the finite plate, the correction factor Ø from (9) is used to redefine (6), when $a \rightarrow W$, is given by

$$\frac{\sigma_c}{\sigma_b} = \frac{2}{\pi} arc \; cos \left[\exp\left(-\frac{\pi K_c^2}{8a\emptyset^2 \sigma_b^2} \right) \right] \quad (7)$$

2.3. Fracture Toughness

Fracture toughness and the behavior of crack propagation mainly depend on the material's thickness [30,31]. The different values of K_I will be produced for specimens with different absolute sizes and standard proportions. This happens because the stress on specimen thickness varies and it continues until the thickness exceeds some critical dimension. After this, K_I becomes relatively constant and this material property is known as the plane-strain fracture toughness "K_{IC}" [34]. The load corresponding to defined unstable crack propagation helps to compute fracture toughness. The stress intensity factor, K, equals the fracture toughness during crack propagation. The stress for LEFM is given by [30].

$$k_c = \sigma_0 \sqrt{\pi a} \times \emptyset\left(\frac{a}{w}\right) \quad (8)$$

In the above expression, "σ_0" represents remote applied stress, "Ø" represents correction factor for center cracked specimen, "a" represents half the crack length, and "w" represents half width of specimen. The geometric correction factor Ø for the center crack specimen is

$$\emptyset\left(\frac{a}{w}\right) = \frac{1 - 0.0025\frac{a^2}{w} + 0.006\left(\frac{a}{w}\right)^4}{\cos^{\frac{1}{2}}\left(\frac{\pi\left(\frac{a}{a}\right)}{2}\right)} \quad (9)$$

Analytical relations for experimental stress vs. crack length are given by Equation below.

$$\sigma_{exp} = \frac{P(a)}{2Bw} \quad (10)$$

In Equation (10), "P" represents the applied force and "a" represents half of the crack length. The specimen configuration used in Equation (10) are presented in Figure 2.

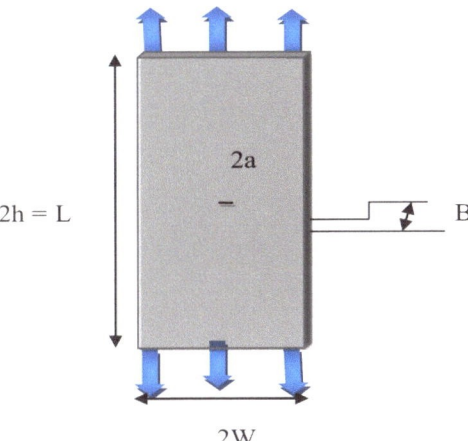

Figure 2. Specimen configuration with a centered crack of length 2a [35].

2.4. Adhesion Behavior

The maximum force required to break the bond is defined as a measure of adhesion. The measure of the strength of an adhesive bond across an interface is the amount of energy needed to break it, i.e., to separate two surfaces. Such a separation may involve the breakage of chemical or van der Waals bonds, and the plastic deformation of one or both of the materials on either side of the interface. The fraction of energy required to break the bonds at the interface is a very small fraction of the total energy necessary for the separation of the two surfaces in all cases where good adhesion is present [36,37]. Most of the mechanical work is used to deform, under stress, the material adjacent to the interface. Therefore, the measured energy of adhesion will be dependent on the ability of the interfacial bonds to sustain stress, as well as on the amount of plastic deformation caused by the above-mentioned stress [38]. In many structures, the use of adhesives is common. To explain the adhesive behavior, separation law can be used. τ_{max} is the maximum stress (damage Initiation in ABAQUS) that the adhesive can take and δ_{max} (damage Evolution in ABAQUS) is the maximum separation of the bulk material before the adhesive breaks and the bulk material is fully separated [27].

3. Experimental Study

The tensile tests were performed using an MTS Qtest100 machine with appropriate clamps manufactured by MTS system corporation, Minneapolis, MN, USA presented in Figure 3. The lower clamp is stationary and the upper one moves up to load the specimen by prescribed displacement. A 2.5 KN load cell was used due to the weight of the clamper. The specimen was placed between the grips and secured properly. The clamped specimen was loaded and extended until it broke or reached a desired displacement. The test speed was adjusted to 10 mm/min. The initial distance between grips was 230 mm.

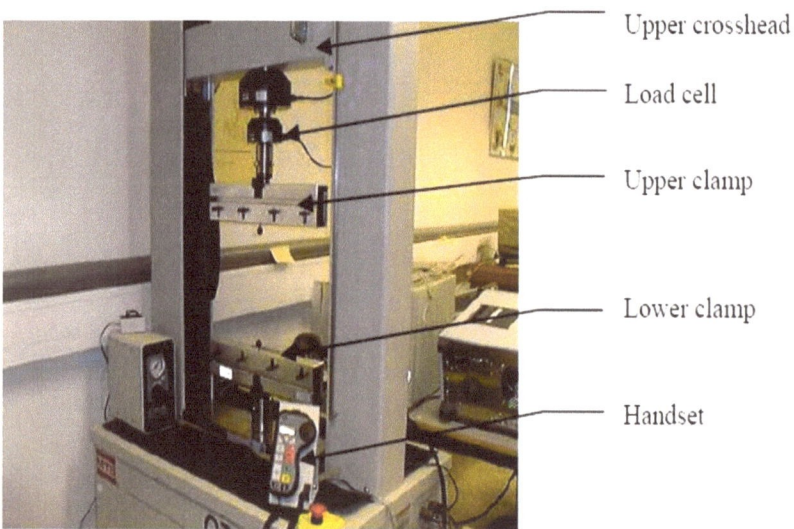

Figure 3. MTS machine at BTH lab.

3.1. Material Preparation

The LET/LDPE laminate was manufactured by spraying melted LDPE onto a PET layer. The LDPE layer was extracted from PET with the help of an extruder, as shown in Figure 4. Al foil, Al/LDPE (without adhesion) and Al/Adh/LDPE (with adhesion) were prepared. A simple paper laminating machine was used to make Al bonded with LDPE. The manual was studied, and it was observed that the machine operated at 120 centigrade. Paper was placed on both sides of the lamina to protect the lamina from having direct contact with the heated roller while laminating. A pre-center crack was introduced after the layers had been laminated. The same procedure was followed for the preparation of Al foil/Adh/LDPE, except that paper gum was used as an adhesive between the layers and then a pre-center crack was introduced after laminating the layers. In all cases for Al and LDPE, Al served as the master layer and LDPE as the slave layer. Both LDPE and Al were assumed to have isotropic behavior in this work, although in practice the materials are anisotropic.

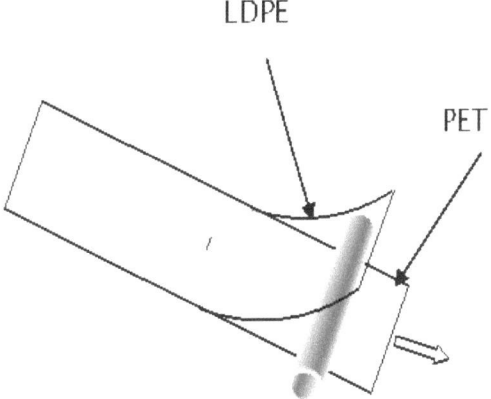

Figure 4. Extrusion of test size LDPE layer from PET laminated with LDPE layers [35].

3.2. Design of Experiment

Sample Preparation: Half width W and height h from the center were measured with the help of a scale. For each material, Al foil and laminated Al foil, at least five test specimens of 95 mm × 230 mm size were prepared, as shown in Figure 5. The width of the specimen (2 W equals 95 mm) was assumed to be infinite in comparison with the crack length in order to remove the effect of finite size. A steel ruler and knife were utilized to make the specimens and cracks. Taking the gripper of the MTS machine into consideration, a height size greater than 230 mm specimen was used. Half crack length was made from the center to both directions in order to ensure similar pre-crack tip shapes. While making the specimen of Al foil, it should be taken into account that no pre-crack was introduced, as Al foil is very sensitive to handle.

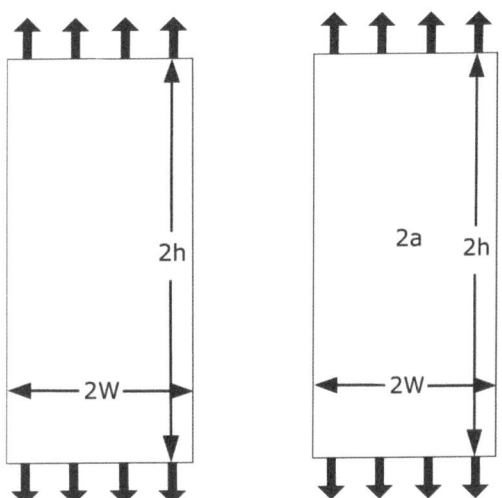

Figure 5. Specimen with center crack and without center crack.

3.3. Experimental Testing for Data Collection

As the test began, the force vs. displacement graph was displayed continually. When there was no pre-crack, the Al foil, Al/LDPE, and Al/Adh/LDPE tear around the lower gripper or fail abruptly from the center; however, when there was a pre-crack, the crack advanced down the width in all circumstances, following the centerline. To investigate the influence of adhesion and crack propagation behavior through the lamina, Al/LDPE and Al/Adh/LDPE specimens with initial cracks of various lengths were manually halted at various displacements. The Al-foil utilized had a thickness of 9 microns, while the test specimens were made of a 27-micron thick LDPE. The test cases are presented in Table 1.

Table 1. Test cases.

Sr. No	Test Cases	Crack	Thickness (μm)	Dimensions (mm × mm)
1	Al Foil	With and without crack	9	230 × 95
2	Al Foil/LDPE/(w/o) Adh	With and without crack	36	230 × 95
3	Al Foil/Adh/LDPE	With and without crack	36	230 × 95

3.4. Experimental Results and Discussion

3.4.1. Case 1: Aluminum Foil

Both materials responded similarly in MD and CD because they are both isotropic. If there was no crack in the aluminum foil, it suddenly failed or tore away from the bottom clamp. As a result, LDPE behaved similarly. When both had pre-cracks, the crack spread along the width of the crack, following the centerline. In the case of no crack, more force was required to induce the fracture. As the length of the initial fracture grew longer, the force required to propagate the crack decreased. At least five test samples of each case were generated and tested, as discussed in Section 3.2. Only the greater displacement and force findings are shown in the main text; the remaining results are presented in Appendix A. The cumulative results of all specimens for Al foil with a crack length of 0 to 45 mm are shown in Figure 6.

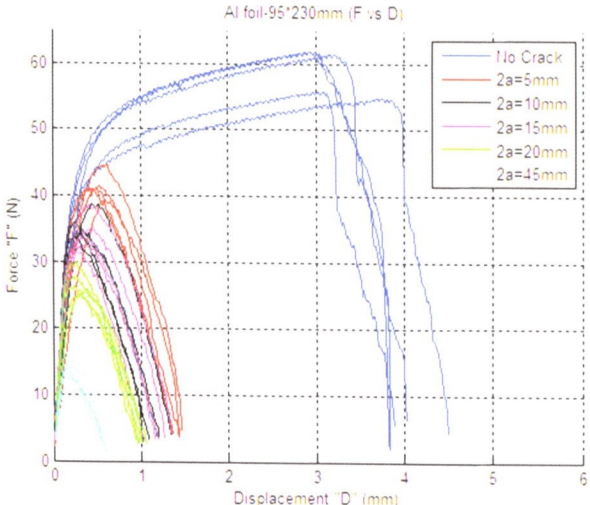

Figure 6. Load vs. displacement of Al-foil (center cracked) of 0–45 mm for all specimens.

As previously stated, for the calculation of mechanical characteristics and numerical analysis of materials, only higher force and displacement values were taken into account in all circumstances. With no crack, the highest displacement measured in Al-foil was 3.218 mm when the applied load reached 61.45 N. In the case of a 45 mm fracture length case with a load of 14.47 N, the minimum extension was 0.16 mm. The findings reveal that a greater force is required to propagate the crack when the crack length is shorter, and that this force reduces as the crack length increases. For fracture lengths ranging from 0 to 45 mm, the results are reported in Figure 7 and Table 2. Experimental results for all the specimens are presented in Appendix A, Figure A1.

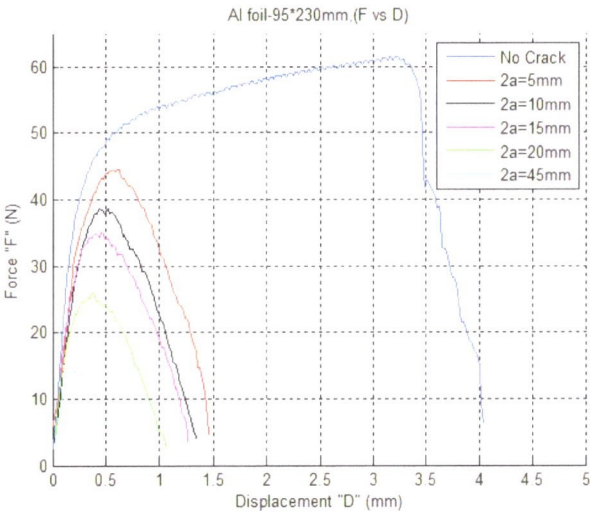

Figure 7. Load vs. displacement of Al foil (center cracked) of 0–45 mm.

Table 2. Load and displacement of fractured Al foil, thickness B of 9 µm and dimensions of 230 mm × 95 mm.

Crack Length (2a) (mm)	Maximum Load (F) (N)	Extension (D) (mm)
No Crack	61.45	3.218
5	44.64	0.606
10	38.62	0.523
15	34.99	0.458
20	25.88	0.38
45	14.47	0.16

3.4.2. Case 2: Aluminum Foil/LDPE (without Adhesion)

In the absence of adhesion, the Al-foil broke faster than the LDPE. As a result, LDPE took longer to fail than Al-foil. This is due to the ductility of LDPE. In the presence of a pre-crack, aluminum had a much lower maximum extension value than LDPE, but in the absence of a crack, both materials failed near the lower clamp. With no crack introduced, the maximum extension observed was 7.553 mm at a load of 53.8 N. However, when the crack length was 20 mm and the peak load recorded was 31.1 N, the minimum extension observed was 2.6 mm. Figure 8 and Table 3 show the maximum displacement and peak values. Test results for all the specimens are reported in Appendix A, Figure A2.

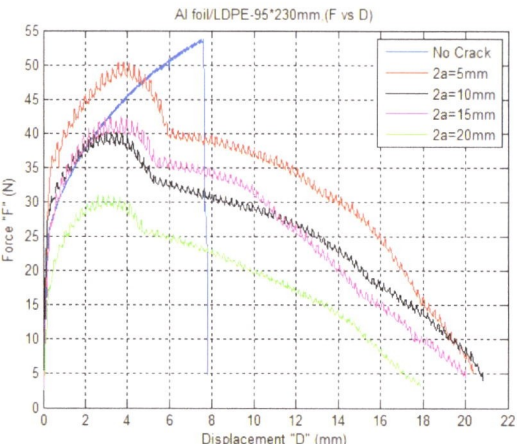

Figure 8. Load vs. displacement of Al foil/LDPE with a center crack of 0–20 mm.

Table 3. Peak load and maximum displacement of fractured Al foil/LDPE, thickness B of 40 µm and dimensions of 230 mm × 95 mm.

Crack Length (2a) (mm)	Maximum Load (F) (N)	Extension (D) (mm)
No Crack	53.8	7.553
5	50.46	3.792
10	40.26	2.7
15	42.51	3.666
20	31.1	2.6

The peak load value for the LDPE with the same specimen size and no crack was recorded as 20.72 N, and the specimen extended 36 mm. The maximum extension observed when the 5 mm crack was introduced was 30.12 mm, and the peak load was 18.6 N, as shown in Figure 9. These values were used to calculate the other mechanical parameters, which are = 8 MPa, E = 126.1 MPa, and KI = 0.75 MPa/m^2 [35]. These variables were then used in the numerical analysis.

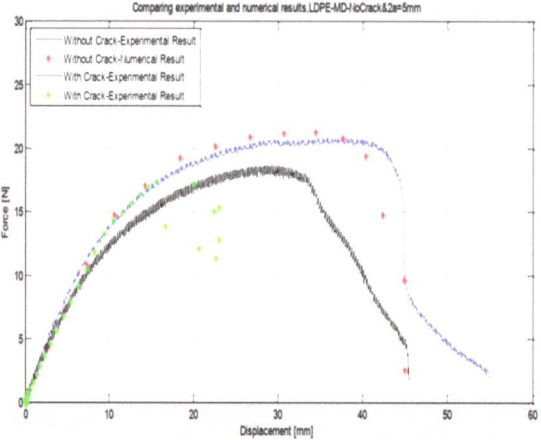

Figure 9. Response of LDPE-MD-95 230 mm experimental and numerical results with crack 2a = 5 mm and without crack.

3.4.3. Case 3: Al-Foil/Adh/LDPE (with Adhesion)

When Al/Adh/LDPE was tested, the maximum load and extension observed were 44.61 N and 4.756 mm, respectively, when there was no crack. The maximum load was 23.59 N and the displacement was 0.653 mm for the initial crack of 20 mm. The single peak in Figure 10 for each case depicts the behavior of materials in which both materials fail at the same point due to a strong adhesion force between them, which causes them to act as one material. Table 4 summarizes the results. Experimental results are presented in Appendix A, Figure A3 for all the specimens.

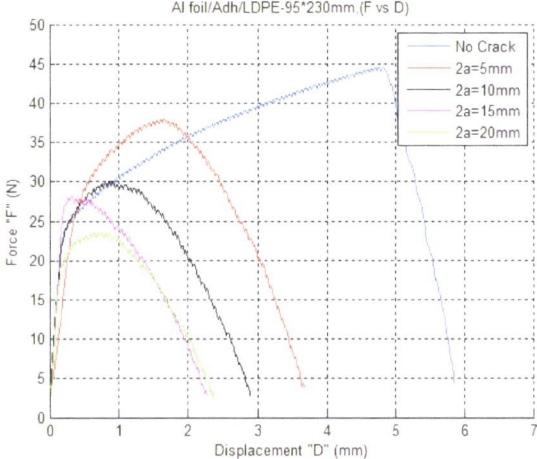

Figure 10. Load vs. displacement of Al foil/Adh/LDPE for 0–20 mm (center cracked).

Table 4. Load and displacement of fractured Al foil/Adh/LDPE, thickness B of 45 μm and dimensions of 230 mm × 95 mm.

Crack Length (2a) (mm)	Maximum Load (F) (N)	Extension (D) (mm)
No. Crack	44.61	4.756
5	38.04	1.653
10	30.07	0.863
15	28.24	0.334
20	23.59	0.653

3.4.4. Comparison between Case 2 and Case 3

Due to the ductility of LDPE, in case 2, the Al broke before the LDPE failed, as shown in Figure 8. For cases with crack lengths ranging from 5 to 20 mm, Al broke before 6 mm, while LDPE did not fail until 20 mm. In case 3, however, due to the strong adhesive bonding between the materials, they acted as one and both broke at the same point, as shown in Figure 10.

4. Numerical Simulation

A finite element simulation was carried out in simulation tool ABAQUS Explicit 6.10 by Dassault Systèmes, Vélizy-Villacoublay, France [39] to understand the material behavior clearly. The subsequent steps were followed to develop the exact tensile test model carried out in Section 3. The following assumptions were made to conduct numerical tests in ABAQUS 6.10. As the thickness is less, plane stress was considered. Shell elements were considered and for polymers, a linear elastic, von Mises isotropic plastic material model together with progressive damage was used as the material model. A trial-and-error calibration of continuum material parameters and calibration of fracture material parameters were used for the modeling procedure to conduct the tensile test in ABAQUS 6.10/Explicit.

4.1. Material Parameters Calibration

The material properties are required in ABAQUS to model the tensile test for Al-foil and LDPE. Experiment results are required to calibrate the data for numerical simulation. Figure 7 depicts a force vs. displacement graph for an Al-foil with no crack. The blue plot was chosen to calibrate the material's properties.

First, the stress vs. strain plot from the experimental results was used to calculate Young's modulus, as shown in Figure 11.

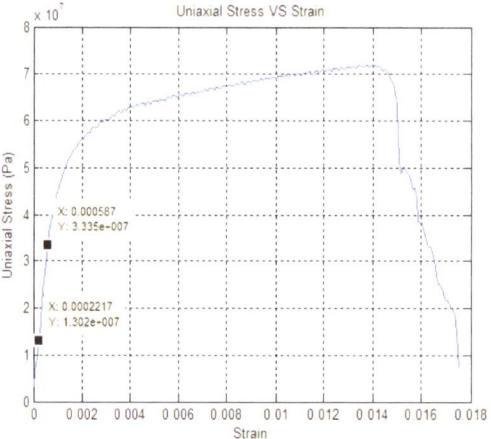

Figure 11. Stress vs. strain graph.

Plasticity is the behavior of a material beyond its elastic limit, which can be calculated using true stress vs. true strain experiments. Analytical methods can be used to calculate true stress and strain. The resulting plot of true stress vs. true strain is shown in Figure 12.

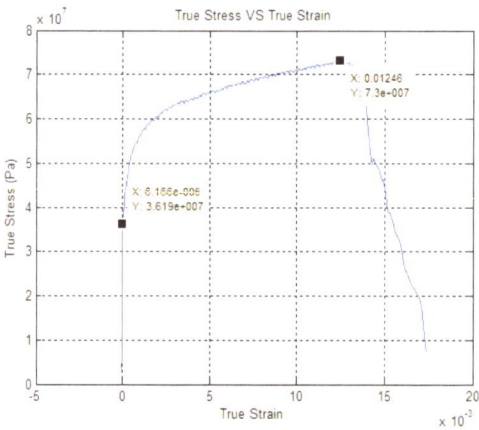

Figure 12. True stress vs. true strain graph for continuum Al-foil.

4.2. Fracture Material Parameters Calibration

The following are the components that define the material: the behavior of material under no damage; the behavior of material under damage (damage initiation (point $D = 0$));

and the behavior of material after damage initiation (damage evolution (D = 0 to ε_f^{pl}). These can be seen in Figure 13.

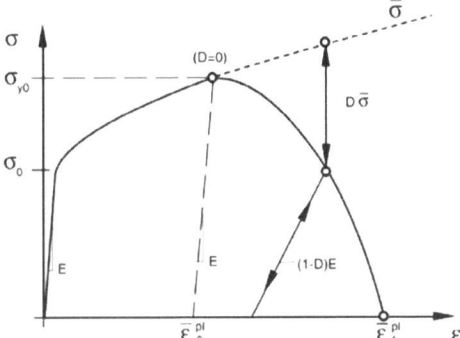

Figure 13. Stress–strain curve under progressive damage degradation.

Damage initiation defines the point at which degradation of stiffness starts. Fracture in a ductile polymer can be caused by two main processes. Nucleation, growth, and coalescence of void cause ductile fracture. Shear band localization causes shear fracture. The behavior of damage growth depends on these above two observations. In ductile metals, the damage caused by nucleation, growth, and coalescence of voids can be predicted by the initiation criterion model. The assumption made in this model is that the plastic strain is a function of stress triaxiality and strain rate for the onset of damage. Damage evolution defines the behavior of a material after damage initiation. It means that once initiation is caused, it defines the rate of degradation of the material stiffness [40].

Figure 13 describe the stress–strain behavior of a material under damage, where, E = Young's Modulus, D = damage variable, σ_{y0} = yield stress when damage onset occurs (at D = 0), ε_0^{pl} = equivalent plastic strain at damage onset (at D = 0), ε_f^{pl} = plastic strain when failure occurs (at D = 1).

All numerical parameters were calibrated by experimental measurement. In Table 5, fracture strains were calculated from the displacement response of the tensile test. As pre-cracked samples were tested, we assumed the strain had fully localized around the cracked surface. The stress triaxiality value was chosen in and of −5 and 5, which made the fracture strain in the material constitutive stress triaxiality independent. Finally, the displacement at failure, which is a FE-simulation material failure parameter, was calibrated by inverse modeling of the force displacement curve from experiment.

Table 5. The ductile damage initiation damage evolution values for aluminum foil and LDPE.

Materials	Fracture Strain	Stress Triaxiality	Strain Rate	Displacement at Failure (mm)
Aluminum Foil	0.0001 0.0001	−5 5	0 0	0.075
LDPE	0.9 0.9	−5 5	0 0	0.9

4.3. Numerical Results and Discussion

By using the Young Modulus equation and (Figure 11), the values within the elastic limit help to calculate Young's Modulus. For aluminum: Young's Modulus (E) = 55.653 GPa, Poisson's ratio (v) = 0.3, thickness (t) = 9 microns. Similarly, Young's modulus calculated for LDPE was (E) = 126.1 Mpa.

4.3.1. Aluminum Foil (No Crack)

Experiment results were used to calculate the material's properties. Plasticity values were calculated analytically based on the results. The results of both numerical and experimental calculations are nearly identical. Figure 14 depicts the response of experimental and numerical results. This leads us to the conclusion that the Al-foil can be modeled in the finite element software ABAQUS plasticity values.

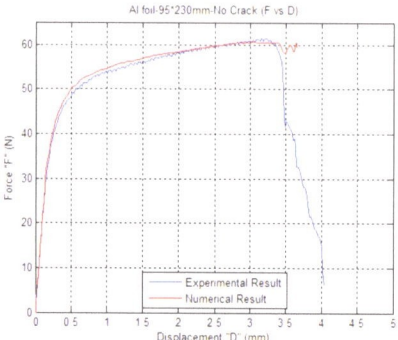

Figure 14. Force vs. displacement graph of aluminum foil (No Crack).

After calibrating fracture material parameters, they were compared with experimental results when the Young's modulus value of 55.63 GPa, and the plasticity values (Appendix B, Table A1) were used; good results were obtained with the damage initiation of 0.0001 and damage evolution of 0.075 mm presented in Table 5. The values of stress triaxiality and strain rate remained fixed in this numerical simulation.

4.3.2. LDPE

The properties of LDPE used in the subsequent theoretical and experimental analysis which are obtained from [35] are presented in Table 5 and Appendix B, Table A2.

4.3.3. Aluminum Foil (5 mm Crack)

All of the continuum and fractsure material parameters mentioned in Section 4.3.1, such as Young's modulus, plasticity, damage initiation, and damage evolution, remained constant for the specimen with a crack length of 5 mm.

When the experimental and numerical results are compared, Figure 15, it is clear that the material behavior is the same within the elastic limit, but there are some discrepancies in the material behavior in the plastic region.

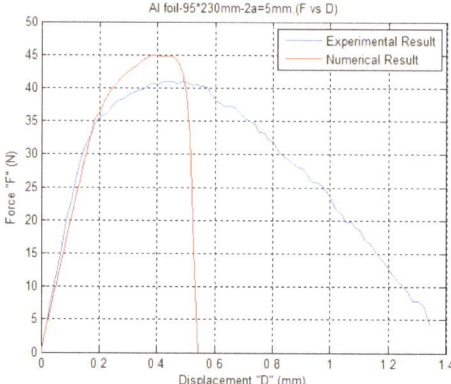

Figure 15. Force vs. displacement of aluminum foil with 5 mm Crack.

4.3.4. Aluminum Foil/LDPE Laminated (No Crack)

To simulate the ABAQUS model, all of the fracture material parameters such as Young's modulus, plasticity, damage initiation, and damage evolution for both Al-foil and LDPE, as mentioned in Sections 4.3.1 and 4.3.2, respectively, were used. When the experimental and numerical results are compared Figure 16, it is clear that the material behavior in both cases differs to some extent.

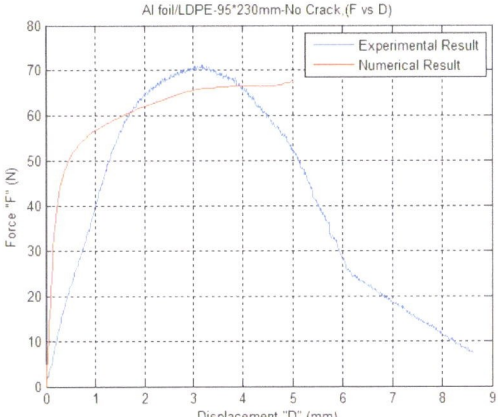

Figure 16. Force vs. displacement of aluminum foil/LDPE without crack.

5. Results and Discussion

Critical stresses were calculated for Al-foil with different crack lengths (5, 10, 15, 20, and 45 mm). The following fracture toughness graph was plotted using the LEFM, Equation (5), the strip yield model, Equation (7), and the experimental results for the Al-foil, which shows the normalized critical stresses vs. normalized crack length.

The strip yield model and LEFM's theoretical calculations show a strong correlation with the experimental data. As depicted in Figure 17, the Al-foil with a thickness of 9 microns has a fracture toughness of 6.39 MPa/m^2 For the Al-foil with a thickness of 6.42 microns, the fracture toughness was calculated using mechanical parameters reported in [23] was KC = 6.1 MPa/m^2.

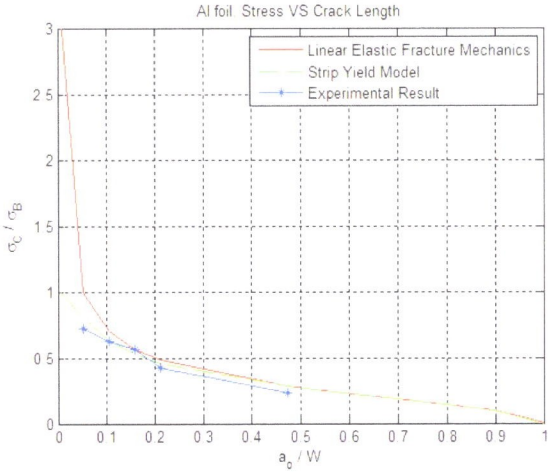

Figure 17. Fracture toughness graph.

When the experimental and numerical findings for Al foil and Al/LDPE were examined Figure 18, it was discovered that the experimental and numerical results for Al foil and Al foil/LDPE exhibited nearly the same behavior until the peak and then altered due to LDPE. However, the results of material laminated in the Blekinge Institute of Technology Lab were not the same as those discussed above. Due to the contact with PET and the heated surface of the machine used for lamination, material properties may alter. Another reason for this is because Al-foil failed first, followed by LDPE. This resulted in oscillation, which influenced the material properties and results.

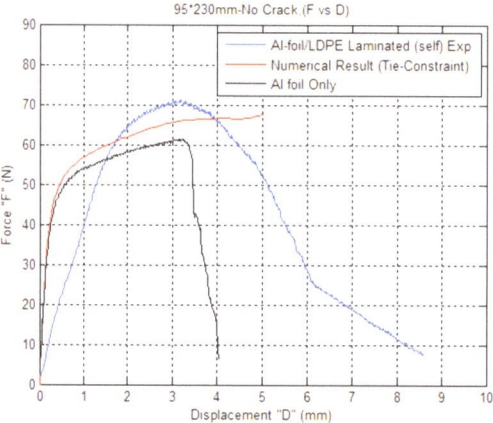

Figure 18. Comparison of experimental and numerical analysis.

When comparing the experimental results for Al foil/LDPE with no and full adhesion, as shown in Figure 19, it was discovered that they showed the same behavior until peak and then changed behavior because the material behaved as a different material with new material parameters during full adhesion. This is due to the strong adhesive bonding between the materials as presented in Figure 16.

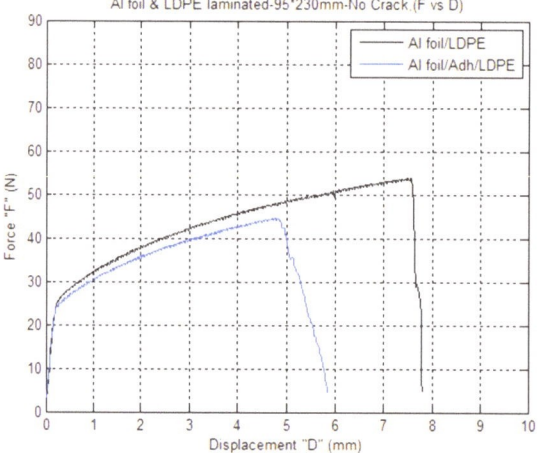

Figure 19. Comparison of experimental results between Al foil/LDPE and Al/Adh/LDPE.

In the case of laminated Al-foil/LDPE (without adhesion), the material's behavior in numerical and physical tests is dynamic along the displacement, which could be due to a number of factors, including the fact that both materials were in contact with Polyethylene Terephthalate (PET) to avoid burning and sticking of materials with paper. This may lead to a change in the material's properties. Second, both materials retained their properties, so that Al-foil broke first, followed by LDPE, which is more ductile than Al-foil. The other reason is that the Al-foil failed first, followed by the LDPE. This resulted in oscillation, which influenced the material properties and results.

The failure occurred at the same force and displacement in both materials while performing experiments on aluminum foil/Adh/LDPE (full adhesion). Due to the strong adhesive bonding between the materials, they behaved as if they were one material, allowing us to conclude that the strong adhesion changes the material parameters.

6. Conclusions

- Different material properties were determined in this study by developing physical test and FE modeling procedures. The experiments were carried out on Al-foil, aluminum foil laminated with LDPE, with and without adhesion. Material parameters were calibrated for use in the numerical test study. The design of experiment technique was used to determine the material's Young's modulus, plasticity, damage initiation, and damage evolution. For Al-foil, the theoretical result using LEFM and modified strip yield model was showing a close relation with the obtained experimental results.
- The fracture toughness K_C obtained for the Al-foil with thickness of 9 microns was 6.39 MPam$^{1/2}$. An earlier study on the same grade Al-foil reported the fracture toughness K_C = 6.1 MPam$^{1/2}$ for thickness of 6.42 microns.
- For the case of laminated Al-foil/LDPE (without adhesion), the behavior of the material in the numerical and physical tests was dynamic along the displacement and there may be number of reasons, such as the fact that Al-foil and LDPE were laminated by a lamination machine and were in contact with PET to avoid the burning and sticking of materials with paper. This may have caused the change in the material properties.
- Failure of Al-foil before the LDPE in case of Al foil/LDPE (without adhesion) cause the specimen oscillation, so vibration can also be a cause of change in material behavior when analyzed numerically.
- In the case of Al foil/Adh/LDPE, both materials broke at the same point so it can be concluded that the strong adhesion force between laminates can change the properties and the materials act as one.

This research can be expanded in the future to calculate the fracture toughness value of laminated materials. Furthermore, the FE modeling strategy can be defined for materials with full adhesion. Under the microscope, bonded material delamination can be defined and studied. Adhesives and their effects on the properties of bonded materials can be studied.

Author Contributions: Supervised all the project and literature review regarding this paper, U.S.; Supervision of the project including literature review and write-up, B.S.; Did all the experimental work and manuplate the data of this paper, M.S.I.; Look after the experimental work, completed illustration and data interpretation, K.M.; Managed the raw materials and give a proper guidance for materials usage and safety, D.S.I.; Helped in paper revisions and data explanations, O.O.A.; Funding acquisition, N.A.; Provided financial help, Z.I.Z.; Helped in data explainations, Z.M.E.-B. All authors have read and agreed to the published version of the manuscript.

Funding: Supported by National Key Research and Development Program of China (Grant No. 2019yFB2006404) and the Taif Researchers Supporting Project (TURSP-2020/42), Taif University, Taif, Saudi Arabia.

Institutional Review Board Statement: Not applicable for this study.

Informed Consent Statement: Not applicable for this study.

Data Availability Statement: Dataset will be provided by coressponding author upon reasonable request. No publicly archived dataset was utilized in this study.

Conflicts of Interest: The authors declare no conflict of interest.

Appendix A

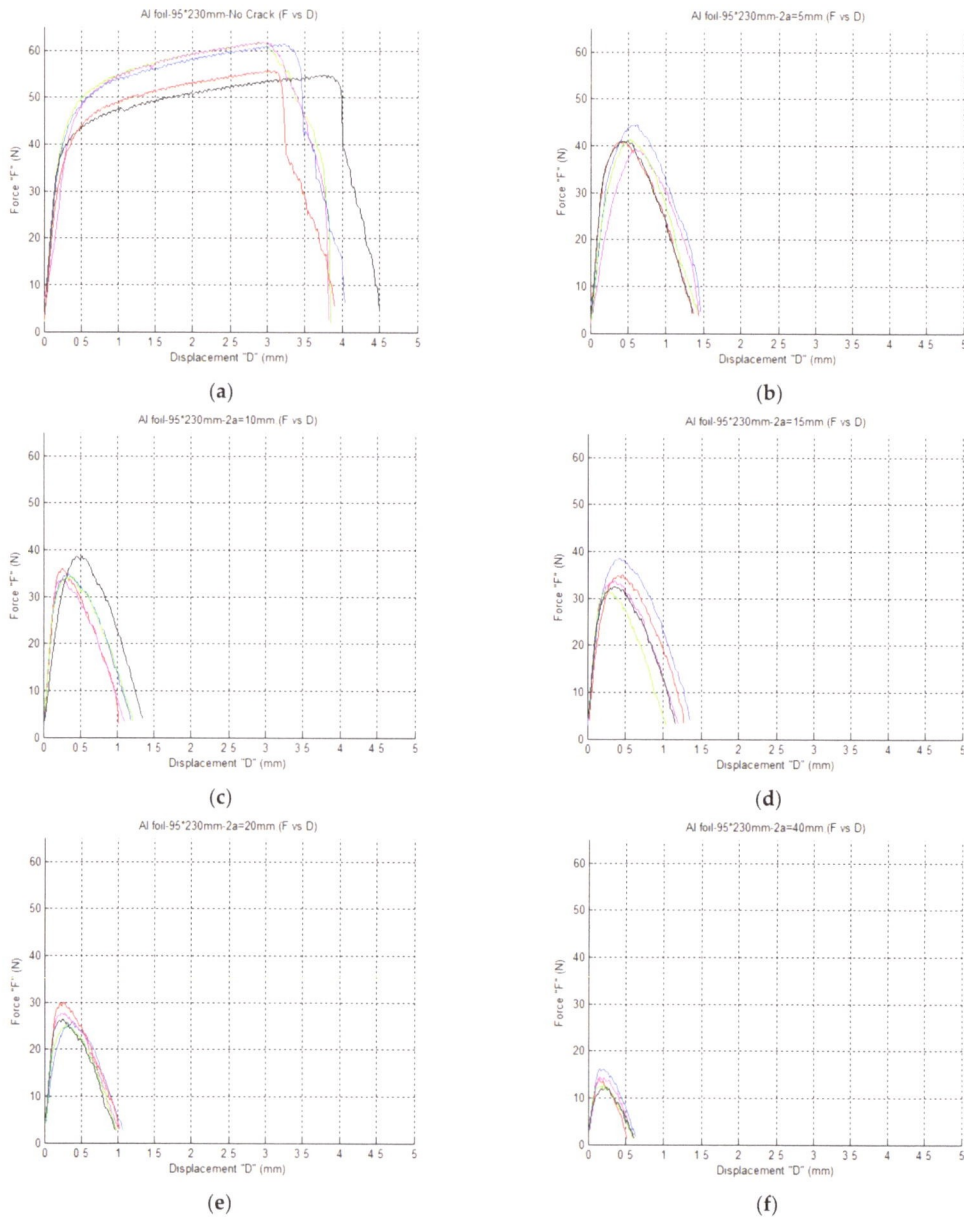

Figure A1. Experimental results for aluminum foil. (**a**) F vs. D for 0 mm crack, (**b**) F vs. D for 5 mm crack, (**c**) F vs. D for 10 mm crack, (**d**) F vs. D for 15 mm crack, (**e**) F vs. D for 20 mm crack, (**f**) F vs. D for 40 mm crack.

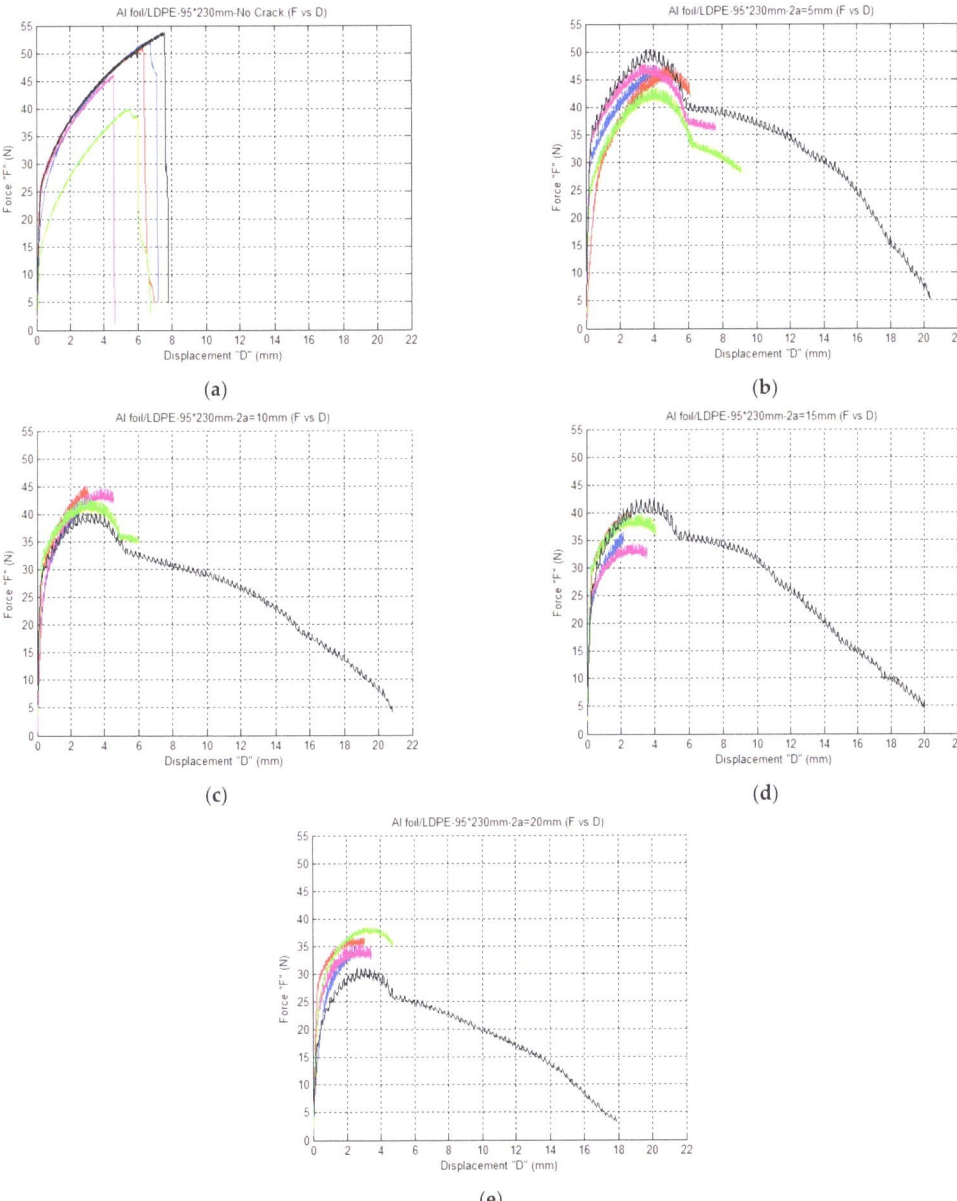

Figure A2. Experimental results for aluminum foil/LDPE. (**a**) F vs. D for 0 mm crack, (**b**) F vs. D for 5 mm crack, (**c**) F vs. D for 10 mm crack, (**d**) F vs. D for 15 mm crack, (**e**) F vs. D for 20 mm crack.

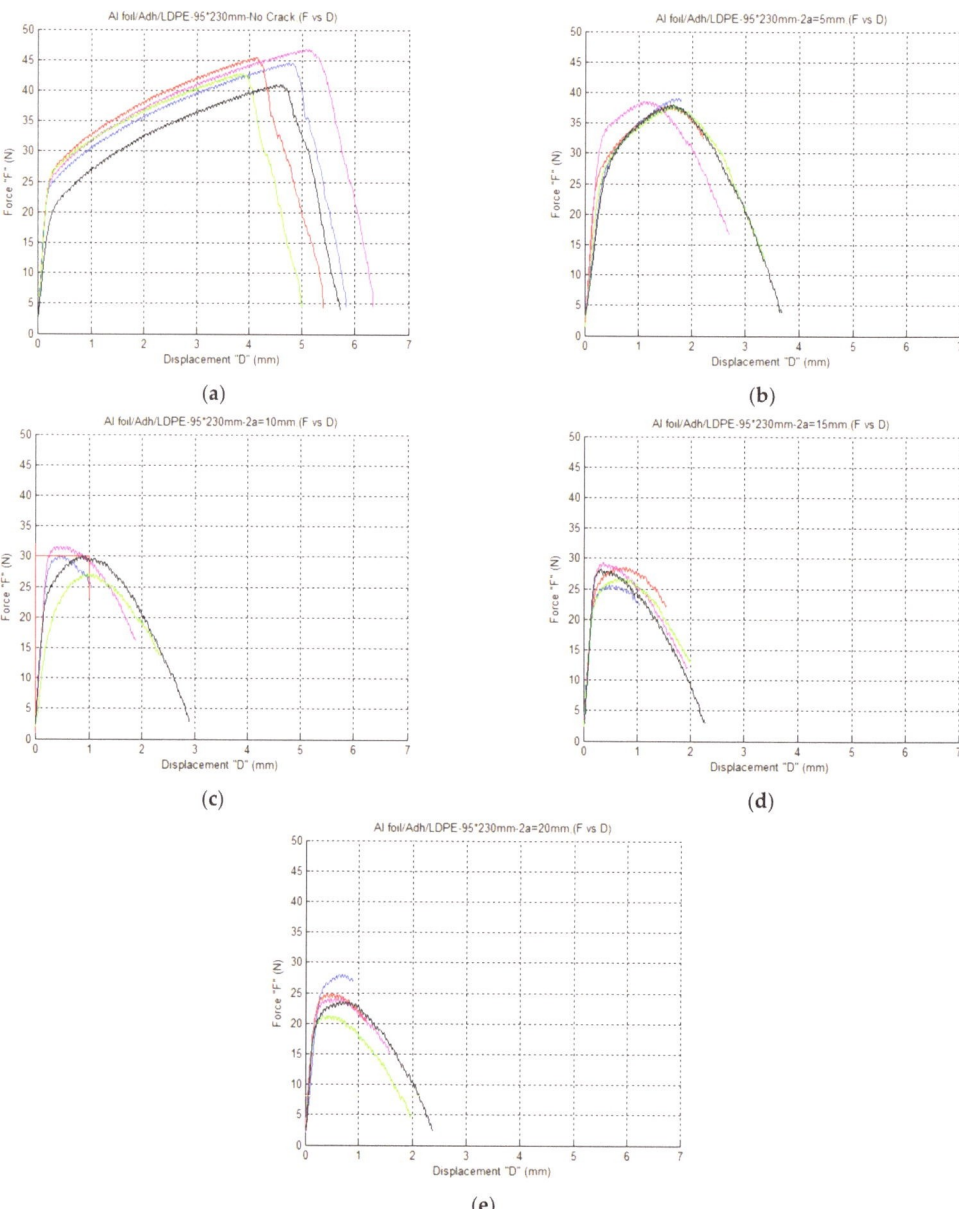

Figure A3. Experimental results for aluminum foil/Adh/LDPE. (**a**) F vs. D for 0 mm crack, (**b**) F vs. D for 5 mm crack, (**c**) F vs. D for 10 mm crack, (**d**) F vs. D for 15 mm crack, (**e**) F vs. D for 20 mm crack.

Appendix B

Table A1. Yield stress and plastic strain of aluminum foil.

Nos.	Yield Stress (MPa)	Plastic Strain
1	36.19	0
2	42.54	0.0001096
3	45.12	0.000211
4	48.13	0.0003004
5	51.16	0.0004633
6	54.11	0.0007017
7	57.38	0.001147
8	60.35	0.00182
9	62.95	0.003004
10	64.77	0.004132
11	66.56	0.005547
12	68.58	0.007468
13	70.63	0.009318
14	72.12	0.01117
15	73	0.01246

Table A2. Yield stress and plastic strain of LDPE.

Nos.	Yield Stress (MPa)	Plastic Strain
1	5.332	0
2	5.444	0.00134
3	5.68	0.0022
4	6	0.00427
5	6.16	0.00569
6	6.3	0.0065
7	6.459	0.0079
8	6.511	0.0083
9	6.62	0.0093
10	6.78	0.01107
11	6.90	0.0124
12	7.08	0.0147
13	7.14	0.0148
14	7.21	0.0164
15	7.328	0.0184
16	7.412	0.0193
17	7.553	0.0227
18	7.638	0.0246
19	7.80	0.0278
20	7.92	0.0313
21	8.02	0.0330
22	8.136	0.0365
23	8.26	0.04
24	8.35	0.0425
25	8.73	0.0696

References

1. Islam, M.S. Fracture and Delamination in Packaging Materials: A Study of Experimental Methods and Simulation Techniques. Ph.D. Thesis, Blekinge Tekniska Högskola, Karlshamn, Sweden, 2019.
2. Islam, M.S.; Alfredsson, K.S. Peeling of metal foil from a compliant substrate. *J. Adhes.* **2021**, *97*, 672–703. [CrossRef]
3. Marsh, K.; Bugusu, B. Food packaging—Roles, materials, and environmental issues. *J. Food Sci.* **2007**, *72*, R39–R55. [CrossRef]
4. Geueke, B.; Groh, K.; Muncke, J. Food packaging in the circular economy: Overview of chemical safety aspects for commonly used materials. *J. Clean. Prod.* **2018**, *193*, 491–505. [CrossRef]
5. Shin, J.; Selke, S.E.M. *Chapter 11 Food Packaging. Food Processing: Principles and Applications*; John Wiley & Sons: Hoboken, NJ, USA, 2014.

6. Brostow, W.; Lobland, H.E.H.; Khoja, S. Brittleness and toughness of polymers and other materials. *Mater. Lett.* **2015**, *159*, 478–480. [CrossRef]
7. Kao-Walter, S. On the Fracture of Thin Laminates. Ph.D. Thesis, Blekinge Institute of Technology, Karlskrona, Sweden, 2004.
8. Tryding, J. In Plane Fracture of Paper. Ph.D. Thesis, Lund Institute of Technology, Lund, Sweden, 1996.
9. Hägglund, R.; Isaksson, P. Analysis of localized failure in low-basis-weight paper. *Int. J. Solids Struct.* **2006**, *43*, 5581–5592. [CrossRef]
10. Isaksson, P.; Hägglund, R.; Gradin, P. Continuum damage mechanics applied to paper. *Int. J. Solids Struct.* **2004**, *41*, 4731–4755. [CrossRef]
11. Islam, M.S.; Zhang, D.; Mehmood, N.; Kao-Walter, S. Study of shear dominant delamination in thin brittle-high ductile interface. *Procedia Struct. Integr.* **2016**, *2*, 152–157. [CrossRef]
12. De-feng, Z.; Md. Shafiqu, I.; Eskil, A.; Sharon, K.-W. Modeling and study of fracture and delamination in a packaging laminate. In Proceedings of the 3rd International Conference on Material Engineering and Application (ICMEA 2016), Shanghai, China, 12–13 November 2016.
13. Li, T.; Suo, Z. Ductility of thin metal films on polymer substrates modulated by interfacial adhesion. *Int. J. Solids Struct.* **2007**, *44*, 1696–1705. [CrossRef]
14. Li, T.; Suo, Z. Deformability of thin metal films on elastomer substrates. *Int. J. Solids Struct.* **2006**, *43*, 2351–2363. [CrossRef]
15. Suo, Z.; Vlassak, J.; Wagner, S. Micromechanics of macroelectronics. *China Particuology* **2005**, *3*, 321–328. [CrossRef]
16. Li, T.; Zhang, Z.; Michaux, B. Competing failure mechanisms of thin metal films on polymer substrates under tension. *Theor. Appl. Mech. Lett.* **2011**, *1*, 041002. [CrossRef]
17. Shahid, S.; Gukhool, W. Experimental Testing and Materialmodeling of Anisotropy in Injection Moulded Polymer Materials. Master's Thesis, Blekinge Institiute of Technology, Karlskrona, Sweden, 2020.
18. Danielsson, M.; Parks, D.M.; Boyce, M.C. Three-dimensional micromechanical modeling of voided polymeric materials. *J. Mech. Phys. Solids* **2002**, *50*, 351–379. [CrossRef]
19. Sedighiamiri, A.; Govaert, L.E.; Van Dommelen, J.A.W. Micromechanical modeling of the deformation kinetics of semicrystalline polymers. *Polym. Sci. Part B Polym. Phys.* **2011**, *48*, 1297–1310. [CrossRef]
20. Chen, S.H.; Wang, T.C.; Kao-Walter, S. A crack perpendicular to the bimaterial interface in finite solid. *Int. J. Solids Struct.* **2003**, *40*, 2731–2755. [CrossRef]
21. Kao-Walter, S.; Ståhle, P.; Chen, S.H. A finite element analysis of a crack penetrating or deflecting into an interface in a thin laminate. *Key Eng. Mater.* **2006**, *312*, 173–178. [CrossRef]
22. Kao-Walter, S.; Ståhle, P.; Hägglund, R. Fracture toughness of a laminated composite. *Eur. Struct. Integr. Society.* **2003**, *32*, 355–364. [CrossRef]
23. Kao-Walter, S. *Mechanical and Fracture Properties of Thin Al-Foil*; Blekinge Tekniska Högskola Forskningsrapport: Karlskrona, Sweden, 2001; ISSN 11031581.
24. He, G.; Li, J.; Zhang, F.; Lei, F.; Guo, S. A quantitative analysis of the effect of interface delamination on the fracture behavior and toughness of multilayered propylene–ethylene copolymer/low density polyethylene films by the essential work of fracture (EWF). *Polymer* **2014**, *55*, 1583–1592. [CrossRef]
25. Iqbal, M.S.; Muhammadi, A.B. Tearing Fracture and Microscopic Analysis of Laminate Toward Sustainable Packaging. Master's Thesis, Blekinge Institute of Technology, Karlskrona, Sweden, 2007.
26. Skanse, H. Investigation of Mechanical Tearing and How it can Be Applied in Package Open Ability Prediction. Master's Thesis, Lund Institute of Technology, Lund, Sweden, 2009.
27. Karmlid, O. Simulation and Testing of Crack Sensitivity in TFA Packaging Material. Master's Thesis, Lund University, Lund, Sweden, 2011.
28. Min, H. *Study of Two Models for Tearing Resistance Assesment Using Essential Work of Fracture Method*; Blekinge Institute of Technology: Karlskrona, Sweden, 2008.
29. Kao-Walter, S.; Dahlström, J.; Karlsson, T.; Magnusson, A. A study of the relation between the mechanical properties and the adhesion level in a laminated packaging material. *Mech. Compos. Mater.* **2004**, *40*, 29–36. [CrossRef]
30. Kao-Walter, S.; Mfoumou, E. *Fracture Toughness Testing of Non Standard Specimens*; Blekinge Institute of Technology: Karlskrona, Sweden, 2004.
31. Anderson, T.L. *Fracture Mechanics Fundamentals and Applications*, 3rd ed.; CRC Press Taylor & Francis Group: Boca Raton, FL, USA, 2005.
32. Kao-Walter, S.; Ståhle, P. Fracture Behaviour of a Thin Al-Foil-Measuring and Modelling of the Fracture Processes. In Proceedings of the Third International Conference on Experimental Mechanics, Beijing, China, 15–17 October 2001; pp. 25–256.
33. Dugdale, D.S. Yielding of steel sheets containing slits. *J. Mech. Phys. Solids* **1960**, *8*, 100–104. [CrossRef]
34. Fracture Toughness. Center, N.E.R. Toughness. 2018. Available online: https://www.nde-ed.org/EducationResources/CommunityCollege/Materials/Mechanical/Toughness.html (accessed on 21 August 2020).
35. Jemal, A.; Katangoori, R.R. Fracture Mechanics Applied in Thin Ductile Packaging Materials-Experiments with Simulations. Master's Thesis, Blekinge Institute of Technology, Karlskrona, Sweden, 2011.
36. Baldan, A. Adhesively-bonded joints and repairs in metallic alloys, polymers and composite materials: Adhesives, adhesion theories and surface pretreatment. *J. Mater. Sci.* **2004**, *39*, 1–49. [CrossRef]

37. Creton, C.A. Materials science of pressure sensitive adhesives, materials science and technology. *Process. Mater.* **1997**, *18*, 434–439.
38. Krenk, S. *Non-Linear Modeling and Analysis of Solids and Structures*; Cambridge University Press: Cambridge, UK, 2009.
39. Systèmes, D. *Introduction to Abaqus Scripting*; Dassault Systèmes: Vélizy-Villacoublay, France, 2009.
40. Shi, Y.; Swait, T.; Soutis, C. Modelling damage evolution in composite laminates subjected to low velocity impact. *Compos. Struct.* **2012**, *94*, 2902–2913. [CrossRef]

Article

Polyvinyl Alcohol and Nano-Clay Based Solution Processed Packaging Coatings

Ali Dad Chandio [1], Iftikhar Ahmed Channa [1,2], Muhammad Rizwan [1], Shakeel Akram [3], Muhammad Sufyan Javed [4,5], Sajid Hussain Siyal [6,*], Muhammad Saleem [7], Muhammad Atif Makhdoom [2], Tayyaba Ashfaq [8], Safia Khan [9], Shahid Hussain [10,11], Munirah D. Albaqami [12] and Reham Ghazi Alotabi [12]

Citation: Chandio, A.D.; Channa, I.A.; Rizwan, M.; Akram, S.; Javed, M.S.; Siyal, S.H.; Saleem, M.; Makhdoom, M.A.; Ashfaq, T.; Khan, S.; et al. Polyvinyl Alcohol and Nano-Clay Based Solution Processed Packaging Coatings. Coatings 2021, 11, 942. https://doi.org/10.3390/coatings11080942

Academic Editor: Yingyi Zhang

Received: 3 July 2021
Accepted: 30 July 2021
Published: 6 August 2021

Publisher's Note: MDPI stays neutral with regard to jurisdictional claims in published maps and institutional affiliations.

Copyright: © 2021 by the authors. Licensee MDPI, Basel, Switzerland. This article is an open access article distributed under the terms and conditions of the Creative Commons Attribution (CC BY) license (https://creativecommons.org/licenses/by/4.0/).

[1] Department of Materials and Metallurgical Engineering, NED University of Engineering and Technology, University Road Karachi, Karachi 75270, Pakistan; alidad@neduet.edu.pk (A.D.C.); iftikharc@neduet.edu.pk (I.A.C.); materialist.riz@gmail.com (M.R.)
[2] Institute of Materials for Electronics and Energy Technology (i-MEET), Friedrich-Alexander Universität Erlangen-Nürnberg, 91058 Erlangen, Germany; atif.ceet@pu.edu.pk
[3] College of Electrical Engineering, Sichuan University, Chengdu 610065, China; Shakeel.Akram@scu.edu.cn
[4] School of Physical Science and Technology, Lanzhou University, Lanzhou 730000, China; safisabri@gmail.com
[5] Department of Physics, COMSATS University Islamabad, Lahore 54000, Pakistan
[6] Department of Metallurgy and Materials Engineering, Dawood University of Engineering and Technology, Karachi 74800, Pakistan
[7] Department of Physics, The Islamia University of Bahawalpur Punjab, Bahawalpur 63100, Pakistan; saleem.malikape@gmail.com
[8] Department of Chemistry, Government College University Faisalabad, Faisalabad 38000, Pakistan; tayyabaa961@gmail.com
[9] Department of Chemistry, Quaid-i-Azam University Islamabad, Islamabad 45320, Pakistan; safiakhan715@gmail.com
[10] School of Materials Science and Engineering, Jiangsu University, Zhenjiang 212013, China; shahid@ujs.edu.cn
[11] Departamento de Quimica Organica, Universidad de Cordoba, Edificio Marie Curie (C-3), Ctra Nnal IV-A, Km 396, E14014 Cordoba, Spain
[12] Chemistry Department, College of Science, King Saud University, Riyadh 11451, Saudi Arabia; muneerad@ksu.edu.sa (M.D.A.); 438202971@student.ksu.edu.sa (R.G.A.)
* Correspondence: sajid.hussain@duet.edu.pk

Abstract: Cost-effective, clean, highly transparent, and flexible as well as a coatable packaging material is envisioned to solve or at least mitigate quality preservation issues of organic materials, originating from moisture interaction under ambient conditions. Liquid phase processing of packaging coatings using nano-clay and polyvinyl alcohol (PVOH) has been developed and reported. Detailed analysis of the developed coating revealed moisture permeability of 2.8×10^{-2} g·cm/m^2·day at 40 °C and 85% relative humidity (RH), which is in close accordance with Bharadwaj's theoretical permeability model. Moreover, the developed coatings are not only more than 90% transparent, when exposed to white light, but also exhibit excellent flexibility and even after going through 10,000 bending cycles maintained the same blocking effect against moisture.

Keywords: flexible barriers; flexible packaging; polyvinyl alcohol; nano-clay; moisture; permeability

1. Introduction

Packaging materials are designed to safeguard the product while it is in storage, transportation, or the distribution phase [1–3]. The basic requirement of this packaging material is to maintain the quality of the product by limiting chemical, physical, or biological changes [4]. Hence, packaging should have a barrier effect against environmental gases, moving in or out, and ultraviolet (UV) light [1,2,5]. In addition to these, packaging materials for food and other organic stuff should also have the ability to increase the shelf life [6] by providing a barrier against moisture and oxygen [7–9]. Miscellaneous materials (e.g., glass, polymers, ceramics, and metals) are in use in contemporary packaging applications. Among these, polymers have found wide application with a market share of 40% in the food industry [10] due to its cost-effectiveness, light-weight, and stability of its chemical and physical properties [10,11]. Transparency to visible light is another additive advantage of the food packaging [10], which polymers can also have. Moreover, these can be easily processed using simple coating techniques (i.e., spin coating, doctor blading, spray coating, etc.) [12–14].

All polymers, with the exception of a few (polystyrene, polycarbonates, etc.,), offer high-quality barrier characteristics [3] that can further be improved against any specific permeating molecule by various factors. In general, there is no single polymer that possesses all of the properties required for packaging applications and hence multifaceted complex types of frameworks are used [7]. Ethylene-vinyl alcohol (EVOH), for example, exhibits excellent barrier properties against oxygen in dry conditions, which further amplifies in humid conditions (>75% RH) due to swelling of polymeric chains [15–17]. Although it is a good barrier against oxygen, it still has to be sandwiched between two hydrophobic polyethylene layers [18,19] for good results. Better barrier properties for packaging applications cannot be achieved via a monolayer of polymers, therefore, a direct mixing of polymers is preferred [20–22] either by blending [21] or by multilayered [23] coatings, but these have not only high production costs but are also difficult to recycle [24]. Therefore, interest has been developed in the recent past to produce novel monolayer packaging films with improved mechanical and barrier properties [24,25]. Due to the advantages of transparency and flexibility of a polymer matrix, quality barrier monolayer films have been developed by incorporating inorganic nanoparticles in them [5,11,24]. This enhances the barrier characteristics against permeating gases by offering a tortuous path [25–27] as the permeating molecules have to travel a long way along the axis of the particle until it finds the polymer or a defect to diffuse to the other side [26,28–31].

The reduction of permeation depends upon aspect ratio, orientation, concentration, compatibility, and uniform distribution of filler within the polymer matrix [28,29,32–34]. Numerous studies [35,36] have been carried out on systems containing nano-clays as gas barrier fillers in polymeric matrices whose results showed a direct effect of nano-clay on the permeation of moisture and oxygen. The hydrophobic nature of nano-clay makes its distribution uniform in water-soluble polyvinyl alcohol (PVOH) [37,38]. Gaume et al. [25] used sodium montmorillonite (MMT-Na$^+$) clay as filling particles and reported a reduction in oxygen and moisture permeation by a factor of 2.7 and 1.7, respectively. However, the influence of clay on haze and the mechanical flexibility of the film is scanty. Therefore, the present research work was planned and carried out using a solution processing route to develop packaging films using PVOH and MMT-Na$^+$ nano-clay coatings. PVOH is one of the well-known biodegradable polymers that can be processed easily [39,40] and exhibits low oxygen permeation and stability under UV irradiations [25,41,42] and hence has wide applications in flexible paper coatings. MMT-Na$^+$ nano-clay, on the other hand, can uniformly disperse in PVOH because of its hydrophilic nature and has a platelet structure with an aspect ratio of 200–1000, which can effectively reduce the diffusion of gases and increase thermal and mechanical stability [25,29]. There have been many studies carried out on PVOH and nano clay. In this study, the processing parameters were optimized to yield the desired results. Not only was the effect of clay concentration studied, but also the optimization of the thickness controlling parameters was carried out to produce a highly bendable/flexible barrier coating for the packaging industry with an economical and single-step easy procedure. Hence, a composite of PVOH and nano-clay (both being environmentally friendly materials) was developed and characterized mainly in terms of barrier characteristics, optical transparency, and flexibility.

2. Experimental

2.1. Materials

PVOH (27,000 g/mol) was obtained from Sigma-Aldrich GmbH (St. Louis, MO, USA) and used as received. Nano-clay (Sodium montmorillonite–MMT-Na$^+$) with a monoclinic structure was purchased from BYK GmbH (Wether, Germany). Polyethylene terephthalate (PET) Melinex ST504 was acquired from DuPont Teijin Films, Chester, VA, USA, and was used as the substrate.

2.2. Processing of Films

Pristine PVOH (10 wt.%) was mixed with de-ionized water and continuously stirred on a hot plate at 90 °C until a clear and homogenous solution was obtained. MMT-Na+ nano-clay was added to this PVOH solution with a concentration ranging from 2–10 wt.% of the matrix and then mixed thoroughly for a few hours at 60 °C followed by ultrasonication for 30 min before coating. It should be noted that the mixing of nano-clay in PVOH solution becomes very hard with an increasing clay content beyond 5 wt.%. Maintaining the uniform distribution of clay within the matrix needs extra effort and extra mixing time. The prepared solution was applied using a doctor blade (Zehntner, Sissach, Switzerland) on a PET substrate, which was maintained at 30 °C followed by drying in an oven at 80 °C. Finally, the dried films were taken off from the PET and characterized as free-standing films.

2.3. Characterization of Films

Microdefects and surface quality of the film were analyzed using the "Olympus MX51" microscope (Olympus, Tokyo, Japan). Transparency of the films under visible region was examined by "Shimadzu UV-1800" spectrophotometer Shimadzu UV-1800 spectrophotometer (Shimadzu Company, Tokyo, Japan). ATR-FTIR (attenuated total reflectance-Fourier transform infrared) analysis of the films was made using "Bruker ALPHA-P" (Bruker, Billerica, MA, USA) at a scan rate of 64 with 4 cm^{-1} resolution. Water vapor transmission rate (WVTR) measurements were performed using the water method with a permeability cup (qualifying ASTM E96 standard, from Thwing-Albert Instrument Company GmbH, West Berlin, NJ, USA) that can measure WVTR values down to 0.1 g/(m^2·day) and following the procedure documented by Wu et al. [43]. WVTR was then calculated from a slop of the weight loss curve using Equation (1).

$$\text{WVTR} = \frac{G}{A \times t} \quad (1)$$

where G/t is the weight loss per day and A is the exposed area of the film.

The flexibility of the barrier films was measured with a bend testing machine equipped with a counter, having one fixed end and another moving back and forth, maintaining a customized bending radius. For this test, a film with a 40 mm width, 80 mm length, and thickness ~100 μm were prepared and for WVTR measurement, the sample was cut from the middle of similar films.

3. Results and Discussion

Figure 1 shows the optical micrographs of PVOH/MMT-Na$^+$ films at various wt.%. The micrographs showed no significant defects in the developed films. However, some tiny black dots were present, which can be referred to as agglomerated nano-clay particles.

The IR spectra of pristine PVOH and PVOH/MMT-Na$^+$ films are shown in Figure 2. The IR spectra of PVOH/MMT-Na$^+$ film showed the summation of characteristic peaks of pristine PVOH and MMT-Na$^+$ nano-clay. The presence of a typical Si–O stretching band of montmorillonite around 970 cm^{-1} was absent in the pristine PVOH spectrum. However, the peak close to 1000 cm^{-1} in pristine PVOH refers to the C–O stretching band, which is in complete alignment with the work done by Reis et al. (2006) [24] in pure PVOH, which appears in PVOH/MMT-Na$^+$ films and whose intensity increases linearly with the percentage of nano-clay (from 2–10 wt.%) in PVOH. These results are in good agreement with the results reported in the literature [25].

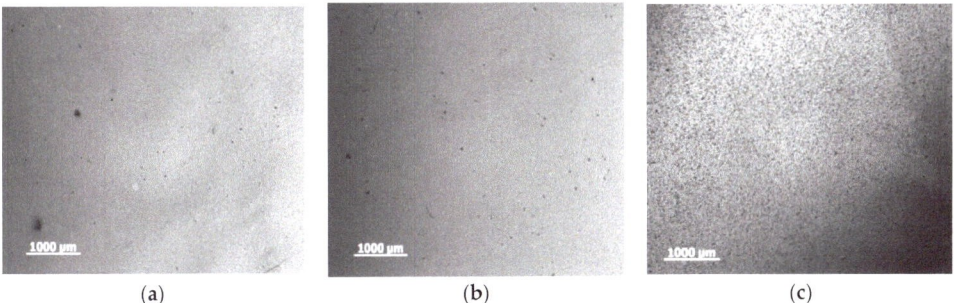

Figure 1. Optical micrographs of PVOH/MMT-Na$^+$ films with (**a**) 2 wt.% nano-clay, (**b**) 4 wt.% nano-clay, and (**c**) 6 wt.%.

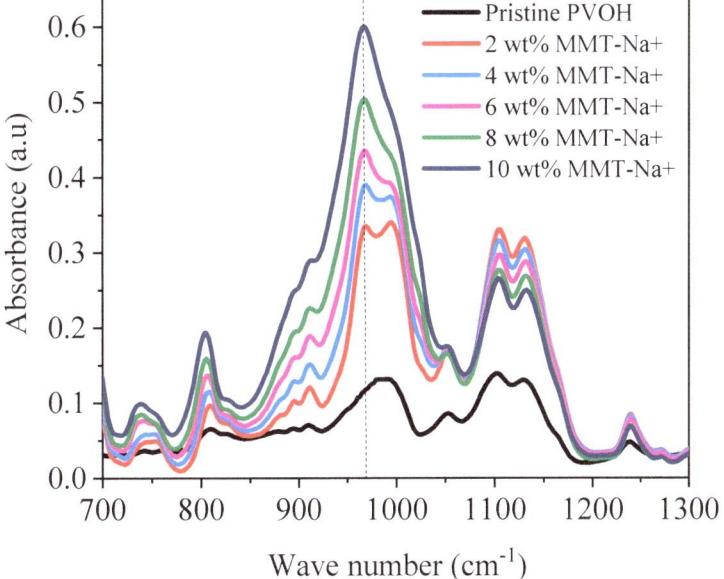

Figure 2. FTIR-ATR spectra of pristine PVOH and PVOH/MMT-Na$^+$ films.

In general, the spectra (Figure 3a) showed that the transmission of white light was not affected by the presence of nano-clay and the transparency level of 92% was almost unchanged. However, the transparency of pristine PVOH (i.e., 92%) slightly dropped to 89% as the wt.% of the nano-clay (Figure 3b) increased in it, which can be attributed to diffused transmittance (Figure 3c). This increase in diffused transmittance suggests that haze is generated, which creates scattering and this effect is more pronounced as we move from the visible to UV region. An increase of 11% in diffuse transmittance, for example, was observed for 6 wt.% PVOH/MMT-Na$^+$ at a wavelength of 400 nm (Figure 3c).

Although pristine PVOH is water-soluble, however, its film shows resistance against moisture and the WVTR exhibited by the PVOH films was comparable to commercially used packaging polymers such as low-density polyethylene (~100 g/(m^2·day)) and polyvinyl butyral (~70 g/(m^2·day)) [3]. However, the PVOH films have several other advantages that include easy processing, biodegradability, and a high barrier against the diffusion of oxygen. Figure 4 shows a decreasing trend in weight loss of water through films of pristine PVOH films by increasing their thicknesses. Using Equation (1) and weight loss measurements, the WVTR values were calculated as shown in Table 1. Table 1 shows that a WVTR of 90 g/m² ·day was observed for a 25 μm thick pristine PVOH film at test conditions of 85% RH and 40 °C, which decreased linearly with thickness.

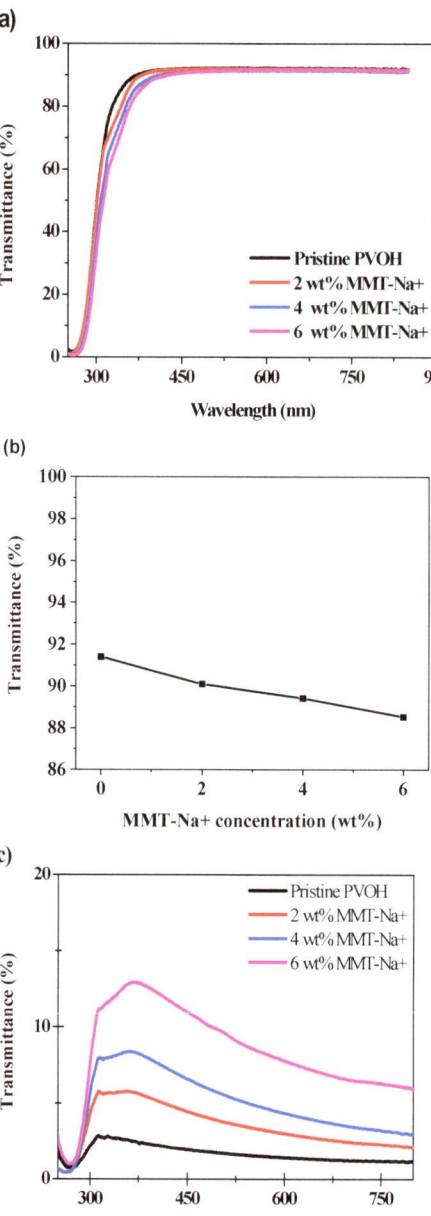

Figure 3. (**a**) UV–Vis spectra of pristine PVOH and PVOH/MMT-Na⁺ films. (**b**) Kinetics of total transmittance decrease for PVOH/MMT-Na⁺ films with a concentration @ 450 nm. (**c**) Diffused transmittance of pristine PVOH and PVOH/MMT-Na⁺ films.

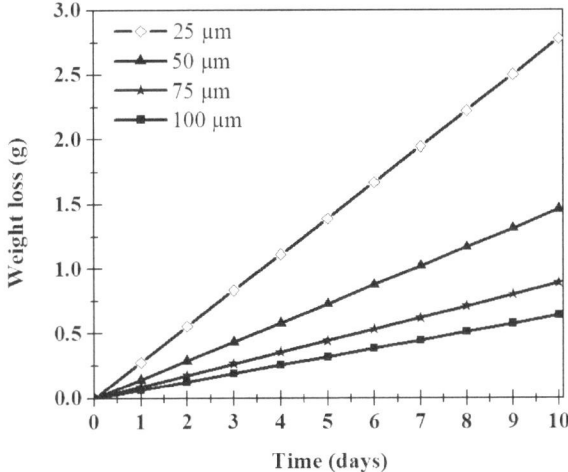

Figure 4. Weight loss versus time using a cup sealed with pristine PVOH films having different thicknesses at 40 °C and 85% RH.

Table 1. Moisture permeation of the pristine PVOH films with different thicknesses. For statistical analysis, at least five samples were tested for each thickness variation.

Film Thickness (μm)	WVTR (g/(m²·Day))
25	90 ± 5.2
50	47 ± 4.2
75	29 ± 3.4
100	20.5 ± 2.5

This can be attributed to the fact that the regular and orderly formation of a defect-free thicker film of the pristine PVOH takes time, taking longer paths for the moisture molecules to diffuse from one side to another. When these WVTR data are plotted against film thickness, a linear graph was obtained (Figure 5), which indicates that barrier properties (inverse of WVTR) of pristine PVOH against moisture is directly proportional to film thickness. Therefore, an improvement of ~77% was observed in the barrier quality of the films by increasing the film thicknesses from 25 to 100 μm.

Figure 6 shows the graphs of the weight loss measurement of moisture through PVOH/MMT-Na$^+$ (2 to 10 wt.%) and pristine PVOH of 100 μm thick films. Using Equation (1) and weight loss measurements, the WVTR values were calculated and tabulated along with the values of permeability, as shown in Table 2. Table 2 shows that the lowest moisture permeability value of 2.8 g/m²·day was reported by the 10 wt.% PVOH/MMT-Na$^+$ film at test conditions of 85% RH and 40 °C, which increases linearly with decreasing wt.% of nano-clay. This means that the moisture permeation value of pristine PVOH can be further reduced by 86% under the same test conditions by the addition of 10% nano-clay. This further reduction in moisture permeability can be attributed to a result of the intercalation of the nano-clay platelets. Moreover, the dispersion of the nano-clay platelets creates hindrances to the diffusing molecules [44], therefore, these hindrances are referred to as the tortuous path, and hence the permeation is decreased.

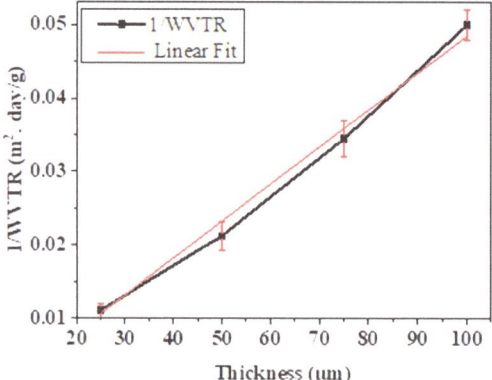

Figure 5. Blocking effect of pristine PVOH films with different thicknesses along with the linear fitting curve.

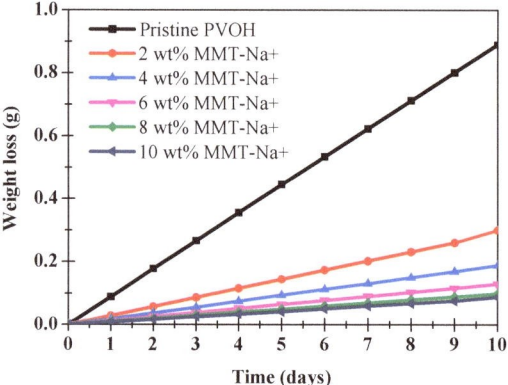

Figure 6. Weight loss versus time using a cup sealed with pristine PVOH and PVOH filled with MMT-Na$^+$ nano-clay films (100 µm thick) at 40 °C and 85% RH.

Table 2. Moisture permeation of pristine PVOH and PVOH/MMT-Na$^+$ films (100 µm thick) at 40 °C and 85% RH. For statistical analysis, at least five samples were tested for each measurement.

Films	WVTR (g/m^2·Day)	Permeability (g·cm/(m^2·Day))
Pristine PVOH	20.5 ± 2.5	2.05×10^{-1}
PVOH/MMT-Na$^+$ (2 wt.%)	10 ± 1.3	1.1×10^{-1}
PVOH/MMT-Na$^+$ (4 wt.%)	6 ± 0.5	6×10^{-2}
PVOH/MMT-Na$^+$ (6 wt.%)	4 ± 0.3	4×10^{-2}
PVOH/MMT-Na$^+$ (8 wt.%)	3.2 ± 0.3	3.2×10^{-2}
PVOH/MMT-Na$^+$ (10 wt.%)	2.8 ± 0.2	2.8×10^{-2}

The current experimental data are in good agreement with Bharadwaj's model of permeation for nanocomposites [45–47]. This permeation model suggests that the tortuous path of gas is influenced by film thickness, fractional volume, and aspect ratio of the particles with the ordered parameter "S". This ordered parameter is based on the orientation of the particles within the polymer matrix. In the present work, MMT-Na$^+$ nano-clay with an aspect ratio of 500 was used as reported in the literature [25,44]. For a semi-exfoliated structure and the calculations based on this aspect ratio using Equation (2), Bhardwaj's model [26] fit well.

$$\frac{P_s}{P_p} = \frac{1 - o_s}{1 + \frac{L}{2W}o_s\left(\frac{2}{3}\right)\left(S + \frac{1}{2}\right)} \quad (2)$$

where P_s and P_p represent the permeability of the composite and permeability of the polymer, respectively. $ø_s$ represents the volume fraction of the nano-clay. L and W represent the length and width of the clay platelets (L/W is the aspect ratio), and S is the order parameter. Experimental results and the theoretical data suggested by Bhardwaj's model are almost similar or in close tolerance, as shown in Figure 7, which corroborate the experimental results.

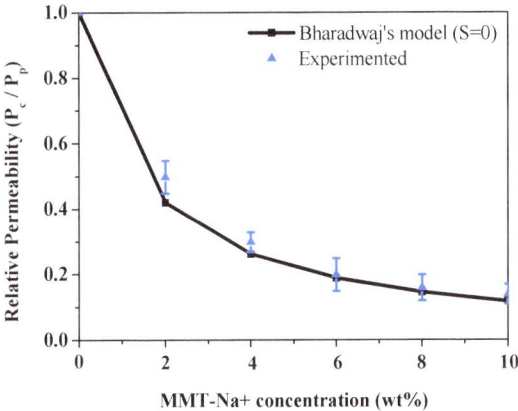

Figure 7. Comparison of the experimental and theoretically calculated data of pristine PVOH and PVOH/MMT-Na$^+$ films.

To check the barrier quality of PVOH/MMT-Na$^+$ films under bending conditions, these were subjected to a maximum of 10,000 bending cycles at a radius of 3 cm and the obtained results are shown in Figure 8. The graph shows that pristine PVOH maintained its barrier properties even after 10,000 bending cycles and a similar trend in barrier properties were demonstrated by respective PVOH/MMT-Na$^+$ films under the same bending cycles. This suggested that the nano-clay platelets, even 10 wt.%, held strongly within the matrix of PVOH, therefore showed no loss in barrier quality [3].

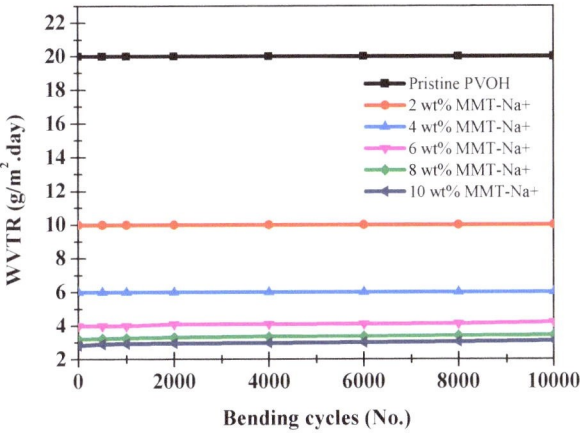

Figure 8. Moisture permeation versus bending cycles of pristine PVOH and PVOH/MMT-Na$^+$ films.

4. Conclusions

A solution-processed route was adopted for the development of a good barrier film using pristine PVOH and MMT-Na$^+$ as dispersants, which is not only a cost-effective route but is also safe for the environment. The developed films exhibited high transparency of ~92% in the white light region and the presence of nano-clay did not affect the transparency of the films and gave similar results to that of the pristine PVOH films. The addition of nano-clay caused the scattering of light due to the agglomeration of particles and found only less than 2% in the visible region. Increasing the percentage of nano-clay from 2 wt.% to 10 wt.% in PVOH resulted in the reduction of moisture permeation to a maximum of 86% when compared with pristine PVOH film, but on the other hand, it becomes very hard to distribute the nano-clay uniformly in the matrix, hence, in that perspective composite, 6 wt.% of MMT-Na$^+$ clay is the best choice. Barriers with 6 wt.% clay are not only easy to process, but also remain transparent and improve the barrier characteristics of PVOH against the diffusion of moisture by over 80%. The experimental data were also validated using Bharadwaj's permeation model and the results were in good agreement with the theoretical calculations. The developed films also depicted excellent flexibility and even after 10,000 bending cycles, the same barrier quality was maintained. All the reported results make PVOH/MMT-Na$^+$ films a potential candidate for various packaging applications such as optoelectronics, foods, etc.

Author Contributions: Conceptualization, I.A.C.; Data curation, T.A.; Formal analysis, S.H.S.; Funding acquisition, R.G.A.; Investigation, S.K.; Methodology, M.R.; Project administration, M.D.A.; Resources, M.A.M.; Software, S.A.; Validation, M.S.J.; Visualization, S.H.; Writing—original draft, A.D.C.; Writing—review & editing, M.S. All authors have read and agreed to the published version of the manuscript.

Funding: This work was funded by the Researchers Supporting Project Number (RSP-2021/267) King Saud University, Riyadh, Saudi Arabia.

Institutional Review Board Statement: Not applicable.

Informed Consent Statement: Not applicable.

Data Availability Statement: Data sharing not applicable.

Conflicts of Interest: The authors declare no conflict of interest.

References

1. Siracusa, V. Food packaging permeability behaviour: A Report. *Int. J. Polym. Sci.* **2012**, *2012*, 302029. [CrossRef]
2. Figura, L.O.; Teixeira, A.A. *Food Physics, Physical Proerties-Measurements and Applicartions*; Springer: Berlin/Heidelberg, Germany, 2008; Volume 39, pp. 561–563.
3. Channa, I.A.; Distler, A.; Egelhaaf, H.; Brabec, C.J. Solution Coated Barriers for Flexible Electronics. In *Organic Flexible Electronics, Fundamentals, Devices, and Applications*; Cosseddu, P., Caironi, M., Eds.; Woodhead Publishing: Sawston, UK, 2020.
4. Channa, I.A.; Distler, A.; Zaiser, M.; Brabec, C.J.; Egelhaaf, H. Thin film encapsulation of organic solar cells by direct deposition of polysilazanes from solution. *Adv. Energy Mater.* **2019**, *9*, 1900598. [CrossRef]
5. Cooksey, K. *Important Factors for Selecting*; Packag: Boston, MA, USA, 2004; pp. 1–12.
6. Geueke, B.; Groh, K.; Muncke, J. Food packaging in the circular economy: Overview of chemical safety aspects for commonly used materials. *J. Clean. Prod.* **2018**, *193*, 491–505. [CrossRef]
7. Channa, I.A. *Development of Solution Processed Thin Film Barriers for Encapsulating Thin Film Electronics*; Friedrich Alexander University of Erlangen Nuremberg: Bavaria, Germany, 2019.
8. Raheem, D. Application of plastics and paper as food packaging materials? An overview. *Emir. J. Food Agric.* **2013**, *25*, 177. [CrossRef]
9. Eustace, I.J. Some factors affecting oxygen transmission rates of plastic films for vacuum packaging of meat. *Int. J. Food Sci. Technol.* **2007**, *16*, 73–80. [CrossRef]
10. Majid, I.; Nayik, G.A.; Dar, S.M.; Nanda, V. Novel food packaging technologies: Innovations and future prospective. *J. Saudi Soc. Agric. Sci.* **2018**, *17*, 454–462. [CrossRef]
11. Silvestre, C.; Duraccio, D.; Cimmino, S. Food packaging based on polymer nanomaterials. *Prog. Polym. Sci.* **2011**, *36*, 1766–1782. [CrossRef]

12. Kim, H.M.; Lee, H.S. Water and oxygen permeation through transparent ethylene vinyl alcohol/(graphene oxide) membranes. *Carbon Lett.* **2014**, *15*, 50–56. [CrossRef]
13. Nazarenko, S.; Meneghetti, P.; Julmon, P.; Olson, B.; Qutubuddin, S. Gas barrier of polystyrene montmorillonite clay nanocomposites: Effect of mineral layer aggregation. *J. Polym. Sci. Part B Polym. Phys.* **2007**, *45*, 1733–1753. [CrossRef]
14. Granstrom, J.; Roy, A.; Rowell, G.; Moon, J.S.; Jerkunica, E.; Heeger, A.J. Improvements in barrier performance of perfluorinated polymer films through suppression of instability during film formation. *Thin Solid Films* **2010**, *518*, 3767–3771. [CrossRef]
15. Müller, K.; Bugnicourt, E.; Latorre, M.; Jorda, M.; Echegoyen Sanz, Y.; Lagaron, J.; Miesbauer, O.; Bianchin, A.; Hankin, S.; Bölz, U.; et al. Review on the processing and properties of polymer nanocomposites and nanocoatings and their applications in the packaging, automotive and solar energy fields. *Nanomaterials* **2017**, *7*, 74. [CrossRef]
16. Mokwena, K.K.; Tang, J. Ethylene vinyl alcohol: A review of barrier properties for packaging shelf stable foods. *Crit. Rev. Food Sci. Nutr.* **2012**, *52*, 640–650. [CrossRef]
17. Hammann, F.; Schmid, M. Determination and quantification of molecular interactions in protein films: A review. *Materials* **2014**, *7*, 7975–7996. [CrossRef] [PubMed]
18. Schmid, M.; Dallmann, K.; Bugnicourt, E.; Cordoni, D.; Wild, F.; Lazzeri, A.; Noller, K. Properties of whey-protein-coated films and laminates as novel recyclable food packaging materials with excellent barrier properties. *Int. J. Polym. Sci.* **2012**, *2012*, 562381. [CrossRef]
19. Schmid, M.; Zillinger, W.; Müller, K.; Sängerlaub, S. Permeation of water vapour, nitrogen, oxygen and carbon dioxide through whey protein isolate based films and coatings—Permselectivity and activation energy. *Food Packag. Shelf Life* **2015**, *6*, 21–29. [CrossRef]
20. Reig, C.S.; Lopez, A.D.; Ramos, M.H.; Ballester, V.A.C. Nanomaterials: A map for their selection in food packaging applications. *Packag. Technol. Sci.* **2014**, *27*, 839–866. [CrossRef]
21. Ahmad, A.; Jini, D.; Aravind, M.; Parvathiraja, C.; Ali, R.; Kiyani, M.Z.; Alothman, A. A novel study on synthesis of egg shell based activated carbon for degradation of methylene blue via photocatalysis. *Arab. J. Chem.* **2020**, *13*, 8717–8722. [CrossRef]
22. Yeo, J.H.; Lee, C.H.; Park, C.-S.; Lee, K.-J.; Nam, J.-D.; Kim, S.W. Rheological, morphological, mechanical, and barrier properties of PP/EVOH blends. *Adv. Polym. Technol.* **2001**, *20*, 191–201. [CrossRef]
23. Siracusa, V.; Ingrao, C.; Giudice, A.L.; Mbohwa, C.; Rosa, M.D. Environmental assessment of a multilayer polymer bag for food packaging and preservation: An LCA approach. *Food Res. Int.* **2014**, *62*, 151–161. [CrossRef]
24. Hahladakis, J.N.; Iacovidou, E. Closing the loop on plastic packaging materials: What is quality and how does it affect their circularity? *Sci. Total Environ.* **2018**, *630*, 1394–1400. [CrossRef] [PubMed]
25. Gaume, J.; Taviot-Gueho, C.; Cros, S.; Rivaton, A.; Thérias, S.; Gardette, J.L. Optimization of PVA clay nanocomposite for ultra-barrier multilayer encapsulation of organic solar cells. *Sol. Energy Mater. Sol. Cells* **2012**, *99*, 240–249. [CrossRef]
26. Strawhecker, K.E.; Manias, E. Nanocomposites based on water soluble polymers and unmodified smectite clays. *Polym. Nanocompos.* **2006**, *206*, 20–33.
27. Ahmad, J.; Bazaka, K.; Anderson, L.J.; White, R.; Jacob, M. Materials and methods for encapsulation of OPV: A review. *Renew. Sustain. Energy Rev.* **2013**, *27*, 104–117. [CrossRef]
28. Atai, M.; Solhi, L.; Nodehi, A.; Mirabedini, S.M.; Kasraei, S.; Akbari, K.; Babanzadeh, S. PMMA-grafted nanoclay as novel filler for dental adhesives. *Dent. Mater.* **2009**, *25*, 339–347. [CrossRef]
29. Nikolaidis, A.K.; Achilias, D.S.; Karayannidis, G.P. Synthesis and characterization of PMMA/organomodified montmorillonite nanocomposites prepared by in situ bulk polymerization. *Ind. Eng. Chem. Res.* **2011**, *50*, 571–579. [CrossRef]
30. Seethamraju, S.; Ramamurthy, P.; Madras, G. Performance of an ionomer blend-nanocomposite as an effective gas barrier material for organic devices. *RSC Adv.* **2014**, *4*, 11176–11187. [CrossRef]
31. Hong, S.I.; Lee, H.; Rhim, J. Effects of clay type and content on mechanical, water barrier and antimicrobial properties of agar-based nanocomposite films. *Carbohydr. Polym.* **2011**, *86*, 691–699.
32. Kurtz, S. *Photovoltaic Module Reliability Workshop 2012: February 28–March 1, 2012*; Office of Scientific and Technical Information (OSTI): Zhenjiang, China, 2013.
33. Carosio, F.; Colonna, S.; Fina, A.; Rydzek, G.; Hemmerlé, J.; Jierry, L.; Schaaf, P.; Boulmedais, F. Efficient gas and water vapor barrier properties of thin poly(lactic acid) packaging films: Functionalization with moisture resistant nafion and clay multilayers. *Chem. Mater.* **2014**, *26*, 5459–5466. [CrossRef]
34. Tsai, T.-Y.; Lin, M.-J.; Chuang, Y.-C.; Chou, P.-C. Effects of modified clay on the morphology and thermal stability of PMMA/clay nanocomposites. *Mater. Chem. Phys.* **2013**, *138*, 230–237. [CrossRef]
35. Dabbaghianamiri, M.; Duraia, E.-S.M.; Beall, G.W. Self-assembled Montmorillonite clay-poly vinyl alcohol nanocomposite as a safe and efficient gas barrier. *Results Mater.* **2020**, *7*, 100101. [CrossRef]
36. Tsurko, E.S.; Feicht, P.; Habel, C.; Schilling, T.; Daab, M.; Rosenfeldt, S.; Breu, J. Can high oxygen and water vapor barrier nanocomposite coatings be obtained with a waterborne formulation? *J. Membr. Sci.* **2017**, *540*, 212–218. [CrossRef]
37. Abdullah, Z.W.; Dong, Y.; Han, N.; Liu, S. Water and gas barrier properties of polyvinyl alcohol (PVA)/starch (ST)/glycerol (GL)/halloysite nanotube (HNT) bionanocomposite films: Experimental characterisation and modelling approach. *Compos. Part B Eng.* **2019**, *174*, 1. [CrossRef]
38. Rangreez, T.A.; Mobin, R. 13-Polymer composites for dental fillings. In *Woodhead Publishing Series in Biomaterials [Internet]*; Woodhead Publishing: Sawston, UK, 2019; pp. 20–24.

39. Tian, Y.; Zhu, P.; Zhou, M.; Lin, Y.; Cheng, F. Effect of Microfibrillated cellulose loading on physical properties of starch/polyvinyl Al-cohol composite films. *J. Wuhan Univ. Technol. Mater. Sci. Ed.* **2020**, *35*, 825–831. [CrossRef]
40. Gao, X.; Tang, K.; Liu, J.; Zheng, X.; Zhang, Y. Compatibility and properties of biodegradable blend films with gelatin and poly(vinyl al-cohol). *J. Wuhan Univ. Technol. Sci. Ed.* **2014**, *29*, 351–356. [CrossRef]
41. Carrera, M.C.; Erdmann, E.; Destéfanis, H.A. Preparation of Poli (Vinylalcohol)/Organoclay Nanocomposites by Casting and in Situ Polymerization. In Proceedings of the 15th European Conference on Composite Materials, Venice, Italy, 24–28 June 2012.
42. Reyes, Y.; Peruzzo, P.J.; Fernández, M.; Paulis, M.; Leiza, J.R. Encapsulation of clay within polymer particles in a high-solids content aqueous dispersion. *Langmuir* **2013**, *29*, 9849–9856. [CrossRef] [PubMed]
43. Wu, P.; Fisher, A.; Foo, P.; Queen, D.; Gaylor, J. In vitro assessment of water vapour transmission of synthetic wound dressings. *Biomaterials* **1995**, *16*, 171–175. [CrossRef]
44. Manias, E.; Touny, A.; Wu, L.; Strawhecker, K.; Lu, B.; Chung, T.C. Polypropylene/montmorillonite nanocomposites. Review of the synthetic routes and materials properties. *Chem. Mater.* **2001**, *13*, 3516–3523. [CrossRef]
45. Bharadwaj, R.K. Modeling the barrier properties of polymer-layered silicate nanocomposites. *Macromolecules* **2001**, *34*, 9189–9192. [CrossRef]
46. Channa, I.A.; Distler, A.; Scharfe, B.; Feroze, S.; Forberich, K.; Lipovšek, B.; Brabec, C.J.; Egelhaaf, H.-J. Solution processed oxygen and moisture barrier based on glass flakes for encapsulation of organic (opto-) electronic devices. *Flex. Print. Electron.* **2021**, *6*, 025006. [CrossRef]
47. Channa, I.; Chandio, A.; Rizwan, M.; Shah, A.; Bhatti, J.; Shah, A.; Hussain, F.; Shar, M.; AlHazaa, A. Solution processed PVB/mica flake coatings for the encapsulation of organic solar cells. *Materials* **2021**, *14*, 2496. [CrossRef]

Review

Microstructure and Oxidation Behavior of Anti-Oxidation Coatings on Mo-Based Alloys through HAPC Process: A Review

Tao Fu, Kunkun Cui, Yingyi Zhang *, Jie Wang, Xu Zhang, Fuqiang Shen, Laihao Yu and Haobo Mao

School of Metallurgical Engineering, Anhui University of Technology, Maanshan 243002, China; ahgydxtaofu@163.com (T.F.); 15613581810@163.com (K.C.); wangjiemaster0101@outlook.com (J.W.); zx13013111171@163.com (X.Z.); sfq19556630201@126.com (F.S.); aa1120407@126.com (L.Y.); L1499923420@163.com (H.M.)
* Correspondence: zhangyingyi@cqu.edu.cn; Tel.: +86-173-7507-6451

Abstract: Mo and Mo-based alloys are important aerospace materials with excellent high temperature mechanical properties. However, their oxidation resistance is very poor at high temperature, and the formation of volatile MoO_3 will lead to catastrophic oxidation failure of molybdenum alloy components. Extensive research on the poor oxidation problem has indicated that the halide activated pack cementation (HAPC) technology is an ideal method to solve the problem. In this work, the microstructure, oxide growth mechanism, oxidation characteristics, and oxidation mechanism of the HAPC coatings were summarized and analyzed. In addition, the merits and demerits of HPAC techniques are critically examined and the future scope of research in the domain is outlined.

Keywords: molybdenum alloys; coating; oxidation; microstructure; mechanism; review

Citation: Fu, T.; Cui, K.; Zhang, Y.; Wang, J.; Zhang, X.; Shen, F.; Yu, L.; Mao, H. Microstructure and Oxidation Behavior of Anti-Oxidation Coatings on Mo-Based Alloys through HAPC Process: A Review. *Coatings* **2021**, *11*, 883. https://doi.org/10.3390/coatings11080883

Academic Editor: Pier Luigi Bonora

Received: 24 June 2021
Accepted: 21 July 2021
Published: 23 July 2021

Publisher's Note: MDPI stays neutral with regard to jurisdictional claims in published maps and institutional affiliations.

Copyright: © 2021 by the authors. Licensee MDPI, Basel, Switzerland. This article is an open access article distributed under the terms and conditions of the Creative Commons Attribution (CC BY) license (https://creativecommons.org/licenses/by/4.0/).

1. Introduction

Mo and its alloys have high melting point, excellent high-temperature mechanical properties, low thermal expansion coefficient and high conductivity and thermal conductivity, which have been widely used in high-temperature structural components in national defense industry, aerospace, and other fields, such as nozzle throat, high temperature electrode, high-temperature heating element, ray shielding material, etc. [1–4]. However, the oxidation resistance of Mo and Mo-based alloys is very poor, and they are easily oxidized to MoO_3 at a temperature of (400–800 °C) [5,6]. With the formation of MoO_3, the volume of molybdenum alloy increases rapidly, and leads to the occurrence of low temperature pulverization phenomenon, namely "Pesting oxidation". In addition, the formation of a large amount of volatile MoO_3 will lead to the catastrophic decomposition of molybdenum and its alloys when the oxidation temperature is greater than 1000 °C [7–10]. At present, alloying and surface coating can be used to improve the oxidation resistance of Mo and its alloys. Alloying is regarded as the preferred method to improve the properties of pure Mo, and Mo-based alloys have better mechanical properties than pure Mo when used at a high temperature above 1000 °C [11–14]. The classification, preparation method, properties, and application fields of molybdenum-based alloys are shown in Table 1 [15,16]. Because of limitations of alloying capability of Mo, its high temperature oxidation resistance cannot be fundamentally improved by alloying. Therefore, surface coating technology is regarded as an ideal method to improve the high-temperature oxidation resistance of molybdenum and its alloys [17–19]. Among them, the HAPC technology is the most widely used. At present, there are many reports about the oxidation behavior of HAPC coatings on Mo and its alloys. However, almost no reviewing of the progress in development of oxidation resistance of Mo has been documented [20].

Table 1. Classification and application of Mo-based alloys.

Alloy Type	Preparation Method	Brand Number	Performance	Application Field	Refs.
Mo-Cu alloy	Co-deposition method; Metal oxide co-reduction, etc.	Mo-Cu	Good conductivity, thermal conductivity, ablation resistance, high hardness and strength	Electrician and electronics, instrumentation, national defense and military industry, aerospace, etc.	[15]
Mo-Ti-Zr alloy	Powder metallurgy; Smelting process	TZM, TZC	Excellent high temperature strength, high recrystallization temperature, good heat conduction and corrosion resistance	It is widely used in aerospace fields, such as rocket nozzles, nozzle throat liners, valve bodies, gas pipelines, etc.	[16]
Mo-Re alloy	Powder metallurgy; Vacuum smelting	Mo-5Re, Mo-41Re, Mo-50Re	Excellent radiation resistance and high tensile strength, good manufacturability and high temperature creep resistance	Aerospace, nuclear energy, chemical, electronics, military and so on.	[17]
Rare earth Mo alloy	Powder metallurgy	Mo-0.5Ti-Y, Mo-La, etc.	Good toughness, high temperature resistance, good bending resistance and tensile strength	High temperature furnace heating elements, nuclear materials, glass melting electrodes, etc.	[18]

In this work, the advantages and disadvantages of the HAPC coatings are summarized and analyzed. The composition, exposure time, exposure temperature and mass change per unit area of the coatings have been given in relevant tables [21,22]. Their oxide growth mechanism and oxidation behavior are emphatically analyzed and summarized. Finally, the oxidation resistance and failure mechanism of the coatings are also summarized, aiming to provide some useful references for researchers in this field [23].

2. Microstructure and Growth Mechanism of HAPC Coatings

Halide activated pack cementation (HAPC) method is to embed the substrate into a mixture (Si powder, B powder, Al powder, NH_4Cl/F, Y, NaF, Al_2O_3 powder, etc.) with a certain particle size and proportion, and carry out thermal diffusion in vacuum or argon atmosphere to prepare diffusion coating. Figure 1 shows a schematic diagram of the HAPC reaction model. It can be seen that the Al_2O_3 crucible is usually used as the reaction device, and the plate on the top of the crucible plays the role of isolating air during the reaction process. The device is placed in a furnace and held at a set temperature for a certain period of time to obtain a coating on the substrate surface [24–29].

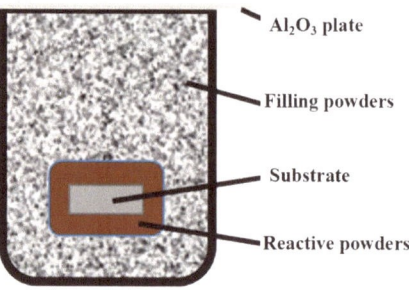

Figure 1. Schematic diagram of the reaction model of the HAPC method. Reprinted with permission from [29]; Reproduced from (Yang et al., 2018).

Table 2 provides a summary of the process parameters, composition, and properties of HAPC coating on molybdenum-based alloys [29–42]. It can be seen that the composition of the mixture and the process conditions have an important influence on the microstructure, phase composition, grain size, and mechanical properties of the coating. Wang et al. [32] successfully prepared $MoSi_2$-MoB coatings with an average hardness of 5.84 GPa on Mo surface, and the hardness of MoB layer is as high as 9.54 GPa. By contrast, the surface hardness of pure $MoSi_2$ coating is only 2.58 GPa [34]. This is attributed to the addition of appropriate amount of B element in the mixed packing, which improves the fluidity of Si and makes the coating structure more dense and uniform [43]. The growth mechanism of HAPC coatings is shown in Figure 2. A $MoSi_2$ layer forms on molybdenum-based alloy, due to the interdiffusion reaction between Si powder and the substrate at high temperature. In the preparation process of coating, adding an appropriate amount of beneficial components (Al, B, YSZ, ZrO_2, Al_2O_3, SiC, MoB, etc.) can significantly improve the oxidation resistance and mechanical properties of the silicide coatings [44–49]. The main reaction equations involved in the above process are shown in Figure 2g.

Table 2. Summary of process, composition, and surface properties of HAPC coatings on molybdenum and its alloy.

Substrate	Composition and Particle Size of HAPC Material		Process Conditions		Composition and Thickness (μm)		Surface Hardness (GPa)	Grain Size of Coating Surface (μm)	Refs.
	Composition (wt.%)	Particle Size (μm)	Atmosphere	Treatment Time and Temperature	Outer Layer	Interface Layer			
Mo	C,Si,NaF	-	Air	1200 °C, 2 h	SiO_2-$MoSi_2$ (55)	Mo_5Si_3 (5)	-	-	[29]
	10Si-10NH_4F-80Al_2O_3	32.77	Ar	1300 °C, 10 h	Al_2O_3-$MoSi_2$ (60)	Mo_5Si_3 (1–2)	-	-	[30]
	10Si-10NH_4F-80SiO_2	17.09	Ar	1300 °C, 10 h	SiO_2-$MoSi_2$ (60)	Mo_5Si_3 (8–10)		10–20	
	10Si-10NH_4F-80SiC	4.87	Ar	1300 °C, 10 h	SiC-$MoSi_2$ (100)	-	-	8–10	
	16Si-4B-4NaF–76Al_2O_3	-	Ar	1200 °C, 5 h	$MoSi_2$ (55–59)	Mo_5Si_3-MoB-Mo_2B (15–20)	-	-	[31]
	20Si-0.8B-5NaF-74.2Al_2O_3	-	Ar	1000 °C, 10 h	$MoSi_2$ (27.2)	MoB (31)	5.84	-	[32]
	16Si-4B-4NaF-2Y-76Al_2O_3	-	Ar	1300 °C, 5 h	$MoSi_2$ (190)	Mo_5Si_3-MoB-Mo_2B (14)	-	-	[33]
TZM	7Si-87Al_2O_3-6NH_4Cl	-	Vacuum	1000 °C, 12 h	$MoSi_2$ (100)	Mo_5Si_3 (2–3)	2.58	-	[34]
	7Al-7Si-10NH_4F-76Al_2O_3	≤75	Ar	1100 °C, 17.5 h	$MoSi_2$ (100)	$Mo(Si,Al)_2$ (10)	-	-	[35]
	12Si-3B-6Al-2Y_2O_3-5NaF-72Al_2O_3	-	Ar	1250 °C, 8 h	$Mo(Si,Al)_2$ (92)	MoB-Mo_2B (2–5)	-	-	[36]
	7Al-7Si-10NH_4F-76Al_2O_3	1–2	Ar	1300 °C, 10 h	$Mo(Si,Al)_2$ (38)	$Mo_3(Al,Si)$-$Mo_5(Si,Al)_3$ (5–7)	-	0.27	[37]
	70Al_2O_3-20Al-10NH_4Cl	-	Ar	1000 °C, 12 h	$Al_2(MoO_4)_3$ (20)	Al_5Mo-Al_7Mo_4 (30)	2.58	-	[38]
	25Si-5NaF-70Al_2O_3	-	Ar	1100 °C, 6 h	$MoSi_2$ (35)	Mo_5Si_3 (2)	-	-	[39]

Table 2. Cont.

Substrate	Composition and Particle Size of HAPC Material		Process Conditions		Composition and Thickness (μm)		Surface Hardness (GPa)	Grain Size of Coating Surface (μm)	Refs.
	Composition (wt.%)	Particle Size (μm)	Atmosphere	Treatment Time and Temperature	Outer Layer	Interface Layer			
	81Al$_2$O$_3$-7Si-7Al-5NH$_4$F	≤75	Ar	800–1000 °C, 8–36 h	Mo(Si,Al)$_2$ (20)	MoSi$_2$-Mo$_5$Si$_3$ (35)	-	-	[40]
Mo-30W	7Al-7Si-5NH$_4$F-81Al$_2$O$_3$	≤75	Ar	1000 °C, 16 h	Al-rich (12)	(Mo,W)Si$_2$ (46)	-	-	[41]
Mo-9Si-8B	34.03Si-0.97B-2.5Na-62.5Al$_2$O$_3$	-	Ar	1450 °C, 8 h	MoSi$_2$-Mo$_5$Si$_3$ (70)	Mo$_5$SiB$_2$-MoB (10–15)	-	-	[42]

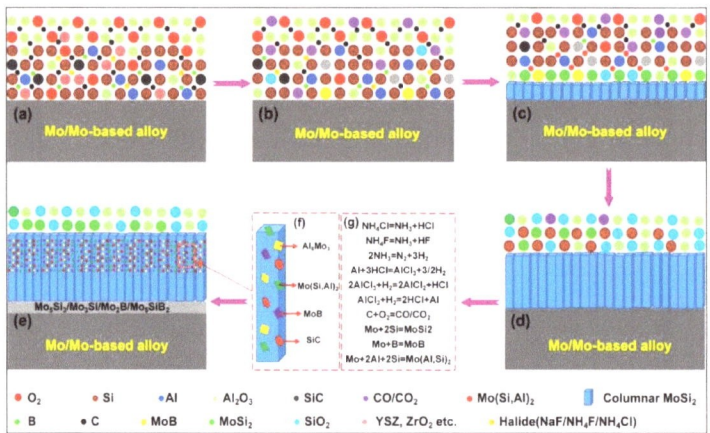

Figure 2. Growth mechanism diagram of HAPC coating on molybdenum and its alloys. (**a**) pre-sedimentation, (**b**) Oxygen consumption in the system, (**c**) The initial phase of the reaction, (**d**) The growth stage of the coating, (**e**) The final structure of the coating, (**f**) Amplification of individual coating grains, (**g**) The reactions involved in the above process.

The typical surface and cross-sectional morphology of the HAPC coatings are shown in Figure 3 [30]. It can be seen that some large particles are accumulated at the folds on the coating surface, which is mainly due to the large particle size of the mixture and uneven mixing, as shown in Figure 3a,c. The formation of micro-cracks is attributed to the thermal expansion mismatch between the coating and the substrate, as shown in Figure 3b,c. In addition, the addition of Al$_2$O$_3$, SiO$_2$, and SiC has great influence on the thickness of coating and interface layer. It is shown in Figure 3e that the SiO$_2$-MoSi$_2$ coating has a thick interface layer with a thickness of 8 to 10 μm. On the contrary, the thickness of interface layer of Al$_2$O$_3$-MoSi$_2$ coating is very thin and is only 1 to 2 μm. It is worth noting that the SiC-MoSi$_2$ coating has the thickest coating thickness, and the interface layer is not observed, as shown in Figure 3f. It can be seen that the addition of SiC significantly improves the deposition efficiency of the coating, which is mainly due to the physical deposition of SiC being faster than thermal diffusion deposition.

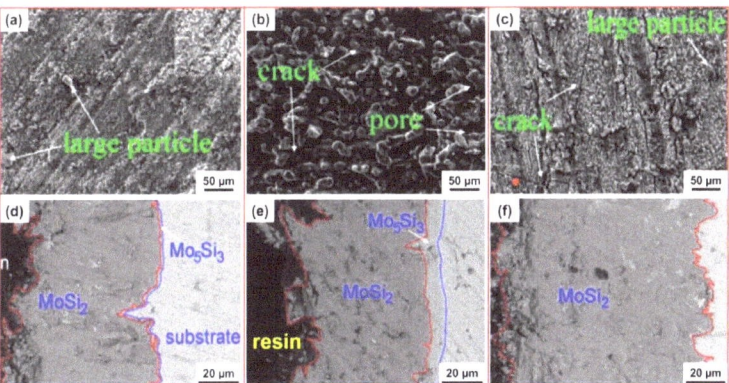

Figure 3. Surface and cross-sectional back scatter electronic (BSE) images of MoSi$_2$ coatings prepared with Al$_2$O$_3$ (**a,d**), SiO$_2$ (**b,e**), and SiC (**c,f**) filler in the pack, respectively. Reprinted with permission from [30]; Reproduced from (Sun et al., 2016).

3. Oxidation Behavior and Failure Mechanism of HAPC Coatings

The oxidation process parameters and oxidation properties of HAPC coatings on molybdenum and its alloy are shown in Table 3. It can be seen that the ingredients of the oxide layer mainly depend on the phase composition of the original coating. The oxide layer of pure MoSi$_2$ coatings is mainly composed of SiO$_2$ and Mo$_5$Si$_3$, however, composite coatings are mainly composed of SiO$_2$, Al$_2$O$_3$, and B$_2$O$_3$. The thickness of the outer coating decreases gradually with the increase of oxidation time. On the contrary, the thickness of the oxide layer and interface coating increases gradually, which is mainly due to the diffusion of oxygen and the internal diffusion reaction of silicon [29–42]. Figure 4 shows the typical surface and cross-section BSE images of oxidized coatings on TZM alloy. A large number of micro-cracks and MoO$_3$ particles are observed on the surface of the oxidized MoSi$_2$ coating, which is mainly due to the volume expansion of MoO$_3$, as shown in Figure 4a. However, the surface structure of Mo(Si,Al)$_2$ coating is complete without obvious cracks and pores after oxidation. This is due to the formation of Al$_2$O$_3$ with small thermal expansion coefficient during oxidation, which inhibits the pulverization of MoSi$_2$, as shown in Figure 5b [39]. The surface of the oxidized Mo(Si,Al)$_2$-MoB coating is very smooth and the oxide layer is clearly visible. This is attributed to the fact that SiO$_2$-Al$_2$O$_3$ generated in the oxidation process inhibits the volatilization of MoO$_3$, and MoB particles are dispersed and distributed inside the MoSi$_2$ coating, which makes the coating have a denser microstructure and better oxidation resistance, as shown in Figure 4c [38].

Table 3. Oxidation process parameters and oxidation properties of HAPC coatings on molybdenum and its alloys.

Substrate	Composition and Thickness of Coatings (μm)		Exposure	Composition and Thickness of Oxidized Coatings (μm)			Mass Gain (mg·cm^{-2}, wt.%)	Refs.
	Outer Layer	Interface Layer		Oxide Layer	Intermediate Layer	Interface Layer		
Mo	SiO_2-$MoSi_2$ (55)	Mo_5Si_3 (5)	1600 °C, 1 h	SiO_2-Mo_5Si_3	$MoSi_2$	Mo_5Si_3	9.86	[29]
	Al_2O_3-$MoSi_2$ (60)	Mo_5Si_3 (1–2)	1200 °C, 110 h	Al_2O_3 (16)	$MoSi_2$ (56)	Mo_5Si_3 (42)	0.15%	[30]
	SiO_2-$MoSi_2$ (60)	Mo_5Si_3 (8–10)	1200 °C, 110	SiO_2 (8–10)	$MoSi_2$ (32)	Mo_5Si_3 (45)	0.05%	
	SiC-$MoSi_2$ (100)	-	1200 °C, 110 h	SiO_2 (25–30)	$MoSi_2$ (64)	Mo_5Si_3 (40)	0.28%	
	$MoSi_2$ (55–59)	Mo_5Si_3-MoB-Mo_2B (15–20)	1250 °C, 100 h	SiO_2-B_2O_3-MoO_3 (100)	Mo_5SiB_2 (20)	Mo_5Si_3-MoB-Mo_2B (50)	3.25	[31]
	$MoSi_2$ (27.2)	MoB (31)	1300 °C, 80 h	SiO_2 (6-8)	$MoSi_2$-Mo_5Si_3 (50)	MoB-Mo_2B (38)	0.34	[32]
	$MoSi_2$ (190)	Mo_5Si_3-MoB-Mo_2B (14)	1000 °C, 100 h, 1 h cycles	SiO_2 (16)	$MoSi_2$ (80)	Mo_5Si_3-MoB-Mo_2B (32)	1.33×10^{-3}	[33]
TZM	$MoSi_2$ (100)	Mo_5Si_3 (2–3)	1200 °C, 55 h	-	-	-	0.15	[34]
	$MoSi_2$ (100)	Mo(Si, Al)$_2$ (10)	1100 °C, 250 h, 0.5 h cycles	SiO_2-Al_2O_3 (5–8)	$MoSi_2$ (45)	Mo(Si,Al)$_2$ (20)	0.08	[35]
	Mo(Si, Al)$_2$ (92)	MoB-Mo_2B (2–5)	1400 °C, 25 h	SiO_2-Al_2O_3 (15–20)	$MoSi_2$-MoB (88)	Mo_5Si_3-Mo_2B (28)	2.38	[36]
	Mo(Si, Al)$_2$ (38)	Mo_3(Al,Si)-Mo_5(Si, Al)$_3$ (5–7)	1100 °C, 10 h	SiO_2-Al_2O_3 (2.5)	Mo(Si,Al)$_2$ (50)	Mo_3(Al,Si)-Mo_5(Si,Al)$_3$ (20)	12.92	[37]
	Al_2(MoO$_4$)$_3$ (20)	Al_5Mo-Al_7Mo_4 (30)	1200 °C, 50 h	-	-	-	0.15	[38]
	$MoSi_2$ (35)	Mo_5Si_3 (2)	1350 °C, 20 h	SiO_2 (1–2)	$MoSi_2$ (20)	Mo_5Si_3 (50)	0.16	[39]
	Mo(Si, Al)$_2$ (20)	$MoSi_2$-Mo_5Si_3 (35)	1300 °C, 72 h	SiO_2-Al_2O_3 (5–10)	-	-	0.694	[40]
Mo-30W	Al-rich (12)	(Mo,W)Si$_2$ (46)	1100 °C, 15 h	-	-	-	-	[41]
Mo-9Si-8B	$MoSi_2$-Mo_5Si_3 (70)	Mo_5SiB_2-MoB (10–15)	1300 °C, 100 h	SiO_2-B_2O_3 (25)	$MoSi_2$-Mo (40)	Mo_5Si_3-MoB-Mo_5SiB_2 (75)	3.82	[42]

In addition, Sun et al. compared the oxidation kinetics curves of the pure Mo and the $MoSi_2$ coatings prepared with Al_2O_3, SiO_2, and SiC filler in the pack, respectively, at different temperatures, as shown in Figure 5 [30]. It can be seen that after oxidizing at 500 °C for 110h, the quality of the deposited coating only changes from −0.18%–0.09%. Among them, the quality of the deposited SiC coating has little change before and after oxidation. However, the mass variation of Mo substrate under this condition is as high as 3%, as shown in Figure 5a,b. Furthermore, in the oxidation experiments at 1200 °C, the Mo substrate rapidly failed in the initial stage of oxidation, and the quality of the deposited Al_2O_3, SiO_2 coatings are Increases of varying degrees, respectively. However, the mass of deposited SiC coating increases first and then decreases under the oxidation conditions. This is caused by the formation and evaporation of oxidized carbon (CO, CO_2) during oxidation, as shown in Figure 5c.

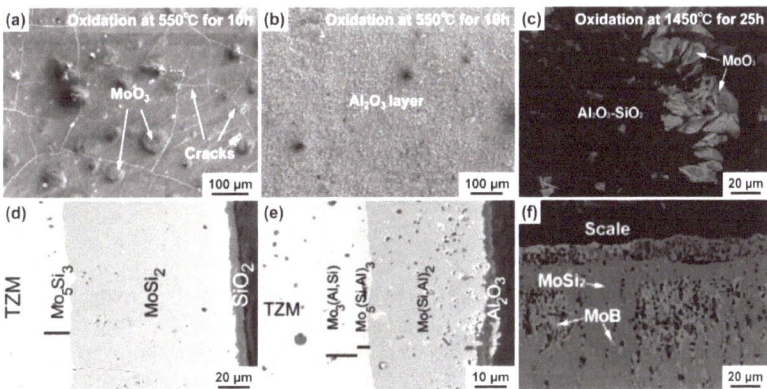

Figure 4. Surface and cross-section images of HAPC coatings at different Oxidation conditions; (**a,d**) MoSi$_2$ coating, (**b,e**) Mo(Si, Al)$_2$ coating. Reprinted with permission from [39]; Reproduced from (Paul et al., 2014). (**c,f**) Mo(Si,Al)$_2$-MoB coating. Reprinted with permission from [38]; Reproduced from (Chakraborty et al., 2008).

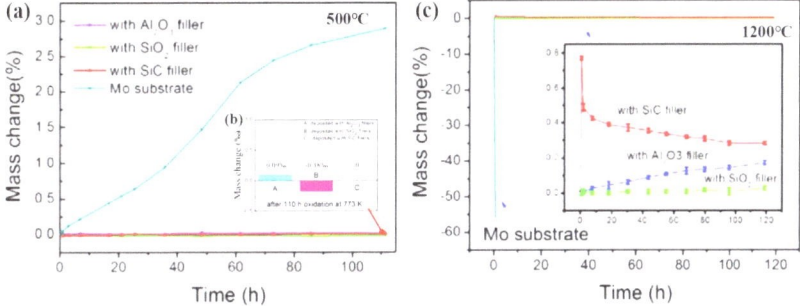

Figure 5. The oxidation kinetics in terms of time of exposure as well as temperature of Mo substrate and coatings at different temperatures, (**a**) 500 °C, (**b**) is the local amplification at 110 h as shown in (**a,c**) 1200 °C. Reprinted with permission from [30]; Reproduced from (Sun et al., 2016).

The microstructure evolution and oxidation mechanism of the coating are shown in Figure 6. Generally, the oxidation of the coating can be divided into transient oxidation and steady oxidation [50]. At the initial stage of oxidation, Mo, Si, B, Al, and other elements are oxidized at the same time. The quality loss of the coating is faster with the formation and volatilization of MoO$_3$, and the SiO$_2$ layer is discontinuous at this stage. At the same time, a large number of pores were observed on the coating surface after the evaporation of MoO$_3$. With the progress of oxidation reaction, continuous and dense oxide films (Al$_2$O$_3$, SiO$_2$, B$_2$O$_3$-SiO$_2$, etc.) gradually form on the coating surface. At last, a large number of pores are closed by SiO$_2$ with a low oxidation rate, and the oxidation process changes to a steady state [51,52].

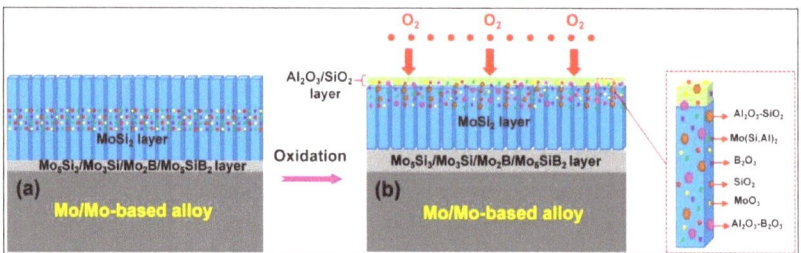

Figure 6. Oxidation mechanism diagram of HAPC coatings on molybdenum and its alloys. (**a**) Initial oxidation stage, (**b**) Late oxidation stage of coating.

4. Conclusions and Prospects

In this work, the growth mechanism, oxidation behavior, and mechanism of HAPC coatings are analyzed and discussed. The process conditions and properties of the coatings are provided. During the process, due to the relatively low process temperature of HAPC methods, the deposition efficiency of the coating is low, and the preparation time is long. However, the application of the processes is not limited by the shape of the substrate, and the coatings prepared have a uniform composition and good adhesion to the substrate. Figure 7 provides a summary of the composition and oxidation characteristics of protective coatings prepared by many researchers on the surface of molybdenum and its alloys by HAPC method.

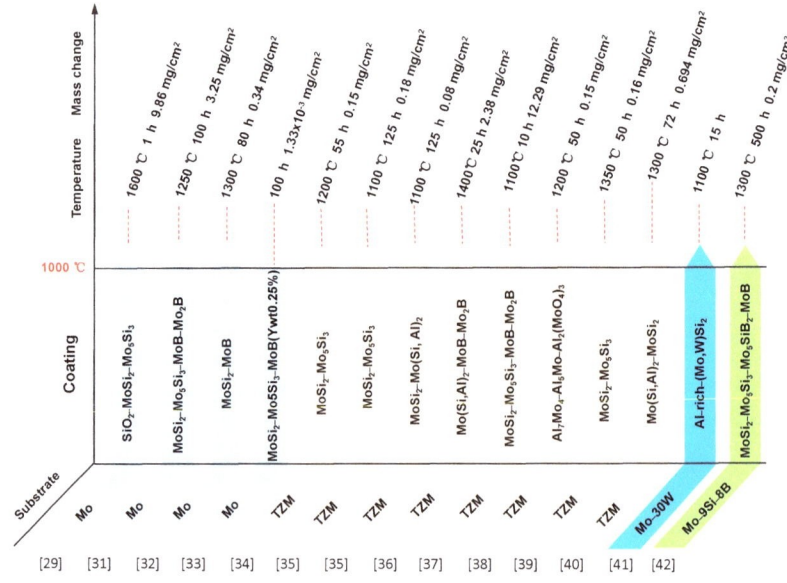

Figure 7. Overview of the composition and oxidation characteristics of HAPC coatings on the surface of Mo and its alloys.

In addition, the addition of beneficial elements and the second phases can not only significantly improve the mechanical properties and high temperature oxidation resistance of the coating, but also can delay the diffusion of silicon to the substrate, inhibit the formation of Mo_5Si_3 with poor oxidation resistance, reduce the formation of volatile MoO_3, and promote the rapid formation of continuous and dense anti-oxidation film on the coating surface. The addition of Al element can significantly inhibit the formation of the pulverization MoO_3 and the occurrence of pulverization, which is mainly due to the formation of Al_2O_3 with low thermal expansion coefficient. The addition of B element can improve the oxidation resistance of the coating mainly in the following two aspects. On the one hand, in the process of high temperature oxidation (above 1400 °C), an appropriate amount of B dissolved into the oxide layer can further improve the fluidity of SiO_2, which promotes the healing of pores and cracks. On the other hand, MoB reacts with Si to produce Mo_5SiB_2 with low diffusion coefficient, which delays the diffusion of Si to the substrate and improves the oxidation life of the coating. In addition, Cr and W elements can also improve the oxidation resistance of the coating, which is mainly due to the formation of Cr_2O_3 with high oxidation resistance, and the formation of (Mo,W) Si_2 phase inhibits the internal diffusion of silicon into the substrate.

In order to improve the mechanical properties and antioxidant properties of the coating, the microstructure and phase composition of the coating must be optimized, and the addition of the second reinforcing phase also plays a positive role. Moreover, the multi-step coating preparation process can organically combine the single coating preparation technology with each other, overcome the limitations of its single application, and effectively extend the oxidation service life of the coating. This will be the direction of future research and development in this field.

Author Contributions: The manuscript was written through contributions of all authors. Y.Z.: Conceptualization, Investigation, and Supervision. Y.Z. and T.F.: Writing original draft and image processing. T.F., K.C., L.Y. and J.W.: Validation, Resources, Investigation, Writing-review & editing. X.Z., H.M. and F.S.: Visualization, Writing-review & editing. All authors have given approval to the final version of the manuscript.

Funding: This research was funded by the National Natural Science Foundation of China, Grant No. 51604049.

Institutional Review Board Statement: Not applicable.

Informed Consent Statement: Not applicable.

Data Availability Statement: The study did not report any data.

Conflicts of Interest: The authors declare that they have no known competing financial interests or personal relationships that could have appeared to influence the work reported in this paper.

References

1. Chaudhuri, A.; Behera, A.N.; Sarkar, A.; Kapoor, R.; Ray, R.K.; Suwas, S. Hot deformation behaviour of Mo-TZM and understanding the restoration processes involved. *Acta Mater.* **2019**, *164*, 153–164. [CrossRef]
2. Tuzemen, C.; Yavas, B.; Akin, I.; Yucel, O.; Sahin, F.; Goller, G. Production and characterization of TZM based TiC or ZrC reinforced composites prepared by spark plasma sintering (SPS). *J. Alloys Compd.* **2019**, *781*, 433–439. [CrossRef]
3. Sharma, I.; Chakraborty, S.; Suri, A. Preparation of TZM alloy by aluminothermic smelting and its characterization. *J. Alloys Compd.* **2005**, *393*, 122–127. [CrossRef]
4. Majumdar, S.; Kale, G.; Sharma, I. A study on preparation of Mo-30W alloy by aluminothermic co-reduction of mixed oxides. *J. Alloys Compd.* **2005**, *394*, 168–175. [CrossRef]
5. Xu, J.; Yang, T.; Yang, Y.; Qian, Y.; Li, M.; Yin, X. Ultra-high temperature oxidation behavior of micro-laminated $ZrC/MoSi_2$ coating on C/C composite. *Corros. Sci.* **2018**, *132*, 161–169. [CrossRef]
6. Cui, K.K.; Fu, T.; Zhang, Y.Y.; Wang, J.; Mao, H.B.; Tan, T.B. Microstructure and mechanical properties of $CaAl_{12}O_{19}$ reinforced Al_2O_3-Cr_2O_3 composites. *J. Eur. Ceram. Soc.* **2021**, in press. [CrossRef]
7. Kong, H.; Kwon, H.S.; Kim, H.; Jeen, G.-S.; Lee, J.; Lee, J.; Heo, Y.S.; Cho, J.-H.; Jeen, H.; Kim, H. Reductive-annealing-induced changes in Mo valence states on the surfaces of MoO_3 single crystals and their high temperature transport. *Curr. Appl. Phys.* **2019**, *19*, 1379–1382. [CrossRef]
8. Peña-Bahamonde, J.; Wu, C.; Fanourakis, S.K.; Louie, S.M.; Bao, J.; Rodrigues, D.F. Rodrigues, Oxidation state of Mo affects dissolution and visible-light photocatalytic activity of MoO_3 nanostructures. *J. Catal.* **2020**, *381*, 508–519. [CrossRef]
9. Smolik, G.; Petti, D.; Schuetz, S. Oxidation and volatilization of TZM alloy in air. *J. Nucl. Mater.* **2000**, *283-287*, 1458–1462. [CrossRef]
10. Majumdar, S.; Kapoor, R.; Raveendra, S.; Sinha, H.; Samajdar, I.; Bhargava, P.; Chakravartty, J.; Sharma, I.; Suri, A. A study of hot deformation behavior and microstructural characterization of Mo-TZM alloy. *J. Nucl. Mater.* **2009**, *385*, 545–551. [CrossRef]

11. Zhang, Y.; Zhao, J.; Li, J.; Lei, J.; Cheng, X. Effect of hot-dip siliconizing time on phase composition and microstructure of Mo-MoSi$_2$ high temperature structural materials. *Ceram. Int.* **2019**, *45*, 5588–5593. [CrossRef]
12. Zhang, Y.; Hussain, S.; Cui, K.; Fu, T.; Wang, J.; Javed, M.S.; Lv, Y.; Aslam, B. Microstructure and mechanical properties of MoSi$_2$ coating deposited on Mo substrate by hot dipping processes. *J. Nanoelectron. Optoelectron.* **2019**, *14*, 1680–1685. [CrossRef]
13. Cui, K.; Zhang, Y.; Fu, T.; Hussain, S.; Algarni, T.S.; Wang, J.; Zhang, X.; Ali, S. Effects of Cr$_2$O$_3$ Content on Microstructure and Mechanical Properties of Al$_2$O$_3$ Matrix Composites. *Coatings* **2021**, *11*, 234. [CrossRef]
14. Liu, L.; Lei, H.; Gong, J.; Sun, C. Deposition and oxidation behaviour of molybdenum disilicide coating on Nb based alloys substrate by combined AIP/HAPC processes. *Ceram. Int.* **2019**, *45*, 10525–10529. [CrossRef]
15. Cui, Y.; Derby, B.; Li, N.; Misra, A. Fracture resistance of hierarchical Cu-Mo nanocomposite thin films. *Mater. Sci. Eng. A* **2021**, *799*, 139891. [CrossRef]
16. Zhang, Y.Y.; Cui, K.K.; Fu, T.; Wang, J.; Shen, F.Q.; Zhang, X.; Yu, L.H. Formation of MoSi$_2$ and Si/MoSi$_2$ coatings on TZM (Mo-0.5Ti-0.1Zr-0.02C) alloy by hot dip silicon-plating method. *Ceram. Int.* **2021**, *47*, 23053–23065. [CrossRef]
17. Mannheim, R.; Garin, J. Structural identification of phases in Mo–Re alloys within the range from 5 to 95% Re. *J. Mater. Process. Technol.* **2003**, *143-144*, 533–538. [CrossRef]
18. Li, W.; Wei, L.; He, J.; Chen, H.; Guo, H. The role of Re in improving the oxidation-resistance of a Re modified Pt-Al coating on Mo-rich single crystal superalloy. *J. Mater. Sci. Technol.* **2020**, *58*, 63–72. [CrossRef]
19. Zhang, Y.Y.; Qie, J.M.; Cui, K.K.; Fu, T.; Fan, X.L.; Wang, J.; Zhang, X. Effect of hot dip silicon-plating temperature on microstructure characteristics of silicide coating on tungsten substrate. *Ceram. Int.* **2020**, *46*, 5223–5228. [CrossRef]
20. Zhang, Y.; Cui, K.; Gao, Q.; Hussain, S.; Lv, Y. Investigation of morphology and texture properties of WSi$_2$ coatings on W substrate based on contact-mode AFM and EBSD. *Surf. Coat. Technol.* **2020**, *396*, 125966. [CrossRef]
21. Zhang, Y.; Li, Y.; Bai, C. Microstructure and oxidation behavior of Si-MoSi$_2$ functionally graded coating on Mo substrate. *Ceram. Int.* **2017**, *43*, 6250–6256. [CrossRef]
22. Zhang, Y.; Cui, K.; Fu, T.; Wang, J.; Qie, J.; Zhang, X. Synthesis WSi$_2$ coating on W substrate by HDS method with various deposition times. *Appl. Surf. Sci.* **2020**, *511*, 145551. [CrossRef]
23. Zhang, Y.; Fu, T.; Cui, K.; Shen, F.; Wang, J.; Yu, L.; Mao, H. Evolution of surface morphology, roughness and texture of tungsten disilicide coatings on tungsten substrate. *Vacuum* **2021**, *191*, 110297. [CrossRef]
24. Fu, T.; Cui, K.; Zhang, Y.; Wang, J.; Shen, F.; Yu, L.; Qie, J.; Zhang, X. Oxidation protection of tungsten alloys for nuclear fusion applications: A comprehensive review. *J. Alloys Compd.* **2021**, *884*, 161057. [CrossRef]
25. Majumdar, S.; Raveendra, S.; Samajdar, I.; Bhargava, P.; Sharma, I. Densification and grain growth during isothermal sintering of Mo and mechanically alloyed Mo-TZM. *Acta Mater.* **2009**, *57*, 4158–4168. [CrossRef]
26. Hu, D.; Fu, Q.; Zhou, L.; Zhang, Y.; Zhang, G. Stress design of a laminated MoSi$_2$/Cr coating under particle impact and high temperature environment. *Ceram. Int.* **2020**, *46*, 10696–10703. [CrossRef]
27. Pang, J.; Wang, W.; Zhou, C. Microstructure evolution and oxidation behavior of B modified MoSi$_2$ coating on Nb-Si based alloys. *Corros. Sci.* **2016**, *105*, 1–7. [CrossRef]
28. Sun, J.; Li, T.; Zhang, G.-P.; Fu, Q.-G. Different oxidation protection mechanisms of HAPC silicide coating on niobium alloy over a large temperature range. *J. Alloys Compd.* **2019**, *790*, 1014–1022. [CrossRef]
29. Yang, K.-M.; Wang, J.-X.; Yang, S.-Y.; Zhang, X.-L.; Luo, P. In-situ synthesis of MoSi$_2$ coating on Mo substrate under carbon protection and its short-term oxidation behavior. *Surf. Coat. Technol.* **2018**, *354*, 324–329. [CrossRef]
30. Sun, J.; Fu, Q.-G.; Guo, L.-P.; Liu, Y.; Huo, C.-X.; Li, H.-J. Effect of filler on the oxidation protective ability of MoSi$_2$ coating for Mo substrate by halide activated pack cementation. *Mater. Des.* **2016**, *92*, 602–609. [CrossRef]
31. Tian, X.; Guo, X.; Sun, Z.; Yin, Z.; Wang, L. Formation of B-modified MoSi$_2$ coating on pure Mo prepared through HAPC process. *Int. J. Refract. Met. Hard Mater.* **2014**, *45*, 8–14. [CrossRef]
32. Wang, Y.; Wang, D.; Yan, J. Preparation and characterization of MoSi$_2$/MoB composite coating on Mo substrate. *J. Alloys Compd.* **2014**, *589*, 384–388. [CrossRef]
33. Tian, X.; Guo, X.; Sun, Z.; Li, M.; Wang, L. Effects of Y$_2$O$_3$/Y on Si-B co-deposition coating prepared through HAPC method on pure molybdenum. *J. Rare Earths* **2016**, *34*, 952–957. [CrossRef]
34. Chakraborty, S.; Banerjee, S.; Sharma, I.; Suri, A. Development of silicide coating over molybdenum based refractory alloy and its characterization. *J. Nucl. Mater.* **2010**, *403*, 152–159. [CrossRef]
35. Majumdar, S. Formation of MoSi$_2$ and Al doped MoSi$_2$ coatings on molybdenum base TZM (Mo-0.5Ti-0.1Zr-0.02C) alloy. *Surf. Coat. Technol.* **2012**, *206*, 3393–3398. [CrossRef]
36. Zhang, P.; Guo, X.; Ren, X.; Chen, Z.; Shen, C. Development of Mo(Si,Al)$_2$-MoB composite coatings to protect TZM alloy against oxidation at 1400 °C. *Intermetallics* **2018**, *93*, 134–140. [CrossRef]
37. Majumdar, S.; Sharma, I. Oxidation behavior of MoSi$_2$ and Mo(Si, Al)$_2$ coated Mo-0.5Ti-0.1Zr-0.02C alloy. *Intermetallics* **2011**, *19*, 541–545. [CrossRef]
38. Chakraborty, S.; Banerjee, S.; Singh, K.; Sharma, I.; Grover, A.; Suri, A. Studies on the development of protective coating on TZM alloy and its subsequent characterization. *J. Mater. Process. Technol.* **2008**, *207*, 240–247. [CrossRef]
39. Paul, B.; Limaye, P.; Hubli, R.; Suri, A. Microstructure and wear properties of silicide based coatings over Mo-30W alloy. *Int. J. Refract. Met. Hard Mater.* **2014**, *44*, 77–83. [CrossRef]

40. Majumdar, S.; Sharma, I.; Raveendra, S.; Samajdar, I.; Bhargava, P. In situ chemical vapour co-deposition of Al and Si to form diffusion coatings on TZM. *Mater. Sci. Eng. A* **2008**, *492*, 211–217. [CrossRef]
41. Majumdar, S.; Sharma, I.; Suri, A. Development of oxidation resistant coatings on Mo–30W alloy. *Int. J. Refract. Met. Hard Mater.* **2008**, *26*, 549–554. [CrossRef]
42. Lange, A.; Heilmaier, M.; Sossamann, T.A.; Perepezko, J.H. Oxidation behavior of pack-cemented Si-B oxidation protection coatings for Mo-Si-B alloys at 1300 °C. *Surf. Coat. Technol.* **2015**, *266*, 57–63. [CrossRef]
43. Paul, B.; Kishor, J.; Majumdar, S.; Kain, V. Studies on growth mechanism of intermediate layer of (Mo,W)5Si3 and interdiffusion in the (Mo,W)-(Mo,W)Si$_2$ system prepared by pack cementation coating. *Surf. Interfaces* **2020**, *18*, 100458. [CrossRef]
44. Burk, S.; Gorr, B.; Christ, H.-J. High temperature oxidation of Mo-Si-B alloys: Effect of low and very low oxygen partial pressures. *Acta Mater.* **2010**, *58*, 6154–6165. [CrossRef]
45. Pan, Y. The structural, mechanical and thermodynamic properties of the orthorhombic TMAl (TM=Ti, Y, Zr and Hf) aluminides from first-principles calculations. *Vacuum* **2020**, *181*, 109742. [CrossRef]
46. Pan, Y.; Guan, W. The hydrogenation mechanism of PtAl and IrAl thermal barrier coatings from first-principles investigations. *Int. J. Hydrogen Energy* **2020**, *45*, 20032–20041. [CrossRef]
47. Zhang, X.; Fu, T.; Cui, K.; Zhang, Y.; Shen, F.; Wang, J.; Yu, L.; Mao, H. Coatings The Protection, Challenge, and Prospect of Anti-Oxidation Coating on the Surface of Niobium Alloy. *Coatings* **2021**, *11*, 742. [CrossRef]
48. Liu, Y.; Shao, W.; Wang, C.L.; Zhou, C.G. Microstructure and oxidation behavior of Mo-Si-Al coating on Nb-based alloy. *J. Alloys Compd.* **2018**, *735*, 2247–2255. [CrossRef]
49. Huang, L.; Pan, Y.; Zhang, J.; Du, Y.; Luo, F.; Zhang, S. CALPHAD-type modeling of the C–Hf–Mo system over the whole composition and temperature ranges. *Thermochim. Acta* **2020**, *692*, 178716. [CrossRef]
50. Qiao, Y.Q.; Kong, J.P.; Xi, Q.L.; Guo, P. Comparison of two kinds of Si-B-Y co-deposition coatings on an Nb-Ti-Si based alloy by pack cementation method. *Surf. Coat. Technol.* **2017**, *327*, 93–100. [CrossRef]
51. Paswan, S.; Mitra, R.; Roy, S. Oxidation behaviour of the Mo-Si-B and Mo-Si-B-Al alloys in the temperature range of 700–1300 °C. *Intermetallics* **2007**, *15*, 1217–1227. [CrossRef]
52. Tian, X.; Guo, X. Structure and oxidation behavior of Si-Y co-deposition coatings on an Nb silicide based ultrahigh temperature alloy prepared by pack cementation technique. *Surf. Coat. Technol.* **2009**, *204*, 313–318. [CrossRef]

Review

The Protection, Challenge, and Prospect of Anti-Oxidation Coating on the Surface of Niobium Alloy

Xu Zhang [1], Tao Fu [1], Kunkun Cui [1], Yingyi Zhang [1,*], Fuqiang Shen [1], Jie Wang [1], Laihao Yu [1] and Haobo Mao [2]

[1] School of Metallurgical Engineering, Anhui University of Technology, Maanshan 243002, China; zx13013111171@163.com (X.Z.); ahgydxtaofu@163.com (T.F.); 15613581810@163.com (K.C.); sfq19556630201@126.com (F.S.); wangjiemaster0101@outlook.com (J.W.); aa1120407@126.com (L.Y.)

[2] School of Civil Engineering and Architecture, Anhui University of Technology, Maanshan 243002, China; L1499923420@163.com

* Correspondence: zhangyingyi@cqu.edu.cn; Tel.: +86-173-7507-6451

Citation: Zhang, X.; Fu, T.; Cui, K.; Zhang, Y.; Shen, F.; Wang, J.; Yu, L.; Mao, H. The Protection, Challenge, and Prospect of Anti-Oxidation Coating on the Surface of Niobium Alloy. *Coatings* 2021, *11*, 742. https://doi.org/10.3390/coatings11070742

Academic Editor: Alexander Tolstoguzov

Received: 25 May 2021
Accepted: 20 June 2021
Published: 22 June 2021

Publisher's Note: MDPI stays neutral with regard to jurisdictional claims in published maps and institutional affiliations.

Copyright: © 2021 by the authors. Licensee MDPI, Basel, Switzerland. This article is an open access article distributed under the terms and conditions of the Creative Commons Attribution (CC BY) license (https://creativecommons.org/licenses/by/4.0/).

Abstract: Niobium (Nb)-based alloys have been extensively used in the aerospace field owing to their excellent high-temperature mechanical properties. However, the inferior oxidation resistance severely limits the application of Nb-based alloys in a high-temperature, oxygen-enriched environment. Related scholars have extensively studied the oxidation protection of niobium alloy and pointed out that surface coating technology is ideal for solving this problem. Based on the different preparation methods of Nb-based alloys' surface coatings, this article summarizes the relevant research of domestic and foreign scholars in the past 30 years, including the slurry sintering method (SS), suspension plasma spraying method (SPS), and halide activated pack cementation method (HAPC), etc. The growth mechanism and micromorphology of the coatings access by different preparation methods are evaluated. In addition, the advantages and disadvantages of various coating oxidation characteristics and coating preparation approaches are summarized. Finally, the coating's oxidation behavior and failure mechanism are summarized and analyzed, aiming to provide valuable research references in related fields.

Keywords: niobium alloy; oxidation resistance; surface coating; growth mechanism; oxidation behavior

1. Introduction

With the human need for space exploration, the development of hypersonic vehicles has attracted wide attention worldwide [1,2]. Due to the nature of long-term hypersonic cruises and the flight of hypersonic aircraft back and forth between the atmosphere and atmospheric reentry [3], the aircraft must face extremely harsh environments, producing high dynamic pressure and aerodynamic heating effects [4,5]. The high-temperature structural materials need to withstand extreme thermal and mechanical loads [6], resulting in large temperature gradients and thermal stresses inside the material, thereby significantly reducing the cycle life of the components. Especially critical parts or components include aircraft nose cones, sharp leading edges, nozzle openings, hot ends of engines [7], etc. Accordingly, the thermal development and high-temperature oxidation resistance of high-temperature structural materials are increasingly required. Traditional steel materials, aluminum alloys, and titanium alloys can no longer meet the extreme environmental requirements of hypersonic aircraft [8]. Niobium and its alloys have become a critical applicant material for high-temperature structural parts in the aerospace and nuclear industries due to their high melting point [9], moderate density, excellent high-temperature strength, and good processability. Niobium-based alloys are expected to replace nickel-based materials and become critical structural materials in the aerospace field by the end of the 21st century [10,11]. Nonetheless, the oxidation resistance of niobium-based alloys is lacking [12], and severe pulverization will occur when exposed to air above

500 °C for a short time, severely restricting its application in high-temperature, oxygen-rich environments. At present, the commonly used methods to inhibit the occurrence of this kind of oxidation include alloying and surface coating technology [13]. Although alloying can improve the corrosion resistance of niobium-based alloys in high temperature and oxygen-enriched environments to a certain extent, this measure often seriously affects the physical properties of the base alloy itself. Surface coating technology is the most effective method for enhancing the oxidation resistance of niobium-based alloys while ensuring the substrate's physical properties [13]. The arrangement of high-temperature, oxidation-resistant defensive coatings on the surface of niobium alloys has developed into a current research hotspot.

At present, there are many reports on the surface coating and oxidation protection of niobium and its alloys, but there are few studies on the growth mechanism and oxidation behavior of the surface coating. This article reviews the main preparation methods of niobium and its alloy surface coatings in recent years (such as the slurry sintering method (SS), suspension plasma spraying method (SPS), halide activated filling cementing method (HAPC), etc.). The latest research status of high-temperature, oxidation-resistant coatings on niobium-based alloys is discussed, and the advantages and disadvantages of various preparation methods are analyzed and summarized. The microscopic morphology, phase composition, and oxidation resistance of coatings prepared by different methods are compared and analyzed. The growth mechanism and oxidation behavior of various coatings are analyzed and summarized. At the same time, the future development direction of niobium and its alloy surface coatings is put forward with the purpose of adding valuable summaries for researchers on this ground.

2. Anti-Oxidation Coating on Niobium Alloys

The comprehensive properties of niobium alloy surface coatings are different due to the different preparation methods. In the following summary, different methods for preparing niobium alloy surface coatings will be described in detail, and the oxidation resistance performance of coatings prepared by different methods will be comprehensively compared.

2.1. Slurry Sintering Method

The slurry sintering method is the most commonly used coating preparation method on the surface of niobium alloys, and the process flow is shown in Figure 1. Firstly, the coating slurry is uniformly mixed with the components of the coating and the binder in proportion, to prepare the coating slurry [14,15], and the slurry is applied to the surface of the substrate by brushing, dipping, or spraying. Then it is solidified by pressurization and heating, and sintered in a vacuum or atmosphere furnace. Finally, a coating is formed on the face of the substrate. The process conditions, composition thickness [16], and the oxidation characteristics of the SS coatings on Nb-based alloys are summarized in Table 1. It can be perceived that the thickness of SS coatings was between 150 and 300 μm under the sintering temperature of 1200 to 1500 °C for 1–4 h in a vacuum or Ar atmosphere.

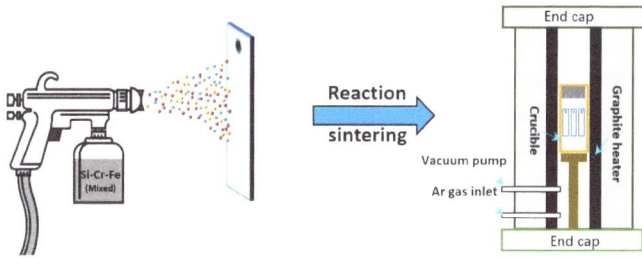

Figure 1. Flow chart of coating preparation by the slurry method.

Table 1. Summary for preparation and oxidation resistance of SS coatings on Nb-based alloys.

Substrate	Slurry Composition (wt.%)	Process Conditions	Coating Composition and Thickness (μm)		Oxidation System		Oxidation Products	Quality Change (mg/cm^2)	References
			Outer Layer	Interface Layer	Oxidation Temperature (°C)	Oxidation Time (h)			
C-103	60Si-15Fe-20Cr-5NaF	1400 °C/2 h vacuum	NbSi$_2$ Cr$_3$Si Fe$_3$Si$_2$ (200)	Nb$_5$Si$_3$ (30)	1300	2.5	Nb$_2$O$_5$ SiO$_2$ Fe$_2$O$_3$	1.5	[17]
R512E	20Si-35Fe-35Cr-10NaF	1250 °C/4 h vacuum	MoSi$_2$ M$_5$Si$_3$ (135)	Nb$_5$Si$_3$ (15)	1200	2	Nb$_2$O$_5$ Cr$_2$O$_3$ SiO$_2$ CrNbO$_4$	1.8	[18]
Nb521	45Mo-45Si-10NaF	1500 °C/1 h Ar	NbSi$_2$ MoSi$_2$ (110)	Nb$_5$Si$_3$ (25)	1700	25	SiO$_2$ MoSi$_2$	−9.5	[19]
Nb-Si	60Si-15Fe-20Cr-5NaF	1400 °C/2 h vacuum	NbSi$_2$ Nb$_4$Si$_5$CrFe$_3$ Fe$_4$Nb$_4$Si$_7$ (250)	Nb$_5$Si$_3$ (50)	1400	6.5	Nb$_2$O$_5$ SiO$_2$	3.5	[20]
Nb-Si-Ti	20Fe-20Cr-50Si-10NaF	1400 °C/2 h vacuum	NbSi$_2$ (Fe, Cr)$_3$Si$_2$ (170)	Nb$_5$Si$_3$ (20)	1400	7	Nb$_2$O$_5$ SiO$_2$	1.9	[21]

The typical surface and corresponding cross-sectional morphologies of SS coatings are shown in Figure 2. It can be noticed that the coating surface was relatively rough with dozens of holes and cracks. This was caused by the considerable particle size of the mixture, uneven mixing, and volatilization of the solvent and binder during the sintering process, as shown in Figure 2a,b. It is worth noting that Xiao et al. used smaller material particles for sintering and obtained a lower surface roughness, and the coating surface was relatively uniform and dense without apparent defects, as shown in Figure 2c. It can be observed from the cross-sectional images that the coating consisted of the outermost layer (NbSi$_2$), the intermediate layer, and the internal interdiffusion layer (IZD). In addition, as there was a mismatch of thermal expansion coefficient between the substrate and the coating during the sintering process, a small number of longitudinal cracks were observed inside the coating [22]. However, the IZD area, where the coating and the substrate were connected, was heavy and uniform, revealing that an excellent metallurgical bond was achieved between the coating and the substrate [23], as shown in Figure 2d–f. Han et al. [24] systematically explained the delamination phenomenon of such coatings. They believed that the growth process of the coating was firstly combined by chemical adsorption and physical adsorption, and then polar groups and substrates diffused between them at high temperatures. Finally, a firm and dense interwoven network coating layer was formed at the interface.

Relevant scholars have researched the oxidation resistance of the coating. After oxidation in the series of 1200–1700 °C for 2–25 h, the mass transition per unit area of the coating was within 10 mg·cm^{-2}, as shown in Table 1. The typical surface and cross-section microstructures of oxidized coatings are shown in Figure 3. It can be perceived that an oxide layer was formed on the oxidized coating surface, and the oxide layer consisted of SiO$_2$ and Nb$_2$O$_5$. Compared with before oxidation, the coating surface was relatively smooth at the initial oxidation stage [25]. This was owing to the formation of SiO$_2$ with a certain fluidity during the oxidation process, which filled up weaknesses such as cracks and gaps on the coating surface, to a certain extent. The inner coating was quite dense without apparent defects [26], as shown in Figure 3b,e. As the oxidation reaction progressed, the thickness of the oxide layer grew gradually. Several longitudinal cracks throughout the coating to the substrate were observed at the coating cross-section. This was caused by the large amount of volatile NbO$_2$ produced during the oxidation process, as shown in Figure 3c. At the same time, a large number of longitudinal cracks throughout the entire cross-section were observed in the coating, and the same composition as the oxide layer was detected in this area, which shows that the coating could no longer provide adequate protection for the substrate, as shown in Figure 3f. Thermal diffusion referred to the spontaneous transition of matter to an equilibrium state in a high temperature and closed environment, significantly affected by temperature. Due to the principle of thermal diffusion [27], the thickness of the interface layer gradually increased throughout the process, as shown in Figure 3d–f.

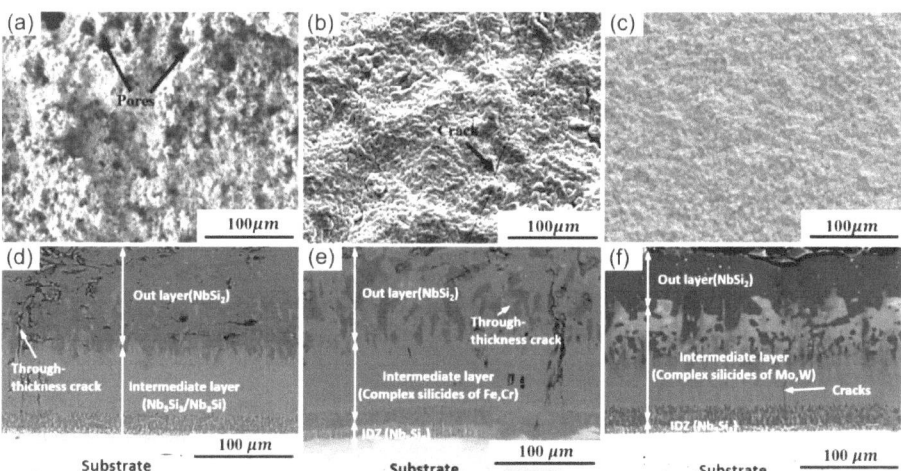

Figure 2. Scanning electron microscope (SEM) images of silicide coatings prepared by SS: (**a**,**d**) of pure $NbSi_2$ coating; (**b**,**e**) $NbSi_2$-(Fe, Cr)$_3Si_2$ composite coating; (**c**,**f**) $NbSi_2$-(Mo,W)Si_2 composite coating.

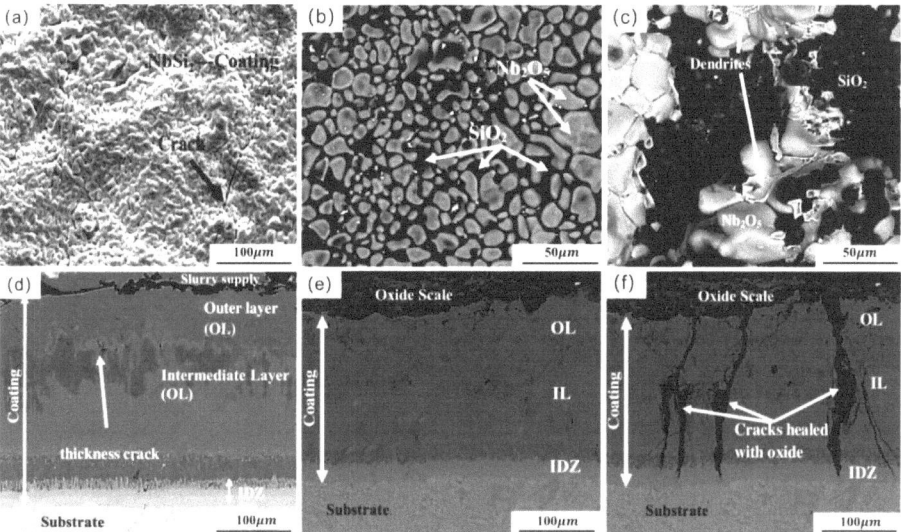

Figure 3. SEM images of SS coating before and after oxidation: (**a**,**d**) Before oxidation; (**b**,**e**) After oxidation for 2 h; (**c**,**f**) Coating morphologies after oxidation failure.

2.2. Thermal Spraying Method

Thermal spraying is also called sputtering technology. Its principle is to spray molten or semi-molten materials on the surface of the substrate to form a coating through high-speed airflow [28,29]. Due to its high surface coating material, short spraying time, and positive bonding performance between the coating and substrate, it is considered one of the coating preparation technologies with the most development potential [30]. The most typical atmospheric plasma spraying (SPS) coating preparation process is shown in Figure 4. Table 2 summarizes the preparation process, coating structure, and oxidation resistance of SPS coatings on the surface of niobium-based alloys. It can be seen that $MoSi_2$ has been favored by more researchers as the primary spray material [31,32]. This is mainly attributed to its high melting point, good thermal stability, and excellent high temperature creep resistance, which will make the material have a higher melting temperature and longer cooling time when sprayed on the substrate, and significantly improve the bonding strength between the substrate and coating particles. In addition, controlling the material size, carrier gas flow, spray distance, and other parameters will substantially impact the coating quality [33,34].

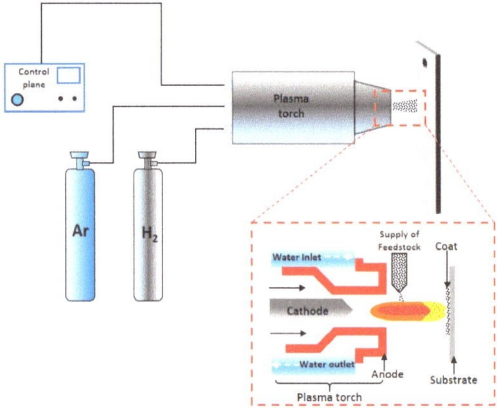

Figure 4. Schematic diagram of plasma spraying coating preparation.

The SEM images of typical SPS coatings are listed in Figure 5. The average surface roughness (R_a) and porosity of the sprayed Mo coating were the same [35,36]. At the same time, apparent holes and cracks were observed at the cross-sectional coating, as shown in Figure 5a,d. However, with the introduction of other phases such as mullite and WSi_2 in the spray material, the surface morphology was significantly improved, and the particle morphology was significantly eased. This was due to the vacancy complementation of the multi-element materials during the spraying process and the self-balance of the thermal conductivity and thermal expansion coefficient during the bonding process, as presented in Figure 5b,c. In addition, a continuous and uniform transition layer was observed between the coating and the substrate, indicating that a positive metallurgical bond was achieved betwixt the substrates and coating, as presented in Figure 5e,f. The author believes that the addition of mullite fills the holes during the spraying process, which can optimize the coating structure, and improves the coating density [37]. The addition of WSi_2 can combine with $MoSi_2$ to form $(W, Mo)Si_2$ and $(W, Mo)_5Si_3$, which effectively inhibits the diffusion of Si elements, maintains the coating morphology, and relieves the internal coating caused by the mismatched thermal expansion coefficients defect.

The researchers have studied the oxidation behavior of SPS coatings on Nb-based alloys, as presented in Table 2. The service life of coatings can reach tens to hundreds of hours in an oxidizing environment from 1200 to 1500 °C. The typical SEM of the oxidized coating is shown in Figure 6. It can be perceived that the large oxide particles appeared on the coating surface after cyclic oxidation at 1500 °C for 43 h. The laser scanning confocal microscopy (LSCM) results show that the R_a of oxidized coating was 23.1 µm, and the main components were SiO_2 and Nb_2O_5 [38]. Moreover, a large number of holes and cracks were observed at the surface, as shown in Figure 6a–c. This was due to surface defects providing channels for the diffusion of oxygen atoms, resulting in excessive oxidation and expansion inside the coating. Meanwhile, the cyclic thermal shock further aggravated the peeling of the outer oxide protective layer from the inner substrate layer. The cross-sectional image shows that the inside of the oxide layer was very loose with a thickness between 20 to 30 µm, and there was a phenomenon of shedding in its local area, as presented in Figure 6d. However, the mass change of the Mo-MoSi$_2$ coating [39] prepared by Zhang et al. after being oxidized at 1500 °C for 140 h was only -2.77 mg·cm^{-2}. This is because the addition of mullite inhibits the crystallization of SiO_2, improves its fluidity, and promotes the formation of a continuous and dense oxide film on the coating surface [40,41].

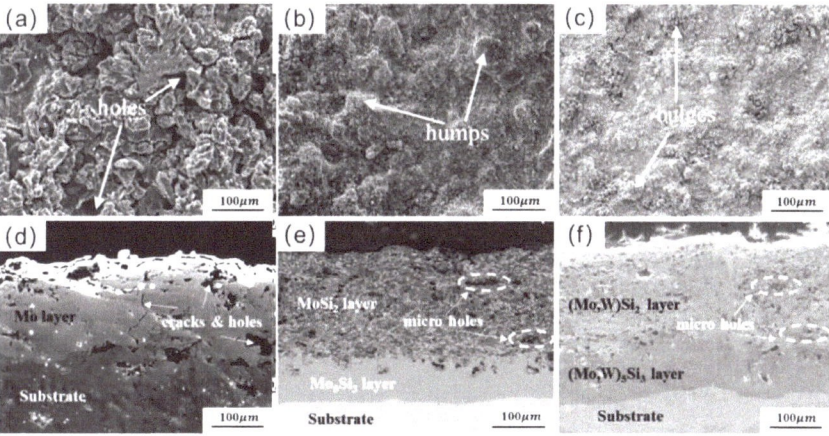

Figure 5. SEM images of coatings prepared by SPS under different procedures: (**a,d**) Pure Mo coating; (**b,e**) MoSi$_2$-mullite coating; (**c,f**) WSi$_2$-mullite-MoSi$_2$ coating.

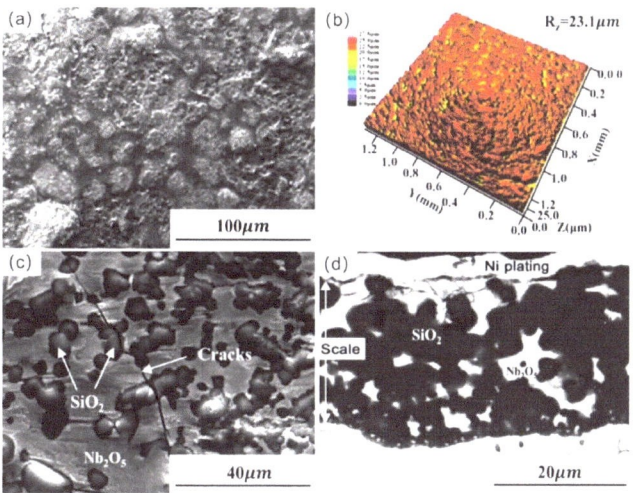

Figure 6. SEM and LSCM images of SPS coatings after oxidation at 1500 °C for 43 h. (**a**) Secondary electron image; (**b**) LSCM image; (**c**) Backscattered electron image; (**d**) Cross-sectional images.

Table 2. Summary for preparation and oxidation resistance of SPS coatings on Nb-based alloys.

Substrate	Material Composition (wt.%)	Process Conditions		Coating Composition and Thickness (μm)		Oxidation System	Oxidation Products	Quality Change (mg/cm^2)	References
		Spray Distance (mm)	Spray Rate (g/min)	Outer Layer	Inner Layer				
Nb521	10mullite-90MoSi$_2$	100	20	h-MoSi$_2$ t-MoSi$_2$ t-WSi$_2$ (150)	Mo$_5$Si$_3$ W$_5$Si$_3$ NbSi$_2$ (50)	1500 °C 500 h	SiO$_2$	4.41	[42]
Nb521	MoSi$_2$ 10mullite-90MoSi$_2$ 30mullite-70MoSi$_2$	100	20	t-MoSi$_2$ h-MoSi$_2$ (155)	Mo$_5$Si$_3$ Mo (45)	1500 °C 140 h	SiO$_2$ Mullite SiO$_2$	21.38 −2.77 −4.06	[43]
Nb521	MoSi$_2$	100	25	b-MoSi$_2$ (98)	Mo$_5$Si$_3$ (23)	1200 °C 94 h	MoSi$_2$	–	[44]
Nb-W	MoSi$_2$	90	20	h-MoSi$_2$ t-MoSi$_2$ SiO$_2$ (170)	Mo$_5$Si$_3$ (34)	1500 °C 43 h	SiO$_2$	5.31	[45]
Nb-Si-Ti	Mo-45Si-45Al	90	25	Mo(Si,Al)$_2$ (100)	Mo$_5$(Si,Al)$_3$ (25)	1250 °C 100 h	SiO$_2$ Al$_2$O$_3$	8.24	[46]
Nb-Si-Ti	2NaF-34Si-B-63Al$_2$O$_3$	100	25	MoSi$_2$ (72)	Mo Mo$_5$Si$_3$ (55)	1250 °C 100 h	SiO$_2$ Borosilicate glass cover	1.28	[47]

2.3. Embedding Method

The embedding method is also known as HAPC, and the principle is to place the substrate in a permeation box containing the halide of the coating element and conduct heat treatment in a vacuum or under the condition of continuous inert gas [48,49]. The required coating is formed through vapor migration and reaction–diffusion; the process flowchart is shown in Figure 7. Because this method has the advantages of an uncomplicated process, significant coating efficiency, and freedom from the limitation of the shape of the workpiece, it is widely used in the oxidation protection of refractory metal surface coatings. The researchers conducted a systematic study on the composition and oxidation mechanism of the HAPC coating on the surface of Nb and its alloys, and the results are shown in Table 3. It can be noticed that if the temperature was in the range of 850–1300 °C for 5–25 h, a coating with a thickness of 40–200 μm was obtained. Qiao et al. [50] reported the significance of the growth law of coating crystals on its surface quality. The consequence is that, during the preparation of HAPC coatings, the contact interface between the substrate and the embedding agent undergo chemical combination and diffusion reactions. When the diffusion rate is greater than the reaction rate, many crystal nuclei are formed on the face of the base metal per unit time, which will cause the surface of the base to form a fine and dense coating. Conversely, when the diffusion rate is less than the reaction rate, there are limited crystal nuclei formed on the surface of the base metal per unit time, the formed crystal grains are coarse, the solid phase accumulation is relatively loose, and the coating pore cracks are more severe [51]. The diffusion rate and reaction rate were closely related to the embedding temperature, which theoretically confirms the decisive influence of the embedding temperature. Further research shows that if the thickness of the outer layer of the coating is controlled to 80–100 μm and the inner layer is controlled to 5–20 μm, the resulting coating is denser, with a porosity of about 5–15%.

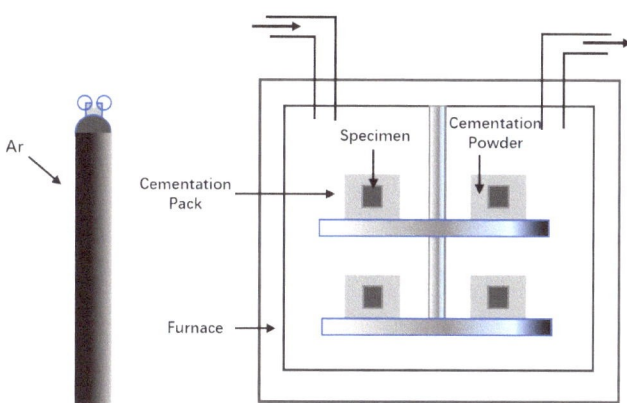

Figure 7. Process flowchart of HAPC method.

Table 3. Summary for preparation and oxidation resistance of HAPC coatings on Nb-based alloys.

Substrate	Experimental Conditions	Embedding Components (wt.%)	Coating Composition and Thickness (μm)		Oxidation System	Oxidation Products	Quality Change (mg/cm^2)	References
			Outer Layer	Interface Layer				
C-103	1100 °C/6 h Ar	25Si-5NaF-70Al$_2$O$_3$	NbSi$_2$ Nb$_5$Si$_3$ (60)	Nb$_5$Si$_3$ (5)	1100 °C/17 h	Nb$_2$O$_5$ SiO$_2$	3	[52]
Nb-Cr	1150 °C/5 h Ar	10Si-10Al-5NaF-75Al$_2$O$_3$	Al$_2$O$_3$ Cr$_3$Si Nb$_5$Si$_3$ (35)	Nb$_5$Si$_3$ Cr$_3$Si (5)	1200 °C/100 h	SiO$_2$ Al$_2$O$_3$	3.38	[53]
	1250 °C/8 h Ar	10Si-2Y$_2$O$_3$-5NaF-83Al$_2$O$_3$	(Nb, Cr)Si$_2$ (Cr, Nb)Si$_2$ (185)	(Nb, Cr)$_5$Si$_3$ (15)	1250 °C/50 h	SiO$_2$ Cr$_2$O$_3$ CrNbO$_4$	13.1	[54]
Nb-Si	1150 °C/10 h Ar	10Si-10Al-5NaF-2Y$_2$O$_3$-73Al$_2$O$_3$	(Nb, X)Si$_2$ (Nb, X)$_5$Si$_3$ (40)	(Nb,Ti)$_3$(Al, X) (5)	1250 °C/50 h	TiO$_2$ Al$_2$O$_3$ SiO$_2$	12.5	[55]
	1300 °C/10 h Ar	16Si-8Ge-Y$_2$O$_3$-5NaF-70Al$_2$O$_3$	(Nb, X)(Si, Ge)$_2$ (180)	(Ti, Nb)$_5$(Si, Ge)$_4$ (Nb, X)$_5$(Si, Ge)$_3$ (12)	1250 °C/100 h	SiO$_2$ GeO$_2$ TiO$_2$ Cr$_2$O$_3$	2.78	[56]
Nb-Ti-Al	850 °C/25 h vacuum	60Al$_2$O$_3$-40Al	NbAl$_3$ (160)		1000 °C/650 h	NbAl$_3$ (α-Al$_2$O$_3$)	1.5	[57]
	1050 °C/25 h vacuum	60Al$_2$O$_3$-40Si	NbSi$_2$ (50)		1000 °C/650 h	SiO$_2$ TiO$_2$	0.4	

The typical SEM of HAPC coatings on Nb-based alloys is shown in Figure 8. It can be perceived that the R_a of the Nb-Si-Mo coating [58] was relatively high, and apparent cracks and large granular filler powder were observed at the facial. The internal porosity of the coating was relatively high, consisting of (Nb, X)$_5$Si$_3$ inner layer and (Nb, X)Si$_2$ outer layer, and apparent cracks were observed at the interface layer, owing to the varying grain size of the embedded material [59] and the difference in thermal expansion coefficient between the substrate and the combined coating, as shown in Figure 8a,d. In order to optimize the coating structure, Majumdar et al. prepared a Ge/Ge-Y modified NbSi$_2$ coating on the Nb substrate. It can be perceived that the surface of the Ge modified coating was relatively flat and smooth [60], but a small amount of pore was observed at the grain boundary, as shown in Figure 8b. The interior coating was relatively uniform, but there were still apparent defects. In addition, a large amount of Al and Cr-rich regions were noticed at the junction of the coating and the substrate, as shown in Figure 8e. Compared with the single-element Ge modified coating, the surface grains of the Ge-Y modified silicide coating after adding Y were smaller, and the average grain size was only 2 to 3 μm, as shown in Figure 8c. It is worth noting that the coating inside was very uniform and dense without any apparent defects [61], and the overall thickness was about 50 μm, as shown in Figure 8f. This is because the addition of the Y element takes advantage of the release of thermal stress and the filling of vacant defects so that the surface grains of the coating are significantly refined, and the gaps between the grains are significantly reduced.

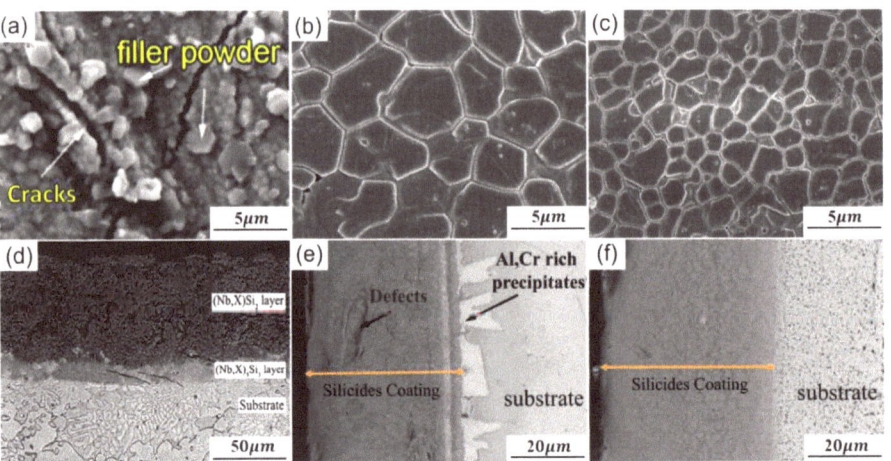

Figure 8. SEM of HAPC coatings: (**a,d**) Al-NbSi$_2$ coating; (**b,c**) Ge/Ge-Y modified coating; (**e,f**) Effect of different binders BaF$_2$/CrCl$_3$ on coating quality.

Related researchers have tested the oxidation resistance of HAPC coatings, as shown in Table 3. It can be noticed that the mass gain of the coating was only 0.4–13.1 mg·cm^{-2} after being oxidized at 1000–1250 °C for 17–650 h. Typical SEM of oxidized coating is shown in Figure 9. Overall, the oxidized Nb-Si-Ti coating surface was smoother, covering a layer of molten oxide film, but a few holes were still observed [62]. The author believes that this is related to the higher surface roughness of the coating before oxidation, resulting in uneven oxidation and the volatilization of Nb$_2$O$_5$ during the oxidation process, as shown in Figure 9a. The oxidation results show that the main component of the oxide film was SiO$_2$-TiO$_2$, and the thickness was about 20 μm. In addition, a small amount of Cr$_2$O$_3$ was also observed at the bottom of the oxide layer [63], as presented in Figure 9c. However, the oxidized (Nb, X)Si$_2$ coating surface was very rough, and a large number of oxide particles mainly composed of SiO$_2$-Al$_2$O$_3$ were observed, as shown in Figure 9b. A large number of holes and longitudinal cracks across the entire cross-section appeared in the coating. At the same time, partial areas of the oxide layer appeared to fall off [64]. It is worth emphasizing that an extended layer of Cr$_2$O$_3$ was observed betwixt the oxide layer and the internal coating, which relieved the further oxidation of the coating to a certain extent, as shown in Figure 9d.

2.4. Other Methods

By analyzing the advantages and disadvantages of the coating preparation process, related scholars appropriately combined different methods to make up for each other's advantages and disadvantages, thereby obtaining a composite coating with superior oxidation resistance [65,66]. Common combination categories, process conditions, coating composition, thickness, and oxidation characteristics are shown in Table 4. It can be seen that the coating system was dominated by Mo-Si-X (B, Ce, etc.), which is related to the high thermal conductivity and good thermal stability of $MoSi_2$. The addition of elements such as B and Ce can further optimize the coating structure and improve its oxidation resistance at high temperatures [67]. The thickness of the coating prepared by the two-step process was 55 to 160 µm. The outer layer mainly consisted of high-priced metal and high silicide, such as $MoSi_2$. Due to the difference in the concentration of internal and external elements, the inner layer was mostly $(Nb, X)_5Si_3$ and $(Nb, X)_3Si$. The typical coating surface and the corresponding cross-sectional morphology are shown in Figure 10. Since most of the later steps of this kind of method are HAPC, the coating surface exhibited a relatively high toughness, and a large amount of granular embedded powder was observed to adhere, as shown in Figure 10a,c. The main component of the coating layer was a mixture of $MoSi_2$ and the second phase was loosely bonded and had many holes and cracks [68,69]. The inner layer mainly consisted of Nb-based metal silicide; its cross-sectional morphology was good, and the structure was dense and uniform. This shows that the coating and the substrate achieved good metallurgical bonding, as shown in Figure 10b,d. This type of coating preparation method realizes the diffusion and filling of the outer layer components to the inner layer components through the latter process so that the structure of the inner coating is further strengthened, and the use performance of the coating is significantly improved. In addition, technologies such as laser cladding technology (LCT), physical vapor deposition (PVD), and hot-dipped silicon (HDS) [70] have also been favored by related scholars in the preparation of Nb surface anti-oxidation coatings. Zhang et al. used HDS [71,72] technology to prepare a WSi_2 coating [73,74] with nano-level roughness on the surface of W. The resulting coating facial was absolutely dense and uniform without defects such as holes, gaps [75], etc. This will help the technology apply the coating to the surface of Nb and its alloys, and provides a helpful reference for the preparation of anti-oxidation coating.

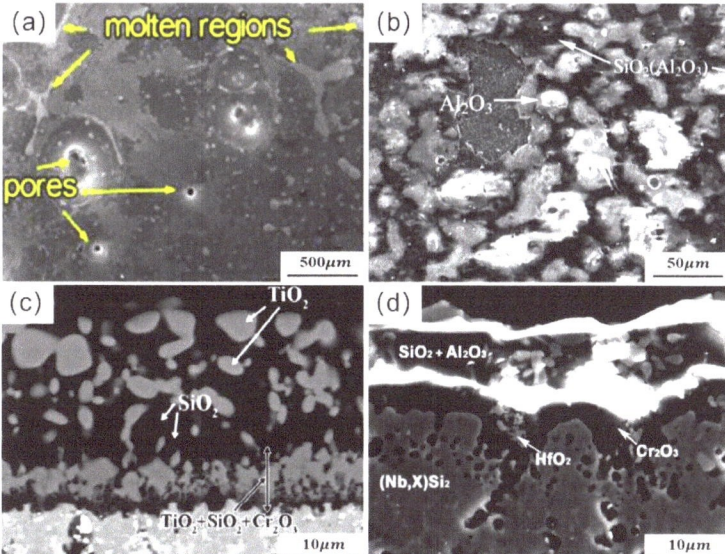

Figure 9. SEM of HAPC coatings after oxidation. (**a,c**) Surface and cross-sectional morphology of Nb-Si-Ti coating after oxidation at 1050 °C for 5 h; (**b,d**) $(Nb, X)Si_2$ coating surface and cross-section morphology after oxidation at 1250 °C for 50 h.

Table 4. Two-step coating preparation procedure and oxidation resistance details on Nb alloy surface.

Substrate	Process Conditions		Coating System	Coating Composition and Thickness (μm)		Oxidation System		Oxidation Products	Quality Change (mg/cm²)	References
				Outer Layer	Interface Layer	Oxidation Temperature (°C)	Oxidation Time (h)			
Nb	SPS SD: 100 mm SR: 20 g/min	HAPC Ar 1000 °C/50 h	Mo-Si-B	$MoSi_2$ NbB_2 $NbSi_2$ (70)	Nb_5Si_3 NbB_2 (10)	1300	24	Nb_2O_5 SiO_2	0.44	[76]
C-103	HAPC 1100 °C/6 h	HAPC 1050 °C/4 h	Si-B	$NbSi_2$ NbB_2 (125)	NbB_2 (13)	1300	100	Nb_2O_5 NbO_2 B_2O_3	1.44	[77]
Nb-Si	SS 1550 °C/2 h	HAPC Ar 1200 °C/5 h	Mo-Si-Ce	$NbSi_2$ $MoSi_2$ (80)	Nb_5Si_3 (4)	1600	24.7	SiO_2	3.57	[78]
Nb-Si-Ti	PVD 300 °C/2 h	HAPC 1450 °C/12 h	Mo-Si-B	$MoSi_2$ $(Nb,Ti)_5SiB_2$ (50)	$(Nb,Ti)Si_2$ $(Nb,X)_5Si_3$ (5)	1300	24	$MoO3$ SiO_2	−0.55	[79]
Nb-Si-Ti	SPS SD: 60 mm SR: 90 g/min	HAPC Ar 1250 °C/4 h	Si-Y-Zr	$(Nb,Ti)_5Si_4$ (110)	$(Nb,X)_5Si_3$ (5)	1250	100	Nb_2O_5 $TiO2$ SiO_2	1.6	[80]
Nb-Si-Ti	SPS SD: 90 mm SR: 20 g/min	HAPC 1000 °C/40 h	Mo-Si-B	$MoSi_2$ MoB (115)	Mo (45)	1250	100	B_2O_3 SiO_2	0.92	[81]

Figure 10. Typical SEM of two-step method coatings. (**a**,**b**) Mo-Si-B coating prepared by SPS (spray distance:100 mm, spray rate: 20 g/min) + HAPC (1000 °C/50 h) process; (**c**,**d**) Mo-Si-B coating prepared by PVD (300 °C/2 h) + HAPC (1450 °C/12 h) process.

Figure 11 present the typical SEM of oxidized coatings prepared by the two-step approach. The mass loss per unit area of the coating was within 4 mg·cm^{-2} after being oxidized in the range of 1300–1600 °C for 24–100 h. The outward layer of the oxidized coating was approximately flat and smooth without apparent cracks and holes. This was due to the formation of aluminoborosilica with a certain fluidity during the oxidation process, filling in the surface defects [82], as shown in Figure 11a,c. $MoSi_2$-SiO_2 dominated the outer phase of the oxidized coating. Due to original defects and continuous consumption during the oxidation process, the partial area of the oxide layer fell off, showing discontinuity and inhomogeneity [83]. However, the inner layer organization of the coating was still dense and compact without apparent flaw, and the transition layer of (Nb, X)$_5$Si$_3$ and NbB$_2$ was observed to grow inward at the interface between it and the substrate, as shown in Figure 11b,d.

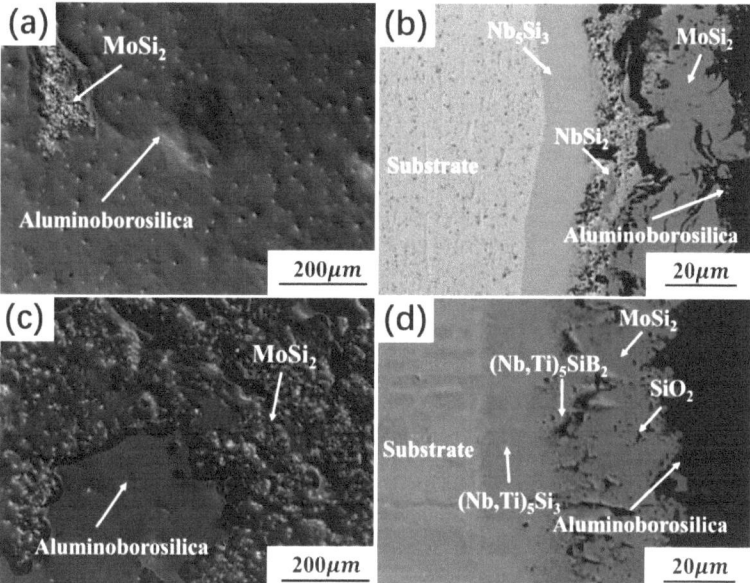

Figure 11. SEM of oxidized coatings prepared by the two-step method, (**a,b**) Mo-Si-B coating after oxidation at 1250 °C for 100 h; (**c,d**) Mo-Si-B coating after oxidation at 1300 °C for 24 h.

3. Oxidation Mechanism and Failure Behavior of Coating

According to the summary of the oxidation characteristics of Nb alloys surface coatings, the oxidation behavior and failure mechanism are summarized, as shown in Figure 12. It can be noticed that oxidation can be divided into two stages. The inner and outer layers of the coating are composed of the Nb$_5$Si$_3$ layer and NbSi$_2$ layer, respectively, as shown in Figure 12a. In the initial stage of oxidation, the oxygen-philic compounds on the coating surface are rapidly oxidized. The oxidation reaction is more severe at weak areas such as cracks, gaps, etc., and the generated oxides such as Nb$_2$O$_5$ and SiO$_2$ are transformed into defects, start to grow, and gradually spread to the entire surface. As the reaction progresses, Nb$_2$O$_5$, NbO$_2$, etc., gradually volatilize, leaving many holes on the surface. In addition, due to the release of thermal stress and the mismatch of thermal expansion coefficients between systems, many cracks sprout on the surface of the coating. At the same time, the addition of some modifying elements (X) improves the fluidity of SiO$_2$, fills up these defects to a certain extent, and forms an Nb$_2$O$_5$-SiO$_2$-X$_2$O$_3$ protective film system on its surface, as shown in Figure 12b. At last, the oxide layer gradually thickens, the NbSi$_2$ layer as the central part of the coating is gradually consumed, and the self-healing ability of the coating gradually deteriorates. However, the Nb$_5$Si$_3$ layer with poor oxidation resistance gradually becomes thicker. With the oxidation process, the low oxidation resistance Nb$_5$Si$_3$ layer is gradually destroyed, resulting in the oxidation failure of the coating, and a large number of holes and cracks are observed, as shown in Figure 12c.

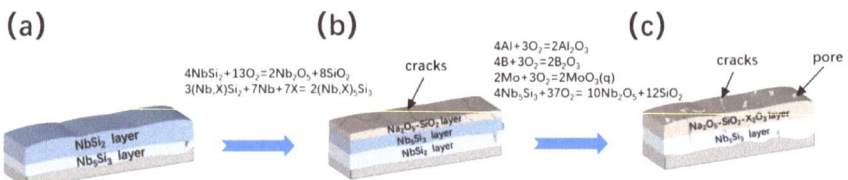

Figure 12. Oxidation and failure mechanism of silicide coatings on Nb-based alloys, (**a**) Before coating oxidation; (**b**) Primary oxidation of coating; (**c**) Deep oxidation of coating.

4. Conclusions and Prospects

In this work, the preparation methods of anti-oxidation coatings on Nb-based alloys are reviewed, and the structure and anti-oxidation performance of coatings obtained by different methods are summarized, as shown in Figure 13. Overall, the high-temperature oxidation resistance of Nb-based alloys has been significantly enhanced by surface coating technology. Through in-depth comparison and analysis of various methods, it can be known that the volatilization of solvents and cement, and the uneven particle size of the mixture during the sintering process, result in poor surface quality and high porosity of the coating prepared by SS. Although the two processes of HAPC and CVD have no volatilization phenomenon and are not limited by the shape of the substrate, their lower deposition temperature makes the growth of the coating slower, and the preparation cycle is longer. In contrast, due to its high diffusion temperature, SPS can deposit coatings of tens to hundreds of microns in a short time. However, due to the uneven melting of the spray paint and a small amount of gas during the spraying process, the porosity of the coating is higher, and the bond with the substrate is poor. In addition, although the two-step coating has a relatively excellent structure, its process is complicated, and the coating preparation efficiency is low. As a new coating preparation process, HDS technology dominates due to short deposition time, high coating preparation productivity, smooth and dense coating surface, etc., and it is expected to protect Nb-based alloys at high temperatures.

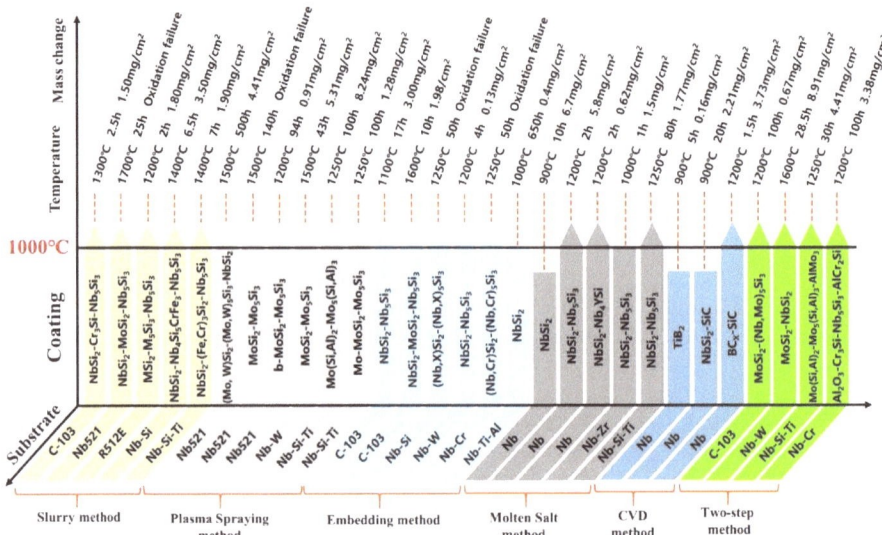

Figure 13. Overview of preparation parameters and oxidation resistance of Nb-based alloys surface coatings.

Summarizing the anti-oxidation mechanism of the coating prepared by the above method, it can be found that the outer protective scale of the coating with better anti-oxidation performance is generally composed of SiO_2 and other inert melt film shielding substances. However, inert molten films such as SiO_2 are formed after oxidation tests, and the oxidation process is challenging to control. Therefore, in order to further improve the high-temperature oxidation resistance of the coating, some beneficial elements are usually added in an appropriate amount during the coating preparation process. Among them, the "selective oxidation type" alloy element X (X = Al, Cr, Mo, Ti, etc.) is added to make it preferentially combine with the O element to form an oxide during the high-temperature oxidation process, setting a dense isolated layer on the surface of the substrate. This blocks the inward diffusion of oxygen atoms, inhibits the formation of Nb_2O_5, and reduces the oxidation rate. The addition of element B and mullite can improve the fluidity of SiO_2 and promote the formation of a uniform and thick oxide film on the coating surface. The addition of Y and Ce elements can refine the coating grains, optimize the coating structure, and significantly improve the strength of the coating at high temperatures so that it can maintain a good shape during the oxidation process. The addition of W and Ge can constrain the diffusion of Si elements into the substrate, slow down the generation of Nb_5Si_3 with poor oxidation resistance, and lengthen the oxidation service life of the coating. The introduction of a proper amount of mullite can fill the pores inside and on the coating surface, optimize the coating structure, increase the density of the coating, inhibit the recrystallization of SiO_2, and promote the thick oxide film on the surface of the coating. In addition, optimizing the coating preparation process and structure can significantly reduce defects produced by thermal expansion coefficient mismatch between coating and substrate, which also plays a crucial role in improving the high-temperature oxidation resistance of the coating.

Author Contributions: Conceptualization, Y.Z.; methodology, Y.Z., X.Z. and T.F.; validation, Y.Z. and K.C.; formal analysis, J.W. and F.S.; investigation, L.Y., H.M. and T.F.; resources, Y.Z.; data curation, Y.Z., X.Z. and T.F.; writing—original draft preparation, Y.Z., X.Z. and T.F.; writing—review and editing, Y.Z., X.Z. and T.F.; visualization, X.Z. and T.F.; supervision, Y.Z.; project administration, Y.Z.; funding acquisition, Y.Z. All authors have read and agreed to the published version of the manuscript.

Funding: This research was funded by the National Natural Science Foundation (51604049).

Institutional Review Board Statement: Not applicable.

Informed Consent Statement: Not applicable.

Data Availability Statement: Not applicable.

Acknowledgments: The authors wish to acknowledge the contributions of associates and colleagues at Anhui University of Technology. The financial support of the National Natural Science Foundation.

Conflicts of Interest: The authors declare no conflict of interest.

References

1. Subramanian, P.; Mendiratta, M.; Dimiduk, D. The development of Nb-based advanced intermetallic alloys for structural applications. *JOM* **1996**, *48*, 33–38. [CrossRef]
2. Wadsworth, J.; Froes, F. Developments in metallic materials for aerospace applications. *JOM* **1989**, *41*, 12–19. [CrossRef]
3. Zhu, Y.; Peng, W.; Xu, R.; Jiang, P. Review on active thermal protection and its heat transfer for airbreathing hypersonic vehicles. *Chin. J. Aeronaut.* **2018**, *31*, 1929–1953. [CrossRef]
4. Ji, T.; Zhang, R.; Sunden, B.; Xie, G. Investigation on thermal performance of high temperature multilayer insulations for hypersonic vehicles under aerodynamic heating condition. *Appl. Therm. Eng.* **2014**, *70*, 957–965. [CrossRef]
5. Jian, D.; Qiuru, Z. Key technologies for thermodynamic cycle of precooled engines: A review. *Acta Astronaut.* **2020**, *177*, 299–312. [CrossRef]
6. Yuan, B.; Harvey, C.M.; Thomson, R.C.; Critchlow, G.W.; Wang, S. A new spallation mechanism of thermal barrier coatings on aero-engine turbine blades. *Theor. Appl. Mech. Lett.* **2018**, *8*, 7–11. [CrossRef]
7. Prasad, V.S.; Baligidad, R.; Gokhale, A.A. Niobium and other high temperature refractory metals for aerospace applications. In *Aerospace Materials and Material Technologies*; Springer: Cham, Switzerland, 2017; pp. 267–288.

8. Inouye, H. Niobium in high temperature applications. In *Proceedings of the Niobium-Proceedings of the International Symposium*; Metallurgical Society of AIME: Warrendale, PA, USA, 1984.
9. Eckert, J. Niobium compounds and alloys. *Int. J. Refract. Met. Hard Mater.* **1993**, *12*, 335–340. [CrossRef]
10. Wojcik, C.C. Processing, properties and applications of high-temperature niobium alloys. *Mrs Online Proc. Libr. (OPL)* **1993**, *322*, 519. [CrossRef]
11. Wadsworth, J.; Nieh, T.; Stephens, J. Recent advances in aerospace refractory metal alloys. *Int. Mater. Rev.* **1988**, *33*, 131–150. [CrossRef]
12. Zhang, S.; Li, X.; Zuo, J.; Qin, J.; Cheng, K.; Feng, Y.; Bao, W. Research progress on active thermal protection for hypersonic vehicles. *Prog. Aerosp. Sci.* **2020**, *119*, 100646. [CrossRef]
13. Wojcik, C.C.; Chang, W. Thermomechanical processing and properties of niobium alloys. In Proceedings of the International Symposium on Niobium 2001, Orlando, FL, USA, 2–5 December 2001; pp. 163–173.
14. Packer, C.M.; Perkins, R.A. Development of a fused slurry silicide coating for the protection of tantalum alloys. *J. Less Common Met.* **1974**, *37*, 361–378. [CrossRef]
15. Streiff, R. Protection of materials by advanced high temperature coatings. *J. Phys. IV* **1993**, *3*, C9–C17. [CrossRef]
16. Kumawat, M.K.; Alam, M.Z.; Kumar, A.; Gopinath, K.; Saha, S.; Singh, V.; Srinivas, V.; Das, D.K. Tensile behavior of a slurry Fe-Cr-Si coated Nb-alloy evaluated by Gleeble testing. *Surf. Coat. Technol.* **2018**, *349*, 695–706. [CrossRef]
17. Alam, M.Z.; Sarin, S.; Kumawat, M.K.; Das, D.K. Microstructure and oxidation behaviour of Fe–Cr–silicide coating on a niobium alloy. *Mater. Sci. Technol.* **2016**, *32*, 1826–1837. [CrossRef]
18. Novak, M.D.; Levi, C.G. Oxidation and Volatilization of Silicide Coatings for Refractory Niobium Alloys. In Proceedings of the ASME 2007 International Mechanical Engineering Congress and Exposition, Seattle, WA, USA, 11–15 November 2007; pp. 261–267.
19. Pan, Y.; Guan, W.M. The hydrogenation mechanism of PtAl and IrAl thermal barrier coatings from first-principles investigations. *Int. J. Hydrog. Energy* **2020**, *45*, 20032–20041. [CrossRef]
20. Geethasree, K.; Satya Prasad, V.V.; Brahma Raju, G.; Alam, M.Z. Cyclic oxidation behavior of Fe-Cr modified slurry silicide coated Nb-18.7Si alloyed with Ti and Zr. *Corros. Sci.* **2019**, *148*, 293–306. [CrossRef]
21. Geethasree, K.; Alam, M.Z.; Raju, G.B.; Prasad, V.V.S. Microstructure and mechanical properties of uncoated Nb-18.7Si and Nb-18.7Si-5Ti alloys and their improved oxidation resistance after application of silicide coating. *Mater. Today Proc.* **2019**, *15*, 36–43. [CrossRef]
22. Li, Y.; Lin, X.; Hu, Y.; Gao, X.; Yu, J.; Qian, M.; Dong, H.; Huang, W. Microstructure and isothermal oxidation behavior of Nb-Ti-Si-based alloy additively manufactured by powder-feeding laser directed energy deposition. *Corros. Sci.* **2020**, *173*, 108757. [CrossRef]
23. Pan, Y.; Pu, D.L.; Yu, E.D. Structural, electronic, mechanical and thermodynamic properties of Cr–Si binary silicides from first-principles investigations. *Vacuum* **2021**, *185*, 110024. [CrossRef]
24. Han, J.; Su, B.; Meng, J.; Zhang, A.; Wu, Y. Microstructure and composition evolution of a fused slurry silicide coating on MoNbTaTiW refractory high-entropy alloy in high-temperature oxidation environment. *Materials* **2020**, *13*, 3592. [CrossRef]
25. Segura-Cedillo, I. Fused Metallic Slurry Coatings for Improving the Oxidation Resistance of Wrought Alloys. Ph.D. Thesis, The University of Manchester, Manchester, UK, 1 August 2011.
26. Sankar, M.; Satya Prasad, V.V.; Baligidad, R.G.; Alam, M.Z.; Das, D.K.; Gokhale, A.A. Microstructure, oxidation resistance and tensile properties of silicide coated Nb-alloy C-103. *Mater. Sci. Eng. A* **2015**, *645*, 339–346. [CrossRef]
27. Ji, P.F.; Li, B.; Chen, B.H.; Wang, F.; Ma, W.; Zhang, X.Y.; Ma, M.Z.; Liu, R.P. Effect of Nb addition on the stability and biological corrosion resistance of Ti-Zr alloy passivation films. *Corros. Sci.* **2020**, *170*, 108696. [CrossRef]
28. Herman, H. Plasma-sprayed coatings. *Sci. Am.* **1988**, *259*, 112–117. [CrossRef]
29. Fauchais, P. Understanding plasma spraying. *J. Phys. D Appl. Phys.* **2004**, *37*, R86. [CrossRef]
30. Stöver, D.; Pracht, G.; Lehmann, H.; Dietrich, M.; Döring, J.-E.; Vaßen, R. New material concepts for the next generation of plasma-sprayed thermal barrier coatings. *J. Therm. Spray Technol.* **2004**, *13*, 76–83. [CrossRef]
31. Zhang, Y.; Hussain, S.; Cui, K.; Fu, T.; Wang, J.; Javed, M.S.; Lv, Y.; Aslam, B. Microstructure and mechanical properties of $MoSi_2$ coating deposited on Mo substrate by hot dipping processes. *J. Nanoelectron. Optoelectron.* **2019**, *14*, 1680–1685. [CrossRef]
32. Pan, Y.; Yu, E. First-principles investigation of structural stability, mechanical and thermodynamic properties of Pt_3Zr_5 compounds. *Phys. B Condens. Matter* **2021**, *611*, 412936. [CrossRef]
33. Odhiambo, J.G.; Li, W.; Zhao, Y.; Li, C. Porosity and its significance in plasma-sprayed coatings. *Coatings* **2019**, *9*, 460. [CrossRef]
34. Hu, D.; Fu, Q.; Zhou, L.; Liu, B.; Sun, J. Crack development behavior in thermally sprayed anti-oxidation coating under repeated thermal-oxygen coupling environment. *Ceram. Int.* **2021**, *47*, 15328–15336. [CrossRef]
35. Sun, J.; Li, T.; Zhang, G.-P. Effect of thermodynamically metastable components on mechanical and oxidation properties of the thermal-sprayed $MoSi_2$ based composite coating. *Corros. Sci.* **2019**, *155*, 146–154. [CrossRef]
36. Hou, Q.; Shao, W.; Li, M.; Zhou, C. Interdiffusion behavior of Mo-Si-B/Al_2O_3 composite coating on Nb-Si based alloy. *Surf. Coat. Technol.* **2020**, *401*, 126243. [CrossRef]
37. Cui, K.; Zhang, Y.; Fu, T.; Wang, J.; Zhang, X. Toughening mechanism of mullite matrix composites: A Review. *Coatings* **2020**, *10*, 672. [CrossRef]

38. Zheng, K.; Wang, Y.; Wang, R.; Wang, Y.; Cheng, F.; Ma, Y.; Hei, H.; Gao, J.; Zhou, B.; Wang, Y. Microstructure, oxidation behavior and adhesion of a CoNiCrAlTaY coating deposited on a high Nb-TiAl alloy by plasma surface metallizing technique. *Vacuum* **2020**, *179*, 109494. [CrossRef]
39. Zhang, Y.; Li, Y.; Bai, C. Microstructure and oxidation behavior of Si–MoSi$_2$ functionally graded coating on Mo substrate. *Ceram. Int.* **2017**, *43*, 6250–6256. [CrossRef]
40. Hou, Q.; Li, M.; Shao, W.; Zhou, C. Oxidation and interdiffusion behavior of Mo-Si-B coating on Nb-Si based alloy prepared by spark plasma sintering. *Corros. Sci.* **2020**, *169*, 108638. [CrossRef]
41. Hu, L.; Li, S.; Li, C.; Fu, G.; He, J.; Dong, Y.; Yang, Y.; Zhao, H.; Qin, Y.; Yin, F. Deposition and properties of plasma sprayed NiCrCoMo–TiC composite coatings. *Mater. Chem. Phys.* **2020**, *254*, 123502. [CrossRef]
42. Zhang, G.; Sun, J.; Fu, Q. Microstructure and oxidation behavior of plasma sprayed WSi$_2$-mullite-MoSi$_2$ coating on niobium alloy at 1500 °C. *Surf. Coat. Technol.* **2020**, *400*, 126210. [CrossRef]
43. Zhang, G.; Sun, J.; Fu, Q. Effect of mullite on the microstructure and oxidation behavior of thermal-sprayed MoSi$_2$ coating at 1500 °C. *Ceram. Int.* **2020**, *46*, 10058–10066. [CrossRef]
44. Sun, J.; Fu, Q.-G.; Huo, C.-X.; Li, T.; Wang, C.; Cheng, C.-Y.; Yang, G.-J.; Sun, J.-C. Oxidation response determined by multiphase-dependent melting degree of plasma sprayed MoSi$_2$ on Nb-based alloy. *J. Alloys Compd.* **2018**, *762*, 922–932. [CrossRef]
45. Pan, Y.; Yu, E.; Wang, D.; Deng, H. Sulfur vacancy enhances the electronic and optical properties of FeS$_2$ as the high performance electrode material. *J. Alloys Compd.* **2021**, *858*, 157662. [CrossRef]
46. Yao, D.; Gong, W.; Zhou, C. Development and oxidation resistance of air plasma sprayed Mo–Si–Al coating on an Nbss/Nb$_5$Si$_3$ in situ composite. *Corros. Sci.* **2010**, *52*, 2603–2611. [CrossRef]
47. Wu, J.; Wang, W.; Zhou, C. Microstructure and oxidation resistance of Mo–Si–B coating on Nb based in situ composites. *Corros. Sci.* **2014**, *87*, 421–426. [CrossRef]
48. Bianco, R.; Rapp, R.A. Pack cementation diffusion coatings. In *Metallurgical and Ceramic Protective Coatings*; Stern, K.H., Ed.; Springer: Berlin/Heidelberg, Germany, 1996; pp. 236–260.
49. Mevrel, R.; Duret, C.; Pichoir, R. Pack cementation processes. *Mater. Sci. Technol.* **1986**, *2*, 201–206. [CrossRef]
50. Goward, G.; Cannon, L. Pack cementation coatings for superalloys: A review of history, theory, and practice. *J. Eng. Gas Turbines Power* **1988**, *110*, 150–154. [CrossRef]
51. Sun, J.; Li, T.; Zhang, G.-P.; Fu, Q.-G. Different oxidation protection mechanisms of HAPC silicide coating on niobium alloy over a large temperature range. *J. Alloys Compd.* **2019**, *790*, 1014–1022. [CrossRef]
52. Alam, M.Z.; Rao, A.S.; Das, D.K. Microstructure and high temperature oxidation performance of silicide coating on Nb-based alloy C-103. *Oxid. Met.* **2010**, *73*, 513–530. [CrossRef]
53. Zheng, H.; Xiong, L.; Luo, Q.; Lu, S. Development of multilayer oxidation resistant coatings on Cr–50Nb alloy. *Appl. Surf. Sci.* **2015**, *359*, 515–520. [CrossRef]
54. Qiao, Y.; Li, M.; Guo, X. Development of silicide coatings over Nb–NbCr$_2$ alloy and their oxidation behavior at 1250 °C. *Surf. Coat. Technol.* **2014**, *258*, 921–930. [CrossRef]
55. Zhang, S.; Shi, X.; Sha, J. Oxidation behaviours of Nb–22Ti–15Si–2Al–2Hf–2V–(2, 14) Cr alloys with Al and Y modified silicide coatings prepared by pack cementation. *Prog. Nat. Sci. Mater. Int.* **2015**, *25*, 486–495. [CrossRef]
56. Wang, W.; Zhou, C. Characterization of microstructure and oxidation resistance of Y and Ge modified silicide coating on Nb-Si based alloy. *Corros. Sci.* **2016**, *110*, 114–122. [CrossRef]
57. Chaia, N.; Cury, P.; Rodrigues, G.; Coelho, G.; Nunes, C. Aluminide and silicide diffusion coatings by pack cementation for Nb-Ti-Al alloy. *Surf. Coat. Technol.* **2020**, *389*, 125675. [CrossRef]
58. Yoon, J.K.; Kim, G.H.; Hong, K.T.; Doh, J.M.; Lee, J.K.; Lee, K.H.; Son, K.H. NbSi-2 Base Nanocomposite Coating and Manufacturing Method Thereof. U.S. Patent 20060029830A1, 29 December 2004.
59. Xiao, L.-r.; Cai, Z.-g.; YI, D.-q.; Lei, Y.; Liu, H.-q.; Huang, D.-y. Morphology, structure and formation mechanism of silicide coating by pack cementation process. *Trans. Nonferrous Met. Soc. China* **2006**, *16*, s239–s244. [CrossRef]
60. Chaliampalias, D.; Vourlias, G.; Pavlidou, E.; Skolianos, S.; Chrissafis, K.; Stergioudis, G. Comparative examination of the microstructure and high temperature oxidation performance of NiCrBSi flame sprayed and pack cementation coatings. *Appl. Surf. Sci.* **2009**, *255*, 3605–3612. [CrossRef]
61. Majumdar, S.; Arya, A.; Sharma, I.; Suri, A.; Banerjee, S. Deposition of aluminide and silicide based protective coatings on niobium. *Appl. Surf. Sci.* **2010**, *257*, 635–640. [CrossRef]
62. Qiao, Y.; Guo, X. Formation of Cr-modified silicide coatings on a Ti–Nb–Si based ultrahigh-temperature alloy by pack cementation process. *Appl. Surf. Sci.* **2010**, *256*, 7462–7471. [CrossRef]
63. Cui, K.; Zhang, Y.; Fu, T.; Hussain, S.; Saad AlGarni, T.; Wang, J.; Zhang, X.; Ali, S. Effects of Cr$_2$O$_3$ content on microstructure and mechanical properties of Al$_2$O$_3$ matrix composites. *Coatings* **2021**, *11*, 234. [CrossRef]
64. Sun, Z.; Tian, X.; Guo, X.; Yin, M.; Zhang, F.; Zhang, X. Oxidation resistance and mechanical characterization of silicide coatings on the Nb-18Ti-14Si-9Al alloy. *Int. J. Refract. Met. Hard Mater.* **2017**, *69*, 18–26. [CrossRef]
65. Pu, R.; Sun, Y.; Xu, J.; Zhou, X.; Li, S.; Zhang, B.; Cai, Z.; Liu, S.; Zhao, X.; Xiao, L. Microstructure and properties of Mo-based double-layer MoSi$_2$ thick coating by a new two-step method. *Surf. Coat. Technol.* **2020**, *394*, 125840. [CrossRef]
66. Sun, J.; Fu, Q.-G.; Guo, L.-P.; Wang, L. Silicide coating fabricated by HAPC/SAPS combination to protect niobium alloy from oxidation. *Acs Appl. Mater. Interfaces* **2016**, *8*, 15838–15847. [CrossRef]

67. Liu, L.; Zhang, H.; Lei, H.; Li, H.; Gong, J.; Sun, C. Influence of different coating structures on the oxidation resistance of MoSi$_2$ coatings. *Ceram. Int.* **2020**, *46*, 5993–5997. [CrossRef]
68. Liu, Y.; Shao, W.; Wang, C.; Zhou, C. Microstructure and oxidation behavior of Mo-Si-Al coating on Nb-based alloy. *J. Alloy. Compd.* **2018**, *735*, 2247–2255. [CrossRef]
69. Majumdar, S.; Mishra, S.; Paul, B.; Kishor, J.; Mishra, P.; Chakravartty, J. Development of MoSi$_2$ coating on Nb-1Zr-0.1 C alloy. *Mater. Today Proc.* **2016**, *3*, 3172–3177. [CrossRef]
70. Zhang, Y.; Cui, K.; Fu, T.; Wang, J.; Shen, F.; Zhang, X.; Yu, L. Formation of MoSi$_2$ and Si/MoSi$_2$ coatings on TZM (Mo–0.5 Ti–0.1 Zr–0.02 C) alloy by hot dip silicon-plating method. *Ceram. Int.* **2021**, in press.
71. Zhang, Y.; Qie, J.; Cui, K.; Fu, T.; Fan, X.; Wang, J.; Zhang, X. Effect of hot dip silicon-plating temperature on microstructure characteristics of silicide coating on tungsten substrate. *Ceram. Int.* **2020**, *46*, 5223–5228. [CrossRef]
72. Zhang, Y.; Zhao, J.; Li, J.; Lei, J.; Cheng, X. Effect of hot-dip siliconizing time on phase composition and microstructure of Mo–MoSi$_2$ high temperature structural materials. *Ceram. Int.* **2019**, *45*, 5588–5593. [CrossRef]
73. Zhang, Y.; Fu, T.; Cui, K.; Shen, F.; Wang, J.; Yu, L.; Mao, H. Evolution of surface morphology, roughness and texture of tungsten disilicide coatings on tungsten substrate. *Vacuum* **2021**, *191*, 110297. [CrossRef]
74. Zhang, Y.; Cui, K.; Fu, T.; Wang, J.; Qie, J. Synthesis WSi$_2$ coating on W substrate by HDS method with various deposition times. *Appl. Surf. Sci.* **2020**, *511*, 145551. [CrossRef]
75. Zhang, Y.; Cui, K.; Gao, Q.; Hussain, S.; Lv, Y. Investigation of morphology and texture properties of WSi$_2$ coatings on W substrate based on contact-mode AFM and EBSD. *Surf. Coat. Technol.* **2020**, *396*, 125966. [CrossRef]
76. Lu–Steffes, O.J.; Su, L.; Jackson, D.M.; Perepezko, J.H. Mo–Si–B coating for improved oxidation resistance of niobium. *Adv. Eng. Mater.* **2015**, *17*, 1068–1075. [CrossRef]
77. Bacos, M.-P.; Landais, S.; Morel, A.; Rimpot, E.; Rio, C.; Sanchez, C.; Hannoyer, B.; Lefez, B.; Jouen, S. Characterization of a multiphase coating formed by a vapor pack cementation process to protect Nb-base alloys against oxidation. *Surf. Coat. Technol.* **2016**, *291*, 94–102. [CrossRef]
78. Xiao, L.; Zhou, X.; Wang, Y.; Pu, R.; Zhao, G.; Shen, Z.; Huang, Y.; Liu, S.; Cai, Z.; Zhao, X. Formation and oxidation behavior of Ce-modified MoSi$_2$-NbSi$_2$ coating on niobium alloy. *Corros. Sci.* **2020**, *173*, 108751. [CrossRef]
79. Su, L.; Lu-Steffes, O.; Zhang, H.; Perepezko, J.H. An ultra-high temperature Mo–Si–B based coating for oxidation protection of NbSS/Nb$_5$Si$_3$ composites. *Appl. Surf. Sci.* **2015**, *337*, 38–44. [CrossRef]
80. He, J.; Guo, X.; Qiao, Y.; Luo, F. A novel Zr-Y modified silicide coating on Nb-Si based alloys as protection against oxidation and hot corrosion. *Corros. Sci.* **2020**, *177*, 108948. [CrossRef]
81. Liu, L.; Lei, H.; Gong, J.; Sun, C. Deposition and oxidation behaviour of molybdenum disilicide coating on Nb based alloys substrate by combined AIP/HAPC processes. *Ceram. Int.* **2019**, *45*, 10525–10529. [CrossRef]
82. Pang, J.; Wang, W.; Zhou, C. Microstructure evolution and oxidation behavior of B modified MoSi$_2$ coating on Nb–Si based alloys. *Corros. Sci.* **2016**, *105*, 1–7. [CrossRef]
83. Li, G.-R.; Yang, G.-J.; Li, C.-X.; Li, C.-J. Force transmission and its effect on structural changes in plasma-sprayed lamellar ceramic coatings. *J. Eur. Ceram. Soc.* **2017**, *37*, 2877–2888. [CrossRef]

MDPI AG
Grosspeteranlage 5
4052 Basel
Switzerland
Tel.: +41 61 683 77 34

Coatings Editorial Office
E-mail: coatings@mdpi.com
www.mdpi.com/journal/coatings

Disclaimer/Publisher's Note: The title and front matter of this reprint are at the discretion of the Guest Editors. The publisher is not responsible for their content or any associated concerns. The statements, opinions and data contained in all individual articles are solely those of the individual Editors and contributors and not of MDPI. MDPI disclaims responsibility for any injury to people or property resulting from any ideas, methods, instructions or products referred to in the content.

www.ingramcontent.com/pod-product-compliance
Lightning Source LLC
LaVergne TN
LVHW072323090526
838202LV00019B/2341